CATCHMENT EXPERIMENTS IN FLUVIAL GEOMORPHOLOGY

Proceedings of a meeting of the International
Geographical Union Commission on Field Experiments
in Geomorphology, Exeter and Huddersfield, UK,
August 16-24, 1981.

Edited by

T.P. BURT
Department of Geography
Huddersfield Polytechnic, UK

and

D.E. WALLING
Department of Geography
University of Exeter, UK

GEO BOOKS
NORWICH

This volume is dedicated to

The memory of Spike Donohoe (1960-1983)

ISBN 0 86094 137 X

© Individual Authors, 1984

Published by Geo Books
Regency House
34 Duke Street
Norwich NR3 3AP
England

Printed in Great Britain at the University Press, Cambridge

CONTENTS

LIST OF CONTRIBUTORS

Chapter

1   Dr. D. Walling
Dept. of Geography,
University of Exeter,
Exeter, Devon, U.K.

2   Dr. M.G. Anderson
Dept. of Geography,
University of Bristol,
Bristol, Avon, U.K.

3   Dr. B. Ambroise
Institut de Geographie,
Universite Louis Pasteur,
67083 Strasbourg, France.

4   Dr. A. Jones
Dept. of Geography,
University College of Wales,
Aberystwyth, Dyfed, U.K.

5   Dr. S. Nortcliff
Dept. of Soil Science,
University of Reading,
Reading, Berks,   U.K.

6   Dr. M. McCaig
Geomorphological Services Ltd,
Old Court House, Trinity Road,
Marlow, Bucks, U.K.

7   Prof. R.B. Bryan
Dept. of Geography,
Scarborough College, West Hill,
Ontario, Canada.

8   Dr. T.P. Burt
Dept. of Geography,
Huddersfield Polytechnic,
Queensgate, Huddersfield, U.K.

9   Dr. A. Armstrong
Field Drainage Experimental Unit,
Ministry of Agriculture,
Fisheries and Food, Anstey Hall,
Trumpington, Cambridge, U.K.

10 Dr. M.P. Mosley
New Zealand Ministry of Works
and Development, Box 1479,
Christchurch, New Zealand.

11 Dr. S.T. Trudgill
Dept. of Geography,
University of Sheffield,
Sheffield, S. Yorkshire, U.K.

12	Dr. A.C. Imeson	Fysisch Geografisch en Bodemkundig Laboratorium, Dapperstraat 115, Amsterdam-Oost, Netherlands.
13	Prof P.D. Jungerius	Fysisch Geografisch en Bodemkundig Laboratorium, Dapperstraat 115, Amsterdam-Oost, Netherlands.
14	Dr. M.J. Haigh	Dept. of Geography, Oxford Polytechnic, Headington, Oxford, U.K.
15	Dr. W. Froehlich	Dept. of Hydrology, Institute of Geography, Polish Academy of Sciences, ul Sw Jana22, 31-018, Krakow, Poland.
16	Dr. D. Balteanu	Institut de Geografie, Str. D. Racovita 12, 70307 Bucuresti 20, Romania.
17	Dr. I. Ichim	Research Station 'Stejaru', Geomorphology Lab., Pingarati, 5648, Neamt, Romania.
18	Dr. G. Leeks	Institute of Hydrology, Plynlimon, Staylittle, Llanbrynmair, Powys, U.K.
19	Dr. D.E. Walling	Dept. of Geography, University of Exeter, Exeter, Devon, U.K.
20	Dr. W.T. Swank	Coweeta Hydrologic Laboratory, Route 1, Box 216, Otto, North Carolina, U.S.A.
21	Prof. H.O. Slaymaker	Dept. of Geography, University of British Columbia, Vancouver, B.C., Canada.
22	Dr. I.D.L. Foster	Dept. of Geography, Coventry (Lanchester) Polytechnic, Coventry, U.K.
23	Dr. A.G. Williams	Dept. of Geography, Plymouth Polytechnic, Plymouth, Devon, U.K.
24	Dr. B.W. Webb	Dept. of Geography, University of Exeter, Exeter, Devon, U.K.

25	Dr. E. Anderson	Dept. of Geography, University of Durham, Science Laboratories, Durham, U.K.
26	Dr. T.P. Burt	Dept. of Geography, Huddersfield Polytechnic, Queensgate, Huddersfield, U.K.
27	Prof. S. Okuda	Disaster Prevention Research Institute, Kyoto University, Uji, Kyoto, Japan.
28	Prof. M. Hirano	Osaka City University, 3-3 Sugimoto-cho, Sumiyoshi-ku, Osaka 558, Japan.
29	Drs. Kang Zhicheng & Zhang Shucheng	Chengdu Institute of Geography, Academia Sinica, Sichuan Province, China.
30	Dr. R.D. Hey	School of Environmental Sciences, University of East Anglia, Norwich, Norfolk, U.K.
31	Dr. A.M. Gurnell	Dept. of Geography, University of Southampton, Southampton, U.K.
32	Dr. A. Werritty	Dept. of Geography, University of St Andrews, St Andrews, Fife, Scotland.
33	Prof. M. Church	Dept. of Geography, University of British Columbia, Vancouver, B.C., Canada.

PREFACE

In August 1981, the International Geographical Union
Commission on Field Experiments in Geomorphology held its
annual meeting in the United Kingdom with a general theme
of 'Catchment Experiments in Fluvial Geomorphology'. Nearly
100 delegates attended the paper sessions held at the
University of Exeter and Huddersfield Polytechnic, and over
40 delegates, primarily those from overseas, participated
in a tour of experimental sites - in South-West England,
Central Wales and the Southern Pennines.

The meeting generated wide-ranging discussion on both
the conceptual framework of field experiments and techniques
and operational details, and in this context considerable
benefit was derived from the international representation.
A common thread running through the discussion held during
the meeting was the need to enhance the scientific status
of fluvial geomorphology through the more rigorous application of scientific method in all its aspects - experimental
design, measurement techniques, and data analysis. In
particular, the need to establish models applicable to a
wide range of locations, and to avoid the 'case study'
approach were strongly emphasised.

Contributions based both on papers presented at the
meeting and on the field experiments visited have been used
to produce this volume. It is the editors' hope that it
will provide a useful view of the current status of catchment studies in fluvial geomorphology, and something of a
follow-up to the volume *Fluvial processes in instrumented
watersheds* published by the Institute of British Geographers
in 1974[1] as a result of another meeting at Exeter. Considerable progress has been made since the early seventies, both
in terms of the questions under investigation and the techniques employed, but it is clear that much still remains
to be done in developing a rigorous scientific method for
catchment experiments. Many, if not most, of the catchment
studies described in this volume do not merit the designation
experiment in the true sense of the word, but there is clear
evidence of the increasing concern of fluvial geomorphologists
to demonstrate a scientific basis for their studies.

In organising the programme for the meeting and in
editing the final collection of manuscripts the editors
experienced some difficulty in producing a meaningful
grouping of the papers into several distinct themes. In
many cases, a paper will embrace a variety of themes and it
is hoped that the groupings finally adopted introduce some
order to the volume, even if they generate some apparent
inconsistencies.

Finally, the editors would like to gratefully acknowledge

the receipt of a grant from the United States Army European Research Office which provided significant support for both the Commission meeting and the production of this volume.

Tim Burt and Des Walling

Thorverton, Devon, August 1982

[1] Gregory, K.J. and Walling, D.E. (eds.), 1974, *Fluvial Processes in instrumented watersheds*, Institute of British Geographers Special Publication no. 6.

PART I

INTRODUCTION

1 Catchment experiments in fluvial geomorphology: a review of objectives and methodology

T.P. Burt and D.E. Walling

'Against the background of relaxed boundaries between the branches of physical geography, geomorphology has emerged as an enquiry devoted to the interpretation of spatial variations in contemporary process in relation to land character, as well as to the temporal variations of process which have significance for the studies of landform evolution'.
K.J. Gregory and D.E. Walling (1974)

I   THE ROLE OF CATCHMENT STUDIES IN FLUVIAL GEOMORPHOLOGY

The Geographical Cycle of W.M. Davis (1909) arranged the elements of the drainage basin into a cycle of landform development. The legacy of Davis's influence was a concern with historical theory in fluvial geomorphology which remained largely unchanged until the 1950's (Chorley, 1978). The major weakness of the Davisian model was that, in concentrating on landform evolution, the study of contemporary landform processes was neglected. This has now been remedied by many detailed investigations of the runoff, erosional and depositional processes operating within the drainage basin. The ability to comprehend and measure accurately the rate of process operation has enabled interrelationships between process and form to be established, using statistical and more recently, deterministic models. Since the 1960's fluvial geomorphological studies have been so dominated by concern with contemporary process that traditional interests in landform evolution, whilst continuing to occur, have provided an almost separate study area (Lewin, 1980). Only recently have these separate interests begun to converge, following the realisation that current process activity must incorporate considerations such as high magnitude - low frequency events (Wolman and Miller, 1960), long-term variations in process operation (Church, 1980), and the concept of geomorphological thresholds (Schumm, 1979). Brunsden and Thornes (1979) have shown that the production of a characteristic form from the regular operation of a constant set of processes may be a valid concept for less resistant systems, but that in many cases, the long relaxation time required for the

characteristic form to develop may not be available due to
environmental changes. In such cases, persistence of prior
forms in the landscape and a limited (spatial) sensitivity
of landform adjustment to a new process domain will be
important. Schumm (1979) concludes that the cyclic and
dynamic equilibrium concepts must be supplemented by additional ideas, such as those noted above. This does not,
however, negate detailed studies of contemporary processes
since such studies will establish the characteristic
response to the forces of change.

It seems likely, therefore, that the use of catchment
experiments in fluvial geomorphology will continue. Such
studies have become firmly established over the last 20
years, and have provided valuable information on the controls of runoff and erosional processes operating in the
drainage basin. Identification and definition of basic
processes continues to be important, as witnessed, for
example, by recent investigations of preferential subsurface
flow through macropores (Beven and Germann, 1981). Increasingly, however, studies seek either directly or indirectly
to provide a rate of process operation in time or space.
Such demands have carried with them a vital interest in the
development and refinement of techniques, to allow ever-
increasing precision over a wide range of measurements
(Goudie, 1981). Indeed, in some cases, our technical ex-
pertise is such that we await the development of analytical
solutions capable of aggregating our detailed findings
(Anderson and Burt, 1981). Such a conclusion requires increased reference to modelling procedures, and to the
careful construction of catchment experiments designed to
test those models.

The drainage basin has provided the fundamental unit
for experiments in fluvial geomorphology. In small catchments it has proved possible to relate catchment characteristics and form to the rates of water and sediment discharged from the outlet of the basin. Through study of
runoff plots, principle process domains have been identified
(eg. Dunne and Black, 1970; Weyman, 1973), although clearly
there has been less success in directly relating the processes operating to the development of the appropriate landform. At the large scale, channel geometry and pattern
show clear (statistical) relationships with position in the
catchment. Spatial variability in the broad controls of
climate, geology and land use have been shown to be associated with variations in stream network density, and in the
spatial variation of solutional denudation rates (Walling
and Webb, 1978). The drainage basin thus provides an excel-
lent example of an open geomorphological system. In considering the system as a set of objects, together with the
relationships between the objects and between their attributes, the systems approach emphasises relationships between
form and process, the multivariate character of many geomorphological processes, and the spatial manifestation of
process operation. The drainage basin provides a clearly
defined functional unit for the study of fluvial processes,
within which many subsystems exist at a variety of scales.
Two major effects of this functional approach may be
detected: firstly, fluvial geomorphology, in adopting a

4

systems approach, provides a broad overlap with many other
environmental science disciplines - ecology, agriculture,
geology and particularly hydrology. Nevertheless, the subject increasingly occupies a distinct niche, providing
valuable interpretations of environmental problems, such as
solute dynamics or sediment erosion, not offered specifically by those other disciplines. The ability to produce
successful predictive models supported by accurate field
and laboratory measurements has significantly increased the
scope of fluvial geomorphologists in this respect.
Secondly, interaction between studies of contemporary
catchment processes and interest in current system operation has determined that much fluvial geomorphology is
now conducted without any strong regard for long-term landform development. A functional view of catchment response,
in which form effectively controls processes, may well not
be linked to rates of landform evolution. As noted above,
a broader view of the term 'contemporary process' may help
bridge the gap between 'evolutionary' and 'contemporary'
fluvial geomorphology. However, since rather short timescales of system operation seem most appropriate for
applied geomorphological studies, the development of an
active applied catchment science (with a firm base in
fluvial geomorphology) may well be at the expense of the
traditional aims of the geomorphologist.

II SCIENTIFIC METHOD IN CATCHMENT EXPERIMENTS

Adoption of the deductive scientific method can be seen as
both a cause and effect of recent changes in fluvial geomorphology. Harvey (1969) has described two alternative
routes which may be followed in order to arrive at a satisfactory scientific explanation (figure 1.1). In the inductive mode, the final explanation is dependent upon the
initial facts available to the investigator and on the
classification procedure adopted. Following the Davisian
model, much fluvial geomorphology before the 1960s was
dominated by a historical inductive approach in which
elements of the landscape were classified and ordered
according to their relative age. Adoption of the deductive
route to scientific explanation was heralded by the adoption of statistical analysis in geomorphology as a means
of investigating process-response systems (Chorley, 1966).
This has led directly to a theoretical approach in fluvial
geomorphology, in which the formulation of an idealised
view or model of reality precedes the establishment of
independent field or laboratory experiments designed specifically to test that model. Production of a successful
model provides a general statement designed to cover all
similar systems, rather than providing a unique explanation
based solely on the facts or events themselves.
     Harvey's model of the deductive route to explanation
does, however, conceal important variations of approach
which must necessarily accompany the development of a
successful model structure. Kuhn (1961) has outlined three
stages of experimentation in science through which all subject areas are seen to progress:

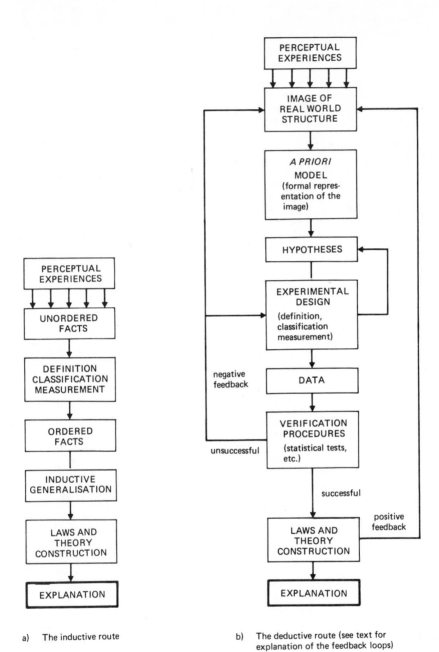

a) The inductive route

b) The deductive route (see text for explanation of the feedback loops)

Figure 1.1. Routes to scientific explanation (after Harvey, 1969).

The inductive stage

This period of study is dominated by informal observations
which allow identification of the processes involved in a
particular system and establishment of functional relationships between variables. Kuhn (1961) sees this stage as
one of essentially 'qualitative experiment', not necessarily
devoid of measurement, but lacking in the rigour and control
which characterises more mature experimentation. The development of ideals and of measurement techniques will begin
at this time. In Harvey's scheme, such case studies accord
to an inductive approach to explanation. Whilst providing
a basis for the construction of theories and models, such
studies do not produce unequivocal conclusions about the
operation and controls of a given process, and must therefore remain merely representative illustrations of the
system. In the study of hillslope geomorphology, defini-
tional studies of runoff processes may be cited as important
descriptive contributions in this regard - for example:
Whipkey (1965), Dunne and Black (1970), Weyman (1973). Such
case studies continue to be valuable, since as Ward (1971)
notes, only a wide range of representative catchment studies
will allow us to begin to integrate all the complexities of
climate, soil and vegetation. In some circumstances, prolonged continuation of such experiments may be justified,
in order to establish the long-term baseline operation of
the system being studied. Nevertheless, protracted attach-
ment to the methodological directive 'Go ye forth and
measure' may well prove only an invitation to waste time
(Kuhn, 1961). Efficient experimental progress must be preceded by the development of a theory-base.

The development of a theory base

The deductive route to scientific explanation depends upon
a clear distinction between the origin and the testing of
theories; only the latter is based upon hard facts and
rigorous logic. Once a theory has been proposed, independent experimentation is required to confirm the theory:
in fact, attempts to falsify or disprove the theory become
the main purpose of experiments. Jevons (1973) points out
that the best established theories are those which have over
a long period withstood a gruelling procedure of testing.
By ruling out what is false, theories approach the truth:
the theory, though supported, is not conclusively proved
because it remains possible that some alternative is the
true, or at least a better, explanation. Such an approach
comes closest to the deductive route proposed by Harvey
(figure 1.1b), involving a progressive development of the
model structure until a 'successful' outcome is achieved.
Even so, the deductive route to explanation is more complex
than Harvey implies. Initial structuring of the model will
in part depend upon the operational constraints known to
apply - what operations are required, what instruments and
measures are to be used, and under what conditions the
observations are to be made; the ideal separation of theory
and fact may not initially be maintained therefore (Anderson
and Burt, 1981). At a later stage of theory development,

7

measurement difficulties may preclude successful experimentation; in this case failure of an experiment to coincide
with the theory will be judged in terms of an inadequate
experiment, rather than indicating a deficient theory. In
this latter case, the 'unsuccessful' feedback loop will
stop at the experimental design stage, requiring perhaps
the development of new measurement techniques before theory
verification becomes possible. It is clear therefore that
neither the theory nor the experimental design need remain
static; indeed several theories may be compared and perhaps
amalgamated until a successful model of reality is produced.

It is at this stage that controlled experimentation
becomes crucial: Church (1983, this volume) describes how
most geomorphological experiments can be seen at best as
'case studies'; organised programmes of field observations
usually do not constitute experiments *sensu stricto*. Church
requires an experiment to include 'a formalised schedule of
measurements to be made in conditions that are controlled
insofar as possible to ensure that the remaining variability
be predictable under the research hypothesis'. Formal
experiments may be of two types: in the initial stages of
theory testing, exploratory experiments which incorporate
simple elements of control may be employed, for example the
use of paired catchments. However, the control may be
widened to include a full range of experimental units in
parallel with the development of the theory. Confirmatory
experiments provide independent verification of the model
already established, and necessarily involve a fixed experimental design and a narrow focus of interest. Church (1983)
provides a full account of the types of experimental control
which may be employed.

Few formal cases of experiments designed specifically
to investigate a given model structure have been reported
in fluvial geomorphology. The most common exploratory
experiments involve the use of contrasting runoff plots
(eg. Dunne and Black, 1970) or contrasting elements of the
catchment (eg. Anderson and Burt, 1977). At the catchment
scale, comparison of catchment response may be achieved
through the use of paired catchments (eg. Hewlett and
Helvey, 1970); in these examples, some specific catchment
treatment, such as clear-cutting or change of tree species,
is usually involved. Comparison of adjacent catchments may
also allow relationships between catchment characteristics
and runoff and sediment response to be postulated (Walling,
1971). However, such experiments may represent little more
than representative case studies: the degree of control is
minimal and often no specific model is being directly
tested. For example, Kirkby (1969) provided simple models
to define the production of throughflow discharge on hillslopes; a number of more recent studies might be said to
provide general but indirect comment on these models, but
have not adopted a rigorous approach to the confirmation
and development of the models. The development of a sound
theoretical base in fluvial geomorphology thus requires a
more formal approach to the testing and development of proposed models than has been the case in the past.

'Normal' scientific studies:

Kuhn (1961) defines 'normal' scientific experiments as those which are required to actualise and consolidate a generally accepted theory. Kuhn views this as 'normal' since most natural scientists are involved in such 'puzzle-solving' studies. The process of actualisation of the potential order defined by the theory, requires production of experiments able to permit quantitative comparison of theory and observation. In this sense, failure of an experiment does not 'infirm' the theory, since this is already accepted, and by definition a successful experiment cannot 'confirm' the theory. All that a successful experiment provides, in essence, is the explicit demonstration of a result which the entire scientific community had anticipated someone would someday achieve (Kuhn, 1961). Nevertheless, the solution of such an experimental puzzle may of course represent a major achievement in itself. This concept of normal science does not accord with Harvey's portrayal of the deductive route to explanation (figure 1.1b); instead the 'unsuccessful' loop feeds into the 'experimental design' stage, denoting the need for further refinements in the experiment. Moreover, as noted above, the successful achievement of 'reasonable agreement' between theory and observation is not indicative of theory confirmation, but of a successful experiment. 'Normal' Science is thus a process of opening up new areas of application for a theory, and of extending and consolidating the implications of the theory. Eventually, of course, persistent anomalies between fact and theory may require a paradigmal shift before further explanatory progress can be made. This brings us full circle in effect, and returns the study to a new period of theory building.

How relevant might Kuhn's tripartite view of Normal Science be to the study of fluvial geomorphology? As noted above, most of the models which have already been proposed remain largely untested; major effort should therefore be directed towards successful verification of theories in fluvial geomorphology. In terms of a process-response model, such as Kirkby's hillslope model (Kirkby, 1971), initial development and calibration of the model must mainly involve observation of current process activity. Once accepted, the theory can then be used to produce simulations of long-term hillslope development in the manner described by Kirkby. The implication of this statement may be that model calibration involves current process observation, whilst model predictions may extend to the long-term development of landforms. The attainment of a successful experimental design to actualise such a model prediction is clearly a difficult, but perhaps not an impossible, task; Kuhn's examples prove that it may be immensely difficult to attain the goal of theory actualisation.

A further implication of the advent of scientific method in fluvial geomorphology, is the need for careful attention to be paid to the whole arena of experimental design. The use of statistical methods in fluvial geomorphology has always been (theoretically) tempered by considerations of the formal requirements of individual

statistical tests, and by the knowledge that any conclusions
drawn will be of little significance if made using data
which do not meet the strict conditions of that test.
Even so, only a few studies have explicitly reported the
use of a statistical test in detail at all stages of analysis (eg. Benson, 1960). The other methodological elements
noted in Harvey's deductive route, also demand our full consideration. For a given model, the objects of interest will
be fairly clear, but careful definition of their attributes
may not be. In many cases, it is necessary to use surrogate
variables perhaps because data on the 'true' variable are not
available or because techniques do not yet exist to provide
such data. The use of micro weight-loss rock tablets by
Burt, Crabtree and Fielder (1983; this volume) illustrates
this problem; in the absence of a direct method of monitoring solute removal from the soil, an indirect indication of
process rates is provided by the weight-loss of rock tablets
buried in the soil. Such measurements may provide a relative indication of solutional losses spatially, but do not
yield absolute erosion rates. It is already clear that
much thought is now being paid to the quality of measurements being made in the field or laboratory, as shown by
the production of recent 'Techniques' volumes (Goudie,
1981; Van Olm, 1981). The need for standard field techniques to be employed to allow comparability of results
should be encouraged by such productions. More attention
still needs to be paid to the errors inherent in any
measurement process. Two sources of error arise: errors
inherent in the use of the instrument itself; and errors
between the measured field or laboratory value and the
actual value. Walling (1978) has shown how the accuracy
of stream sediment and solute load calculations is greatly
reduced when continuous monitoring techniques are not used.
Even so, such attention to data quality is rare; but as
Kuhn (1961) has noted, the presence of major inaccuracies
in data collected can serve only to invalidate the experiment, not the theory. The representativeness of data collected also requires further consideration: Church (1983)
shows that conditions of experimental control demand a
strict sampling procedure if control is to be achieved by
statistical means. The need for a large enough sample size
to predict accurately the numerical characteristics of a
given variable are also too often forgotten, or ignored,
although economic constraints may also contribute to such
negligence. Too frequently, a single storm event, soil
sample, or laboratory measurement is used as the basis for
major conclusions. Freeze (1978) has discussed the limitations imposed by the scarcity of data on the mathematical
modelling of hillslope hydrological processes. He notes
that a physically-based approach requires complete spatial
specification of relevant hydrological parameters, but even
where such properties are uniformly distributed there are
seldom enough data available. Where the hydrological patterns are heterogenous, the use of simplified 'representative' values to produce an idealised view of reality
introduces much uncertainty into the model predictions.
Infiltration capacity is a central variable in any hillslope runoff model; even so, single measurements are often

employed (Beven and Kirkby, 1979), and it is rare that a
sample as large as 26 is reported (Sharma, Gander and Hunt,
1980), even though this is probably still statistically
unacceptable. The representativeness of data deserves
greater emphasis in all fields of fluvial geomorphology than
it has received in the past.

Successful experiments will lead to the production of
effective models. But of what type will such models be?
At present in fluvial geomorphology, there is a clear division between statistical explanations, which stress interrelationships, and physically-based explanations based on
process mechanics. This situation may be viewed as competition between rival theories in a developing science, rather
than as a more fundamental paradigmal shift (Kuhn, 1962).
It seems that increasingly fluvial geomorphology will be
tackled in terms of process mechanics. This can be com-
patible with traditional aims of landform study, if aggregation of microprocess studies proves possible, since much
fundamental process work must be undertaken at the smaller
scale. However, since in some areas, process-based studies
may be inapplicable in the foreseeable future, statistical
relationships must continue to play an important role. The
analogy with developments in runoff modelling may prove
valuable; traditional hydrological models relied on a black-
box approach where fixed relationships (sometimes using
strict regression analysis) were important. The inadequacy
of this approach has led to new methods of approach:
physically-based models have been implemented at the scale
of individual hillslopes (eg. Beven, 1977) although aggregation of such data to the scale of the drainage basin has
not proved possible due to modelling and computational
limitations. Because of this, conceptual catchment models
have been developed (eg. Beven and Kirkby, 1979); calibration is achieved using field or map measurements. It
seems likely that a similar progression of approach will
also occur in fluvial geomorphology. This may entail some
narrowing of the scale of interest, but not necessarily.
Aggregation of results in space, as well as time, may become
an important component of theory actualisation therefore,
with the theory structure tending increasingly towards a
firm base in process mechanics.

III   THE DRAINAGE BASIN AS A FUNDAMENTAL UNIT?

*'Although the river and the hillside waste sheet do not resemble each
other at first sight, they are only the end members of a continuous
series, and when this generalisation is appreciated, one may fairly
extend the 'river' all over its basin and up to its very divides'.*
                                                    W.M. Davis (1899)

The drainage basin is now firmly accepted as the basic
erosional landscape element (Leopold, Wolman and Miller,
1964; Chorley, 1969). The basin provides a convenient, and
usually clearly defined, topographic unit which may be
viewed as an open physical system in terms of inputs of
precipitation and solar radiation, and outputs of discharge,
evaporation and reradiation (Chorley, 1969). The topographic,

hydraulic and hydrological unity of the basin was recognised
in the 'Laws' of Drainage Composition proposed by R.E.
Horton (1945), which identified the morphometric unity of
nested hierarchies of basins on the basis of stream order.
Much subsequent work has been devoted to the investigation
of this geometric regularity, although the integration of
basin geometry and process has not been generally achieved.
The range of drainage basin studies which developed directly
from the work of Horton, have only provided a unified
approach within the context of the hydrological cycle: the
cascade of water, sediment and solutes through a catchment
system has encouraged an integrated investigation of the
whole runoff system. However, as noted above (Section I),
this systems approach is essentially 'functional' in
character, involving the dependence of process upon basin
form and characteristics, and being 'contemporary' in out-
look, long-term landscape evolution is disregarded.

Areas of study which have involved landforms as the
central unit of interest, have tended to develop independently of one another. With the exception of studies of
network topology, the fact that such landforms occur within
a drainage basin system has generally been of little direct
consequence. In the 1950's and 1960's, study of hillslope
forms, channel patterns, hydraulic geometry and spatial
morphometry were all characterised by the use of statistical
methods, providing explanations in the form of 'relationships' rather than being based upon physical principles.
More recently, such landform studies have tended to become
process-based, but even so their essential independence has
been preserved. Whilst the independent variables -
discharge and sediment characteristics - are themselves
controlled by their relative position within the drainage
basin, the interactions between these discrete strands of
fluvial research have not been explicitly examined. This
independent approach was not envisaged by Horton (1945)
whose work on the erosional development of streams en-
compassed several aspects of fluvial processes. Recent
theoretical approaches have, however, indicated that integration of the various facets of fluvial research may be
favoured by a process-based approach. For example, Kirkby
(1978) developed a storage model to describe the generation
of surface and subsurface hillslope runoff; the model was
then used to predict the slope forms and drainage densities
which would develop under given sets of conditions. The
results suggest that a knowledge of hillslope hydrology may
help in the understanding of differences in drainage density
in humid areas, a topic which has hitherto proved intr-
actible (Kirkby, 1978, p.354). Dunne (1980) has also
examined the dependence of channel network initiation on
hillslope runoff processes. He notes that it is difficult
to isolate the hydrological controls of channel formation
and drainage density using statistical analysis because of
the high degree of correlation among the variables analysed.
He argues that deterministic modelling supported by field
measurement will allow isolation of controlling factors.
These two examples illustrate the important conclusion that
the use of physically-based process-response models may
allow integration of the various disparate elements of

fluvial geomorphology, so returning the subject to the
essential unity envisaged by Davis and Horton.

An integrated approach to drainage basin analysis is
now firmly established in the science of hydrology, with
the increasing use of physically-based distributed runoff
models. The example of hydrology suggests that aggregation
of results from the small to larger scale is possible,
although the use of deterministic models at the hillslope
scale (eg. Beven, 1977) must be replaced by a physically-
based conceptual framework at the catchment scale (eg. Beven
and Kirkby, 1979). The ability to model the runoff system
in a catchment is a prerequisite to modelling the erosional
system; the example of Kirkby (1978) has shown that a combination of a hillslope runoff model with a sediment erosion
model can provide valuable insight into landform development. Such results must encourage the continued integration
of hydrology and fluvial geomorphology.

Hydrological modelling thus provides a means of characterising the complete catchment system: such knowledge will
allow, in particular, the downstream effects of land-use or
other changes in part of the basin to be evaluated. Such
predictions would prove valuable at both contemporary and
longer-term timescales: in the short-term, downstream
changes in flood discharge and associated sediment yield
might be examined, for example; in the longer-term, the
effects of changes in flow duration on channel morphology
might be investigated. This strategy suggests that the
period of theory development will be essentially functional
in nature, as noted in Section II. Successful modelling
will then provide both an operational end in itself, as
well as a foundation for predictions of landform evolution.

Catchment experiments in fluvial geomorphology must
increasingly aim to calibrate and test geomorphological
models of catchment response. With a firm base in hydrological theory, such models allow integration of fluvial
landform studies, provide a context within which specific
landforms may be examined, and should allow aggregation of
results to the larger scale. It is suggested that experiments in small catchments (both at the plot and complete
basin scale) will continue to serve a useful and representative purpose, since they allow the 'land phase' of the
hydrological cycle to be defined (Hewlett, Lull and
Reinhart, 1969). Observation of channel response (both
discharge and morphology) must on the other hand, clearly
encompass a wider range of locations. Successful hydrogeo-
morphological modelling will enable fluvial geomorphologists
to play a valuable functional role in the management of
drainage basins, as well as providing a basis for explaining
landform character and evolution.

IV   THEORY AND PRACTICE - THE CURRENT STATE OF
     CATCHMENT RESEARCH IN FLUVIAL GEOMORPHOLOGY

This review provides an introduction to a volume which may
be viewed as a reflection (possibly an unrepresentative
sample?) of the current state of catchment research in
fluvial geomorphology. It is therefore of some interest to

13

assess how the ideals discussed here equate with the papers which follow.

It is clear that studies of 'contemporary' processes provide the major focus of attention at present. Most of the papers are essentially concerned with establishing (short-term) rates of process-operation; few relate to the longer-term variation of processes, nor to the linkages between process and form. Thus, only one side of the 'timescales' schism identified by Lewin (1980) is displayed here. Such preoccupation with contemporary processes is justifiable on two grounds: firstly, concern with applied issues in fluvial geomorphology invariably requires only the definition of current system operation, with a longer-term view of landform evolution often being much less relevant; secondly, successful calibration of process-form models must occur at the small scale (in both time and space) before such models can be applied to larger scales. It is worth recalling that criticism of the Davisian cycle centred upon its lack of concern with process; unless our models are firmly based in process-mechanics, we may not have gained much despite nearly a century of 'progress'. Even so, it is to be hoped that the scale division will disappear and that contemporary process studies may be applied both to current land management issues and to con- siderations of landform evolution. This is not to imply that concern with the larger scale should be neglected until the theory actualisation stage. Interest in topics such as the spatial variation of landform characteristics, the dating of events, and the response of landforms to climatic change, provide both the 'image' of the real world from which our models are derived, as well as the data base for longer-term landform development studies. However, it is the modelling of process-form relationships in the short- term which must remain our first objective. If such models can then be applied over a range of spatial and temporal scales, this can only serve to make fluvial geomorphology more worthwhile and productive as a science.

The major divisions of this volume reflect a trend towards integration of research within the drainage basin. Whilst many studies of runoff processes and erosion dynamics continue to be made without much reference to the overall catchment system, others do reflect a more integrated approach. Thus topics such as runoff processes, erosion rates, and landform character may all be discussed within one paper (eg. Hey and Thorne), and in several other papers, the breadth of interest goes well beyond a single process- form domain. Thus the sections have been made deliberately broad, given the difficulty of providing narrower boundaries across which a number of papers would straddle uncomfort- ably.

It is to be regretted that so much of our fieldwork remains empirical, at the level of the case study, attempt- ing simply to describe the response of yet another small catchment. What is required instead is a programme of planned experiments designed to generate the critical data needed for model evaluation. The lack of dialogue which currently exists between the theoretician and the field scientist must therefore be replaced by active collaboration.

14

From such planned research projects useful generalisations will develop to broaden our understanding and to improve our ability to predict. As Church (1983, this volume) notes, few catchment studies have constituted experiments in the strict sense. Even so, a few papers in this volume do pay strict attention to the requirements of sampling in space and time (Anderson and Cox; Armstrong; McCaig; Webb and Walling). Production of effective models depends upon successful experimentation in the widest sense. Undoubtedly, we have made much progress in terms of measurement techniques and statistical procedures, but such advances do not provide the whole answer. A formal and rigorous approach is required in all rather than just some aspects of the model-building process.

It would be unwise to conclude by attempting to assess the current health of catchment research in fluvial geomorphology, or by considering the likely future progress of the subject. One final observation may be offered, however: fluvial geomorphology is now becoming established as a scientific discipline; even so, without continued strict attention to all aspects of scientific methodology, the development of models applicable to catchment landforms at a variety of temporal and spatial scales may prove intractable. It is salutory to end with a quotation by Dunne (1981): his assessment suggests that fluvial geomorphology remains very much an infant science (perhaps even 'pre-scientific' in Kuhn's terms), and that, as we have claimed here, there is the need for a fundamental change in our style of doing research in catchments if the subject is to progress:

'Science progresses through the making of generalizations in the face of the complexity of nature. But if Isaac Newton had reported his reaction to a falling apple in the manner that we commonly use he would have described the gauging station by which he was sitting, the uniqueness of weather patterns during the preceding three years, the particular apple, and his plans to spend the next three years sitting there to observe other apples, in the hope that at the end of his data collection program, he or someone else would be able to decide what it all meant! ..... We need to plan our next research projects with the express intention of developing some useful generalizations that will expand the theoretical framework of the science. More emphasis needs to be placed on planning field measurement programs that will generate the critical data required for modelling rather than just the data that are easy to obtain. Such planning requires that from the outset the study should be designed either by someone skilled in both theory and fieldwork or by a partnership of such interests.'

REFERENCES

Anderson, M.G. and Burt, T.P., 1977, Automatic monitoring of soil moisture conditions in a hillslope hollow and spur, *Journal of Hydrology*, 33, 27-36.

Anderson, M.G. and Burt, T.P., 1981, Geomorphological Techniques - Part One: Introduction, in: *Geomorphological Techniques*, ed. Goudie, A.S., (George Allen and Unwin, London), 3-24.

Benson, M.A., 1960, Areal flood frequency analysis in a humid region, *Bulletin of the International Association of Scientific Hydrology*, 19, 5-15.

Beven, K.J., 1977, Hillslope hydrographs by the finite element method, *Earth Surface Processes*, 2, 13-28.

Beven, K.J. and Kirkby, M.J., 1979, A physically-based, variable contributing area model of basin hydrology, *Hydrological Sciences Bulletin*, 24, 43-69.

Brunsden, D. and Thornes, J.B., 1979, Landscape sensitivity and change, *Transactions of the Institute of British Geographers*, 4, 463-484.

Burt, T.P., Crabtree, R.W. and Fielder, N., 1983, Patterns of hillslope solutional denudation in relation to the spatial distribution of soil moisture and soil chemistry over a hillslope hollow and spur, in: *Catchment Experiments in Fluvial Geomorphology*, ed. Burt, T.P. and Walling, D.E., (GeoBooks, Norwich), this volume.

Chorley, R.J., 1966, The application of statistical methods to geomorphology, in: *Essays in Geomorphology*, ed. Dury, G.H., (Heinemann, London), 275-388.

Chorley, R.J., 1969, The drainage basin as a fundamental geomorphic unit, in: *Water, Earth and Man*, ed. Chorley, R.J., (Methuen, London), 77-100.

Church, M.J., 1980, Records of recent geomorphological events, in: *Timescales in Geomorphology*, ed. Cullingford, R.A., Davidson, D.A. and Lewin, J., (Wiley, London), 13-30.

Church, M.J., 1983, On experimental method in geomorphology, in: *Catchment Experiments in Fluvial Geomorphology*, ed. Burt, T.P. and Walling, D.E., (GeoBooks, Norwich), this volume.

Davis, W.M., 1899, The geographical cycle, *Geographical Journal*, 14, 481-504.

Davis, W.M., 1909, *Geographical Essays*, (Dover, Boston).

Dunne, T., 1980, Formation and controls of channel networks, *Progress in Physical Geography*, 4, 211-239.

Dunne, T., 1981, Concluding comments to the Christchurch Symposium on 'Erosion and Sediment Transport in Pacific Rim Steeplands', *Journal of Hydrology N.Z.*, 20, 111-114.

Dunne, T. and Black, R.D., 1970, An experimental investigation of runoff processes in permeable soils, *Water Resources Research*, 6, 478-490.

Freeze, R.A., 1978, Mathematical models of hillslope hydrology, in: *Hillslope Hydrology*, ed. Kirkby, M.J., (Wileys, London), 177-225.

Goudie, A.S., (ed.), 1981, *Geomorphological Techniques*, (George Allen and Unwin, London).

16

Gregory, K.J. and Walling, D.E., 1974, The geomorphologist's approach to instrumented watersheds in the British Isles, in: *Fluvial Processes in Instrumented Watersheds*, ed. Gregory, K.J. and Walling, D.E., (Institute of British Geographers Special Publication no.6), 1-6.

Harvey, D., 1969, *Explanation in Geography*, (Arnold, London).

Hewlett, J.D., Lull, H.W. and Reinhart, K.G., 1969, In defense of experimental watersheds, *Water Resources Research*, 5, 306-315.

Hewlett, J.D. and Helvey, J.D., 1970, Effects of forest clear-felling on the storm hydrograph, *Water Resources Research*, 6, 768-782.

Horton, R.E., 1945, Erosional development of streams and their drainage basins; hydrophysical approach to quantitative morphology, *Bulletin of the Geological Society of America*, 56, 275-370.

Jevons, F.R., 1973, *Science Observed*, (George Allen and Unwin, London).

Kirkby, M.J., 1969, Infiltration, throughflow, and overland flow, in: *Water, Earth and Man*, ed. Chorley, R.J., (Methuen, London), 85-97.

Kirkby, M.J., 1971, Hillslope process-response models based on the continuity equation, *Institute of British Geographers Special Publication*, no.3, 15-30.

Kirkby, M.J., 1978, Implications for sediment transport, in: *Hillslope Hydrology*, ed. Kirkby, M.J., (Wiley, London), 325-363.

Kuhn, T.S., 1961, The function of measurement in modern Science, *Isis*, 52, 161-190.

Kuhn, T.S., 1962, *The structure of Scientific Revolutions*, (Chicago University Press, Chicago).

Leopold, L.B., Wolman, M.G., and Miller, J.P., 1964, *Fluvial Processes in Geomorphology*, (Freeman, San Francisco).

Lewin, J., 1980, Available and appropriate timescales in geomorphology, in: *Timescales in Geomorphology*, ed. Cullingford, R.A., Davidson, D.A. and Lewin, J., (Wiley, London), 3-10.

Schumm, S.A., 1979, Geomorphic thresholds: the concept and its applications, *Transactions of the Institute of British Geographers*, 4, 485-515.

Sharma, M.L., Gander, G.A. and Hunt, G.C., 1980, Spatial variability of infiltration in a watershed, *Journal of Hydrology*, 45, 101-122.

Van Olm, P.W., 1981, *Compendium of Geomorphological Methods*, (KNGMG, Amsterdam), 3 volumes.

Walling, D.E., 1971, Sediment dynamics of small instrumented catchments in south-east Devon, *Transactions of the Devonshire Association*, 103, 147-165.

Walling, D.E., 1978, Reliability considerations in the evaluation and analysis of river loads, *Zeitschrift fur Geomorphologie*, Supplementband 29, 29-42.

Walling, D.E. and Webb, B.W., 1978, Mapping solute loadings in an area of Devon, England, *Earth Surface Processes*, 3, 85-99.

Ward, R.C., 1971, Small watershed experiments: an appraisal of concepts and research developments, *University of Hull Occasional Papers in Geography*, 18.

Weyman, D.R., 1973, Measurements of the downslope flow of water in a soil, *Journal of Hydrology*, 20, 267-288.

Whipkey, R.Z., 1965, Subsurface stormflow from forested slopes, *Bulletin of the International Association of Scientific Hydrology*, 10, 74-85.

Wolman, M.G. and Miller, J.P., 1960, Magnitude and frequency of forces in geomorphic processes, *Journal of Geology*, 68, 54-74.

PART II

RUNOFF PROCESSES
AND EROSION DYNAMICS

2 Controls on overland flow generation

M.G. Anderson, D. Bosworth and P.E. Kneale

INTRODUCTION

In the study of overland flow there are two principal
approaches that can be taken. Firstly, there is the application of theories that are applicable to specified boundary
and hydraulic conditions of natural flow on hillslopes.
Wooding (1965), for example, showed that kinematic wave
theory was applicable to gradually varied unsteady flow if
the Froude number was less than 2.0. Secondly, field
investigations indicate the problems of calibrating models
of overland flow. For example, the resistance to overland
flow has been examined by many workers showing the variation in resistance to be very large. Rovey et al. (1977)
show that for a given Reynolds number, the Darcy Weisbach
values range over five orders of magnitude dependent upon
the surface conditions of concrete (Yu and McNowan, 1964),
or grass Prairie (Morgali, 1970). Emmett (1970) has provided perhaps the most comprehensive field measurements of
overland flow in which the varying slope sectors (convex,
straight, and concave) were monitored. The apparent dilemma
is that while various theories can be shown to be suitable
for specific conditions of overland flow, the hydraulic
parameters vary so greatly that it is especially difficult
in the case of overland flow to design relatively simple
general models for inclusion in hydrological models.
General support for this assertion comes from the findings
of Naef (1981) who, having tested out a range of rainfall-
runoff models, deduced their collective failure to lie in
the fact that the rainfall-runoff process is inadequately
represented.

One source of possible misspecification of process
parameters is that of incorrect identification of the
dominant process controls, which need subsequently to be
incorporated into the hydrologic model. In any modelling
procedure there is not only the overriding necessity to
incorporate such controls, but also to ensure that the correct effective parameter values are being utilized, given
the natural variability already referred to.

Table 2.1   Site details

(A)   Site 1 - 1000 m² Scale for Overland Flow Analysis

Catchment characteristics

Location: Winford, Bristol, England
Geology: Red Keuper Marl
Soils: Clay loams, Worcester Series
Land Use: Pasture

Saturated hydraulic conductivity at 80 cm depth: 1×10^{-3} to 8×10^{-4} cm s^{-1}

Subcatchment characteristics

Subcatchment	Predominant Slope Angle (°)	Slope Form
Source to D	4	Straight
D to C	2-6	Spur and hollow
C to B	6-10	Straight
B to A	6-10	Spur and hollow

(see figures 2.1 and 2.2)

Field equipment: electrical stage recorders, automatic tensiometer system

(B)   Site 2 - 10 m² Scale for Overland Flow Analysis

Location:	Bristol, England
Geology:	Portishead beds; consolidated sandstone
Soils:	Weathered Keuper Marl
Land Use:	Forest clearing
Saturated hydraulic conductivity:	Surface 7×10^{-3} cm s^{-1}
	25 cm depth 1×10^{-4} cm s^{-1}

Slope angle 32°

8 individual plot studies (see figure 2.6)

Field equipment: tipping bucket/data logger overland flow recorders, 'jetfill' tensiometers.

The general contention in this paper is that in order to meet these two fundamental model requirements, further experimental work needs to be undertaken to isolate dominant overland flow controls at scales from plot studies (10 m²) to that of identifiable field topographic units of divergence or convergence (1000 m²). In this context it is the variability in overland flow that requires examination so that stronger empirical guidelines than hitherto can be provided to specify model design at these scales. There have been a number of previous empirical studies of overland flow which have sought to identify the process controls (eg. Lundgren, 1980) but none have had, as an explicit element in their experimental design, the potential for isolating the variation in dominant controls as the scale

22

Figure 2.1. Site 1 - Winford catchment near Bristol, U.K.
View looking across A-B subcatchment.

of study is increased. Accordingly, for this study two
sites were selected (one of the order 1000 m^2, the other
10 m^2). Site details are given in table 2.1, and the
following sections detail variability/scale effects in
overland flow, monitored and analyzed for these two contrasting locations.

MACROTOPOGRAPHIC CONTROLS ON OVERLAND FLOW - SITE 1

Given the premise that there is a size of area above which
overland flow will be generated in sufficiently uniform a
manner for local microtopographic or vegetation feature
effects to be overriden, this section examines the overland
flow response from side-slope areas greater than 600 m^2.
At the small basin scale, the precipitation characteristics
may be assumed to be constant across the basin and therefore
the major element of variation in the case considered here
is the hillslope topography.
    The 0.8 km^2 basin chosen for study was divided into
four subcatchments on the basis of the side-slope topography. The catchment and subcatchment characteristics are
summarized in table 2.1 and figure 2.1 shows the spur and
hollow topography in the A to B subcatchment.
    Given the size of the basin and the more long-term aim
to model the overland flow phase of the stream hydrograph,
the use of troughs to intercept overland flow was impractical. Therefore the overland flow discharge was

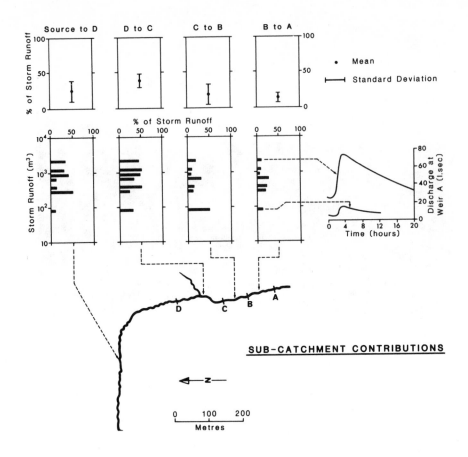

Figure 2.2. Total storm runoff apportioned by subcatchment
for seven winter storm events.

monitored using electrical stage recorders (Anderson and
Burt, 1978) at each of the four weirs which bounded the
subcatchments. The recorders gave a continuous and syn-
chronous record of stage which allowed the individual
catchment discharges to be calculated by subtraction. In
this paper all baseflow and hillslope discharge-controlled
recession flow has been subtracted so that the overland
flow hydrographs can be examined in isolation.
   Figure 2.2 shows the total storm runoff from each subcatchment for a range of winter storms. It is important to
note that the source to D area (60% of the catchment) does
not contribute the largest volume of runoff. This occurs
from the D to C subcatchment where the spur and hollow
topography is particularly marked. Here the shallow slopes
are frequently saturated with depression storage occurring
during the winter months which causes the very marked overland flow response to precipitation.

24

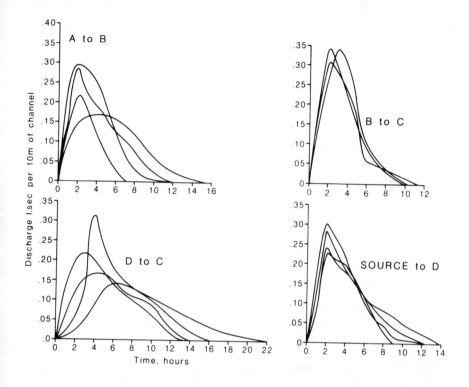

Figure 2.3. Typical individual unit hydrographs for the
subcatchments (5 mm storm used for derivation purposes).

In order to look in detail at the runoff characteristics
of each area, individual subcatchment hydrographs were
standardized to control for channel bank (side slope)
length and then unit hydrographs were derived (Anderson
and Kneale, 1982; Sherman, 1932; Schulz, 1976). It is not
practical here to present all the storms analyzed. However,
typical forms of subcatchment response are illustrated in
figure 2.3, which shows the individual unit hydrographs for
a 5 mm storm for the four subcatchments. The major point
to note is the consistency of the hydrographs in the two
straight slope cases (B-C and Source to D). On the contrasting hollow and spur slopes the peak discharge and
length of recession are both extremely variable. However,
for each slope form the mean peak discharge is greatest
from the steeper sloped areas.

Deriving mean unit hydrographs allows predictions to be
made of the hydrograph morphology at the catchment outflow,
given modified hillslope topography. Figure 2.4 gives two

25

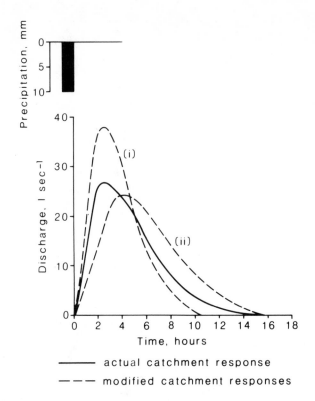

Figure 2.4. Simulated catchment hydrographs used when catchment topography is simulated to be (i) 10° slopes, rectilinear and (ii) 4° slopes with hollow-spur topography.
(After Anderson and Kneale, 1982).

examples of this for a 10 mm storm where (1) the side slopes are 10° and straight, and (2) the side slopes are 4° with spur and hollow topography. Clearly even over this relatively short channel length, minor changes in the slope topography can have a significant effect on the peak discharge and timing of the overland flow discharge.

The principal causes for these variations relate to convergence of soil water. Figure 2.5 illustrates changes in the saturated areas, plotting the air entry soil water potential contour (-70 cm) as it is seen to move upslope during precipitation. The contrast between response in the hollow topography and that for the rectilinear slope is particularly marked. On the rectilinear side slopes the size of the hillslope 'saturated' area is relatively constant and the distance for flow across the slope to the stream is seen to vary little. This leads to a near constant response in the unit hydrograph in terms of both peak and timing.

On the hollow and spur topographies the size and position of the 'saturated' area varies much more widely during

26

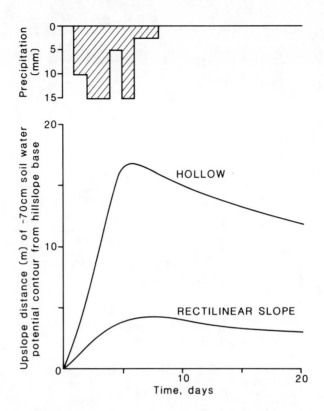

Figure 2.5. Variation in the upslope position of the
-70 cm soil water potential contour in (1) hillslope
hollow and (2) rectilinear slope locations, during
precipitation.

and following precipitation (Anderson and Kneale, 1980;
Kneale, 1980). Consequently, the overland flow paths can
be more variable both spatially and temporally. This leads
to the much greater variability in the hydrographs peak
volumes and runoff times observed in figure 2.3.
    Generally it seems that for straight side-slope topo-
graphies a relatively constant overland flow hydrograph
response can be anticipated, and that steeper slopes
generate larger peak discharges when channel length, precipitation, and topography are held constant. Thus at this
small basin side-slope scale, some of the microtopographic
elements are overridden.

SMALL-SCALE CONTROLS ON OVERLAND FLOW - SITE 2

The first part of this work, at site 1, dealt with overland
flow development at a relatively large scale - that of
hillslope spurs and hollows. At that scale the effect of
small-scale controls is not felt since within any one large
unit there is a sufficient variety of small-scale features

27

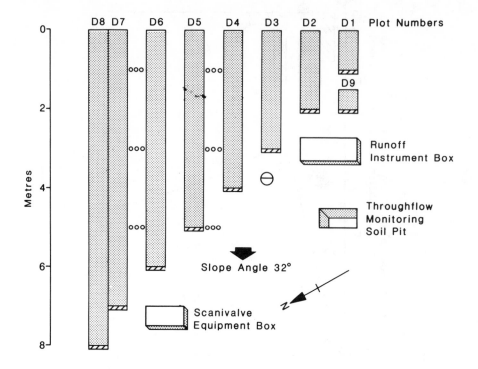

Trough with tipping bucket

Tipping bucket rain gauge

o   Tensiometer connected to scanivalve system

SURFACE RUNOFF EXPERIMENTAL SITE, ABBOTS LEIGH, BRISTOL

Figure 2.6. Site 2 - Abbots Leigh near Bristol, U.K.
Instrumentation layout and plot configuration.

to ensure that discharge variations are cancelled out and
that the slope gives a near constant response.
   On the other hand, details of work from smaller scale
studies show a marked variability in results. Overland flow
discharges from small experimental plots are notoriously
variable. The four plots of Bryan, Yair, and Hodges (1978)
produced very different responses in a semiarid area where
surface vegetation is absent. Their plots varied from 15
to 50 m^2 and the controlling variables involved the properties of the surface materials. Emmett (1970) included
a study of seven field plots of area 28 m^2 in his study.
Using artificial rainfall on the plots until steady conditions were reached, this experiment gave very detailed
measurements of the surface vegetation and topographic
controls on the distribution of flow lines in the thin sheet
flow that occurred. The data show that overland flow is
typically something between uniform sheet flow and rill
flow, whereby the whole surface is covered with standing

28

water but most of it moves in a number of flow concentrations. A further example of variability is that of
Morgan (1980) whose sediment traps were set up in pairs on
unbounded plots on a hillslope. The discharge and sediment
yield to each trap within a pair was extremely variable, so
much so that the traps could not be regarded as replicates
but were lumped to give an average yield for the slope
length they represented.

This aspect of the study deals with a set of small
bounded plots of varying sizes to see if individual control
variables can be identified on each plot and to what extent
they are common to all. The object is to ascertain if
these variables are amenable to simple and quick assessment
in the field. Our argument states that the large-scale
controls are of this type, while those controls that act on
the smaller scale are, because of their intrinsic variability, not amenable to simple assessment.

The instrumentation of the site is shown in figure 2.6,
and site details are given in table 2.1. Data were collected from natural rain-storms for the winter of 1980/81,
from October to March. Sixteen storms produced overland
flow on the sites and these comprise the data set. The
overland flow discharge for each plot was monitored by a
trough installed at the soil surface which drains to a
tipping bucket. The soil surface was defined as being at
the top of the organic layer but below the undecomposed
litter. Troughs were 0.5 m long and the system recorded an
event on a clockwork chart recorder each time a bucket
tipped. For the period of study the soil moisture state
was monitored by a set of eight manually operated tensiometers which were read at each visit to the site.

Surveys of the surface vegetation and topography were
completed during the winter for each plot. These data are
used with respect to between-plot variability.

A total of 16 variables were included in the analysis
as follows:

Y, total storm runoff per   X8, average slope of plot
unit area

X1, total storm precipi-   X9, standard deviation of slope
tation   angles within the plot

X2, maximum 2-hr precipi-   X10, percent bare ground
tation

X3, maximum 1-hr precipi-   X11, percent litter cover
tation

X4, average intensity   X12, percent plant cover

X5, precipitation in the   X13, percent straight sections
last 6 hr   in plot

X6, precipitation in the   X14, percent concave sections
last 12 hr   in plot

X7, precipitation in the   X15, percent convex sections
last 24 hr   in plot

These can be split into two groups. X1-X7 are the time-
varying group which represents various measures of the

Residuals from regression of total storm precipitation (X_2) and σ slope angle (X_9) with observed overland flow (Y)

$$Y = -16.22 + 1.06X_2 + 2.17X_9$$

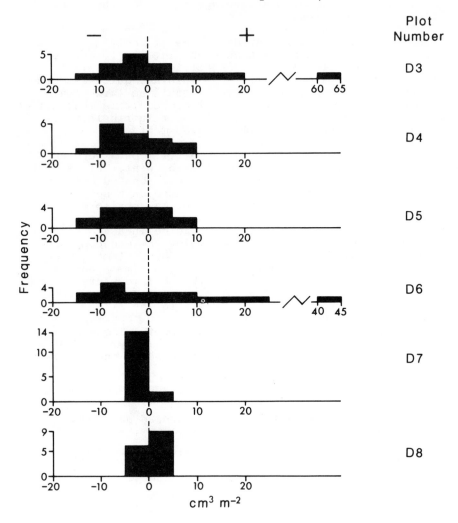

Figure 2.7. Plot of regression residuals apportioned by plot size.

rainfull input to the slope. X8-X15 represent measurements
taken of the plot surface and are both spatially variable
and fixed for the winter period. The former group is
termed the 'precipitation' group and the latter the 'topographic' group. The work which follows represents the
initial stage of the study. It serves to indicate the lines
further investigation should take and that progress can be
made in this manner.

Multiple regression

An initial inspection of the data collected indicated that
the two smallest plots did not respond to any of the events.
These were then left out of the analyses that followed. A
correlation matrix was then produced for the remaining data
including all 15 variables. It was found that only X1, X3,
X9, and X10 were significantly correlated with the overland
flow discharge. Further analysis, by regression, only uses
these four variables. It was not possible to use more than
a two variable multiple regression since, within the identified groups, the variables were multicollinear. Details of
the regression models are as follows:

Model 1: $Y = -16.2 + 1.1X_1 + 2.2X_9$ $r = 0.36$ $F = 7.2$

Model 2: $Y = -9.7 + 1.1X_1 + 0.4X_{10}$ $r = 0.36$ $F = 7.2$

Model 3: $Y = -16.2 + 2.3X_3 + 2.2X_9$ $r = 0.35$ $F = 6.5$

Model 4: $Y = -9.7 + 2.3X_3 + 0.4X_{10}$ $r = 0.35$ $F = 6.5$

($n = 96$; all F ratios significant at 99.9%)

It can then be said that the linear models applied are
useful as a first step in explaining the overland flow
events observed. However, the amount of explanation
achieved with two variables is low and allows no firm conclusions to be drawn from the data.
    The next line of enquiry is to study the residuals from
the regressions to see if any particular clusters occur,
perhaps relating to other controlling variables. Figure
2.7 shows one of the distributions and it is a typical case.
No particular grouping of residuals is apparent either in
the figure or in the scatter plot for the data. However, we
note that the distribution of residuals is nearly normal
about the regression and that the accuracy of prediction
increases as the plot size increases. This is an indication
of the tendency of large plots to be less variable and
therefore more susceptible to predictive approaches.

Individual plot regression models

It is only possible to use the 'precipitation' variables
for individual plots since the other group is constant for
each plot. Testing with linear regression on each plot we
find that only three plots now produce significant relationships; these are plots D5, D7, and D8, as follows:

31

Table 2.2. Summary of overland flow analyses

Site	Control Variables*		Resulting Overland Flow		Remarks
(See table 2.1)	Scale	Topography	Analysis	Predictions	
1	1000 m²	Rectilinear	Overland flow unit hydrographs consistent	Good predictions using unit hydrograph methods	See figure 2.4 - empirical predictions based on unit hydrograph possible with inclusion of topographic factors
		Hollow/spur	Overland flow unit hydrographs variable dependent upon soil water convergence	Poor using unit hydrograph	
2	3-5 m²	Rectilinear	Overland flow dominated by precipitation alone for individual plots	Good using precipitation alone in regression	Confirmed by results from plots D5, D7, and D8 (figure 2.7)
	<3 m²	Rectilinear	Overland flow dominated by suite of controls including roughness	Poor using regressions methods	Predictions necessitate laboratory controlled calibration of theoretical models or detailed field use of such models

* More minor controls are given in table 2.1

32

Plot D5: $Y = -3.4 + 1.6X_1$ $r^2 = 0.363$ $F = 7.98$ (99%)

Plot D7: $Y = -2.5 + 0.6X_1$ $r^2 = 0.708$ $F = 33.88$ (99.9%)

Plot D8: $Y = -0.2 + 0.5X_1$ $r^2 = 0.262$ $F = 4.98$ (95%)

(n = 16)

These single plot and single variable regressions have a higher level of explanation than those for the whole data set. This is to be expected since we are now only dealing with one set of controlling factors; for each plot the topographic factors are constant. The various regression coefficients for each plot indicate how the static controls vary from plot to plot, since rainfall input was the same on all plots and the responses should be similar.

CONCLUSIONS

Table 2.2 summarises the principal findings from this investigation. In two cases it has been determined that overland flow is primarily a function of precipitation alone - plots > 3 m^2 (D5, D7, and D8 - site 2) and rectilinear slopes approximately 1000 m^2 (site 1). For smaller plots (site 2) and diverse topographic conditions at the larger scale (site 1), it is shown that overland flow cannot be construed as a simple precipitation function alone.

This study has thus begun an attempt to delimit those conditions under which it may prove possible to make straightforward empirical estimations of overland flow responses, and those conditions demanding a much greater detail of investigation for such predictions to be successful. We have demonstrated that such a distinction cannot usefully be made according to a scale criterion alone, as table 2.2 and figure 2.4 illustrate. Rather at the 'larger' scale, topography is shown to be an important determinant regarding the feasibility of estimating overland flow using unit hydrograph techniques (figures 2.3-2.5), while at the plot scale the detailed topographic group of variables (X8-X15) is confirmed to be of overriding importance.

ACKNOWLEDGEMENTS

The work reported in this paper was supported in part by the US Army through its European Research Office, London. Typing of the manuscript was completed by the US Army Engineer Waterways Experiment Station, Vicksburg, Mississippi.

REFERENCES

Anderson, M.G. and Burt, T.P., 1978, Time synchronised stage recorders for the monitoring of incremental discharge inputs in small streams, *Journal of Hydrology*, 34, 141-159.

Anderson, M.G. and Kneale, P.E., 1980, Topography and hill-slope soil water relationships in a catchment of low relief, *Journal of Hydrology*, 47, 115-128.

Anderson, M.G. and Kneale, P.E., 1982, The influence of low angled topography on hillslope soil water convergence and stream discharge, *Journal of Hydrology*, 57, 65-80.

Bryan, R.B., Yair, A. and Hodges, W.K., 1978, Factors controlling the initiation of runoff and piping in Dinosaur Provincial Park, Badlands, Alberta, Canada, *Zeitschrift fur Geomorphologie*, 29, 151-168.

Emmett, W.W., 1970, The hydraulics of overland flow on hillslopes, *US Geological Survey Professional Paper* 662A.

Kneale, P.E., 1980, Soil water processes in low permeability soils with reference to hillslope hydrology and slope stability. *Unpublished Ph. D. thesis*, University of Bristol.

Lundgren, L., 1980, Comparison of surface runoff and soil loss from runoff plots in forest slopes, *Geografiska Annaler*, 62A, 113-148.

Morgan, R.P.C., 1980, Field studies of sediment transport by overland flow, *Earth Surface Processes*, 5, 307-316.

Morgali, J.R., 1970, Laminar and turbulent overland flow hydrographs, *Journal of the Hydraulics Division, ASCE*, 96, HY2, 441-460.

Naef, F., 1981, Can we model the rainfall-runoff process today? *Hydrological Sciences Bulletin*, 26, 281-289.

Rovey, E.W., Woolhiser, D.A. and Smith, R.E., 1977, A distributed kinematic model of upland watersheds, *Colorado State Hydrology Paper*, no. 93.

Schulz, E.F., 1976, *Problems in applied hydrology*, (Water Resource Publications, Colorado State University).

Sherman, L.K., 1932, Streamflow from rainfall by unit-graph method, *Engineering News Record*, 108, 501-505.

Wooding, R.A., 1965, A hydrological model for the catchment-stream problem (1) kinematic wave theory, *Journal of Hydrology*, 3, 254-267.

Yu, Y.S. and McNowan, J.S., 1964, Runoff from impervious surfaces, *Journal of Hydraulics Research*, 2, 1, 3-23.

3 Spatial variability of soil hydrodynamic properties in the Petite Fecht catchment, Soultzeren, France: preliminary results

Bruno Ambroise, Yves Amiet and Jean-Luc Mercier

INTRODUCTION

Properties of superficial formations and soils vary from one point to another, both laterally and vertically. This spatial heterogeneity has been recognised for a long time in many geomorphologic, pedologic and agronomic works, but since the systematic investigations of Rogowski (1972) and Nielsen et al. (1973), an increasing number of field studies have been devoted to this subject, in order to estimate the lateral and vertical variations of soil physical properties and particularly of the soil water characteristics (eg. Keisling et al., 1977; Cameron, 1978; Wright and Wilson, 1979; Topp et al., 1980; Imbernon, 1981).

For many scientific and practical applications, it is clearly necessary to have a good knowledge of the spatial pattern and statistical distribution of the hydrodynamic properties which control water retention and movement in soils. They strongly influence water redistribution processes of runoff, infiltration or evapotranspiration, and thereby the hydrologic behaviour of a watershed. They also affect the vegetation pattern and dynamics, and many geomorphologic processes (solifluction, creep, solute transport). Such information is therefore needed for water management, land use planning and agricultural development (irrigation, drainage, fertilization), but also to elaborate and calibrate spatially-distributed hydrologic models at the watershed level (Ambroise et al., 1980).

Such a study has been undertaken in the Petite Fecht catchment (12.3 km^2) and the small adjacent Ringelbach catchment (0.36 km^2) near Soultzeren in the Vosges Mountains, France (figure 3.1). Climatological, hydrological and geomorphological processes, and also their spatial distributions and interactions, have been studied there since 1975 (ERA 569 CNRS, 1982). These basins have been incised in a granitic bedrock under glacial and periglacial conditions; their altitudes vary between 585 m and 1303 m (on the Vosges main ridge); the mean slope is $18°9'$ and the modal aspect is south; 60% of the catchment area is

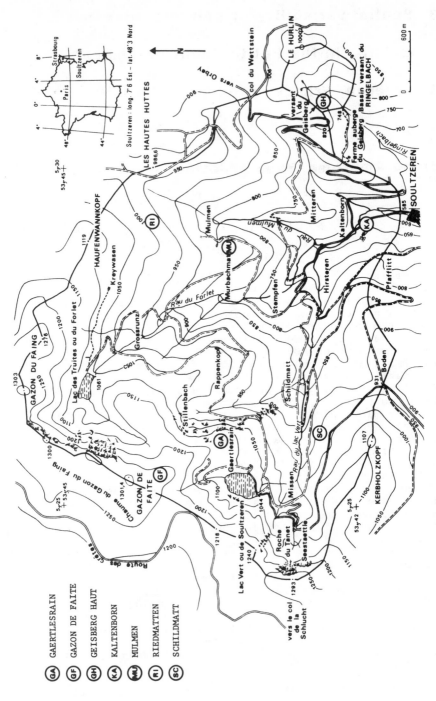

Figure 3.1.   The Petite Fecht and Ringelbach catchments at Soultzeren – Sites of undisturbed soil core sampling.

(GA) GAERTLESRAIN

(GF) GAZON DE FAITE

(GH) GEISBERG HAUT

(KA) KALTENBORN

(MU) MULMEN

(RI) RIEDMATTEN

(SC) SCHILDMATT

36

covered by forests, the other 40% mostly by grassland (Humbert *et al.*, 1982).

The experimental design will be presented here, together with the results of the preliminary investigations. Undisturbed small soil cores have been sampled in representative profiles. Their water retention curves (desorption branch) have been measured in the laboratory (Amiet *et al.*, 1982). Van Genuchten's function (1980) has been fitted to these experimental data in order to deduce estimations of the corresponding relative conductivity curves (Ambroise, 1982).

MAPPING AND SAMPLING METHOD

Water fluxes into a soil layer are controlled by two hydro-dynamic properties, which both depend on the soil's pore size distribution and therefore on its texture and structure:
- the water retention curve $\theta(\psi)$ or pF curve $\theta(pF)$, which relates the water content θ to the water tension ψ (or $pF = \log_{10}|\psi|$) and expresses the ability of the soil to retain water as a function of its energy status;
- the hydraulic conductivity curve $K(\psi)$ or $K(\theta)$, which expresses the ability of the soil to transmit water as a function of its degree of saturation.

A map of these two characteristics could be made by sampling the study area systematically, but to define soil units with sufficient precision a very close sampling framework would be needed. The soil unit characteristics are in fact the results of the entire geomorphological history of the soils and superficial formations: it is therefore useful to make a preliminary subdivision of the area into 'genetically homogeneous units' using geomorphologic and pedologic maps, and then to apply a less dense sampling routine on each unit. A statistical analysis will establish:
- whether the genetic units are significantly different as hydrodynamic units (inter-unit variability);
- the residual variation in each unit because of local heterogeneities (intra-unit variability).

Such a sampling design was applied at Soultzeren. Maps of superficial formations and soil types have been surveyed at a 1:10 000 scale (Amiet, 1980). The formations are mainly tills, moraines and periglacial slope deposits, which are all more or less disturbed by solifluction. The main soil types belong to the podzolic series (from acid brown earth to podzol). Their texture is generally very heterogeneous, with a high stone content, a sand or sand-silt matrix and some accumulation of organic matter in the surface horizon. As a first step, seven sites representative of the main units were chosen to determine and compare their soil water properties (figure 3.1, table 3.1).

Sites	Altitude a.s.l. (m)	Orientation	Slope angle	Vegetation	Superficial formation constitution	Soil type	Number of samples by layer A	B	B/C,R	sum
GEISBERG Haut "GH"	750	S-E	20°	grassland	Blocks imbedded in finer material	brun de culture	4	6	1	11
MULMEN "MU"	870	E-S-E	15°	forest	ibid.+ removal of some interstitial matrix	ocre-podzo-lique	1	4	-	5
SCHILDMATT "SC"	940	N	17°	forest	ibid.	brun ocreux	1	6	-	7
GAERTLESRAIN	1100	S	10°	grassland	ibid.	ocre-podzo-lique	2	3	-	5
KALTENBORN "KA"	620	N-E	5°	fir nursery	alluvial deposits	brun de culture	2	1	-	3
RIEDMATTEN "RI"	990	S-E	17°	grassland	Blocks imbedded in finer material	brun ocreux	1	1	-	2
GAZON DE FAITE "GF"	1260	E-N-E	15°	grassland	in situ weathered granite	ranker crypto-podzo-lique	2	-	-	2

Grain Size Distribution and organic matter at 5 sites

Sampled depth (cm)	Organic matter (%)	Material > 2mm (%)	Sand (%)	Material ≤ 2mm Silt (%)	Clay (%)
GEISBERG Haut					
0-4	9.2				
4-15	5.4				
7-10	-	51.1	67.0	20.2	12.8
15-30	2.7				
25-30	-	51.3	68.5	19.8	11.7
30-70	0.5				
35-40	-	48.4	69.9	17.8	12.3
85-95	-	41.3	76.1	15.7	8.2
MULMEN					
0-5	26.7	35.6	55.5	24.5	20.0
4-10	5.6				
10-30	4.0				
15-25	-	35.3	57.1	23.1	19.8
30-65	1.8				
30-35	-	47.4	54.2	25.3	20.5
50-55	-	46.8	53.7	24.3	22.0
65-100	1.6				
70-75	-	31.9	63.6	21.4	15.0
SCHILDMATT					
0-5	42.0				
4-8	-	33.9	55.3	31.0	13.7
5-13	9.2				
10-15	-	4.7	61.7	26.3	12.0
13-35	11.1				
23-30	-	50.3	60.2	28.3	11.5
35-60	6.6				
60-90	3.7				
60-95	-	48.5	60.6	26.5	12.9
GAERTLESRAIN					
6-9	-	16.4	55.9	27.9	16.2
17-23	-	41.2	63.3	22.9	13.8
KALTENBORN					
10-20	-	11.2	51.1	31.0	17.9
30-40	-	18.6	48.0	30.7	21.3
50-60	-	30.6	57.8	25.0	17.2

Table 3.1.　Description of the sampled profiles at Soultzeren.

38

EXPERIMENTAL DETERMINATION OF THE WATER RETENTION CURVES

The determination of retention curves $\theta(\psi)$ is possible in the field, by measuring simultaneously for each soil layer the water tension ψ with tensiometers, and the water content θ with a neutron probe or, for more precise measurements, 'heat pulse' probes (Pouyaud, 1975; Hamburger and Mercier, 1978). Such a method, which requires well-equipped sites, is much too demanding to be used over a watershed to study its spatial variability, but it gives us a good reference to verify the validity of any other simpler methods. The four sites - Geisberg Haut, Mulmen, Schildmatt and Gaertlesrain - have been equipped for such field control.

A commonly used and simpler method is to determine $\theta(\psi)$ in the laboratory on soil samples, taken at any point in a watershed. For the high tension range ($|\psi| \geqslant 1000$ cm, $pF \geqslant 3$), water retention depends more on the soil texture than on its structure, and $\theta(\psi)$ measurements can be performed on small disturbed soil samples with the pressure-membrane apparatus (Richards, 1965). But for lower tensions, nearer saturation, the retention is much more influenced by the soil structure and it is necessary to use undisturbed soil cores.

For this low tension range, we have used the sand-box apparatus (Stackman *et al.*, 1969; Vàrallyay, 1973), which is cheap, reliable and also very suitable for serial analyses (as needed for such a study). The undisturbed soil cores are placed in a ceramic or plexiglas box, which is filled up with a calibrated porous medium. The water tension is fixed by the level of a bottle connected to the drainage tube of the box. For the range pF 0-2 this medium is a very fine sand, and for the range pF 2-2.7 a layered mixture of sand and kaolin is used. At each chosen tension ψ, once the equilibrium is reached, the cores are weighed. The corresponding water contents θ are then obtained by oven-drying them after the last measurement.

For the seven chosen sites, a first series of 35 undisturbed cores was sampled in 250 cm^3 (ϕ 8 cm, H 5 cm) stainless steel cylinders. Table 3.1 gives the sampled depths and layers for each site and some textural data. The retention curves were determined at nine tension values: six with the sand-box apparatus (pF 0.5, 1., 1.3, 1.5, 1.8, 2.) and three with the pressure-membrane apparatus (pF 2.5, 3.5, 4.2) to complete the curves up to wilting point. The results are presented in table 3.2 and figure 3.2.

ANALYSIS OF THE RETENTION CURVES

A break in the retention curves at pF 2.5 is clearly evident in the case of very organic samples (SC1-2,GA1-2). In such layers with high macroporosity the structural retention is still important at pF 2.5 and even pF 3.5 and the use of disturbed samples leads to underestimated retention values: the 4 values at pF 2.5 for these samples should therefore be disregarded.

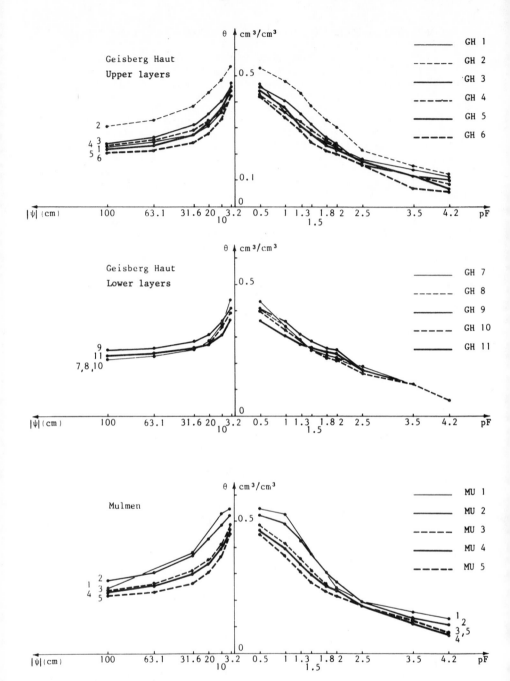

Figure 3.2. Experimental water retention curves for seven soil profiles at Soultzeren.

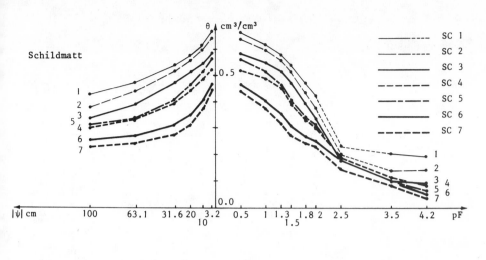

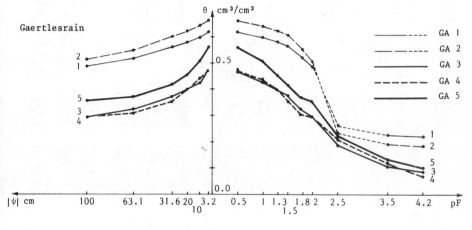

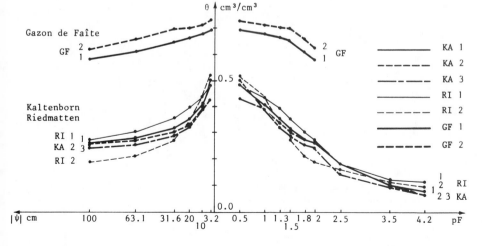

Name	N°.	Z(cm)	S	cm 3.2 / pF 0.5	10.0 / 1.0	20.0 / 1.3	31.6 / 1.5	63.1 / 1.8	100.0 / 2.0	316.2 / 2.5	3162 / 3.5	15849 / 4.2
GEISBERG HAUT												
GH1	22	1-6	1.124	.470	.358	.308	.277	.243	.227	.178	.140	.117
GH2	23	1-6	1.223	.531	.481	.435	.386	.331	.307	.216	.156	.127
GH3	20	6-11	1.320	.461	.407	.357	.312	.261	.240	.160	.120	.104
GH4	21	6-11	1.309	.429	.362	.323	.292	.253	.236	.179	.122	.085
GH5	18	35-40	1.433	.444	.372	.318	.274	.236	.219	.173	.114	.067
GH6	19	35-40	1.380	.425	.336	.279	.246	.215	.203	.160	.069	.056
GH7	24	55-60	1.339	.440	.346	.286	.258	.231	.220	.179	.120	.060
GH8	26	55-60	1.444	.415	.342	.285	.258	.233	.222	.190	.120	.060
GH9	16	75-80	1.461	.412	.358	.311	.285	.261	.251	.176	.123	.059
GH10	17	75-80	1.349	.408	.335	.281	.255	.230	.219	.167	.124	.059
GH11	25	100-105	1.526	.368	.309	.279	.262	.244	.235	.176	.116	.059
MULMEN												
MU1	07	1-6	0.891	.550	.529	–	.382	–	.248	.198	.155	.131
MU2	36	7-12	1.108	.525	.487	.433	.376	.307	.276	.195	.133	.107
MU3	08	15-20	1.161	.486	.413	.357	.314	.267	.246	.187	.119	.085
MU4	12	35-40	1.333	.469	.398	.338	.301	.263	.246	.185	.117	.076
MU5	11	48-53	1.246	.458	.371	.309	.272	.236	.222	.179	.126	.081
SCHILDMATT												
SC1	29	2-7	0.611	.671	.617	.579	.542	.475	.426	.236	.209	.196
SC2	10	10-15	0.743	.646	.598	.559	.520	.440	.382	.206	.139	.147
SC3	31	20-25	0.983	.586	.549	.514	.472	.391	.342	.183	.102	.091
SC4	13	45-50	1.152	.522	.489	.448	.396	.333	.306	.194	.119	.085
SC5	32	60-65	1.112	.566	.518	.468	.403	.340	.313	.186	.106	.065
SC6	46	65-70	1.246	.472	.405	.352	.311	.273	.255	.189	.104	.052
SC7	09	85-90	1.362	.442	.376	.314	.277	.246	.232	.145	.086	.038
GAERTLESRAIN												
GA1	27	4-9	0.768	.619	.598	.582	.567	.521	.488	.261	.225	.220
GA2	35	4-9	0.828	.666	.643	.627	.608	.550	.512	.235	.192	.184
GA3	33	12-17	1.292	.472	.428	.399	.374	.323	.294	.190	.111	.088
GA4	05	23-28	1.238	.470	.433	.399	.352	.309	.291	.207	.122	.071
GA5	06	44-49	1.008	.560	.509	.455	.421	.377	.357	.223	.137	.100
KALTENBORN												
KA1	30	5-10	1.245	.484	.409	.353	.316	.281	.264	.185	.103	.079
KA2	34	5-10	1.122	.500	.391	.336	.305	.275	.261	.188	.107	.068
KA3	14	45-50	1.340	.431	.393	.325	.290	.255	.243	.146	.093	.064
RIEDMATTEN												
RI1	01	10-15	1.084	.482	.436	.399	.357	.302	.274	.184	.127	.119
RI2	02	30-35	1.114	.512	.437	.345	.273	.212	.192	.164	.114	.093
GAZON DE FAITE												
GF1	03	5-10	0.542	.694	.681	.668	.653	.612	.584	–	–	–
GF2	04	5-10	0.509	.730	.714	.707	.700	.662	.620	–	–	–

Table 3.2. Laboratory measured water contents (cm^3 cm^{-3}) for nine tensions or pF.

The 11 GH curves obtained for the Geisberg profile under grassland have very similar shapes. For this little differentiated profile with a coarse texture, each retention curve decreases rapidly when the tension increases (very low air entry pressure). If we compare the values at each tension, some slight retention decrease with increasing depth is observable, owing to a related small decrease in fines and organic matter.

The Schildmatt site is quite different. For this well-differentiated profile under a forest canopy, the seven SC curves indicate a regular and sizeable decrease in retention with increasing depth, and also a clear decrease in their curvatures. In this case the content of fines and organic matter, which is high in the top layers, decreases rapidly in the deeper layers.

The other sites sampled may be related to these two basic types. The retention curves for Mulmen MU, Kaltenborg KA and Riedmatten RI, which have rather homogeneous profiles, look like the Geisberg ones, although the soil types are different. The Gaertlesrain profile GA, which is also well differentiated, and the upper layer at Gazon de Faîte GF are much more similar to the Schildmatt profile. The neutron probe and tensiometric measurements confirm this distinction: the Geisberg and Mulmen sites are much drier than those at Gaertlesrain and Schildmatt, where the superficial layers are always moist. Thus, the main difference lies between the 'organic' layers with high retention (SC1-2,GA1-2,GF1-2) and the 'mineral' layers having a lower retention (GH3-11,MU3-5,SC7). When the content of fines and organic matter decreases, the bulk density δ_S increases, so that the porosity and therefore the retention decreases (figure 3.3).

Finally, it should be noted that local variability of the retention curve for any layer seems to be rather low, according to the results for replicate samples (GA1-2, KA1-2,GF1-2); the two corresponding curves are very similar. Moreover, the nine curves GH3-11 of the mineral layers at Geisberg are so near that a mean retention curve can be sufficient to describe all these layers (table 3.3).

In conclusion, it seems that the differences in water retention between the soils in this region are not very important for the deeper layers, whose coarse textures are very similar; the mineralogy and geomorphologic evolution of the soil material are rather homogeneous over the whole watershed. Much greater differences are observed in the upper layers because of important differences in their content of fines and organic matter. In future studies, it will clearly be necessary to sample more frequently from upper than from deeper soil layers on the watershed.

FITTING OF THE RETENTION CURVES

Experimental $\theta(\psi)$ curves are defined by points. They are not easy to analyse, to compare, or to introduce into run-off simulation models. It is therefore vital to fit a carefully chosen mathematical function which will synthesize in a few parameters all the available information. Various

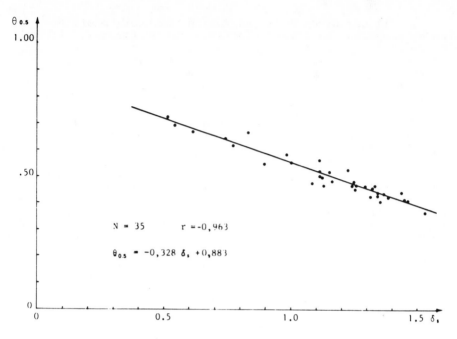

Figure 3.3. Linear regression between bulk density δ_S and water content $\theta_{0.5}$ at pF 0.5 ($|\psi| = 3.2$ cm).

Table 3.3. Mean and standard deviation of the nine GH3-11 retention curves at Geisberg Haut.

Pf	0.5	1.0	1.3	1.5	1.8	2.0	2.5	3.5	4.2
Mean	0.422	0.352	0.302	0.271	0.240	0.227	0.173	0.114	0.068
Standard Deviation	0.266	0.028	0.027	0.021	0.016	0.014	0.010	0.017	0.016

functions have been used since the pioneering work of Gardner *et al.* (1970), but systematic comparisons of them have shown that the best results are obtained with functions having a sufficient number of parameters or degrees of freedom (Vàrallyay and Mironenko, 1979).

To adjust the experimental retention curves obtained at Soultzeren, we have chosen the function recently proposed by van Genuchten (1980), which has five parameters and gives generally very satisfactory results:

$$\theta(\psi) = \theta_r + \frac{\theta_s - \theta_r}{\left[1 + (\frac{\psi}{\psi_o})^\alpha\right]^\beta} \quad \text{or} \quad S_e = \frac{\theta - \theta_r}{\theta_s - \theta_r} = \frac{1}{\left[1 + (\frac{\psi}{\psi_o})^\alpha\right]^\beta} \tag{1}$$

where S_e is the effective saturation rate, and where all

Table 3.4. Values of the fitted van Genuchten's parameters for 32 soil samples from Soultzeren.

NAME	θ_s θ_r (cm^3/cm^3)		ψ_o (cm)	α	β	NAME	θ_s θ_r (cm^3/cm^3)		ψ_o (cm)	α	β
GEISBERG HAUT						SCHILLMATT					
GH1	0.667	0.096	1.00	1.329	0.248	SC1	0.674	0.143	19.74	1.375	0.273
GH2	0.553	0.089	10.40	1.338	0.253	SC2	0.640	0.111	29.08	1.517	0.341
GH3	0.483	0.084	9.38	1.407	0.289	SC3	0.586	0.058	29.86	1.488	0.328
GH4	0.486	0.018	3.04	1.223	0.182	SC4	0.537	0.041	16.75	1.353	0.261
GH5	0.525	0.021	2.51	1.255	0.263	SC5	0.589	0.001	12.27	1.310	0.237
GH6	0.577	0.000	1.07	1.241	0.194	SC6	0.512	0.000	5.38	1.253	0.202
GH7	0.582	0.018	1.00	1.232	0.188	SC7	0.483	0.000	5.32	1.280	0.219
GH8	0.536	0.000	1.00	1.197	0.165						
GH9	0.444	0.000	4.92	1.214	0.176	GAERTLESRAIN					
GH10	0.530	0.000	1.00	1.201	0.167	GA1	0.606	0.199	74.70	1.648	0.393
GH11		0.000	4.92	1.195	0.163	GA2	0.653	0.161	77.09	1.663	0.399
						GA3	0.478	0.006	15.85	1.267	0.211
MULMEN						GA4	0.486	0.000	12.38	1.257	0.204
MU1	–	–	–	–	–	GA5	0.573	0.013	13.10	1.272	0.214
MU2	0.539	0.097	16.07	1.499	0.333						
MU3	0.539	0.042	4.53	1.291	0.225	KALTENBORN					
MU4	0.548	0.064	2.58	1.229	0.187	KA1	0.541	0.000	3.71	1.235	0.190
MU5	0.630	0.043	1.00	1.261	0.207	KA2	0.667	0.000	1.00	1.221	0.181
						KA3	0.457	0.023	8.69	1.315	0.240
Imposed values are underlined						RIEDMATTEN					
						RI1	0.486	0.098	17.57	1.480	0.324
						RI2	0.541	0.103	9.46	1.683	0.406

the parameters have a physical meaning: θ_s is the saturation water content, θ_r the residual water content, ψ_o a tension value between the air entry pressure and the abscissa of the θ(pF) curve inflexion point; α and β, which control the slope at this inflexion point, are related by equation (2) which reduces the degrees of freedom of the chosen function to four.

$$\beta + \frac{1}{\alpha} = 1 \quad , \quad 0 < \beta < 1 \tag{2}$$

The incomplete MU1 and GF1-2 retention curves were not analysed; values of the five parameters θ_s, θ_r, ψ_o, α and β were determined for the 32 complete (nine or eight point) curves by a non-linear regression taking into account some other physical and computational conditions:

$$0 \leqslant \theta_s \leqslant 1; \ 0 \leqslant \theta_r \leqslant 1; \ 1 \leqslant \psi_o; \ 1 < \alpha < 10; \ 0.01 < \beta < 1 \tag{3}$$

The resulting values are summarized in table 3.4.
This fitting procedure gave very good results for the retention curves obtained at Soultzeren. Figure 3.4 shows the fitted curves and experimental points for typical samples. Most of the fitting errors,

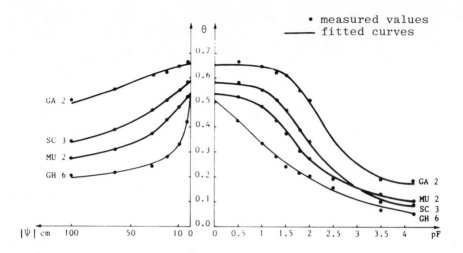

Figure 3.4. Examples of retention curve adjustment with van Genuchten's function.

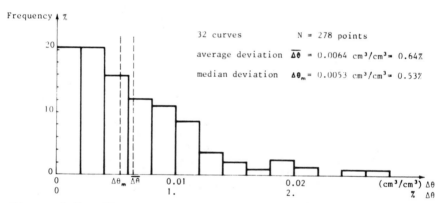

Figure 3.5. Histogram of adjustment errors
[$\Delta\theta$=|θmeasured - θcomputed|] with van Genutchen's function.

[$\Delta\theta$ = |θmeasured - θcomputed|] are low and of the same size order as the experimental errors (figure 3.5). The few errors greater than 0.02 cm^3 cm^{-3} may be related to dubious measurements or to poor linking between the two parts of the experimental curves.

The parameter values confirm the layers identified on the experimental curves. The estimated values of θ_s and θ_r are much greater for the 'organic' layers (SC1-2, GA1-2) than for the 'mineral' horizons (GH6-10,KA2,MU5). The higher values of ψ_0 are obtained for organic layers where

46

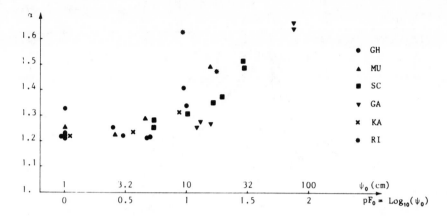

Figure 3.6. Graph of the fitted parameters α versus ψ_0.

the air entry pressure is relatively high, with lower values
for the mineral layers which have a very low water retention
capacity. The values of α are also very low: most of the
curves have no evident inflexion point because of the
rather heterogeneous grain size distribution of the soils
material.

The parameters ψ_0 and α, which determine the shape of
the retention curve, permit us to classify and to compare
globally the samples according to their macroscopic behaviour: on figure 3.6 the differentiation between the
organic layers (high ψ_0 and α) and the mineral ones (low ψ_0
and α) appears clearly. We can also observe, for instance,
the very regular retention decrease in the Schildmatt
profile, in contrast to the sharp decrease at Gaertlesrain
between the upper and the lower layers.

ESTIMATION OF THE RELATIVE HYDRAULIC CONDUCTIVITY CURVES

The unsaturated hydraulic conductivity curve is much more
difficult to measure than the water retention curve, both
in the field (Vachaud et al., 1978), especially on slopes,
and in the laboratory on undisturbed soil cores (Arya
et al., 1975; Boels et al., 1978). However, many workers
have demonstrated that there is some relationship between
these two soil water properties, since both depend on the
size distribution of the pores and on the geometry of their
connections. The θ(ψ) function proposed by van Genuchten
has a major advantage in that the fitted retention curve
may be used to provide a good estimation of the corresponding hydraulic conductivity curve K(ψ).

Mualem (1976) has proposed a statistical model of K(ψ)
where the microscopic characteristics of the porous medium
are assumed to be globally expressed by its macroscopic
retention curve θ(ψ), which is easier to determine:

47

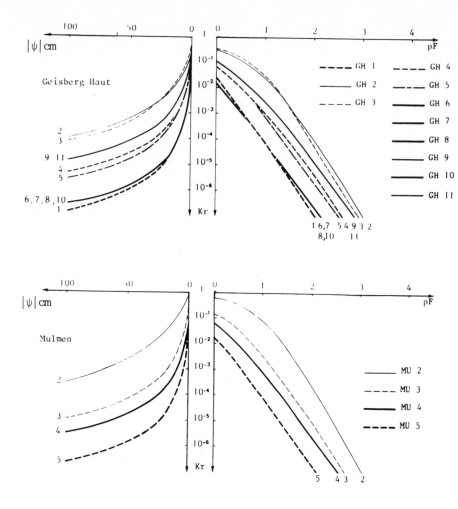

Figure 3.7. Estimated relative hydraulic conductivity
curves at Soultzeren.

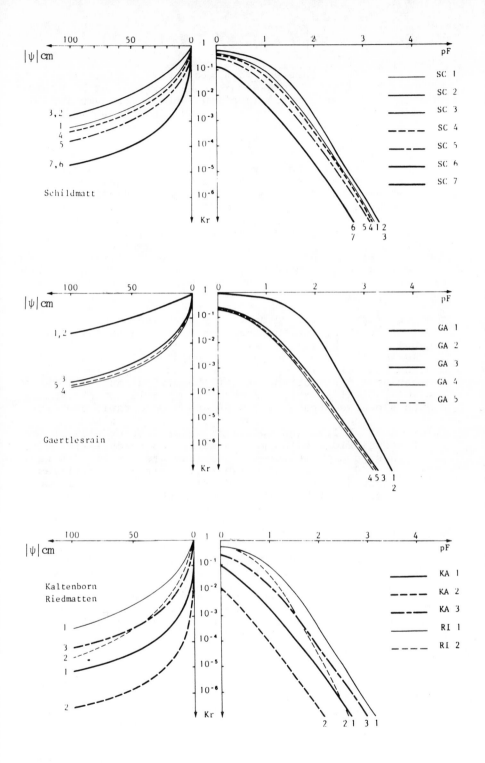

$$K(\psi) = -K_S \, S_e^{1/2} \left[\int_0^{S_e} dS_e/\psi \Big/ \int_0^1 dS_e/\psi \right]^2 \qquad (4)$$

where K_S is the saturated hydraulic conductivity. The integration of this expression can be carried out by using the van Genuchten's $S_e(\psi)$ function, when the assumptions of equation (2) for the parameters α and β are respected:

$$K(\psi) = K(S_e) = K_S \, S_e^{1/2} \left[1 - (1 - S_e^{1/\beta})^\beta \right]^2 \qquad (5)$$

The saturated conductivity K_S has not been measured on the first series of samples at Soultzeren. Equation (5) has been applied merely for predicting the relative conductivity curves $K_r = K/K_S$ from the fitted retention curves (figure 3.7). The two parameters ψ_o and α control the shape of these $K_r(\psi)$ curves at, respectively, low and high pF:
- the lower ψ_o, the more rapid is the conductivity decrease near saturation: at pF 0 or $|\psi| = 1$ cm, for GH6 ($\psi_o = 1.07$) $K_r = 10^{-2}$ and $K \approx 0.01 \, K_S$, but for GA1 ($\psi_o = 74.7$) $K_r = 1$ and $K = K_S$;
- the higher α, the steeper is the $K_r(pF)$ curve slope and the more rapid is the conductivity decrease at high pF: this can be seen, for instance, on curves GA1($\alpha = 1.648$) and GA4 ($\alpha = 1.257$).

As for the retention curves, the mineral layers (eg. GH6,MU5,KA2) with low ψ_o and α have rapidly but regularly decreasing $K_r(pF)$ curves, whilst the more organic layers (eg. GA1,SC1,RI1) exhibit a more curved relationship with a much slower decrease at low pF and a more rapid decrease at high pF than for the mineral layers (for instance, $K_r = 0.1$ at pF 1.75 and $K_r = 10^{-7}$ at pF 3.5 for GA1).

In conclusion, although the saturated conductivities K_S of these coarse-textured formations are certainly high (more than 40 cm day^{-1} at Geisberg), their conductivities decrease rapidly with a small decrease in their water content: eg. at pF 2, with a rather high water content ($\theta = 0.20$), K_r is already as low as 10^{-7} for GH6. The water fluxes in the mineral layers are very high when the soils are close to saturation, but become very low even when the soils are only a little more unsaturated. In layers rich in fines and organic matter, the saturated conductivities K_S are lower, but their conductivities remain relatively greater until a higher pF level is reached.

To analyse further the hydrodynamic behaviour of the soils at Soultzeren, it will be necessary to compare not only the shapes of the relative conductivity curves K_r but also the absolute conductivity values $K = K_S.K_r$.

CONCLUSIONS

The preliminary results obtained at Soultzeren with a limited sampling framework are in good agreement with field observations obtained elsewhere in the granitic Vosges. In such superficial formations and soils with coarse texture and high macroporosity, the water retention $\theta(\psi)$ and the unsaturated conductivity $K(\psi)$ decrease rapidly during the

desorption phase. The main differentiation between the
soil units relates to the variable content of fines and
organic matter in their top layers; the vertical anisotropy
is at least as important as the lateral one. Spatial variability of soils at Soultzeren will be detailed by further
sampling of each soil unit. The curve values at saturation
θ_S and K_S will also be systematically measured on the
sampled cores, in order to complete the experimental curves
and simplify their adjustment.

Field determination of $\theta(\psi)$ and $K(\psi)$, and also laboratory determination of $K(\psi)$ are rather difficult and time-
consuming. The method used here, which is suitable for a
statistical analysis of the fitted curves and parameters,
limits these tedious measurements to a few control sites or
cores. Serial determination of water retention curves can
be made with sand-box apparatus on undisturbed soil cores,
and the van Genuchten's $\theta(\psi)$ function gives good estimations
of the corresponding conductivity curves; the fitted
parameters also permit a global classification of the
samples to be made. The sampling work would be further
simplified if it proved possible to correlate the hydrodynamic characteristics with other soil physical properties
which are easier to sample and measure, for instance bulk
density and texture (Bloemen, 1980).

In conclusion, the experimental procedures described
(soil mapping, core sampling, laboratory determination of
retention curves, curve fitting, conductivity estimation,
field controls, statistical analysis) provide an efficient
and simple tool for assessing the spatial variability of
the soil water properties, and thereby the hydrologic
structure and behaviour of a watershed.

ACKNOWLEDGEMENTS

We wish to express our thanks to W.P. Stackman (ICW,
Wageningen) and to the Laboratoire de Science du Sol
(INRA, Rennes) for their essential contribution to this
work.
  This project has been supported by:
  - le Ministere de l'Environnement (Convention no.77-124)
  - le Centre National de la Recherche Scientifique
    (ASP PIREN no.2180)

REFERENCES

Ambroise, B., 1982, Première caractérisation hydrodynamique
        des formations superficielles et des sols des
        bassins de la Petite Fecht et du Ringelbach:
        II - Ajustement des courbes de rétention hydrique
        et estimation des courbes de conductivité
        hydraulique relative, Recherches Géographiques à
        Strasbourg, 19/20/21, 139-146.

51

Ambroise, B., Gounot, M. and Mercier, J.L., 1980, Hydrologic cycle modelling of ecosystems at the watershed level, in: *Symposium NSF-CNRS 'Disturbance and Ecosystems - Components of response'*, Stanford, USA, July 14-16, 52p.

Amiet, Y., 1980, Méthode d'approche de la spatialisation des caractéristiques hydrodynamiques des formations superficielles - Exemple du haut bassin de la Petite Fecht (Haut-Rhin), Thèse 3e cycle, ULP, Strasbourg, France, 127p.

Amiet, Y., Ambroise, B. and Mercier, J.L., 1982, Première caractérisation hydrodynamique des formations superficielles et des sols des bassins de la Petite Fecht et du Ringelbach: I - Détermination des courbes de rétention hydrique, *Recherches Géographiques à Strasbourg*, 19/20/21, 129-138.

Arya, L.M., Farrell, D.A. and Blake, G.R., 1975, A field study of soil water depletion patterns in presence of growing soybean roots: I - Determination of hydraulic properties of the soil, *Proceedings of the Soil Science Society of America*, 39, 424-430.

Bloemen, G.W., 1980, Calculation of hydraulic conductivities of soils from texture and organic matter content, *Zeitschrift für Pflanzenernährung und Bodenkunde*, 143, 581-615.

ERA 569 CNRS, 1982, Structure et fonctionnement d'un bassin versant de montagne - Exemples des bassins de la Petite Fecht et du Ringelbach à Soultzeren, Hautes Vosges, France, *Recherches Géographiques à Strasbourg*, 19/20/21, 276p.

Gardner, W.R., Hillel, D. and Benyamini, Y., 1970, Post irrigation movement of soil water: I - Redistribution, *Water Resources Research*, 6, 851-861; II - Simultaneous redistribution and evaporation, *Water Resources Research*, 6, 1148-1153.

Genuchten, M.T. van, 1980, A closed-form equation for predicting the hydraulic conductivity of unsaturated soils, *Proceedings of the Soil Science Society of America*, 44, 892-898.

Hamburger, J. and Mercier, J.L., 1978, Mesure in-situ de la diffusivité thermique d'un milieu poreux non saturé - Application à la mesure de la teneur en eau, *6e Congres International sur le Transfert de Chaleur*, Toronto, Canada, 3, 99-104.

Humbert, J., Najjar, G., Ambroise, B. and Amiet, Y., 1982, Caractéristiques morphométriques et hydrographiques des bassins de la Petite Fecht et du Ringelbach. *Recherches Géographiques à Strasbourg*, 19/20/21, 53-64.

Imbernon, J., 1981, Variabilité spatiale des caractéristiques hydrodynamiques d'un sol du Sénégal - Application au calcul d'un bilan sous culture, Thèse 3e cycle, INPG, Grenoble, France, 152p.

Keisling, T.C., Davidson, J.M., Weeks, D.L. and Morrison, R.D., 1977, Precision with which selected soil physical parameters can be estimated, *Soil Science*, 124, 241-248.

Mualem, Y., 1976, A new model for predicting the hydraulic conductivity of unsaturated porous media, *Water Resources Research*, 12, 513-522.

Nielsen, D.R., Biggar, J.W. and Erb, K.T., 1973, Spatial variability of field-measured soil-water properties, *Hilgardia*, 42, 215-259.

Pouyaud, B., 1975, La mesure de l'humidité du sol par chocs thermiques, *Cahiers ORSTOM, série Hydrologie*, 12, 259-284.

Richards, L.A., 1965, Physical condition of water in soil, in: *Methods of soil analysis - Part 1*, ed. Black, C.A. et al., American Society of Agronomy, Monograph 9, Madison, USA, 128-152.

Rogowski, A.S., 1972, Watershed physics: soil variability criteria, *Water Resources Research*, 8, 1015-1023.

Stackman, W.P., Valk, G.A. and Harst, G.G. van der, 1969, Determination of soil moisture retention curves: I - Sand-box apparatus - Range pF 0. to 2.7, *Institute for Land and Water Management Research Note* no. 81, Wageningen, The Netherlands, 19p.

Topp, G.C., Zebchuk, W.D. and Dumanski, J., 1980, The variation of in-situ measured soil water properties within soil map units, *Canadian Journal of Soil Science*, 60, 497-509.

Vachaud, G., Dancette, C., Sonko, S. and Thony, J.L., 1978, Méthodes de caractérisation hydrodynamique in situ d'un sol non saturé - Application à deux types de sol du Sénégal en vue de la détermination des termes du bilan hydrique, *Annales d'Agronomie*, 29, 1-36.

Vàrallyay, Gy., 1973, A talaj nedvesség potenciàlja és ùj berendezés annak meghatàrozàsàra az alacsony (atmosféra allati) tenziótartomànyban, *Agrokémia és Talajtan*, 22, 1-22.

Vàrallyay, Gy. and Mironenko, E.V., 1979, Soil-water relationships in saline and alkali conditions, *Agrokémia és Talajtan*, 28(Suppl.), 33-82.

Wright, R.L. and Wilson, S.R., 1979, On the analysis of soil variability with an example from Spain, *Geoderma*, 22, 297-313.

4 Pipeflow and pipe erosion in the Maesnant

experimental catchment

J.A.A.Jones and F.G. Crane

THE RESEARCH DESIGN

The current experimental work was begun in 1979 and aimed primarily at determining the hydrological role of natural piping in a humid upland catchment, using the Maesnant basin in mid-Wales as a study area (figure 4.1). Preliminary work had suggested that pipeflow may call for a significant modification of the theory of streamflow generation advocated by Hewlett and Nutter (1970), and, in particular, that it may provide the much-needed speed of transmission for throughflow that has been a major source of criticism of the theory following the field observations of Dunne and Black (1970) and the computer simulations of Freeze (1972). An outline for such a conceptual model has been described elsewhere by Jones (1979).

The research has involved three major areas of enquiry, namely, i) the spatial extent, connectivity and general nature of the pipe networks, ii) the relative importance of pipeflow to stream discharge vis-à-vis other drainage pathways such as overland flow and diffuse seepage, and iii) the factors responsible for generating pipeflow, as a first step towards developing a computer model of pipeflow contributions. During the progress of the work it became evident that many of the pipes were shifting much larger sediment loads than had previously been supposed in the British environment (Jones, 1981) and a bedload monitoring scheme was set up at four pipe outlets with contrasting flow regimes.

In devising a field experiment to test the conceptual model, a number of problems were encountered which are common to many such schemes, although perhaps somewhat exaggerated by the special nature of piping networks. In the first instance, no available maps could possibly provide an adequate basis for devising a sampling scheme, nor do these subsurface networks have uniquely recognisable vegetative or topographic expression. Although, for example, Conacher and Dalrymple (1977) have associated piping with specific units in their nine-unit land surface model and have advocated the use of this model as a frame-

work for instrument location in a stratified sampling
design (Conacher and Dalrymple, 1978), it is clear first
that piping is not <u>necessarily</u> present in any specific
unit, and secondly that the hydrological or erosive importance of pipes is not uniquely determined by the landsurface unit. By the same token, an experimental plot
approach, whilst economical in terms of effort, was clearly
inappropriate given the spatial variability in response
inherent in the piping process.

A necessary prerequisite for sampling was, therefore,
an extensive and detailed mapping of the pipe networks.
This was begun from pipe outlets near to the stream, since
it is these outlets which must be monitored in order to
answer the questions of volume and timing of pipe discharge
entering the stream, which are central to the main proposi-
tion that pipeflow is a major source of storm runoff in the
basin. It was soon apparent that the number of recognisable
outlets increased during storms, and extra monitoring
points were set up to cover these high flow outlets. Upslope tracing and monitoring of the pipe networks was
needed to answer questions of the sources of pipeflow and
the factors causing variations in response.

A second set of problems centre on the variability in
response encountered between different pipes. In order to
devise a sampling scheme of known accuracy, one needs
reliable estimates of the variance in the phenomena to be
measured, estimates which are commonly lacking for hydrological and geomorphological processes. This deficiency
may be overcome either by a pilot sampling scheme to
estimate population parameters or, as in this instance, by
abrogating sampling altogether and measuring the complete
population. Although it proved possible to do this with
the pipes, competing drainage processes such as overland
flow, diffuse throughflow and saturated flow from riparian
seepage hollows were not susceptible to a complete monitoring scheme nor, with the possible exception of overland
flow, is it feasible to carry out a pilot sampling scheme
on these prior to running an experiment. There are two
main problems here, the disruption caused and the cost.
The interference with the natural environment, caused by
excavating sampling pits and inserting instruments, is so
great that sites have to be minimised; this is an extreme
case of the act of measurement interfering with the phenomena being measured. Two categories of riparian seepage
hollow were recognised in the field, namely, 1) those small
enough ($\leq$ 5 m wide) to permit a trench to be dug completely
across their outfall down to the impermeable clayey parent
material, and lined with damp-proofing grade polythene
directing drainage to a sharp-crested weir, and 2) larger
ones where just 5 m of the complete seepage zone could be
thus measured and total yield was estimated on a geometric
basis, assuming that 5 m sample to be representative.
Similarly, the investment in time, effort and relatively
expensive instrumentation, such as a Scanivalve tensiometer
system, at any one site also demands that only a small
number of sites are selected.

Given these constraints, the investigation was designed
to provide for total and continuous monitoring of all pipe-

flow inputs to the stream and to estimate inputs from other subsurface sources by a combination of measurement and extrapolation, using seepage hollow weirs and a scanivalve system. In the case of overland flow, pilot surveys with crest stage gauges have indicated that, outside the seepage hollows, overland flow is closely associated with pipe networks, and spatially restricted. Because of this, and in view of limited resources, overland flow contributions are not being continuously monitored, but may be estimated from the level of 'unexplained contributions' from the monitored sources.

The field constraints also mean that classical forms of experimental design (cf. Cox, 1958) are inappropriate here. Nevertheless, the research design contains the six distinctive characteristics of an experiment outlined by Church (this volume). It also provides the framework for many 'sub'-experiments, for example, on dye injection and movement, in which the element of control is more perceptible. At the same time, the pipe bed load survey has indicated that further geomorphological experiments must be designed within the context of the hydrological programme to determine processes and relative rates of erosion within the basin system, and extending the earlier work of Lewin, Cryer and Harrison (1974).

THE EXPERIMENTAL CATCHMENT

The Maesnant catchment covers an area of 0.54 km^2 falling from 752 m O.D. on the summit of Plynlimon to 470 m O.D. at the Welsh Water Authority sharp-crested weir. It is underlain by Ordovician greywacke, mudstone and grits with partial coverings of soliflucted material forming 4-5 m high terrace scars flanking the stream, and areas of peat 1-2 m deep, especially in the headwater zone fringing the scree slopes which fall directly from the summit of Plynlimon. Soils are generally poorly developed, ranging from rankers on the upper slopes through peaty soils and peaty gleyed podzols of the Hiraethog Association (Rudeforth, 1970). Measured annual rainfall lies in the range 1700-2100 mm, with a tendency towards a slight rainfall peak in early winter and an annual water surplus of about 1450 mm. Median stream discharge at the outfall is 0.015 m^3 s^{-1}.

PIPEFLOW RESPONSE

Mapping of the pipe network revealed a much more extensive network than expected. The longest perennially-flowing pipes with typical diameters of 8-10 cm extend over 330 m across the 4° river terrace slope and the ephemeral feeder pipes (c. 5 cm diameter) on the steeper, 15° slopes above extend a further 250 m or more. Dye tracing experiments have proven direct connections between many of these ephemeral pipes and the perennial network. However, baseflow in the perennial pipes can be traced back to an origin in the saturated zones at the upper end of the river terrace

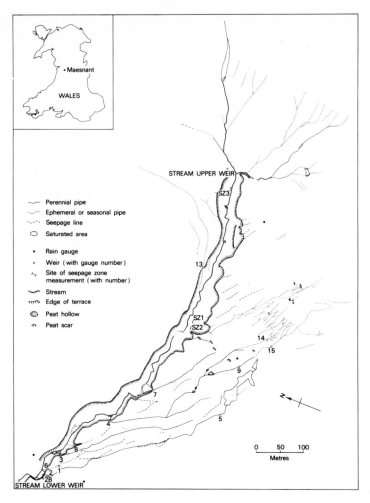

Figure 4.1.   The Maesnant Experimental Catchment: pipe
networks and monitoring sites.

which act as 'secondary source areas'. Pipeflow sampling
stations have been established at three categories of site.
Weirs have been inserted on all pipe outlets near the
stream to quantify pipeflow contributions to runoff. Two
pipeflow weirs have also been established near the lower
end of the ephemeral network (14 and 15 on figure 4.1) and
two at the upper end of the perennial network near the outfall from the saturated zones (5 and 9 on figure 4.1), to
provide data on the source of pipeflow contributions, the
function and response of the upper ephemeral network and
the transmission of stormflow across the slope. The data
are being used in a computer modelling experiment to predict contributions. The records suggest that the storm
hydrograph is recognisably established at the head of
perennial pipes 4 and 2, approximately 4 hr before it

reaches the outlets at the stream edge, but that there is a slight smoothing of the response and a marked increase in discharge over the 300 m distance, much of which can be explained by inflow from tributary pipes. The storm response appears to be fed by the equivalent of 'channel precipitation' on the saturated zones, by ephemeral pipe discharge often channelled through the saturated zones, together with short distance overland flow captured by 'blow-holes' in the pipe network, and possibly some augmentation of groundwater seepage. Routing of discharge from the upper ephemeral pipes through the saturated zones could explain the large difference in peak lag times between 1 hr at the lower end of the ephemerals' slope and 4-5 hr at the terrace scar, when dye tracing suggests travel times of only 1-2 hr through the perennial pipes alone.

In addition to the pipeflow monitoring sites, diffuse throughflow contributions are being estimated from three weirs installed across seepage areas in hollows near to the stream and from a scanivalve tensiometer system on a percoline at site 1 (figure 4.1), together with two phreatic surface float gauges. The extent and frequency of overland flow, as estimated from a combination of crest stage gauges and the extent of flattened vegetation after a storm, would appear to suggest that overland flow occurs principally in association with the pipe network, either as a supplier of pipeflow or as overflow from pipes when their storm capacity is exceeded. Overland flow tends only to contribute directly to stream discharge as localised saturation overland flow from riparian seepage zones or from the outlets of pipes on the terrace scar. In these cases it is basically a form of seepage zone or pipeflow contribution.

The hydrological response of both pipes and seepage zones appears to be remarkably similar in form and timing (figure 4.2). Indeed, there seems to be at least as much variation within these groups as between them. Two groups of riparian seepage zone can be distinguished in terms of response, broadly, but not entirely coincident with the two types of sampling sites referred to above. These are i) zones occupying marked hollows with steep head-walls, some evidence of landslipping and of feeding from short distance piping, which display a flashy response above a baseflow (figure 4.2; SZ1 and SZ2 on figure 4.1), and ii) broader, generally drier seepage zones in less well defined topographic hollows with more subdued response and discontinuous baseflow (SZ3 on figure 4.1). A combination of ground survey and interpretation of hang-glider photography is being used to estimate the extent of these seepage zones, to classify them and to extrapolate from the three sample sites that we have been able to establish. The pipes seem to fall into four clear groups, namely, i) perennially flowing pipes with flashy responses (cf. pipes 4 and 7 in figure 4.2), ii) perennially flowing pipes with less peaked response due to more limited flow capacities and overspilling into distributary pipes at higher flows (cf. pipe 3 in figure 4.2), iii) ephemerally active or seasonally active overflow pipes with flashy responses which do not necessarily react to every storm

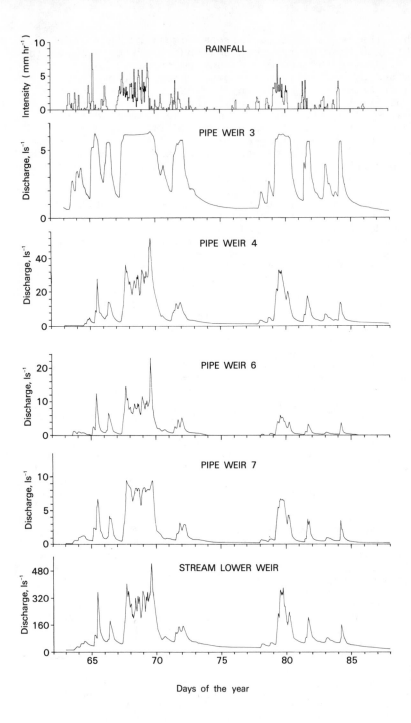

Figure 4.2. Rainfall and runoff for the stream and selected pipes during early spring 1981.

Table 4.1. Drainage Pathway Response in the Maesnant Catchment

(Time in hr)

Weir	Start lag time	Rise time	Peak lag time	Timing relative to stream lower weir Start	End	Total stormflow volume (m³) $\bar{x}$	max	Maximum peak discharge ($l\ s^{-1}$)
2	4.6	13.2	7.0	-1.4+	-4.8	156	431	4.9
2B	6.2	11.5	7.4	-1.7	-29.0	88	260	3.7
3	4.8	13.2	11.8	0.5	-7.2	474	736	6.5
4	4.8	8.6	7.7	2.2	2.2	535	2281	25.8
8	7.4	12.7	11.0	5.3	-12.0	40(52)*	273	10.0
SZ1	1.9	11.0	5.8	-2.4	-19.7	49	138	2.9
Stream, upper	7.9	8.4	7.9	2.2	0.5	2794	10523	82.2
Stream, lower	4.8	10.1	7.0	-	-	4658	17196	141.3

+ Negative times indicate response before lower stream weir.

* Using only storms in which pipe responded in parenthesis.

which appears on the hydrographs of pipes in categories
(i) and (ii) (cf. pipe 6 in figure 4.2), and (iv) ephemeral
headwater pipes, which also sometimes fail to react.

The diversity of response is illustrated in table 4.1
in terms of lag times, rise times and mean discharges for a
few selected examples from the major groups of pipe, calculated from the record of autumn and early winter 1980. Two
salient points emerge in the table. First, pipes and
seepage zones tend to begin to respond within a very short
time of the response at the lower stream weir. Many start
to respond before the lower stream weir and all except
extreme examples of ephemeral overflow pipes (pipe 8) begin
to respond before the upper stream weir. Presumably the
pipes and seepage zones are therefore responsible for the
average 3 hr reduction in mean response time between the
upper and lower stream weirs. Secondly, all except the
most extreme perennial case (pipe 4) have finished their
response well before the stream. Much of the late recession
drainage in the stream seems to be coming from above the
upper weir, from pools and bogs in the headwaters. The
mean discharges show the wide differences between pipes,
with total stormflows reaching a maximum in pipe 4 and a
minimum in the overflow pipes 2 and 8. However, pipe 8 has
relatively high peak discharges.

Flow duration curves show the different categories of
pipe response very clearly (figure 4.3). The winter flow
duration curve for pipe 3 shows a marked convexity towards
the origin which is typical of the pipes with constricted
flows during high discharges in group (ii). Flow levels are
lower in summer and the convexity does not appear.
Similarly, no such constriction is apparent in the flow
duration curves for group(i)pipes, as illustrated by the
annual curve for pipe 4 in figure 4.3B. Nor, indeed, does
pipe 8 show any such limitation to flow. However, pipe 2B,
the overflow distributary to pipe 2, does reach capacity in
winter, and plotting discharge in 2 against discharge in
2B has shown a very interesting relationship. Flow increases rapidly in 2 up to 1.5-2 l s^{-1} when 2B begins to
flow. Flow then rapidly rises in 2B up to 2.25 l s^{-1} with
a parallel rise of only O.5-1 l s^{-1} in pipe 2. Above this
level there is another rapid rise in discharge in pipe 2 to
5 l s^{-1} or above with virtually no increase in flow in 2B,
presumably largely achieved by flow under pressure in
pipe 2. Although the existence of pressure flow in pipes
has been doubted by Gilman and Newson (1980),it clearly
exists on pipe 3, where a 5 cm fountain can form during
high flows.

PIPE EROSION

These contrasts in flow regime seem to be reflected in the
erosive activity of the different pipes. This might well
explain the curious contrast in physiographic expression
between the outlets of pipes 3 and 8. Pipe 8 occupies a
major hollow in the terrace scar implying the removal of
190 m^3 of material, whilst pipe 3 has no physiographic or
vegetative expression (figure 4.4). Dye tracing indicates

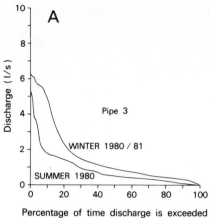

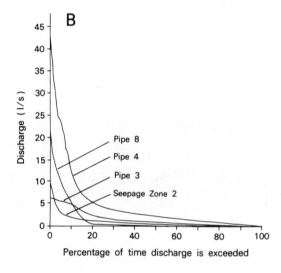

Figure 4.3.   Flow duration curves: A, pipe 3 winter and
    summer half years; B, comparison of annual curves.

that pipe 8 is at least partly activated by overspill
from 3.   Whilst pipe 3 has a perennial but 'constricted'
flow regime with discharges generally below 7 l s^{-1} (group
(ii)), pipe 8 is ephemeral but flashy.   Pipe 8 does not
respond to every stormflow recorded in pipe 3 (table 4.1),
but when it does it is capable of delivering over 400 m^3 of
water with peak discharges in excess of 20 l s^{-1}.   This is
clearly illustrated in the annual flow duration curves in
figure 4.3B.   The evidence now tends to refute an earlier
hypothesis that the contrasting topographic expressions of
pipes 3 and 8 may have been due to subsurface drainage
capture by pipe 3 depriving pipe 8 of a more continuous
flow regime.

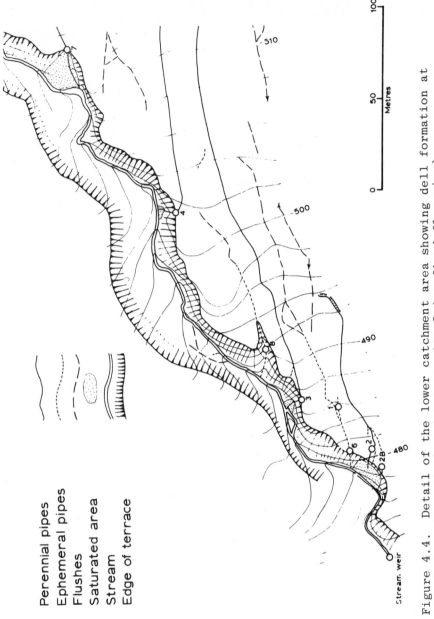

Perennial pipes
Ephemeral pipes
Flushes
Saturated area
Stream
Edge of terrace

Figure 4.4. Detail of the lower catchment area showing dell formation at site 8 (overflow pipe) and not at site 3 (confined-flow pipe).

64

Table 4.2. Bedload movement in selected pipes

A. Mean weekly bedload sediment by season (kg week^{-1})

Pipe	Winter 1979/80	Summer 1980	Winter 1980/81	Summer (partial) 1981
2		0.557	1.022	·0.459
3	-	0.210	0.474	0.052
4	6.168	1.680	3.852	0.267
7	-	0.774	2.152	0.169

B. Product moment correlation coefficients between pipeflow
parameters and rates of bedload transport

Pipe	Flow levels at % points of flow duration curve				Maximum discharge
	25%	15%	5%	2%	
2	0.64	0.66	0.71	0.61	0.70
4	0.63	0.79	0.93	0.95	0.55
7	0.88	0.90	0.98	0.98	0.78
2+3*+4+7	0.71	0.83	0.95	0.94	0.78

*Insufficient values to merit separate treatment

Bulk bedload sediment yields have been estimated for
pipes 2, 3, 4 and 7 from traps set in the weir pools (table
4.2). These show winter loads approximately twice as large
as summer loads, which is closely commensurate with discharge durations. In all cases the relatively low discharge
levels during the early summer of 1981 are matched by the
lowest weekly sediment yields recorded. The main contrasts
between pipes are between the flashy regimes of pipes 4 and
7 and the 'constricted' regimes of 2 and 3. Not surpris-
ingly, pipe 4 which has the highest discharges and highest
total stormflow volumes of all has the greatest sediment
yield, shifting of the order of·6 kg per week in winter and
nearly 2 kg per week in summer. The most marked feature is
the overall volumes recorded. The traps on pipe 4 have
collected 290 kg during 21 months of monitoring. Over the
18 month period prior to June 1981 the other samplers
together collected 140 kg. Extrapolating for unmeasured
pipes suggests a total sediment yield of over 300 kg yr^{-1}
or approximately 15% of the 2 t yr^{-1} total sediment yield
for the catchment estimated by Lewin, Cryer and Harrison
(1974) in 1971-72.
    Approximate though this estimate is, it is considerably
greater than suggested by previous measurements in this
country. For example, Newson (1976) reported 0.5 kg,
mainly peat nodules, trapped in a small ephemeral pipe
during the wet winter of 1974-75. The contrast appears
even greater when it is remembered that most of the eroded
peat passing through the Maesnant pipes is not being

65

collected as bedload; only 1-2% of the samples is organic material. The sediment is predominantly coarse sand and gravel (65-90%) derived mainly from bed erosion, with median grain sizes also showing a seasonal trend, for example, on pipe 4 from a D_{50} in winter of 1.6 mm to one of 0.7 mm in summer. The seasonal contrasts are greatest on pipe 4 commensurate with the greater seasonal contrasts in flow durations.

There is a clear relationship between rate of sediment movement and the rarer storm events, as illustrated in table 4.2B. The product-moment correlation coefficient between sediment weight and discharge levels at specific flow duration points during the collection period show a steady, monotonic increase from the 25% to the 5% duration levels for each pipe, and for all pipes taken together (correlations significant at 5% level). The pattern is best shown on pipe 4 where the improvement in correlation between the 25% and 5% points is significant at the 1% level. So too is the subsequent fall in correlation between the 2% flow duration point and maximum discharge levels. The only exception to the pattern appears to be pipe 2, where the confined flow features seem to be responsible. Confined flow is encountered at c. 2 l s^{-1} which is a common highflow level in summer; hence, 5% and 2% flow levels are very similar in summer. However, in winter high flow levels are greater, often exceeding the capacity of the overflow (2B), and probably generating pressure flow in pipe 2. In winter 5% duration levels are c. 3 l s^{-1} compared with 4 l s^{-1} at the 2% point; hence there is a different relationship between the 5% and 2% discharges according to season.

VOLUMETRIC SIGNIFICANCE OF PIPEFLOW TO STREAMFLOW

Both pipeflow discharges and total volumetric yields during stormflow conditions appear, like sediment yield, to be much higher than suggested by any previous work in this country (cf. Ward, 1975; Jones, 1978; Gilman and Newson, 1980; Jones, 1981). Table 4.3 takes the records for September 1980 to March 1981 and bulks together the best sets of records for each category of pipe regime to arrive at an overall estimate that 46% of the increment in streamflow across the study reach can be explained as contributions from pipeflow. The bulk of this comes via perennial outlets, although much of this will have passed through ephemeral tributary pipes. On average only c. 6% is actually derived from ephemeral overflow pipes near the terrace edge. However, this overflow contribution can be considerably more significant in certain storms, and indeed, the same can be said for the main perennial arteries. Figure 4.5 plots contributions from the main perennial pipes against 7-day antecedent rainfall totals and total storm rainfall. The scattergram adds weight to earlier evidence that pipeflow contributions are of reduced significance both under very wet antecedent conditions and in the heavier rainstorms (Jones, 1978; Jones and Crane, 1982). There is also some additional evidence here for another

Table 4.3. Mean percentage contribution to stormflow increment between stream weirs

Pipeflow regime	Site identifier	No. of events employed	% contribution		Combined into categories No. of events	% contribution
Group(i), perennial	4	84	15.5	)	17	15.7
"	7	17	2.4	)		
Group(ii), perennial	2	49	6.5	)	40	16.8
"	3	70	10.3	)		
Group(iii), ephemeral	2B	26	1.0	)	7	6.0
"	6	37	3.2	)		
"	8	26	1.7	)		
Seepage zone, piped	SZ1	43	2.6	)	41	5.6
"	SZ2	41	3.0	)		
Seepage zone	SZ3	38	0.2			

Estimated mean total pipe contribution (excluding seepage zone 3) = 46.2%

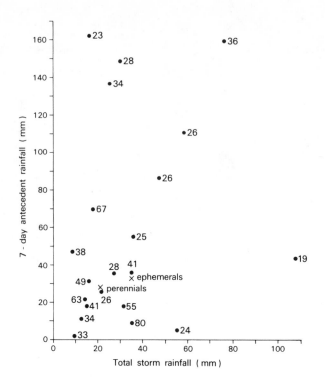

Figure 4.5. Percentage contribution of pipeflow to streamflow compared with 7-day antecedent rainfall totals and
total storm rainfall. Points marked 'perennials' and
'ephemerals' indicate mean positions for the six
highest contributions from group (i) and group (iii) pipes
respectively.

fall-off in percentage contributions in drier antecedent
conditions and in the lighter storms (cf. Jones and Crane,
1982).

The implication is that under these circumstances other
drainage processes are more effective. We now know, from
more than a year of monitoring the ephemeral overflow pipes,
that part of the answer is that the overflow pipes are
activated on wetter occasions, so that the 'fall-off' along
the principal diagonal of the scattergram is delayed somewhat when these are included. For clarity, the mean position of the six top-ranking contributions out of 16 events
in these overflow pipes has been plotted on figure 4.5 and
compared with a similar value for the main perennials; the
mean percentages here are 14% and 57% respectively. It
seems likely that the most important alternative contributor to streamflow is drainage from riparian seepage
zones.

FACTORS RESPONSIBLE FOR PIPEFLOW GENERATION

Two approaches have been adopted to determine the factors controlling pipeflow. In the first instance, statistical analyses are being made of the relationships between pipe-flow parameters and parameters of rainfall and antecedent moisture conditions. Secondly, both qualitative and quantitative dye injection experiments have been begun in order to obtain positive, physical evidence for the sources of pipe water. The latter approach needs considerably more work and will be a matter for a future report.

Table 4.4 summarises the statistically significant factors identified by multiple regression for one of each of the four categories of pipeflow regime for a number of pipeflow parameters. The factors are listed in order of significance, as determined by F-values in a stepwise procedure. Correlation analyses were performed using the variables described in table 4.4 in the most logical groupings for the different dependent parameters. To overcome problems induced by multicollinearity, the regressions quoted here contained only one predictor variable from any significantly cross-correlated group, i.e. the one most closely correlated with the dependent variable.

In most cases a clear distinction can be made between the factors governing flow in the perennial pipes and those for the ephemeral pipes. Total volume of stormflow is here defined as the integrated flow volume occurring above a long-term baseflow level (where the pipes have baseflow), with the recession curve at the end of each storm extrapolated linearly to baseflow - in a sequence of storms extrapolated volumes are added to the recession or subtracted from the rising limb as appropriate. There appears to be a distinction here between the main perennial pipe, 4, and the upslope ephemeral tributary, 15. For the ephemeral pipe, short-term antecedent conditions are more significant than longer-term antecedent conditions, whilst the reverse is true for the perennials - in fact, the 7-day antecedent rainfall total is a reasonable substitute for the pre-storm flow levels in table 4.4 giving a multiple correlation of 0.84 based on 84 cases. This would seem to suggest that near-surface wetness is more significant in determining the severity of a flow event in the shallow, upslope ephemeral pipes which run at a depth of about 15 cm, whereas in the lower slope perennials running at depths of 30-40 cm phreatic surface levels are more important.

Again, peak discharges in the perennial pipes 2 and 4 are closely related to flow levels immediately prior to the storm, whereas the response in the ephemeral network is probably, like pipe 15, controlled more by rainfall intensities before the hydrograph peak. Another point of contrast is that the perennial network, including the overflow pipes, responds more closely to total rainfall than to mean rainfall intensities. Since lag times are markedly shorter in the ephemeral network and the feeder routes perhaps more direct and singular in nature, rainfall intensities might well be expected to be an important determinant of flow concentration. The perennial network, however, is slower to react and must integrate flow inputs

Table 4.4. Factors affecting pipe response as suggested by stepwise multiple regression

Flow Parameter	Pipe	Variables and Beta Coefficients in full equation[†]						Multiple Correlation Coefficient	Sample Size
Total volume of stormflow	2	R_{tot}	.69*					.69	55
	4	R_{tot}	.91*	Q_b	.13			.93	94
	8	R_{tot}	.58*					.58	31
	15	A_1	.46	R_{tot}	.44			.58	20
Peak discharge	2	Q_b	.49*	R_{pp}	.44			.67	55
	4	R_{pp}	.63*	Q_b	.27*	I_{max}	.25*	.85	94
	8	R_{pp}	.58*					.58	31
	15	I_{pp}	.78*					.78	20
Rate of rise	2	I_{pp}	.39*					.39	55
	4	I_{max}	.56*	A_1	.35*			.68	84
	8	Q_b	.67*	I_{pp}	.27			.73	31
	15	none significant							20
Start lag time	2	I_{ps}	-.37*					.37	55
	4	I_{ps}	-.30					.30	74
	8	none significant						-	31
	15	I_{ps}	-.43					.43	20
Peak lag	2)							-	55
	4)	none significant						-	84
	8)							-	31
	15	I_{pp}	-.51					.51	20

[†]Variable names: R_{tot} - total storm rainfall

R_{pp} -   "   "   "   prior to pipeflow peak

I_{pp} - mean rainfall intensity prior to pipeflow peak

I_{ps} -   "   "   "   "   "   " start

I_{max} - maximum "   "

Q_b - discharge at start of stormflow

A_1, A_4 - one and four day total antecedent rainfall

Significance levels indicated by analysis of variance: 1% starred, otherwise 5%.

from a number of sources. Presumably, this blurs the
effects of shorter term rainfall intensities. The overflow
pipe, 8, differs from the main perennial network only
insofar as flow levels prior to the storm do not generally
affect peak response because there is no baseflow. Nevertheless, when there is still some recession flow in pipe 8
at the start of a storm, this does appear to favour a
steeper hydrograph rise.

Although mean rainfall intensities prior to maximum
discharge do not appear to be significant to peak discharge
levels in the perennial pipes, maximum rainfall intensity
certainly is. It is a reasonable substitute for the total
rainfall in the equation for pipe 2 ($R = 0.61$, $n = 55$) and
falls just short of the conventional level of significance
on the overflow pipe (8). However, pre-peakflow rainfall
intensities are important for group(ii) pipes and their
overflows (group(iii)) in terms of the rate of rise of the
hydrograph.

Mean intensity plays a similar role in the ephemeral
pipes, but the factors controlling rate of hydrograph rise
in the group(i) pipes are not so easily explained. Here
the importance of maximum rainfall intensities and short-
term antecedent rainfall suggests that steeper rising limbs
are associated with well wetted surface layers and intense
rainfall that is rapidly transmitted across, around or
through those layers. Even so, any such effects could not
be found in terms of peak lag times on these pipes. On the
other hand, peak lag times could only be satisfactorily
'explained' in the case of the ephemeral pipes in group(iv).

Rainfall intensities again figure in the equations for
start lag time, the time between the start of rainfall and
the start of pipe response, indicating shorter lag times
with greater average rainfall intensities prior to the
start of stormflow in all groups of pipes.

CONCLUSIONS

The work clearly demonstrates that pipes are a major source
of stream runoff in this catchment and it substantiates the
conceptual model of runoff processes under examination. In
so doing, it has revealed much greater variation between
pipes than hitherto suspected and the categories of flow
regime need to be expanded considerably beyond a simple
division into ephemeral and perennial. In studying these
differences, some progress has been made towards explaining
the causes of this variation and the next phase of the
research will be moving beyond the stage of field monitoring
into computer simulation experiments and field experiments
with tracer dyes. In fact, the computer modelling will be
a consummation of the present field experiment insofar as
physical prediction is the highest form of explanation.

It is hoped that erosion studies can follow a parallel
course. The bedload traps have revealed that pipeflow on
Maesnant is not quite as insignificant an agent of erosion
as has previously been thought in this country. Bulk loads
are very closely related to discharge levels in rare events,
i.e. events with frequencies of 5% or less, but a continuous

monitoring scheme is needed to better understand the thresholds and to truly model the processes.

REFERENCES

Church, M., 1983, On experimental method in geomorphology, in: *Field Experiments in Fluvial Geomorphology*, ed. Burt, T.P. and Walling, D.E., (GeoBooks, Norwich).

Conacher, A.J. and Dalrymple, J.B., 1977, The nine unit landsurface model : an approach to pedogeomorphic research, *Geoderma*, 18, 1-154.

Conacher, A.J. and Dalrymple, J.B., 1978, Identification, measurement and interpretation of some pedogeomorphic processes, *Zeitschrift für Geomorphologie*, Supp. Band 29, 1-9.

Cox, D.R., 1958, *Planning of Experiments*, (Wiley, New York).

Dunne, T. and Black, R.D., 1970, An experimental investigation of runoff production in permeable soils, *Water Resources Research*, 6, 478-490.

Freeze, R.A., 1972, Role of subsurface flow in generating surface runoff - 2. Upland source areas, *Water Resources Research*, 8, 609-623.

Gilman, K. and Newson, M.D., 1980, *Soil pipes and pipeflow - a hydrological study in upland Wales*, British Geomorphological Research Group Monograph no.1, (GeoBooks, Norwich).

Hewlett, J.D. and Nutter, W.L., 1970, The varying source area of streamflow from upland basins. *Proceedings of Symposium on Interdisciplinary Aspects of Watershed Management*, Montana State University, 65-83.

Jones, J.A.A., 1978, Soil pipe networks : distribution and discharge, *Cambria*, 5, 1-21.

Jones, J.A.A., 1979, Extending the Hewlett model of stream runoff generation, *Area*, 11, 110-114.

Jones, J.A.A., 1981, *The nature of soil piping : a review of research*, British Geomorphological Research Group Monograph no.3, (GeoBooks, Norwich).

Jones, J.A.A. and Crane, F.G., 1982, New evidence of rapid interflow contributions to the streamflow hydrograph, *Beiträge zür Hydrologie*, Sonderheft 3, 219-232.

Lewin, J., Cryer, R. and Harrison, D.I., 1974, Sources for sediments and solutes in mid-Wales, in: *Fluvial Processes in Instrumented Catchments*, ed. Gregory, K.J. and Walling, D.E., Institute of British Geographers Special Publication No.6, (GeoBooks, Norwich).

Newson, M.D., 1976, Soil piping in upland Wales : a call for more information, *Cambria*, 1, 33-39.

5 Floodplain response of a small tropical stream

Stephen Nortcliff and John B. Thornes

INTRODUCTION

In 1977 and 1978 we instrumented a simple hillslope and
floodplain in a small catchment near Manaus in Amazonas,
Brazil. In earlier papers we have outlined a cup analogy
for micropore and macropore differentiation of flows
(Nortcliff and Thornes, 1978), discussed the implications
of this model for solute and water fluxes (Nortcliff,
Thornes and Waylen, 1979) and modelled and estimated the
subsurface storm fluxes at various points and depths on the
hillside (Nortcliff and Thornes, 1981).
    It is evident that in addition to differentiating
between micropore and macropore drainage in the soil, we
can identify in the environment as a whole zones in which
residence times are short and throughput relatively rapid
and those in which residence times are longer and volumetric turnover correspondingly slower. In the same way
that the micro-macropore difference must be significant to
solute budgeting, so too must be the broad configuration
of the catchment. In particular, the performance of the
floodplain environment in these tropical catchments may be
especially critical for the maintenance of sufficient
nutrients in river water, in the form of macroscopic
organic matter, and hence for aquatic life, especially for
fish stocks. The floodplain zones are also critical in
the response to tree felling in these environments. Hillslope channel coupling for both water and sediment usually
takes place across the floodplains, and to this extent the
control of particulate matter in the channels, if not soil
erosion on the hillsides, is largely operational through
the floodplain. In this paper we shall show that, contrary
to widely held beliefs among conservationists and forest
managers, the runoff dynamics may also very largely be
related to activity on the floodplain rather than to hillslope hydrology. In this sense the paper may be regarded
as a contribution to the understanding of agricultural
practice. In another sense, it is relevant to the debate
concerning the relative contribution of subsurface stormflow to catchment quickflow in different climatic

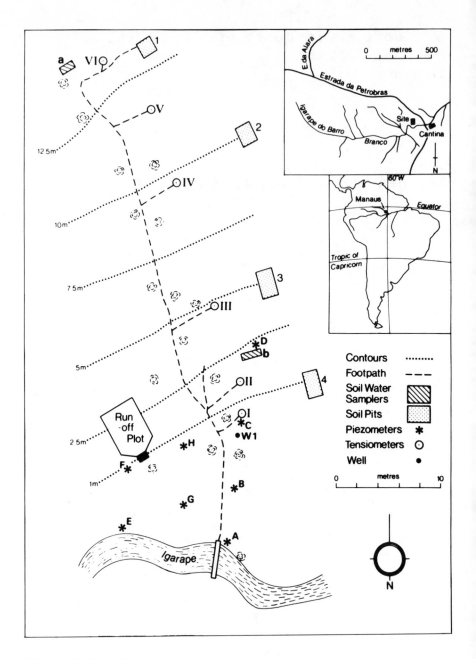

Figure 5.1. Map to show instrumentation of experimental slope. The inset shows the general location and the location within the reserve of the experimental site.

environments and under different soil conditions (Freeze, 1972; Dunne, 1978; Mosley, 1979; Bonnell, Gilmour and Sinclair, 1981). We do not hold the view that a dichotomy exists between alternative models, but rather that a variety of flow-generating mechanisms will occur in different environments and in the same environment at different times.

The catchment investigated is described in detail in Nortcliff and Thornes (1978), and only a brief description will be given here. It lies in Reserva Ducke, a tropical forest reserve some 26 km northeast of Manaus (figure 5.1). The area is underlain by Tertiary deposits of the Barreiras Series which have a wide textural range, and is covered by three forest types, 1) Riverine Forest, 2) Carrasco Forest and 3) Terra Firme Forest. The experimental hillslope and floodplain are located in the second of these types which has canopy heights of 22-32 m and is quite heterogenous, with an understory of numerous seedlings and saplings and some herbaceous plants and palms. The soils of the hillslope are plinthic haplorthox (figure 5.2), composed predominantly of kaolinitic clays and quartz, with few weatherable minerals and having hardened plinthitic nodules at depth within the profile. The dominant features are a strongly developed root mat down to 15 cm, poorly expressed soil horizons with merging boundaries, and a persistent nodular laterite layer between 100 and 140 cm, running more

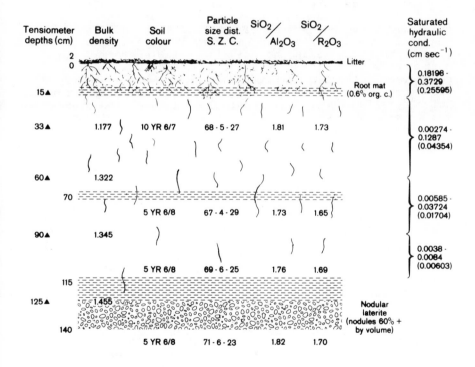

Figure 5.2.   Schematic soil profile showing structural and
    hydrological parameters.

75

or less parallel to the surface. The soils have sandy clay
loam textures throughout, but have very high porosities
(c.50%) because of the well developed, very stable micro-
aggregates which proved very difficult to disperse. The
average values of saturated hydraulic conductivity reveal
a strong vertical gradient, with values of 2.5×10^{-1} cm s^{-1}
in the organic mat and values of 6.0×10^{-3} cm s^{-1} below
1 m depth. Even the latter are quite high, however. At
the foot of the slope and across the floodplain the soils
are developed on white sands with a thin layer of down-
washed latosolic material at the slope foot. These soils
are strongly gleyed with ochreous mottles and stainings
around root channels within a greyish matrix. The floodplains continue well up into the headwater areas of the
catchment and, as in many other areas, are separated from
the hillslopes by a sharp break of slope. The slopes
usually about 25° and 10-40 m long lead to broad gently
undulating interfluves. The channels are sandy bedded and
almost rectangular in section, reaching a maximum of 1.5 m
wide at the gauging station. We were not able to determine
the base of the sands and there is the possibility that we
have a 'leaky' basin.

Typically the temperatures range from 14.3 to 37°C, the
average daily maximum being about 33°C. The wettest months
are March and April (about 400 mm) and the driest is
September (about 50 mm). The rainfall distributions for
the two periods investigated shows strong seasonal contrast.
The first period was characterised by regular and frequent
rainfalls with four storms of appreciable size. In
contrast, the second period had very little rain, with long
gaps between the falls and only two notable storms. In
both periods the storms tended to be relatively concentrated, occurring mainly in the early morning hours.
The storms marked with arrows (figure 5.3) are those for
which detailed records of tension, moisture and piezometric data have been collected and form the basis of the
discussions in the succeeding sections.

HYDROGRAPHS

A preliminary commentary on the stage-hydrographs for the
Reserva Ducke catchment has been published by Franken
(1980). In our analysis we have followed the terminology
used by Mosley (1979). The hydrographs are based on two
gauging points. For 1977 they are based on an autographic
gauge installed by Franken and corrected by readings
collected near the experimental site. In 1978 a new autographic gauge was installed by the site to avoid the backwater effects of a downstream weir which influenced some of
the highest flows recorded in the previous year.

Analysis of the hydrographs has focussed on the question
of what is the major source of quickflow. Recent investigations of forest areas (Mosley, 1979; Bonnell et al., 1981)
both suggest large and significant subsurface stormflow
inputs to the quickflow, in apparent contrast to Freeze
(1972). However, in both cases there is evidence of a
strong impeding layer within the profile and short steep

76

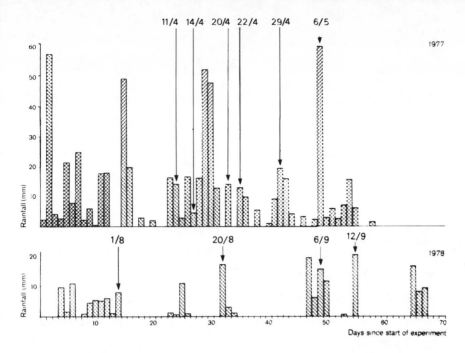

Figure 5.3. Daily rainfall totals for the two observation
   periods with instrumented storms marked with arrows.

slopes leading directly to the channel, i.e. there is
direct external coupling with no feedback. The existence
of a floodplain implies a rather different set of controls.
   The basis of our investigation is that the floodplain
receives direct inputs from hillslope subsurface stormflow,
hillslope groundwater flow, direct rainfall and overflow
and subsurface inflow from the channel (figure 5.4). Losses
occur by evaporation (neglected in the situation considered
here), vertical percolation and subsurface inflow and overflow to the channel.
i) We have estimated the total volume of discharge in quickflow as a percentage of the total rainfall input from a
storm and the results for all storms reveal a remarkably
consistent result. The mean percentage is 5.04 with a
standard deviation of 1.69%. Unfortunately the small
number of dry season storms does not enable a formal
statistical comparison of means to be made, however it
seems clear that the dry season mean (5.6%) and the wet
season mean are only marginally different.
ii) The range of times from the modal value of rainfall to
the peak flow (i.e. time to peak), is 50-260 min but the
values are strongly clustered about the mean of 126 min.
These values are in many cases quite close to the times to
rise, since the hydrographs typically show very sharp rises
(figure 5.5). For example the storm of 25/03/77 shows a
quite steep rise. Some storms do however have broad convex
summits. We regressed the ratio of peak discharge to time
to peak against intensity and an index of antecedent

77

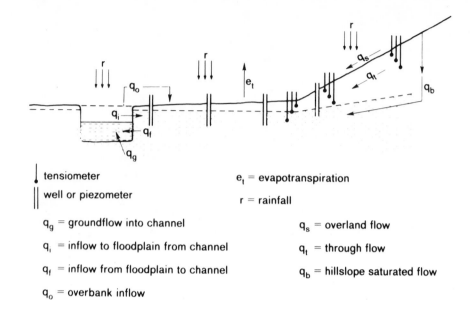

↓ tensiometer	e_t = evapotranspiration
‖ well or piezometer	r = rainfall

q_g = groundflow into channel q_s = overland flow

q_i = inflow to floodplain from channel q_t = through flow

q_f = inflow from floodplain to channel q_b = hillslope saturated flow

q_o = overbank inflow

Figure 5.4.   Diagrammatic representation of the slope-
  floodplain-river system showing inputs and outputs.

precipitation on the assumption that higher intensities
coupled with lower soil moisture deficits would inevitably
lead to a much more rapid rise.   The results show only a
modest correlation with intensity, and the antecedent precipitation did not enhance the correlation significantly.
The total explained variance was only 64%.
iii) Typically the quickflow component of the storms is of
relatively short duration, the range is from 10-42 hr for
both seasons with a mean value of 17 hr.   This is close to
the mean of 15.3 hr for the wet season.   Since baseflows
are consistently higher in the wet season than the dry
season, the difference between the dry (22.6 hr) and the
wet season means may be more apparent than real and reflect
simply the decision as to where to draw the separation line.
iv) All the storms have rapid recessions.   Calculating
$q_t/q_o \times 100 = e^{-kt}$, k has values in the range of 0.08-0.55
with a mean value of 0.263.   The highest values occurred
as expected, with the very largest flows.
    These results imply that quickflow is consistently
coming from a small part of the catchment, with very short
travel distances.   The fact that they are not strongly
differentiated seasonally implies that storage is not

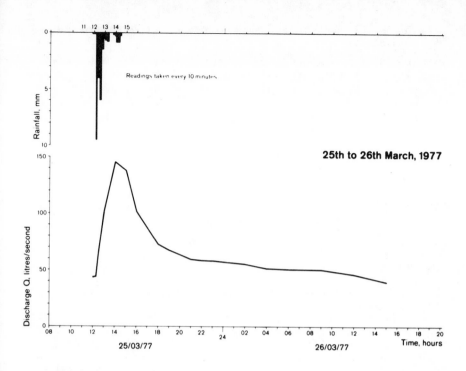

25th to 26th March, 1977

Readings taken every 10 minutes

25/03/77   26/03/77   Time, hours

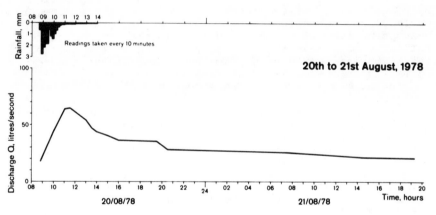

20th to 21st August, 1978

Readings taken every 10 minutes

20/08/78   21/08/78   Time, hours

Figure 5.5.   Typical wet (25/03/77) and dry (20/08/78)
season hydrographs.

highly variable and the correlation of rate of rise
principally with intensity implies that the storage is
fairly constant and filled as a function of input rate.
All this lends support to the hypothesis that the quickflow
is generated principally on the wetlands adjacent to the
channel, in fact, from the floodplain, with little direct
coupling between the slope and the channel during the
quickflow period. Further support for this thesis could
be found from:

- The available storage in the floodplain and its temporal variability.

- The relative amounts of direct inflow from subsurface stormflow and rainfall into this store, and

- The relative magnitude of vertical fluxes on the hill-slope and its role in groundwater recharge. This recharge enables the level of water in floodplain storage to be sustained.

FLOODPLAIN INPUTS

The available 'space' in the floodplain is determined by the distribution of soil moisture above the piezometric level. The latter has been obtained by observations of tensiometers and piezometers, and the soil moisture by neutron probe. These sources indicate that in the dry periods the upper bound of the saturated zone is charac-teristically 44 cm, 32 cm and 20 cm below the surface as one moves from the back to the front of the floodplain. The central piezometers tend to show relatively little vari-ation through time, whereas the bank and slope piezometers fluctuate by about 10 cm and 30 cm respectively. The former responds to, and sometimes lags behind, the rise in the stream hydrograph (see below), indicating the 'internal' type of coupling in the sense of Pinder and Sauer (1971) and Freeze (1972). The slope piezometers are controlled by a combination of groundwater flow (along the piezometric surface) and subsurface storm flow. Prior to a storm, with mean porosities (55%) and actual/saturated soil moisture ratios (0.85-0.95), the available storage at the back of the floodplain could be as high as 24 mm, in the centre 18 mm and on the stream bank as little as 6 mm.

In the wet season between storms the piezometric levels are typically of the order of 10, 14 and 30 cm (respec-tively A, B and C in figure 5.1), indicating that sustained groundwater inflow is keeping the water table about 15 cm nearer the surface, except near the channel, which is again controlled by stream level. Even above the saturated zone the soils are very close to saturation, the ratio of θ/θ_{sat} being between 95 and 97%. With these depths, porosities and moisture contents the available storage might be 3-5 mm in the middle and 7 mm in the rear. Under these conditions, virtually all wet season storms could be expected to pro-duce a response. A large storm such as that of 20/03/77 with 56.9 mm of rainfall should be yielding of the order of 830 1 m^{-1} length of channel. Under these circumstances the slope of the rising limb of the hydrograph should be closely related to intensity and storage. The rather im-perfect relationship referred to earlier (r = 0.8) must in part reflect the inadequacy of the storage measure used, as well as the subjectivity involved in hydrograph separation.

The second input source to the floodplain is from sub-surface stormflow. This can be estimated from moisture, tension and head differences between sites on the slope and floodplain sites. The details of these procedures have been published elsewhere (Nortcliff and Thornes, 1981). Average

Table 5.1. Slope flux values from Site II
to Site I at different depths (cm hr^{-1})

Depth (cm)	Wet Season	Dry Season
15	1.37	3.41
33	1.02	1.06
66	0.07	0.0121
90	0.14	0.00001

computed fluxes in a downslope direction from the lowest
slope site to the back of the floodplain are shown in table
5.1.   In both the wet and dry seasons these means are
heavily biased towards periods immediately following storms
and will therefore generally represent storm recession
periods when the fluxes are much higher than usual.   In
the wet period the downslope fluxes (with depth) in cm hr^{-1}
are, from top to bottom respectively 1.37, 1.02, 0.07 and
0.14.   For a section 1 m wide and 1.4 m deep, which is the
typical section available between the two stations, these
would be yielding subsurface stormflow to the floodplain of
2.5, 3.4, 0.2 and 0.77 l m^{-1} hr^{-1} respectively.   Assuming
these rates prevailed for the average storm recession
period, they would supply to the floodplain less than 25%
of the water being supplied to the channel by saturated
overland flow on the floodplain.   Under the very strongest
assumption (hillslope saturated at all levels) the downslope components at different depths could be as high as
47.26 (<15 cm), 24 (15-60 cm), 6.2 (60-90 cm) and 2.6
(90-140 cm) l hr^{-1} giving a total yield of 80 l hr^{-1} m^{-1}
width of slope.   In practice such rates could not be
sustained for more than a very short period of time and
the figures obtained above are considered to be more
realistic, though even these are probably on the high side.
     The question arises then as to where the water falling
on the non-floodplain areas is going?   It might be suggested
that much of it is flowing at the boundary between the root
mat and the mineral soil.   In fact we monitored 'surface'
runoff including that from the vegetal mat by setting a
corrugated metal strip into the mineral soil to bound a
plot 4 m long (upslope) and 2 m wide.   During the entire
observation period in the wet season, runoff from this
plot was negligible.   We are forced to conclude therefore
that most of the losses were due to deep percolation and
to evapotranspirational losses.
     A comparison of the computed midslope vertical and
downslope fluxes indicates that in the wet season, the
vertical fluxes were typically 2 to 3 times greater than
the lateral fluxes.   Vertical fluxes near the surface are
between 1 and 11 times greater than those at depth (125 cm)
on the hillslope sites.   In the dry season the vertical
fluxes are comparable to the lateral fluxes only in the
shallowest layer.   At other depths the vertical fluxes are
much higher and below about 60 cm both vertical and lateral

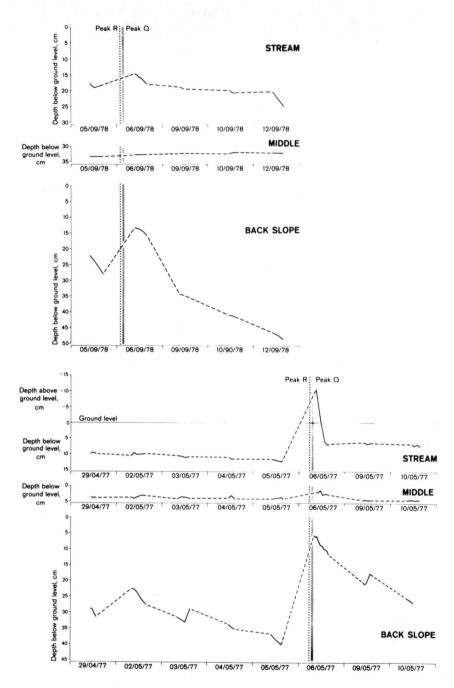

Figure 5.6.   Piezometric levels for a dry season (06/09/78)
and a wet season (06/05/77) storm.

fluxes become negligibly small. This indicates that on the
hillslope, in contrast to the floodplain, the dominant movement is strongly vertical in both seasons. In the wet
season there is a direct feeding to the saturated zone. In
the dry season most of the water is used in replenishing
storage and in evapotranspiration. This inference is confirmed by the greatly depleted baseflow component in the
dry season.

Our final line of evidence relates to the behaviour of
piezometric levels during storms. We illustrate the point
with reference to the wet season storm of 06/05/77 and the
dry season storm of 06/09/78. The piezometric responses
for bank, middle and slope foot positions are shown in
figure 5.6. In the first storm, one of the largest in the
observational period, 46 mm of rain fell between 0530 and
0600 hours and by 0530 rapid response of the river was
already apparent. Peak discharge was reached and overbank
flow probably occurred by 0630 and recession from this peak
was not evident (because of overbanking) until 0830. The
bankside piezometer (E) remained flooded until approximately
1200 and thereafter remained more or less constant. The
piezometer in the middle position (B) remained fairly
constant reaching a maximum height at about 1120, presumably as a result of direct irrigation from the overbank
water. This probably also affected the slope foot
piezometer (C) which reached a maximum between 0800 and
0930, presumably due to flooding and then fell quite consistently to its pre-storm level after 4 days. There is
no evidence of a delayed strong slope input during this
period, the dynamics being largely controlled by the
dissipation of the overbank input which in turn sustained
the middle piezometric levels. We observed several times
during the wet season the inflow of water from the channel
into the floodplain either by overbank flow or by lateral
throughflow from the channel.

In the dry season storm, 4 mm of rain occurred between
2320 and 2340 the previous evening and a further 15 mm fell
between 0130 and 0200. Peak flow occurred at about 0330,
the hydrograph having a broader more convex peak than in
the wet season. The stream side piezometer had already
begun to fall by 0935 and fell exponentially keeping pace
with the recession limb of the hydrograph. The middle
piezometer remained almost constant through the storm,
whereas the slope foot piezometers both showed an exponen-
tial decay consistent with progressive diminution of a
groundwater hump produced through vertical percolation over
the hillslope. There is no evidence to suggest a rapid
influx of subsurface stormflow.

CONCLUSIONS AND IMPLICATIONS

In this environment quickflow hydrographs appear to be
almost completely produced by saturated overland flow
derived from the floodplain areas immediately adjacent to
the channel. Even though there is a strong seasonal contrast in precipitation, the floodplain retains sufficient
moisture in the dry season for the response times and the

volumetric yield (in % of rainfall on the catchment) to
remain very similar to those in the wet season. The floodplain is fed by two dominant mechanisms: lateral inflow
from the channel, though this probably affects only a
relatively small area, and groundwater inflow (as opposed
to subsurface stormflow) from beneath the interfluves. A
third mechanism, replenishment from overbank flow, is important in the wet season.

ACKNOWLEDGEMENTS

The authors gratefully acknowledge the help they have
received from several persons and institutions whilst
carrying out this research. The work was sponsored in 1977
by the Royal Society and in 1978 by the Organisation of
American States Regional Scientific and Technological
Development Program. In the field logistic support was
provided by the Instituto Nacional de Pesquisas da Amazonia,
Manaus and the Max Planck Institute of Limnology. The
diagrams were drawn by the Drawing Office, Geography
Department, London School of Economics. Our personal
thanks go also to Dr. Wolfram Franken and especially to
Dr. M.J. Waylen.

REFERENCES

Bonnell, M., Gilmour, D.A. and Sinclair, D.F., 1981, Soil
    hydraulic properties and their effect on surface
    and subsurface water transfer in a tropical rain-
    forest catchment, *Hydrological Sciences Bulletin*,
    26, 1-18.

Dunne, T., 1978, Field studies of hillslope flow processes,
    in: *Hillslope Hydrology*, ed. Kirkby, M.J., (Wiley,
    Chichester).

Franken, W., 1979, Untersuchen im Einzugsgebeit des zentral-
    amazonischen Urwalkbaches 'Barro Branco' auf der
    'terra firme'. I. Abflussverhalten des Baches,
    Amazoniana, 6, 459-466.

Freeze, R.A., 1972, Role of subsurface flow in generating
    surface runoff - 1. Base flow contributions to
    channel flow, *Water Resources Research*, 8, 609-623.

Mosley, M.P., 1979, Streamflow generation in a forested
    watershed, New Zealand, *Water Resources Research*,
    15, 795-806.

Nortcliff, S. and Thornes, J.B., 1978, Water and cation
    movement in a tropical rainforest 1. Objectives,
    experimental design and preliminary results, *Acta
    Amazonica*, 8(2), 245-258.

Nortcliff, S., Thornes, J.B. and Waylen, M.J., 1979,
    Tropical forest systems: a hydrological approach,
    Amazoniana, 6, 557-568.

Nortcliff, S. and Thornes, J.B., 1981, Seasonal variations in the hydrology of a small forested catchment near Manaus, Amazonas, and the implications for its management, in: *Tropical Agricultural Hydrology*, ed. Lal, R. and Russell, E.W. (Wiley, Chichester).

Pinder, G.F. and Sauer, S.P., 1971, Numerical simulation of flood wave modification due to bank storage effects, *Water Resources Research*, 7, 63-70.

6 The pattern of wash erosion around an upland stream head

Mike McCaig

INTRODUCTION

One of the main purposes of field measurement in geomorphology is to collect data with which to develop and
test process-based models of sediment transport. Measurement studies in pursuit of this purpose can be divided into
3 complementary types, these are studies of:

i)   Process mechanisms

ii) Relationships between process rates and dynamic
     variables (eg. rainfall intensity, flow depth etc)

iii) Relationship between processes and landforms

The first two study types require experimental measurements
(as defined by Ahnert, 1978) made under laboratory or
experimental conditions where control of individual
parameters can be maintained. The third type of study
relies, in contrast, on observational measurements of
process rates and patterns under natural conditions. This
paper describes the sampling design and methods used to
determine the pattern of wash erosion around a simple landform: an upland stream head in the Central Pennines of
Yorkshire.

STUDY CATCHMENT DESCRIPTION

Work was carried out in the 4.3 ha catchment of the
Slithero Clough, near Ripponden, Yorks. The catchment
covers the south-facing side of Joiner Stones Hill (grid
reference SE 009102), rising from 301 m OD to 360 m at the
highest point in the catchment. The bedrock geology is
massive Namurian Millstone Grit of the Kinderscout series
(Wray, et al., 1930). The slope itself is convex-linear
in profile with a small lithologically produced scarp just
below its wide, flat, summit surface. Above the scarp the
catchment is truncated below the slope divide by a water
supply catchwater (built in the 19th century) which follows
the 360 m contour (figure 6.1).

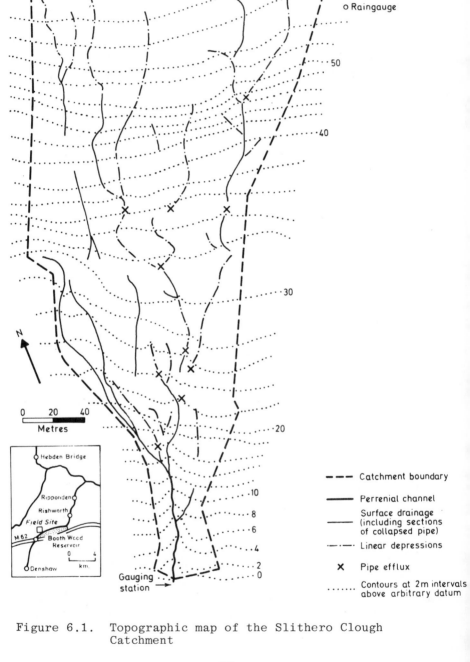

Figure 6.1.   Topographic map of the Slithero Clough
               Catchment

On the hillside the Slithero Clough has formed a valley
4 m deep by 7 m wide at its outflow. Stream flow is
perennial to 60 m upslope of this point. The perennial
flow limit occurs at the junction of two major lines of
drainage, beyond which the topographic expression of the
surface drainage becomes relatively indistinct. Drainage
from the easterly part of the catchment reaches the
perennial stream through a network of natural pipes, the
largest pipe being 0.3 m in diameter. The pipeflow source
area is the flatter part of the slope above the scarp.
This area has a typical 'wet moorland' *Erica sp.* vegetation
community (Moss, 1913), developed over a peat cover which
reaches a depth of 2 m in places. On the mid-slope, below
the scarp, a mixed grass vegetation community dominated by
Molinia sp. is found on a thin (13-30 cm) sandy <u>ranker</u>
type of soil. The westerly part of the catchment contains
only a few small pipes. The main drainage feature in this
part of the catchment is a wet boggy area just below the
scarp. This bog drains to the perennial channel through a
wide *Juncus* flush (the term flush is used here as defined
by Ingram, 1967). Rainfall, evapotranspiration and runoff
totals for the study year Nov.1977-Nov.1978 were 1130 mm,
540 mm and 606 mm respectively.

CONSIDERATIONS FOR A SAMPLING DESIGN AND
APPROPRIATE METHODS

The following sections describe the considerations which
determined the sampling design and methods used in the study
catchment:

Approach to producing a process map

One method of describing the pattern of wash erosion on a
landform is through the construction of a map of wash and
sediment transport rates. It would be impossible, however,
to measure wash rates at a sufficient number of points to
produce any sort of accurate map. The only way a process
map can be produced is if some empirical relationship
between location (topographic position) and the process
being studied can be established. For the case of surface
wash, the 'saturation excess' model of runoff generation
described by Kirkby and Chorley (1967) (figure 6.2),
suggests which variables will determine, for a given
effective rainfall distribution, the volume and frequency
of overland flow generated. These are: i) inflow and
ii) outflow subsurface flow rates, iii) soil moisture
storage capacity and iv) losses to groundwater. Saturated
subsurface discharge rates can generally be described by
Darcy's law (Darcy, 1856):

$$V = K\frac{d\phi}{dx} \qquad\qquad (1)$$

where   V = subsurface flow rate

        K = soil constant

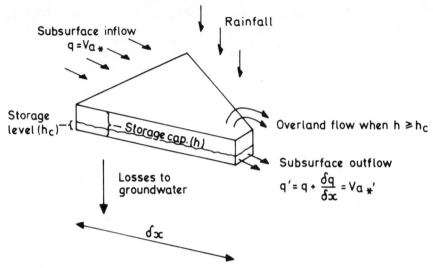

For both inflow and outflow $V = K \dfrac{d\psi}{dx} \simeq K' \cdot s$

where s = tangent of topographic slope

ψ = total potential gradient (including both gravitational and moisture potential)

K = soil constant

a_* = cross-section area

Figure 6.2. The saturation excess model of runoff generation.

$\dfrac{d\phi}{dx}$ = total potential gradient, including gravitational and moisture potential (this is equivalent to topographic gradient alone in a completely saturated soil mass, assuming that hydrostatic conditions prevail)

For the slope element drawn in figure 6.2, therefore, the equation of continuity between inflow, outflow and moisture storage is:

$$V_{a_*} - V_{a'_*} + \frac{\delta q}{\delta x} = \frac{ds}{dt} \qquad (2)$$

where $\dfrac{ds}{dt}$ = change in storage level

$a_* a'_*$ = inflow and outflow cross section areas

$\dfrac{\delta q}{\delta x}$ = increase in flow due to drainage from element

The storage deficit (rainfall required to generate overland flow) is least for those parts of a slope where, in the downslope direction, either $\dfrac{d\phi}{dx}$ decreases, or area drained per unit contour length increases, or both. The effect of increasing gradient for points draining the same area is to decrease overland flow frequency and volume, since more rapid drying occurs due to increased drainage rates. Conversely, increases in area drained per unit contour width for the same gradient increases overland flow frequency and volume. An appropriate combination of the two overland flow controlling variables is, therefore, as the ratio: a/s (where a = area drained per unit contour width, s = tangent of topographic slope). This ratio has large values in hollows and becomes small on spurs. Taking natural logarithms of the ratio for convenience, maps of $\ln(a/s)$ can be produced to determine patterns of surface and subsurface drainage. The map produced for the Slithero Clough (figure 6.3) was originally drawn from a field survey of 1:1000 scale. The small scale was found necessary because of the complex drainage patterns resulting from the presence of natural pipes in the catchment.

The spatial pattern of the variable $\ln(a/s)$ should (according to the saturation excess model of runoff generation) correspond to the pattern of local overland flow production. If empirical relationships between field observations of wash frequency, volume and $\ln(a/s)$ values can be established, relatively detailed 'process maps' can be produced. However, these maps will only be accurate if the following assumptions are satisfied:

i) The saturation excess (storage controlled) model of overland flow generation is applicable to the study site

ii) Systematic variations in soil moisture storage capacity (ie. soil depth and hydraulic conductivity) with topography do not significantly affect relationships between $\ln(a/s)$ and wash processes.

If a further empirical relationship between wash and sediment transport can be established from field data the maps produced can be used to determine the pattern of wash erosion around the stream head.

Determination of empirical relationships between $\ln(a/s)$ and wash frequency and volume

The standard technique for establishing empirical relationships between variables is through statistical analysis. It is important, therefore, that the data set (sample) on which analysis is to be performed should be representative of the population from which it is drawn. Given that the purpose of this study is to establish a relationship applicable to the complete range of $\ln(a/s)$ values encountered, the crucial questions of sampling design to be answered are:

i) How many sample points are required?

ii) What sort of layout of these sampling points is required?

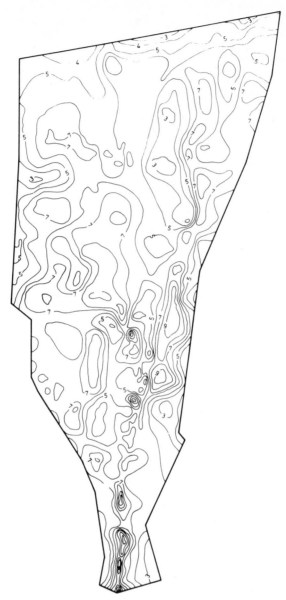

Figure 6.3. Contour map of ln(a/s) in the Slithero
    Clough Catchment.

Sample size determination

The sample size required to determine the true population
characteristics of a variable depends upon the variability
within the sampled population. Estimation of sample sizes
requires, therefore, that a variables' population distribution and an estimate of its standard deviation be known

(information that is usually obtained by a pilot survey).
Such is the degree of variability in natural landscapes,
however, that even the size of pilot survey required to
give a realistic estimate of a variables standard deviation
is often beyond the scope of the research project. In this
project insufficient time was available for a pilot survey
to be undertaken and so the sample size used was determined
only by the maximum amount of field effort that could be
maintained. In effect this approach assumes that the
variability of the variables being monitored is so great
that a degree of 'undersampling' will occur with any sample
size. Sampling theory shows, though, that it is sample
size rather than sample fraction which determines the precision of the sample mean as an estimate of the population
mean. That is, the sampling variance of a sampling mean
is related to the true population variance as:

$$V_s = V_p \cdot \frac{1}{n} \left(\frac{N - n}{N} \right) \tag{3}$$

where V_s = variance of sampling mean

V_p = population variance

N = number of individuals in population

n = number of individuals in sample

Since $N \gg n$ equation 3 can be simplified to

$$V_s = V_p \cdot \frac{1}{n}$$

In a case where no estimate of the population variance is
available increasing sample size to the maximum possible
appears, therefore, to be a reasonable solution to the
sample size problem. Another consideration that determines
the sample size taken is whether the number of samples is
sufficient for certain statistical techniques to be applied
to the data set. In the present study empirical relationships between $\ln(a/s)$ and wash characteristics will be
sought through linear regression analysis.
    The linear model is:

$$y = ax + B + \varepsilon$$

and assumes that the error terms, ε, are independent of x
and are normally distributed with a mean of zero and a
constant variance. In order to fit an unbiased regression
line a true estimate of the mean error, $\bar{\varepsilon}$, is required.
According to the central limit theorem, the sampling
distribution of error terms will approximate a normal
distribution and will therefore have a mean close to the
population mean when sample size exceeds 30.

Sampling pattern

To produce a realistic map using the approach outlined, the
sampling pattern adopted must draw samples from a 'representative' range of locations.

Sample site type	Area (m²)	% of Catchment area	Number of sediment traps
DLA	1172.7	2.7	10
DLI	598.7	1.39	6
S	37800.0	87.7	8
SS	2324.0	5.3	7

The assumed wide variability of wash processes and the
limited number of sampling points to be set up suggest that
a simple random distribution of sampling locations is inappropriate. An appropriate stratification for the present
study is on the basis of site morphology. Four site types
were recognised which together cover the entire surface
area of the catchment. These were:

  i) Drainage line - active (DLA)
     A surface depression formed presumably by fluvial
     erosion, with signs of that type of erosion between
     clumps of vegetation, there being conspicuous bare
     soil areas between individual grasses etc.

  ii) Drainage line - inactive (DLI)
     A similar surface depression to (i) above, but
     without any sign of current erosive activity, and
     only limited bare ground between vegetation.

iii) Slope (S)
     The undissected slope draining to the drainage
     lines.

  iv) Drainage line side slopes (SS)
     Those slopes facing normal to the flow direction
     of drainage lines.

Besides acknowledging the functional significance of different morphologies, these divisions fulfil (as far as is
possible) the stratification principle of strata variance
minimization and mean difference maximization with respect
to sample attributes such as wash frequency and volume. A
further question concerning the layout of sample points is
the number of sample points taken from each strata. Table
6.1 shows the areas covered by each site type in the catchment. For a sample size of, say, 30 a constant sample
size:area ratio would result in the majority of sample
points being placed in the slope (S) class. Sampling
theory shows, however, that the best estimate of population
mean and variance is obtained when the most variable strata
is 'oversampled'. Intuitively, ephemeral drainage lines
can be expected to show a large variability (in terms of
annual wash total for instance) between adjacent or similar

sites, and so it is appropriate to take the largest sampling
fraction from this strata. Subject to the practical
constraints of the measurement techniques employed (see
below), these general principles of sample point layout,
have been used to determine the distribution of sample
sites around the study stream head.

Choice of observation/sampling method

The method of sampling chosen for a field study is determined by the time resolution required from the data set
produced. For this study, valuable information could be
obtained if the field methods adopted allowed the relationship between $\ln(a/s)$ and wash frequency/volume to be
established for a number of different time durations. Data
from individual storms would, for example, indicate the
significance of dynamic variables, eg. rainfall totals and
catchment moisture storage. Four useful time bases for a
study of wash processes are:

  i)   Within a storm

 ii)   Single Storm

iii)   Seasonal

 iv)   Annual

Appropriate methods for these time bases are:

  i)   Multi-probe continuous monitoring of runoff gener-
       ation (flow detection and measuring)

 ii)   Runoff/sediment traps and detectors (flow collection
       in Gerlach type troughs)

iii)   Erosion markers (pins or tracers)

 iv)   Erosion markers and photographs

Since the main interest of the project was in the processes
of hillslope hydrology the field methods used were the
first two of the list above. Cost and maintenance factors
determined that the main part of the field instrumentation
comprised a network of simple runoff traps and a supplemen-
tary network of qualitative runoff detectors. A limited
continuous monitoring installation was set up, part of
which recorded soil water levels and runoff generation at
a total of 11 sites.
   The choice of methods determined the sample sizes for
the sediment trap and runoff detecting networks, because of
the field and laboratory effort required to collect and
analyse samples. In total it proved possible to service
on a weekly basis 31 sediment traps and 52 runoff detectors.
The total of 83 sampling points represents an average
sampling density of one sample point per 520 m^2. The
numbers of sample points in each sampling strata are shown
in table 6.1.

DESIGN OF FIELD EQUIPMENT

The runoff detectors and sediment traps used in the study
were extremely simple, easy to install and cheap to replace.

Runoff detectors

These detectors were used to record the occurrence or
otherwise of surface runoff in the different parts of the
catchment. The design has previously been described by
Kirkby *et al*. (1976) and termed a 'crest stage saturation
tube' (CST). The tubing resting in contact with the ground
surface has a series of holes drilled in its side so that
only flow intercepted by the tube and not rainwater may
enter it and eventually flow down into the vertical
reservoir. The installation is shown in figure 6.4a. The
recorder is checked and emptied by removing the top bung
in the T piece and draining out the collected flow.

Runoff/sediment traps

The design of the runoff/sediment traps used is shown in
figure 6.4b. They consisted of a polypropelene 4 litre con-
tainer connected via a 1 m length of 1 cm diameter PVC
tubing to a flow collecting strip. The flow collector was
constructed from two strips of 75 cm x 5 cm aluminium
strip inserted approximately 3.5 cm into the ground surface.
These strips were placed so as to form an obtuse angled
'V' with the apex of the 'V' pointing down the direction
of maximum local slope and the open upslope facing having
a cross-contour width of approximately 1 m. At the apex
of the 'V' the strips were connected to a 1.5 cm diameter
plastic 'T' piece; the third entry to the 'T' was occupied
by the tube connecting to the collecting tub. The strips
were cut to fit into the 'T' piece and riveted into place.
The junction was held in position by setting its back in a
small lump of cement. The entire flow collecting assembly
was then sealed using Bitumen paint to prevent flow leakage
or infiltration and any possible additional sediment entrainment due to flow deflection along the strip.
   The collecting tub itself was normally placed 1 m downslope of the catching strip, in a position so that flow
from the strip would quickly drain into it via the tubing.
One centimetre diameter tubing was used to prevent any
possible blocking. To ensure rapid drainage, it was frequently necessary to excavate a small pit in which to place
the trap. The traps were fitted with an overflow pipe (in
the lid) and had plastic cups placed inside them (to act
as flow slowing baffles) so that if the volume of flow
caught during a week was greater than 4 l at least the
majority of sediment draining to the tub would be trapped.
   In practice, only 4 of the 31 sites consistently
recorded flow volumes greater than 4 l per flow event.
After each week, when there had been flow, the tubs were
replaced and transported back to the laboratory for
analysis. The traps proved quite robust over the period
of study, requiring only seasonal resealing and, on
occasions replacement when pulled up by sheep or humans.

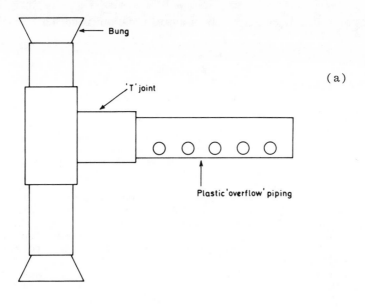

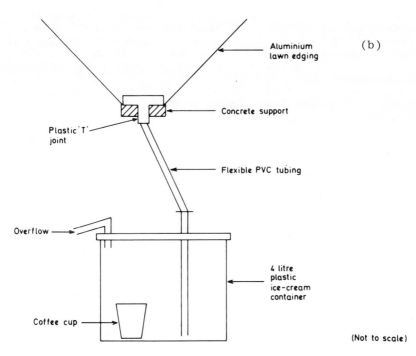

Figure 6.4. Field equipment designs a) runoff detector b) sediment trap.

The approximate cost of the materials required to construct each trap was only £1 (1977 price).

RESULTS

The relationship between ln(a/s) and runoff source areas

It was assumed above that the saturation excess model of overland flow generation was applicable to the study catchment. If this assumption is not valid, the variable ln(a/s) cannot be used as a basis for mapping the pattern of wash frequency/volume. Examining the field results, several relationships suggest that the saturation excess model may be valid. First, the continuously monitored sites around the perennial stream head showed that during a storm there was a spread of surface saturation from high to low ln(a/s) as discharge and catchment water storage increased (table 6.2). This indicates that discharge and saturated area are related. A second way of demonstrating this is through a peak discharge/saturated area rating curve (figure 6.5). The curve shown is constructed from sediment trap and runoff records data for several storms. This form of rating curve could be used with streamflow records to roughly estimate the frequency of wash at all points in a catchment. However, errors in the estimated frequencies might arise since a corollary of the dynamic contributing area concept is that similar areas of saturation will generate different peak discharges for different rainfall intensities. A more accurate relationship would be obtained if catchment water storage were substituted for discharge.

The relationship between wash generation and hydrometeorological variables

A third indication that wash is generated by the saturation excess mechanism, which also demonstrates the importance of catchment water storage levels, is given by the correlations found between wash volumes and hydrometeorological variables.

Table 6.2. Timing and occurrence of surface saturation around the perennial stream head hollow during a storm

Site type	ln(a/s)	Timing of saturation after initial stage rise (hr)	Discharge at saturation (1 s^{-1})	Catchment storage* at saturation (mm)
DLA	11	0.5	2.42	+ 2.3
DLA	10.5	1.6	2.97	+ 2.5
DLE	6	3.5	3.8	+ 6.7
SS	3.0	4.5	5.7	+ 8.8
SS	2.7	7.0	8.1	+ 13.81

* relative to arbitrary zero at beginning of storm

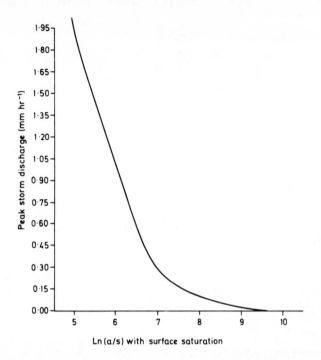

Figure 6.5. Relationship between ln(a/s) associated with surface saturation and storm discharge rate.

Pearson product-moment correlations between weekly wash volumes collected by sediment traps and the following variables were calculated:

i)   Maximum total daily discharge during sample week

ii)   Maximum 30 minute rainfall intensity during sample week

iii) Maximum total daily effective rainfall during sample week

iv)   Maximum catchment water storage level during sample week

Correlation coefficients calculated between sampling group totals and these variables are shown in table 6.3. If wash were generated by the classic Horton 'infiltration excess' mechanism (Horton, 1933) the highest correlation coefficients could be expected to occur with the rainfall variables. However, this is not the case: wash totals are generally significantly correlated with both discharge and catchment storage, once more suggesting that runoff is produced by the saturation excess mechanism.

Table 6.3.   Correlation matrix: sample site type wash
totals with hydrometeorological variables

Sampling site type	Max. daily discharge	Catchment water storage	Max. daily effective rainfall	Max. 30 min rainfall intensity
DLA	0.51	0.51	0.47	0.11
DLI	0.46	0.42	0.17	0.18
S	0.339	0.42	0.25	0.12
SS	0.422	0.57	0.22	-0.02

— = significant at 1% level

The relationship between ln(a/s) and wash frequency
and volume: annual results

The data set produced from the field instrumentation has
indicated that discharge, saturated areas and catchment
storage are interrelated. In particular, wash volumes are
significantly correlated with maximum discharges and
storage levels attained during sampling weeks. These
results confirm that the saturation excess model of runoff
generation is applicable to the study catchment, and so to
map annual wash frequencies and volumes from a ln(a/s) map
is justified. To produce the map, empirical relationships
between ln(a/s) and these variables have been established
through linear regression analysis (figures 6.6a, 6.6b).
Scatter in these plots is due to the effects of:

  i)   Isolated occurrence of infiltration excess runoff
 ii)   Natural pipes
iii)   Occurrence of both channelled and non-channelled
       overland flow in drainage lines

The first source of scatter is only significant for sites
of infrequent flow (small ln(a/s)). The second and third
factors are the more important, being most significant in
drainage line (large ln(a/s)) sites. The occurrence of
natural pipes in the catchment made accurate determination
of ln(a/s) values difficult. Despite careful mapping some
anomalies have clearly arisen in the data set. These are:
i) sites with large ln(a/s) underlain by natural pipes
that therefore record low wash volume/frequencies; ii) sites
with small ln(a/s) values downslope of pipe efflux's that
record high wash volume/frequencies.
     The occurrence of channelled flow in drainage lines is
the most difficult to include in a map of surface wash.
With increasing drainage area channelled flow rapidly becomes the most important component of total runoff.
Channelled flow at the study site was observed along
drainage lines (ln(a/s) > 9) in indistinct anastomising
courses formed between clumps of *Molinia sp.* The volumes
of flow collected by sediment traps placed in drainage
lines, therefore, depended on whether the trap collecting

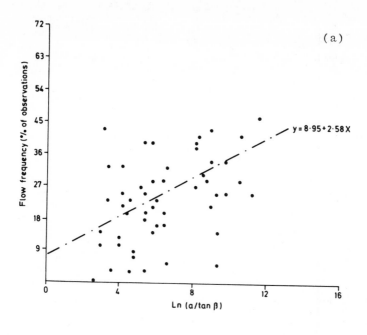

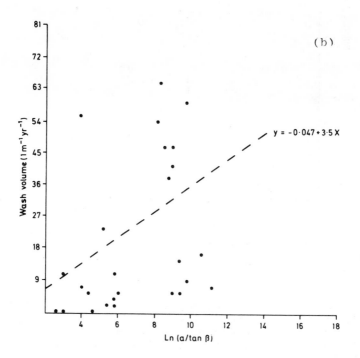

Figure 6.6. a) Relationship between ln(a/s) and wash frequencies. b) Relationship between ln(a/s) and wash volumes.

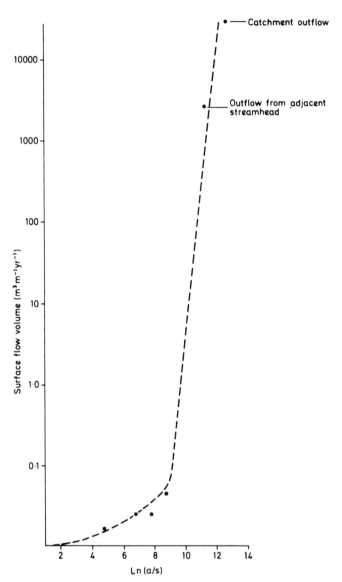

Figure 6.7. Relationship between ln(a/s) and total annual
surface runoff.

strip intercepted one of these lines of flow. As well as
accounting for the wide scatter of points at the upper end
of the ln(a/s) range the failure of the sediment traps to
consistently collect channelled runoff means that the
linear relationship fitted in figure 6.6b will not be valid
for ln(a/s) values $\geqslant 9$.

An estimation of the relationship between total surface
runoff (ie. wash and channelled flow) and ln(a/s) can,

however, be obtained by replotting figure 6.6b with runoff totals recorded by streamflow gauges at the catchment outflow and in an adjacent stream head included in the data set (figure 6.7). It is not unreasonable, however, to use the linear relationship obtained from the sediment trap data to map wash frequencies and volumes for ln(a/s) < 9. The contour maps produced are shown in figures 6.8a, 6.8b.

Flow-sediment relationships

To estimate sediment transport and erosion totals around the stream head it is necessary to establish a relationship between wash volumes and sediment totals. Once more the main problem is to include sediment transport by both channelled and unchannelled flow.

Examining the wash-sediment relationships for the sediment trap data set first, regression analysis of annual totals gives the result:

$$S_{(mg)} = 194 \ Q_{(1)}^{0.66} \tag{4}$$

(r = 0.806; significant at 0.001 level)

Inclusion of slope gradient as an independent variable in a multiple regression did not significantly improve the correlation coefficient obtained.

The exponent for wash volume computed above contrasts markedly with results obtained in other studies (see for example discussion by Carson and Kirkby, 1972, p 212). Generally other studies have shown that:

$$S \propto Q^2$$

In seeking an explanation of why the wash volume exponent is so low for this data set, an important factor is the failure of the traps to consistently collect channelled flow. It might be expected, for example, that the exponent of the sediment-flow relationship will increase with increasing flow depth (since the sediment transporting capacity of channelled flow is much greater than that of unchannelled wash). This trend is confirmed from analyses of weekly group mean wash and sediment totals, wash volume exponents for the sampling groups decreasing in the order DLA>DLI>SS>S. The low exponents obtained (0.71, 0.63, 0.55, 0.5 respectively) might, therefore, be explained by the limited transporting capacity of the flows sampled.

In contrast, however, there is considerable field evidence to suggest that sediment transport in both channelled and unchannelled runoff is detachment rather than transport limited. The sediment trap data set, for example, shows that the highest sediment concentrations recorded in wash samples collected by each trap occur when sample volumes are small (≤ 250 ml). This feature of the data set is interpreted as indicating that there is a rapid exhaustion of a 'new flow' available sediment supply when overland flow begins. It is suggested that an appropriate wash-sediment relationship for unchannelled overland flow is of the form:

103

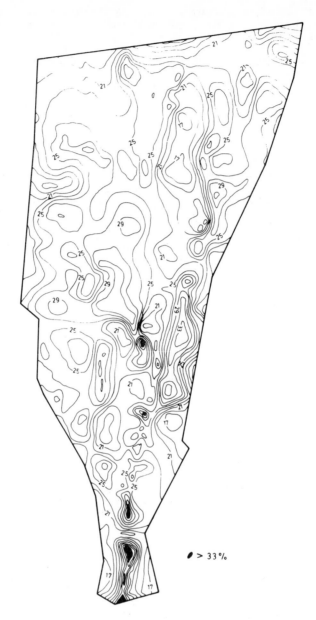

Figure 6.8a. Contour map of annual wash frequency
              (% of weekly observations).

$$S = aQ^b + \Omega$$

where Ω = a new flow sediment contribution.
Regeneration of the available or 'new flow' sediment supply
is related to time between flow events. Consequently for
samples of overland flow collected on, say, consecutive
days, only the first sample will contain the 'new flow'

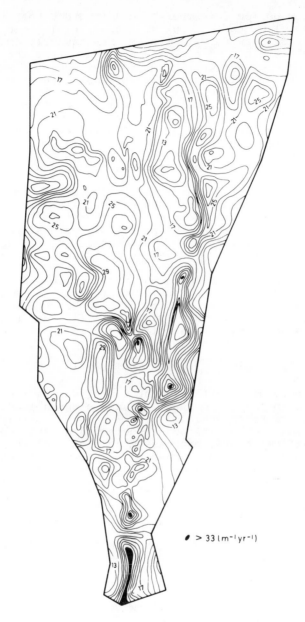

Figure 6.8b. Contour map of annual wash volumes ($m^{-1}yr^{-1}$).

sediment contribution. In effect, therefore, the data set contains samples of two different wash-sediment relation-ships, one of the form:

$$S = aQ^b + \Omega \qquad (5)$$

and another:

$$S = aQ^b \qquad (5a)$$

The winter period (November-March) recorded both the
greatest volume (around 80% of annual totals) and greatest
frequency of flow for all traps. Consequently, most samples
where relationship 5a is valid are drawn from the upper end
of the range of collected flow volumes and most samples
where equation 5 is applicable are drawn from the lower end
of this range. The effect of this mixture in the data set
on the parameters of the wash-sediment relationship (of the
form $S = aQ^b$) estimated using annual sediment/wash totals
is to decrease the exponent (b) and increase the constant
(a). Thus the method of sampling used in the study and the
importance of an availability limited sediment supply
appear to have combined to produce the unusual result
found in equation 4.

The parameters of the flow (discharge)-sediment relationship for channelled flow can be established from
records of suspended sediment concentrations and discharge.
Turbidity records show that pulses of sediment occur independently of variations in discharge, suggesting that
sediment transport by channelled flow is also detachment
rather than transport limited. Given the poor correlation
between suspended sediment concentrations and discharge,
the flow-sediment relationship for the stream is best
described simply by the median of the observed data. For
400 samples taken in the $0.01\text{-}25 \ \mathrm{ls}^{-1}$ flow range the median
suspended sediment concentration is $9 \ \mathrm{mg} \ \mathrm{l}^{-1}$, or expressed
in the same terms as equation 4:

$$S = 9Q^1 \qquad\qquad (6)$$

The problem of establishing a single flow/sediment
relationship for both channelled and unchannelled surface
flow can now be conveniently resolved. The parameters of
the flow/sediment relationship determined from the analysis
of streamflow records can (assuming that sediment transport
in both channelled and unchannelled flow is detachment
rather than transport limited) be substituted into equation
5 and used to estimate the new flow sediment constant Ω.
Making this substitution a calculated mean new flow sediment constant for sediment trap samples of < 250 ml volume
is $9 \ \mathrm{mg} \ \mathrm{m}^{-2}$ of runoff generating area drained per unit
contour length. The flow-sediment relationship used to
estimate sediment transport by both channelled and unchannelled flow in the following sections is therefore:

$$S = 9Q^1 + \Omega \qquad\qquad (7)$$

where $\Omega = 9 \ \mathrm{mg} \ \mathrm{m}^{-2}$ per stormflow event if flow begins
during that event. This flow/sediment relationship is
included in the computer simulation model described below.

The distribution and pattern of erosion

Using the sediment-flow relationship suggested above the
distribution of erosion and its relationship to $\ln(a/s)$
can be examined. Combining equation 7 above with the wash
frequency/volume - $\ln(a/s)$ relationships shown in figures
6.6a and 6.7, allows the determination of sediment source

Table 6.4. Distribution of sediment transport
and erosion for ln(a/s) classes

ln(a/s) class	Class downslope contour outflow width	Sediment Transport (kg year^{-1})	Erosion (inflow-outflow) (kg year^{-1})
1- 1.99	5.89	0.001	-
2- 2.99	37.3	0.0037	0.003
3- 3.99	405.2	0.1216	0.118
4- 4.99	506.0	0.4554	0.334
5- 5.99	545.7	1.527	1.132
6- 6.99	311.6	2.488	0.960
7- 7.99	134.6	3.360	0.870
8- 8.99	51.9	3.630	0.270
9- 9.99	19.85	4.560	0.930
10-10.99	7.35	11.02	6.46
11-11.99	2.73	60.06	49.04
12-12.99	1.0	260.0	150.8

areas around the stream head, provided that the total
contour outflow width for drainage from each ln(a/s) class
is calculated from the ln(a/s) map. This is calculated as:

$$\frac{\text{area of elements in ith ln(a/s) class}}{a_i}$$

where a_i = area drained per unit contour width which for
the average slope gradient of the ith class
produces the class mean ln(a/s) value

Using the widths calculated, sediment transport total and
erosion for each ln(a/s) class can be calculated. The
result of this calculation (table 6.4) shows that erosion
from ephemeral drainage lines (ln(a/s) 11-12) accounts for
around 80% of the total erosion occurring upstream of the
perennial stream head position. Drainage lines forming
this ln(a/s) class occupy only 2-7% of the total catchment
area. Of the total solid-sediment discharge from the
stream head recorded during the study year (260 kg,
60.4 kg ha^{-1}), the results indicate that 58% is accounted
for by erosion of the perennial channel head.

EXTENSION OF RESULTS

Thus far, the results discussed relate only to a single,
possibly atypical, stream head. A question which should be
considered is how the distribution and pattern of erosion
around stream heads which have different hydrological
characteristics compares with that recorded here. A second

important question is how different rainstorm event magnitude and frequencies might affect the observed patterns. The only convenient approach to examining these questions is through the development and use of a computer simulation model.

A brief description of, and some limited results produced by, a simple model are therefore included in the rest of this paper to demonstrate how the field results can be extended to a consideration of these questions.

A model of overland flow generation

The results discussed previously have shown that the saturation excess model of runoff generation is valid for the study catchment. Kirkby (1975) has suggested that an appropriate relationship between soil moisture storage levels and discharge is:

$$q = q_0 \, e^{h/m} \cdot s \tag{8}$$

where q = discharge

q_0 = maximum (saturated) subsurface discharge rate

h = soil moisture storage level

m = storage 'recession' parameter, determined by soil characteristics

s = slope gradient

Referring back to figure 6.2, the relationship between discharge and soil moisture is:

$$\frac{1}{w} = \frac{\partial(q_0 \cdot w \cdot e^{h/m} \cdot s)}{\partial x} + \frac{\partial h}{\partial t} = i \tag{9}$$

where w = element width

i = instantaneous rainfall rate

Equation 9 contains two parts, a discharge due to changing drainage area and a discharge due to the depletion of the storage (by drainage) over time, ie.

$$q = q_{(x)} \cdot q_{(t)}$$

$$h = h_{(x)} + h_{(t)} \tag{10}$$

substituting equation 10 into equation 9 and introducing the constant α simply as a mathematical device gives:

$$\frac{1}{w} \frac{d \, (q_0 \cdot w \cdot e^{h_{(x)}/m} \cdot s)}{dx} = \left[i - \frac{dh_{(x)}}{dt} \right] e^{-h_{(x)}/m} = \alpha \tag{11}$$

Solving for $h_{(x)}$;

$$q_{(x)} = q_0 \cdot e^{h_{(x)}/m} \cdot s = \frac{\alpha}{w} \int w \, dx = \alpha a$$

where a = area drained per unit contour width

$$\therefore \; h_{(x)} = m \ln (a/s) + m \ln(\alpha) - m \ln(q_o) \qquad (12)$$

During a period of steady continuous rainfall, soil moisture storage attains a steady state, $(q_{(t)} = i)$, and so the right hand side of equation 11 reduces to:

$$e^{h_{(t)}/m} = i/\alpha$$

Solving for $h_{(t)}$;

$$h_{(t)} = m \ln(i) - m \ln(\alpha) \qquad (13)$$

Finally combining equations 9 and 10:

$$h = h_{(x)} + h_{(t)} = m \ln(a/s) - m \ln(q_o) + m \ln(i) \qquad (14)$$

which shows that for spatially uniform q_o and i, $\ln(a/s)$ is the critical determinant of overland flow generation, ie. the criterion for overland flow generation at any point is:

$$a/s > q_o/i \; e^{h/m}$$

For non-steady state conditions $q_{(t)}$ should be substituted for (i) in equation 14 above. For the purposes of a simulation model, values of $q_{(t)}$ can be approximated by the mean catchment runoff rate. This is calculated from the solution to the storage model originally presented by Kirkby (1975):

$$q_* = \frac{i}{1 - e^{-i/m} + i/q_{*_o} \; e^{-i/m}}$$

where q_* = runoff rate at end of the period

$\quad$ i $\;$ = rainfall during time period

$\quad q_{*_o}$ = runoff rate at beginning of time period

$\quad$ m $\;$ = storage 'recession' parameter

The volume of overland flow generated per unit time period from the ith slope element is simply determined as:

$$q_o f_{(i)} = q_o \; e^{h_{(i)}/m} \cdot s_{(i)} - q_o \quad \text{or } \phi, \text{ which ever is greater}$$

Model results

The model described above has been used to simulate runoff and erosion from the study catchment under observed and hypothetical conditions. A comparison of simulation results with observed data is shown in figure 6.9. The results, shown in figure 6.9 were produced using 'best fit' estimates of the model parameters m and q_o together with the wash-sediment relationship described in chapter 5.

The effect of different catchment hydrological charac-teristics on the distribution of sediment transport and

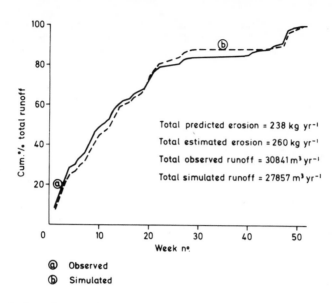

Figure 6.9.   Comparison of observed data with model
   results: Seasonal distribution of runoff generation
   and predicted runoff totals.

erosion can be assessed by introducing different parameter
values into the model.   The model parameters m and q_O
define the rapidity of hydrograph recession and maximum
subsurface flow rates respectively.   The effect of increasing values of both is to simulate an increase in soil
depth and saturated hydraulic conductivity.   Predicted
wash frequencies and sediment transport totals for two
contrasting soil types (shallow, low saturated hydraulic
conductivity; deep, high saturated hydraulic conductivity)
are shown in figure 6.10a, b.   The differences in predicted
sediment transport rates are greatest for low $\ln(a/s)$
values.   Assuming that a certain annual sediment transport
rate is required to form/maintain ephemeral channels, the
result indicates that channels will tend to be more numerous
around a stream head in, say, a clay covered catchment than
one with a sandier brown earth type of soil.   The result
also suggests that in the former case the divideward limit
of ephemeral channel development will respond more sensi-
tively to small changes in annual sediment transport rates
(magnitude and frequency effects) since the gradient of
the $\ln(a/s)$-sediment transport curve is generally less
steep.
   The importance, in terms of at-a-point erosion, of
different rainstorm event magnitudes in the four different
sample site types identified in the study can be assessed
through use of the model.   Table 6.5, for example, shows
the relative amounts of erosion predicted for the largest
storm recorded during the study year.   The relative

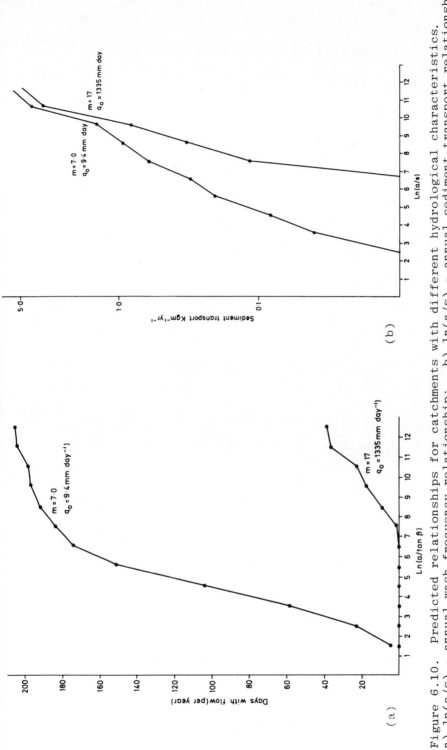

Figure 6.10. Predicted relationships for catchments with different hydrological characteristics.
a) ln(a/s) – annual wash frequency relationship; b) ln(a/s) – annual sediment transport relationship.

111

Table 6.5. Contributions of rainstorm/erosion events to total annual sediment transport rates

Sample site type	Predicted flow frequency (day year^{-1})	Total annual sediment transport due to largest observed storm* (56 mm)	Sediment transport for 10 mm storm-dry**	Sediment transport for 10 mm storm-wet	Sediment transport for 10 mm storm-dry	Sediment transport for 10 mm storm-wet
DLA	282	9	10.2	14.8	15.6	20.1
DLI	130	13	15.1	22.7	19.6	25.7
S	37	26	20.0	40.0	50.0	70.0
SS	46	23	25.0	37.5	50.0	75.0

* all figures as percentages of annual total

** dry conditions = initial q rate 0.5 mm day^{-1}

  wet conditions = initial q rate 5.0 mm day^{-1}

importance of this single event clearly decreases with increasing flow frequency. A similar trend is shown by results produced using artificial rainfall sequences; indicating that a more extreme type of rainfall distribution will produce, relatively, more widespread erosion on the slopes around the stream head - a similar effect to that produced by decreases in subsurface flow rates.

CONCLUSIONS

Studies in instrumented catchments over the last decade or so have concentrated on recording the occurrence and rates of geomorphological processes. As these processes become better understood and models are established, it seems a natural development for catchment studies to concentrate more on the relationships between process and landform patterns. In this study the pattern of wash erosion around a stream head has been determined using the surrogate mapping variable ln(a/s). The appropriateness of using ln(a/s) as a basis for the map is indicated by a consideration of the saturation excess model of runoff generation. By employing this variable (which can be mapped directly from topography) the problems of producing a wash map are reduced to those of establishing empirical relationships between ln(a/s), wash frequency, volume and sediment transport. The key sampling problems are, therefore, the determination of i) sample size, and ii) sampling pattern. The use of surrogate mapping variables identified from consideration of process models is suggested as a simple yet general solution to the problem of mapping process-landform relationships.

ACKNOWLEDGEMENTS

I would like to thank Mike Kirkby and Pam Naden for their helpful comments on an earlier draft of this paper. The work was funded by a NERC research studentship.

REFERENCES

Ahnert, F., 1980, A note on measurements and experiments in Geomorphology, *Zeitschrift für Geomorphologie*, Sp Bd 35, 1-10.

Carson, M.A. and Kirkby, M.J., 1972, *Hillslope form and process*, (Cambridge University Press).

Darcy, H., 1856, *Les Fontaines publiques de la Ville de Dijon*, (Dalmont, Paris).

Horton, R.E., 1933, The role of infiltration in the hydrological cycle, *Transactions American Geophysical Union*, 14, 446-460.

Ingram, H.A.P., 1967, Problems of hydrology and plant distribution in mires, *Journal of Ecology*, 55, 711-724.

Kirkby, M.J., 1975, Hydrograph modelling strategies, in: *Processes in Physical and Human Geography*, ed. Peel, R.F., Chisholm, M. and Haggett, P., (Heinemann), 69-90.

Kirkby, M.J., Callen, J.C., Weyman, D.R. and Wood, J., 1976, Measurement and modelling of dynamic contributing areas in very small catchments, Working Paper 142, School of Geography, University of Leeds.

Kirkby, M.J. and Chorley, R.J., 1967, Throughflow, overland flow and erosion, *Bulletin of the Association of Scientific Hydrology*, 12, 5-21.

Moss, C.E., 1913, *Vegetation of the Peak district*, (Cambridge).

Stuart, A., 1962, Basic ideas of scientific sampling, Griffiths statistical monographs and courses, 4, London).

Wray, D.A., Stephens, J.U., Edwards, W.N. and Bromehead, C.E.N., 1930, *The geology of the country around Huddersfield and Halifax*, (Memoirs of the Geological Survey of England and Wales - explanation of sheet 77).

7 Runoff and sediment transport dynamics

in Canadian badland micro-catchments

Rorke B. Bryan and William K. Hodges

INTRODUCTION

For the last 300 km of its course above the Saskatchewan border, the Red Deer River in Alberta is fringed by the most extensive area of badland topography in Canada, which covers some 800 km^2. Although part of the Red Deer may follow pre-glacial channels, virtually the complete badland development post-dates the Wisconsin glaciation. The badlands occur on a number of Cretaceous units of shallow marine and lagoonal facies, but the most impressive topography coincides with strata of the Upper Cretaceous Oldman formation which outcrop in the Deadlodge Canyon sector of the Red Deer, which forms the core of Dinosaur World Heritage Park. This lies entirely in the semi-arid sector of Alberta where the annual moisture deficit is high and precipitation averages only 325 mm year^{-1}. About 30% falls as snow and the remainder as sporadic summer rainstorms, usually of rather low intensity, and often associated with considerable wind turbulence.

Although the badlands form only about 2% of the complete Red Deer drainage basin, they contribute about 80% of the total suspended sediment load (Campbell, 1977) and probably most of the bed load as well. Most of the sediment load from the badland area enters the Red Deer through ephemeral and some perennial streams, but the pattern of sediment delivery is complex and varied. Most rainstorms are highly localized but track through the area at varying rates. The incidence and volume of runoff from any tributary basin and the synchronization of inputs to the Red Deer are markedly affected by the storm trajectory and the rate at which the storm centre travels (Bryan and Campbell, 1982).

Runoff and sediment delivery patterns are further complicated by variations in surface lithology, vegetation and microtopography. Quite extensive grassed areas occur on Pleistocene till remnants and on alluvial flats near the Red Deer. Otherwise the surface is formed of bare bedrock exposed in a wide variety of topographic forms. All lithologic units are lagoonal or deltaic and are formed of thin, irregular or lenticular beds of muddy sandstone or shale, interbedded with occasional bands of massive sandstone or clay ironstone, and thin carbonaceous shales. Most of the shales are almost pure montmorillonite and adsorbed sodium is abundant increasing shrink-swell capacity so that in dry conditions the surface is densely fragmented by desiccation cracks (figure 7.1). Lithologic diversity has been increased by differential sheetwash, rillwash, rainsplash and by various forms of mass wasting. Amongst the more common of the features resulting are talus slopes of shale debris beneath steep faces, and low-angle transport on depositional features resembling miniature pediments. The dynamics of runoff and sediment production are further complicated by very extensive developments of piping and tunnel erosion features.

The combination of highly variable storm activity with surfaces which differ greatly in infiltration characteristics and in dynamics during rainfall results in an extremely complex pattern of runoff generation. This makes it very difficult to establish a clear set of interrelationships between rainfall, runoff and sediment transport which can be used to understand the water and sediment budgets of the complete badland area. Even within a single tributary catchment of 0.5 km^2, initial attempts at budgeting were frustrated by inadequate understanding of localized controls on runoff generation and sediment production (Bryan and Campbell, 1980). These showed that any large scale budgeting must be based on detailed appreciation of surface behaviour under rainfall at microcatchment level. Initial experiments were carried out using simulated rainfall on a small range of surface types (Bryan, *et al.*, 1978). Although limited in scope these did provide considerable information about runoff generation and demonstrated fundamental differences in surface response. On this basis a more comprehensive set of experiments was established using closely controlled simulated rainfall on a more representative selection of surfaces. These provided comprehensive information on runoff generation and sediment entrainment on a range of surfaces under different antecedent moisture conditions. They also permitted precise monitoring of hydraulic patterns in sheet and rillwash (Hodges, 1982) and detailed study of the behaviour of surface materials under controlled rainfall (Hodges and Bryan, 1982).

EXPERIMENTAL DESIGN

Ten experimental microcatchments were selected which covered 33 lithological units and examples of most small-scale topographic forms (table 7.1). Detailed geological and topographic maps were drawn for each microcatchment (eg. figures 7.1 and 7.2). Simulated rainstorm experiments were carried out at all sites in dry antecedent moisture conditions, and at eight sites in wet antecedent moisture conditions. The simulator used had two pole-mounted spray units which produced approximately 80% of natural kinetic energy from a 5 m fall height. Rain was applied at a constant rate but wind disturbance caused some intensity variation. The average intensity was 29 mm hr^{-1} with a usual duration

Table 7.1.   Characteristics of microcatchments
and simulated rainstorms

Micro-Catchment No.	Unit No.	Lithology	Avg. Slope Angle (°)	Drainage Area (m²)	Storm Duration (hr)		Storm Intensity (mm hr⁻¹)	
					Dry	Wet	Dry	Wet
1	1	claystone	17.5					
	2	sandyshale	21.0					
	3	claystone	20.0	20.1	0.58	NA	40.7	NA
	4	sandstone	19.0					
	5	claystone	52.0					
2	debris	shale reg.	39.5	15.7	0.62	0.53	24.0	14.0
3	6	shale	12.5					
	7	shale	8.5	36.8				
	8	shale	23.0		0.46	0.47	28.6	28.2
	9	shale	40.0					
4	10	shale	14.0					
	11	shale	6.5	44.3	0.44	0.44	20.2	25.2
5	12	shale	7.5					
	13	shale	8.0	31.5				
	14	sandstone	19.5		0.42	0.45	26.5	24.2
	14p	sandy-silt	4.5					
6	13	shale	29.5					
	15	shale	40.5					
	16	shale	35.5					
	17	shale	47.0	26.0	0.41	NA	21.9	NA
	18	shale	25.0					
	19	shale	9.5					
	20	shale	21.5					
	21	shale	37.5					
7	22	shale	16.5					
	23	sandstone	32.5					
	24	shale	33.5					40.4
	25	sandstone	37.0	34.2	0.50	0.40	19.3	
	26	shale	20.5					
	27	shale	46.5					
	28	shale	45.0					
8	29	shale	14.5					
	30	shale	25.5	10.6	0.30	0.38	47.0	34.8
	32	sandstone	N.A.					
9	29	shale	30.0					
	30	shale	27.5	30.2	0.23	0.46	23.7	23.4
	31	sandyshale	23.5					
	32	sandstone	20.0					
10	33	loam	22.5	14.9	0.42	0.38	38.0	43.5

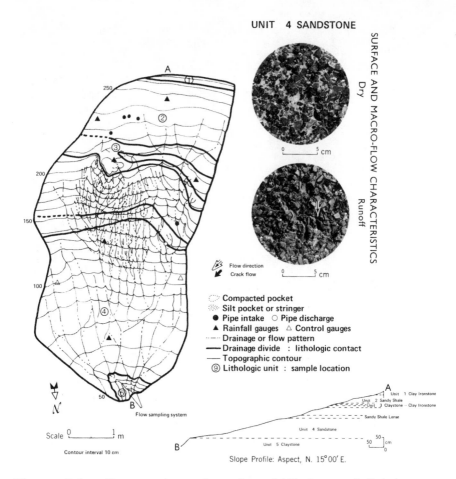

Figure 7.1. Topography and surface lithology of Catchment 1; clay ironstone debris over sandstone.

of 30 min. These rainstorm characteristics recur in the
area every 2-5 years. During each simulated storm, water
and sediment loads were sampled at regular intervals, rainfall distribution was measured with standard gauges, flow
conditions were monitored with dye tracers and regolith
moisture content was measured at different stages. At the
end of each test splash detachment was assessed by the
weight of material collected in special traps.

EXPERIMENTAL RESULTS

Experimental data were collected from all ten microcatchments, but for this paper only four are discussed, selected
to provide the widest range of lithology and surface
characteristics. The only type not represented was the
grassed alluvial surface (Catchment 10) which produced
neither runoff nor sediment transport during experiments.

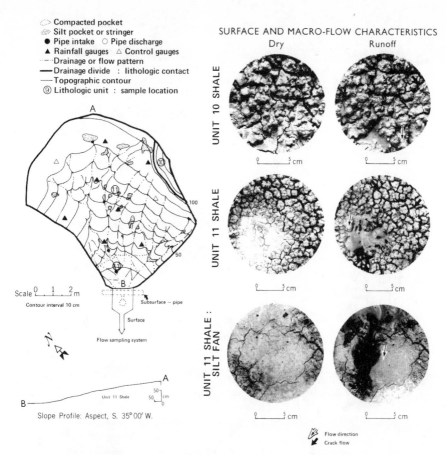

Figure 7.2.   Topography and surface lithology of Catch-
    ment 4.

The detailed morphology and lithology of the selected
microcatchments are shown in figures 7.1-7.4. Experimental
results have been summarized in hydrographs and sediment
transport curves in figure 7.5, while details of rainfall
characteristics are given in table 7.1.

RUNOFF GENERATION

Catchment 1 (figure 7.1) is formed of sandstone and sandy
shale, armoured by clay ironstone fragments. The sandstone
seals and yields runoff almost instantly while the fragments
are impermeable and shed water onto the surrounding surface.
This is a miniature analogy of the situation described by
Yair and Lavee (1974) on talus slopes in Sinai. The result
is almost instantaneous runoff response with a high peak
discharge. The recession is fairly gentle, probably reflecting a slight delay in return flow from small-scale
pipes. Some of the discharge was also 'lost' as no outlet

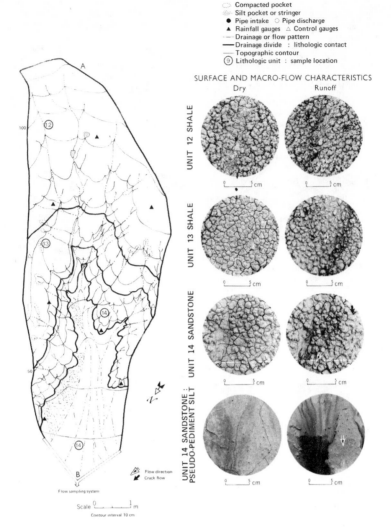

Figure 7.3. Topography and surface lithology of Catchment 5.

could be found in the catchment.   Adverse weather conditions prevented a 'wet' run.

Several shales outcrop in Catchment 4 (figure 7.2) and hydrographs (figure 7.5) illustrate the reduced runoff and pronounced response lag associated with these surfaces.   In dry antecedent moisture conditions runoff is very limited, coming mainly from rills and compacted areas (Bryan, *et al*., 1978).   In wet antecedent moisture conditions complex changes occur in the character of the shale surface (Hodges and Bryan, 1982) which reduce response time and increase runoff discharge.   The contributing area of the catchment expands slightly, but is still largely confined to the vicinity of rills.

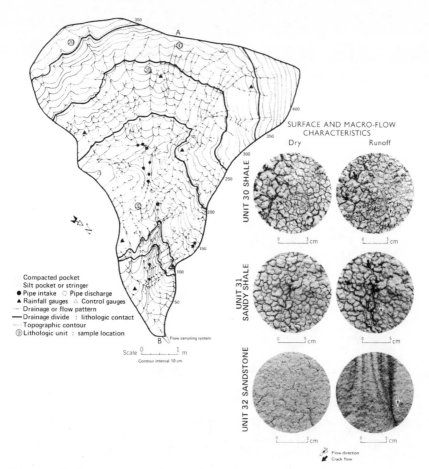

Figure 7.4.   Topography and surface lithology of
    Catchment 9.

Hydrographs for both Catchment 5 and 9 (figure 7.6)
both resemble that for the armoured surface and reflect the
dominant influence of a few lithologic units. In Catchment
5 runoff is generated very rapidly on silt deposits of the
pseudopediment (figure 7.3), and runoff from the upper
shale units becomes significant only near the end of the
tests.   The irregular fluctuation of discharge in the 'dry'
test reflects variations in precipitation inputs due to
wind disturbance. In the 'wet' test approximate equilibrium
is reached for the silt area after about 7 min, and the
slight subsequent rise in discharge marks the contribution
of runoff from the shales. Recession is almost instantaneous.
     The hydrograph for Catchment 9 reflects the dominance
of lower sandstone units throughout the tests. The 'dry'
test was abbreviated due to pump failure, but in the 'wet'
test, equilibrium was again established after about 7 min,
though the latter part of the test and the start of the
recession was disturbed by wind.

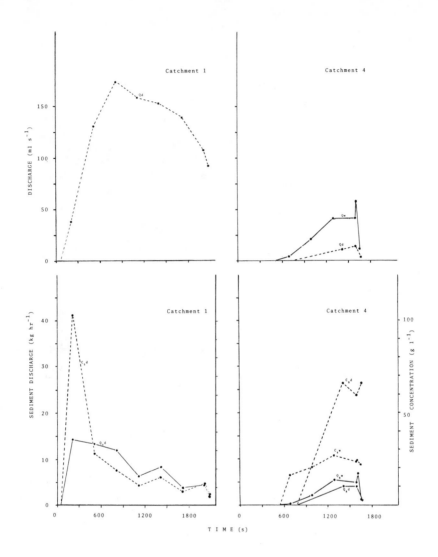

Figure 7.5. Runoff discharge (Q), sediment concentration
(Cs) and sediment discharge (Qs) during simulated rainstorm experiments in dry (d) and wet (w) antecedent
moisture conditions. Catchments 1 and 4.

Bryan *et al*. (1978) first showed the critical importance
of lithology in runoff generation in this area and this is
demonstrated clearly by the hydrographs in figures 7.5 and
7.6. The differences are so marked that even a very small
area of rapidly responding surface, such as sandstone, can

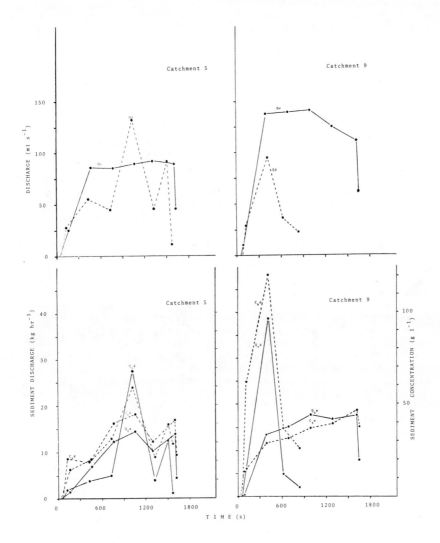

Figure 7.6. Runoff discharge (Q), sediment concentration
(Cs) and sediment discharge (Qs) during simulated rainstorm experiments in dry (d) and wet (w) antecedent
moisture conditions. Catchments 5 and 9.

totally transform the discharge pattern for a complete
microcatchment. In natural rainfall, differences are
probably even more marked as most rainstorms are short and
fall on dry surfaces (Hodges and Bryan, 1982). While
microcatchments are often dominated by a single lithologic

unit, the mesocatchments which carry tributaries of the Red Deer include a complex of surface types. Discharge characteristics therefore tend to be equalized, but a disproportionate amount of runoff comes from rapidly responding surfaces. Some mesocatchments are very largely dominated by pseudopediments and sandstones (Bryan and Campbell, 1982). It appears that as the badland landscape evolves, these will tend to assume progressively larger prominence, and so discharge per unit area of catchment may tend to increase.

SEDIMENT TRANSPORT

The four microcatchments selected show quite different patterns of sediment concentration and discharge during rainfall tests, and quite different relationships with runoff discharge. The patterns shown in figures 7.5 and 7.6 are complex but description is not necessary. Only one of the catchments (Catchment 1) clearly conforms to the pattern described by Emmett (1970) and Mosley (1974) with a rapid rise of sediment concentration and discharge immediately after runoff initiation, with maxima prior to peak runoff discharge, and thereafter steady decrease. It is possible that the abbreviated 'dry' test in Catchment 9 is also essentially of the same pattern. It is clear, however, that although this pattern may be common in thin overland sheetflow under intense rainfall, it is certainly not universal.

The decline of sediment concentration and discharge from an initial peak regardless of runoff discharge described by Emmett (1970) and Mosley (1974) and shown in Catchment 1 may result from a number of interacting processes and the dominance in any particular circumstance varies. The simplest and most common explanation is that the surface is progressively protected by the evolution of an erosion pavement of coarse particles (Shaw, 1929; Lowdermilk and Sundling, 1950). This may evolve either by selective entrainment and transport of finer particles by splash or rillwash or by washing of fines beneath the surface to form a 'filtration pavement' below the coarse surface lag deposit (Bryan, 1973). The evolution of such pavements has been demonstrated in a number of studies (Swanson and Dedrick, 1965; Bryan and Luk, 1981). On the other hand, in detailed studies of a selection of materials, Poesen (1981) has shown that a protective erosion pavement does not always develop, and in some cases the surface is progressively enriched in fines. The processes are complex, but Poesen and Savat (1980) have shown that this may result from selective transport of coarser particles by terminal or afterflow, and this has certainly been observed in residual pipeflow in the Dinosaur badlands.

Emmett (1970) rejected erosion pavement protection as an explanation for his results because he found that in successive tests the initial sediment concentration in a test always exceeded the terminal concentration in preceding tests. He suggested that drying between tests produces a loose, erodible surface mulch vulnerable to entrainment in the next storm. There is no question that drying can

124

markedly affect erodibility and Hodges and Bryan (1982) have shown that the duration and intensity of drying is a critical control on behaviour of surface materials during rainstorms in the Dinosaur badlands. However, the effect noted by Emmett (1970) has also been observed in laboratory flume experiments on very thin sheetwash over cohesive soils. Bryan (unpublished data) has found that an initial 'flush' of sediment concentration invariably follows intervals in the flow, some as short as 5 min, during which no significant drying takes place. It appears that in these circumstances intertest preparation must relate primarily to very minor changes in cohesion and porewater pressures. In turn this means that a pattern of declining sediment concentration and discharge during a rainstorm may not reflect changes in the surface composition but merely changes in the state of cohesive materials. This was demonstrated clearly by Karcz and Shanmugam (1974) in flume tests without rainfall, while Poesen (1981) has also shown exactly similar time-dependent changes in removal of material in flume tests with simulated rainfall. In this case entrainment by splash is particularly affected by porewater pressures and Sloneker et al., (1976) have shown that changes in porewater pressure can cause splash entrainment of sand to fluctuate by 50%.

Rainsplash plays a vital role in entrainment when sheetwash velocities fall below the critical threshold, and several studies (eg. Bryan, 1974; DePloey et al., 1976) have shown that this mixed rainwash process is a very important component in hillslope erosion. Time dependent changes in the effectiveness of rainsplash entrainment occur regardless of any change in the nature or resistance of the material because of changes in the depth of the surface water film. Palmer (1965) suggested that an accumulating depth of water would first enhance and then reduce rainsplash entrainment, but recent work by Savat (1981), Poesen (1981) and Poesen and Savat (1981) indicates that the water film is entirely protective. In situations where rainsplash entrainment dominates, a progressive decline in sediment concentration during the test would occur, simply due to progressive increase in the water film thickness.

Observations of the plot, and the sediment collected from Catchment 1 show that all material entrained comes from the underlying sandstone, and the coarse clay ironstone debris is not affected. As the sandstone is essentially homogeneous it is unlikely to form an erosion pavement, and as it has little cohesion, it should not be vulnerable to the effects noted by Karcz and Shanmugam (1974). It seems therefore that the decline in sediment concentration and discharge towards the end of the test must primarily reflect reduced splash entrainment.

Sediment concentration and discharge patterns in Catchment 4 are quite different but rainsplash still appears to be highly significant. In both tests most of the flow is generated on or near rill channels which are compacted and silt-enriched. The high sediment concentrations in the 'dry' test suggest a flushing of 'prepared' material which is absent in the 'wet' test, but it may simply reflect

deeper flow in the rilled areas and reduced rainsplash
entrainment. The slightly higher sediment discharge in the
'wet' test reflects both greater runoff discharge and a
greater area over which entrainment occurred.

Rillwash is the dominant entrainment process in Catchment 5. In the 'dry' test sediment concentration and discharge are closely correlated with runoff discharge, but in
the 'wet' test both peak while flow on the pseudopediment
is in equilibrium before there is significant contribution
from the upper shales. Maximum sediment movement in both
tests coincides with the scouring of relatively deep chutes
in the pseudo-pediment surface, with abundant sediment
supply. These filled rapidly during recession and after
flow could be recognized only by peripheral bands of clay
deposition (Hodges, 1982). At no stage during either test
did rainsplash appear as a significant agent in entrainment,
reflecting the relatively deep water film on the low-angle
pseudo-pediment.

On the sandstone dominated slopes of Catchment 9, the
greatest sediment concentration and discharge appear to be
related to the availability of a thin weathering rind of
loose material (Hodges and Bryan, 1982) which varies in
thickness from 1 to 5 mm. The exact process of formation
is not clear but test observations show that it disperses
almost instantly on wetting and is readily entrained by
either rainsplash or sheetwash. The very high initial
sediment concentrations and discharge in the 'dry' test,
which slightly precede the runoff peak indicate rapid
stripping of this rind. In the 'wet' test little of this
vulnerable rind is still available and so sediment movement
rises rather gradually as the sandstone is gradually
dispersed by wetting, reaching a maximum after peak runoff
discharge just as the test ends.

The data from the four catchments described indicate
that sediment entrainment and transport patterns in the
Dinosaur badlands are highly complex and variable. They
depend not only on lithologic variations and the presence
or absence of protective vegetation or coarse clastic
debris, but also on the precise interrelationships of storm
and infiltration characteristics and particularly on the
interval since the last runoff-producing storm. Although
the pattern of sediment concentration and discharge
described by Emmett (1970) and Mosley (1974) exists, it is
clearly not universal in its occurrence.

SPLASH DETACHMENT

While the experiments and the preceding discussion have
focused particularly on transport by overland flow, it is
clear that at least on some surfaces rainsplash is a very
important process. It was not possible to make detailed
observations of the interaction of rainsplash and sheetwash
during the experiments but some measurements of splash
detachment were made with simple traps set outside the
catchment peripheries. These consisted of 20 x 20 cm
collectors set above and below 'plots' measuring 20 x 25 cm.
Although sufficiently high to catch all material detached

Table 7.2. Splash detachment in microcatchments
during rainstorm experiments

Catchment	Unit	Slope	Moisture Condition	Splash Detachment ($mg\ cm^{-2}\ mm^{-1}$)			
				Upslope	%<62μm	Downslope	%<62μm
1	4	19°	Dry	.006	75	.021	73
			Dry	.023	27	.093	40
4	11	6.5	Dry	.029	85	.024	87
			Dry	.040	85	.042	93
			Wet	.025	90	.020	88
			Wet	.025	68	.021	75
5	13	8.0	Dry	.049		.031	
			Wet	.038		.051	
	14p	4.5	Dry	.050		.067	
			Wet	.019		.152	
9	30	27.5	Dry	.037	74	.029	60
			Wet	.033	65	.026	90
	32	20	Dry	.119	82	.240	23
			Wet	.109	32	.906	14

by splash (apart from that landing on the plot) they give
no indication of transport distance and therefore provide
at best only an index of splash detachment. Nevertheless
the data in table 7.2 are of some interest, demonstrating
the marked influence of lithology on vulnerability to
splash. On shale surfaces all detachment rates are similar
and rather low ($\bar{x}$ = 0.033 mg cm² mm⁻¹) confirming the
observations of Yair et al., (1980) that detachment processes on montmorillonitic shales are not closely related
to the kinetic energy of rainfall. Antecedent moisture
conditions influence detachment rates only slightly and
inconsistently.

Contrary to the observations of numerous workers (eg.
DePloey and Savat, 1968; Poesen and Savat, 1981) slope
angle does not markedly influence the ratio between upslope
and downslope catch. However, because of wind turbulence
and driving rain, the distribution of splash material is
rather unreliable. On unit 4 the effect of slope angle is
more impressive but again only very small amounts of
material are detached. The small detachment and the variation between splash traps reflect the high, but variable
protection of the surface by clay ironstone fragments.
Substantially larger amounts of material are detached from
the unprotected sand or silty surfaces of unit 14p and
unit 32, and in each case slope angle and antecedent
moisture conditions play an important role. The high

vulnerability of sands to splash confirms the spectrum of
vulnerability observed by Poesen (1981) and Poesen and
Savat (1981) but it is notable that in unit 32 the material
is stable and coherent in dry condition but loses resistance
dramatically on wetting reflecting weak cementation by
montmorillonite and dispersion in the presence of high
sodium concentrations. Heavier particles are more resistant
to wind drifting of splash, which may account for the
curious enrichment of fines in the upslope trap (table 7.2).

Although they raise interesting questions the splash
data are too scant to provide a good basis for incorporation
in a general sediment budget. More detailed studies of
time-dependence of splash vulnerability during runoff and
the interaction between rainsplash entrainment and transport
in sheet and rillwash, and the appearance of erosion and
filtration pavements are clearly essential, and these are
now in progress. In the meantime it is clear that direct
transport by splash is a potentially significant erosional
process only on steep sandstone surfaces, particularly in
wet antecedent moisture conditions.

MASS WASTING

General observations suggest that mass wasting is quite
significant in the overall sediment budgets of mesoscale
tributaries, particularly where steep loose shale surfaces
predominate. Of the microcatchments studied only Catchment
2 could be classified as highly vulnerable, and in all
tests carried out, viscous mudflows occurred (Bryan *et al.*,
1978) followed by rilling. Raincreep unquestionably occurs
on some surfaces, and also unstable rolling of aggregates
early in the test. Measurements of marked aggregates by
theodolite comparing the start and end of the field season
were, however, inconclusive. On most surfaces movement
was close to the limits of measurement accuracy, and on
gentle slopes differential movement in response to wetting
and drying exceeded any downslope component. It appears
that creep is a significant process only on steep shale
surfaces, but even these measurements do not confirm the
impression of consistent general significance, conflicting
somewhat with Schumm's (1956) observations in the South
Dakota badlands.

CONCLUSIONS

Although the experiments described were constrained in
duration and in the variety of antecedent moisture conditions, they do permit a general statement of the major
components of sediment budgets. Rainsplash plays an
important role in sediment entrainment, but it is not
significant in direct sediment transport. Likewise,
although it can dominate certain surfaces in some rainstorms, mass-wasting is not of widespread importance. The
dominant mode of sediment transport on all lithologies and
surfaces is surface or subsurface runoff and the timing,
duration and volume of runoff discharge are critical. As

the tests demonstrate, these vary widely on different
lithologic surfaces, and change markedly with antecedent
moisture conditions. Although the experiments show very
complex patterns of water and sediment transport they provide a good basis for understanding the variable response
of mesoscale tributary catchments to typical storm events,
and the fluctuating contribution of sediment and water to
the Red Deer river.

ACKNOWLEDGEMENTS

This work was made possible by the co-operation of the
Alberta Provincial Parks Department and particularly by the
assistance of Mr. Jim Stomp, Chief Ranger of Dinosaur World
Heritage Park. It was supported by research grants to
Professor Bryan from the Natural Sciences and Engineering
Research Council of Canada.

REFERENCES

Bryan, R.B., 1973, Surface crusts formed under simulated
rainfall on Canadian soils, *National Research
Council of Italy, Laboratory for Soil Chemistry,
Pisa, Conferenze 2*, 30 p.

Bryan, R.B., 1974, Water erosion by splash and wash and the
erodibility of Albertan soils, *Geografiska Annaler*,
56, 159-181.

Bryan, R.B., Yair, A. and Hodges, W.K., 1978, Factors
controlling the initiation of runoff and piping in
Dinosaur Provincial Park Badlands, Alberta, Canada,
Zeitschrift für Geomorphologie, Suppl. Bd., 29,
151-168.

Bryan, R.B. and Campbell, I.A., 1980, Sediment entrainment
and transport during local rainstorms in the
Steveville Badlands, Alberta, *Catena*, 7, 51-65.

Bryan, R.B. and Luk, S-H., 1981, Laboratory experiments on
the variation of soil erosion under simulated rainfall, *Geoderma*, (in press).

Bryan, R.B. and Campbell, I.A., 1982, Surface flow and
erosional processes in semi-arid mesoscale channels
and drainage basins, in: *Recent Developments in
the Explanation and Prediction of Erosion and
Sediment Yield*, ed. Walling, D.E., International
Association of Hydrological Sciences Publication
no.137, 123-133.

Campbell, I.A., 1977, Stream discharge, suspended sediment
and erosion rates in the Red Deer River basin,
Alberta, Canada, *International Association of
Hydrological Sciences Publication* no. 122, 244-259.

DePloey, J. and Savat, J., 1968, Contribution à l'étude de
l'érosion par le splash, *Zeitschrift für Geomorphologie*, 12, 174-193.

DePloey, J., Savat, J. and Moeyersons, J., 1976, The differential impact of some soil loss factors on flow, runoff creep and rainwash, *Earth Surface Processes*, 1, 151-161.

Emmett, W.W., 1970, The hydraulics of overland flow on hillslopes, *United States Geological Survey Professional Paper*, 662-A.

Hodges, W.K., 1982, Hydraulic characteristics of a badland pseudo-pediment slope system during simulated rainfall experiments, in: *Badland Geomorphology and Piping*, ed. Bryan, R.B. and Yair, A., (GeoBooks, Norwich), 127-152.

Karcz, I. and Shanmugam, G., 1974, Decrease in scour rates of fresh deposited mud, *Journal of the Hydraulics Division, ASCE*, 100, 1735-1738.

Lowdermilk, W.C. and Sundling, H.L., 1950, Erosion pavement, its formation and significance, *Transaction American Geophysical Union*, 31, 96-100.

Mosley, M.P., 1974, Experimental study of rill erosion, *Transaction American Society of Civil Engineers*, 17, 909-913, 916.

Palmer, R.S., 1963, The influence of a thin water layer on waterdrop impact forces, *International Association of Scientific Hydrology Publication* no. 65, 141-148.

Poesen, J. and Savat, J., 1980, Particle size separation during erosion by splash and runoff, in: *Assessment of Erosion*, ed. DeBoodt, M. and Gabriels, D., (Wiley, Chichester).

Poesen, J., 1981, Rainwash experiments on the erodibility of loose sediments, *Earth Surface Processes and Landforms*, 6, 285-307.

Poesen, J. and Savat, J., 1981, Detachment and transportation of loose sediments by raindrop splash, Part II Detachability and transportability measurements, *Catena*, 8, 19-41.

Savat, J., 1981, Work done by splash: laboratory experiments, *Earth Surface Processes and Landforms*, 6, 275-284.

Schumm, S.A., 1956, The role of creep and rainwash on the retreat of badland slopes, *American Journal of Science*, 254, 693-706.

Shaw, C.F., 1929, Erosion pavement, *Geographical Review*, 19, 638-641.

Sloneker, L.L., Olson, T.C. and Moldenhauer, W.C., 1976, Effect of porewater pressure on sand splash, *Proceedings of the Soil Science Society of America*, 40, 948-951.

Swanson, N.P. and Dedrick, A.R., 1967, Soil particles and aggregates transported in runoff under various slope conditions using simulated rainfall, *Transaction American Society of Agricultural Engineers*, 10, 246-247.

Yair, A. and Lavee, H., 1976, Runoff generative processes and runoff yield from arid talus mantled slopes, *Earth Surface Processes*, 1, 235-248.

Yair, A., Bryan, R.B., Lavee, H., Adar, E., 1980, Runoff and erosion processes and rates in the Zin Valley Badlands, Northern Negev, Israel, *Earth Surface Processes*, 5, 205-225.

8 Runoff and sediment production in a small peat-covered catchment: some preliminary results

T.P. Burt and A.T. Gardiner

INTRODUCTION

The blanket peat deposits of the Pennine uplands represent
an important land resource - as catchment supply areas for
many reservoirs, for agriculture and for recreation. The
peat soil system is extremely delicate and processes of
runoff and erosion must be fully understood if potential
effects of disturbance are to be properly predicted
(Phillips et al., 1981). In particular those factors which
control the extension of the stream network require detailed
examination. The drainage density of peat moorlands is
very high compared to most other areas of Britain (Burt and
Gardiner, 1981). Two types of network occur - a very dense
branching network on the deep peat of the flat interfluves,
with a more open network of unbranching tributary streams
on the thinner peat of slightly sloping ground (up to 5°).
The origin and development of these contrasting erosional
zones has been discussed by several workers (Johnson, 1958;
Bower, 1961; Radley, 1962; Mosley, 1972). Tallis (1965)
argued that the start of the network extension coincided
with the loss of sphagnum moss species from the peat
surface, due to air pollution emissions from nearby industrial areas, from about 1750 onwards. Other factors
such as moor burning, overgrazing and footpath erosion may
also have contributed to the erosion. Certain areas of the
peat moorland appear to be eroding rapidly. It is therefore of some importance to study stream runoff and erosion
in peat-covered catchments. This may allow definition of
current rates of peat erosion as well as help interpretation
of the differential development of the two erosional zones.
  Little attention has been paid to runoff and erosion
processes operating in peat-covered catchments. Conway and
Millar (1960) discussed general differences in the storm
runoff response of two contrasting erosional areas: heavily
eroded subcatchments were shown to be very sensitive to
rainfall inputs, producing peak flows within 30 minutes of
peak rainfall intensity; in contrast, uneroded peat subcatchments exhibited a smoother storm hydrograph and a
greater lag between peak rainfall and peak runoff.

Similarly there has been little detailed research into peat erosion. Crisp (1966) and Crisp and Robson (1979) have shown that individual runoff events account for the bulk of peat transported, but their use of bulk samples collected over a period of hours has allowed comparison only between mean discharge and mean peat transport rates. Crisp and Robson provide only one example of a storm where instantaneous discharge and peat concentration data was available; thus no detailed analysis of rating curve relationships has been possible (cf. Walling, 1979). It is clear, therefore, that there has been little previous discussion of runoff generation processes in peat-covered catchments, or of the links between storm runoff and peat erosion. In addition, contrasts in runoff and erosional processes in the two contrasting peat zones remain largely undefined.

The aim of this paper is to describe the production of runoff and sediment in two contrasting peat-covered catchment areas. By defining the hydrological processes operating at those sites, and by relating those processes to the movement of sediment in the streams, it is hoped to provide a process-based explanation of the contrasting types of stream network development found on the blanket peat moorlands.

THE STUDY AREA

Observations of runoff processes have been made since May 1978 at Shiny Brook, a peat-covered catchment at the headwaters of the River Colne, West Yorkshire, England (NGR 40604072). The catchment covers an area of 2 km^2, and lies on the Millstone Grit plateau of the Pennine Hills at an altitude of 450 m. The underlying bedrock consists of interbedded sandstones and shales. All slope gradients are below 5°, except close to incised tributaries, and the interfluve zones have a minimal slope gradient. The overall drainage density of the catchment is high (11.15 km km^{-2}), being typical of peat moorland catchments (Burt and Gardiner, 1982), although as noted below clear differences exist between network density in different parts of the catchment.

On the interfluves, the peat is heavily eroded and incised, and in many places a bare peat surface is exposed. Peat depths are variable but reach 6 m in places. Because of the deep channel incision, some areas of peat are now relatively well drained and, where vegetated, are dominated by Crowberry (*Empetrum nigrum*). The dense stream channel network is supplemented by a system of vegetated 'peat flush' channels (Burt and Gardiner, 1982), where Cotton Grass (*Eriophorum vaginatum*) dominates. Much of the interfluve zone consists of bare peat with little or no vegetation cover. The peat flush channels and bare peat areas remain saturated right to the surface; in contrast the Crowberry-dominated ridges are quite dry with the water table well below the surface.

Over the remainder of the catchment, the slightly steeper slopes (up to 5°) are associated with peat depths of 1-3 m. There are few areas of bare peat except on the

stream channel banks. Cotton Grass is the main species
present. In this zone the peat is in a stable, uneroded
state and the drainage density is much lower than in the
heavily eroded interfluve zone.

The general climatic regime of the catchment is severe.
Long term average annual rainfall is estimated at 1500 mm
(Burt, 1980). Rainfall is distributed throughout the year
with an autumn maximum; one day in three receives at least
0.1 mm rainfall (Burt, 1980). Mean annual temperature is
low - about 6°C - with January and February having mean
monthly temperatures close to zero. Because of the low
temperatures and high rainfall, rainfall exceeds evapotranspiration in every month of the year. Frosts are
prevalent throughout the winter - half of the year, with
16% of the year experiencing subzero temperatures. Snowfall
is a common occurrence, with 45 days having snow lying on
average. Wind exposure is also severe at this altitude,
with speeds in excess of 50 km hr^{-1} being experienced for
2% of the time. All these climatic factors[1] may be significant to the erosion and general instability of the peat
moorlands.

Stream discharge has been measured at several sites in
two subcatchments of Shiny Brook using thin-plate 90°
v-notch weirs and Ott stage recorders. Rainfall has been
recorded at three locations using Casella 12 inch autographic gauges, with standard 5 inch gauges as a check on
recorded totals. Moisture conditions in the peat have been
monitored using manual tensiometers (Burt, 1978) and manual
piezometers. Overland flow has been recorded using crest-
stage tubes. The concentrations of suspended sediment and
solutes have been measured from samples collected using
Rock and Taylor pump water samplers.

The two subcatchments chosen illustrate the contrast
between heavily dissected, eroded peat (Bower's Type I
erosion) and uneroded peat (Bower's Type II erosion).
Because of the contrasting nature of the erosional zones,
the experimental design has not been identical at the two
sites. In the heavily eroded subcatchment, the dense network of channels, separated by well drained, vegetated
ridges, has not allowed the use of a regular grid of soil
moisture monitoring sites; nor was it possible to use
throughflow troughs since a 'typical' length of stream
channel bank does not exist. In the uneroded subcatchment,
much less spatial variability exists, and it was possible
to use a regular grid of monitoring sites and to use
throughflow troughs to categorise the hillslope runoff
response. The area of the eroded subcatchment is 0.09 km^2,
and the uneroded subcatchment 0.11 km^2.

HYDROLOGICAL RESPONSE OF THE TWO SUBCATCHMENTS

Infiltration rates

Cylinder infiltrometers were used to determine infiltration
rates at several sites in both subcatchments. In the eroded
subcatchment, measurements were only possible on the ridges,

[1] except where stated, climatic data have been taken from Phillips
   et al. (1981).

since the peat flush channel areas remain saturated. On the ridges, infiltration rates varied from 2.19×10^{-4} cm s^{-1} (7.66 mm hr^{-1}) up to 0.0296 cm s^{-1} (1069 mm hr^{-1}), with a mean value of 0.00508 cm s^{-1} (183 mm hr^{-1}). Since maximum observed rainfall intensities have not exceeded 15 mm hr^{-1}, Hortonian overland flow will be rare on the ridges, in sharp contrast to the saturated channels which can be expected to produce direct runoff even in low intensity rainfall events.

In the uneroded subcatchment, infiltration rates varied from 9.35×10^{-5} cm s^{-1} (3.36 mm hr^{-1}) down to 2.06×10^{-7} cm s^{-1} (0.007 mm hr^{-1}) with a mean of 4.27×10^{-5} cm s^{-1} (1.54 mm hr^{-1}). This implies that infiltration-excess runoff will be produced on the hillslopes of the uneroded subcatchment during most storm events, and that such runoff will be produced over the entire hillslope area. This is in contrast to the eroded subcatchment where the contributing areas for surface runoff should be confined to the peat flush channels and bare peat areas. Nevertheless, the generally low infiltration capacities existing at both sites implies that much surface runoff will be produced, resulting in a large and rapid stream discharge response.

Evidence of overland flow production

The use of crest-stage tubes allowed the frequency of overland flow at various points within the two subcatchments to be investigated. The overland flow data, described in table 8.1, shows the frequency that crest-stage tubes were found to be full. The time periods between emptying are variable and are not restricted to individual storm events. Even so, the relative occurrence of overland flow at different sites can be evaluated. The results given in table 8.1 suggest that overland flow is a frequent occurrence on the vegetated hillslopes of the uneroded subcatchment, but is only frequent on the peat flush channel and bare peat zones of the eroded subcatchment. On the ridges of the eroded site, as would be expected from the infiltration data, overland flow is much less common. In addition, the lower water table and interception by the Crowberry vegetation and litter layer will help to restrict infiltration-excess runoff on the ridges. The implications of the infiltration experiments are therefore confirmed by the crest-stage tube data, namely a dominance of surface runoff processes.

Soil water conditions

In addition to infiltration-excess runoff, the permanent near-saturation of the peat implies that saturation overland flow and throughflow may also be produced during periods of storm runoff generation. Once again, clear differences exist between the two subcatchments. Figures 8.1 and 8.2 show the distributions of soil water and total potential for slope profiles in the eroded and uneroded subcatchments. The soil water potential profile for the uneroded catchment (figure 8.1a) shows the water table to be very close to the soil surface across the entire slope.

Table 8.1.   The frequency of overland flow
at the two subcatchments

Uneroded site

Vegetated hillslopes	Site 1	84.4%
	Site 2	78.2%
	Site 3	78.2%

Eroded site

Unvegetated peat surface	88.2%
Vegetated peat flush channel	87.5%
Vegetated hummock	47.1%
Vegetated hummock	17.7%

(The data indicate the percentage of occasions that crest-stage tubes were full)

The total potential distribution (figure 8.1b) indicates that the hydraulic gradient increases close to the stream, due to the lateral drainage effect caused by the incised channel. Away from the stream, hydraulic gradients are lower and flow direction is essentially vertical, implying fairly stagnant conditions within the peat. Given the low hydraulic conductivity of the peat ($k = 10^{-5}$ cm s^{-1}), and the low hydraulic gradient, the amounts of throughflow generated are likely to be low, even close to the stream bank. In the eroded subcatchment the soil water conditions are more variable spatially (figure 8.2a). The greater density of 'channels' has provided better drainage of the surface peat, though the water table is still at most only 50 cm below the surface. Even so, the bare peat flush channel remains saturated at the surface for most of the year and will generate direct runoff immediately during rainstorm events. The total potential distribution indicates flow towards all three channels (figure 8.2b). Once again, the influence of the main channel induces a high hydraulic gradient in the adjacent bank. Figure 8.3 emphasises the contrast between the two subcatchments in terms of the average water table levels at the two sites. This underlines the fact that the surface peat is generally much better drained in the eroded subcatchment, except for specific sites such as the peat flush channels, whereas in the uneroded catchment the water table remains at or close to the peat surface at all times except during very dry periods such as the drought period of April and May 1980.
    The observations of soil water conditions confirm that surface runoff generation at the eroded site is likely to be limited to the permanently saturated zones. In the uneroded subcatchment, in addition to Hortonian overland flow, saturation overland flow will be produced during most events, since the water table remains so close to the soil surface. Even though infiltration losses will remain small when infiltration is able to occur, there will be some

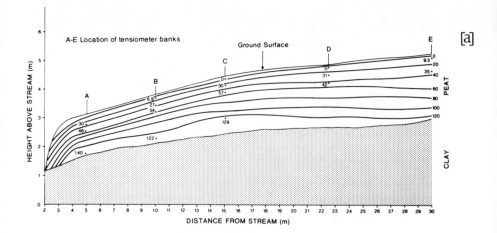

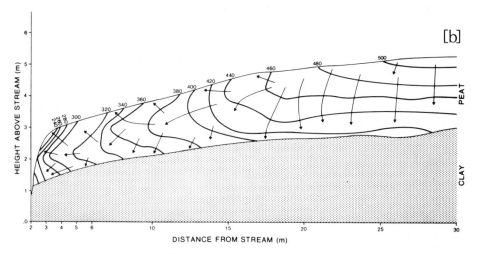

Figure 8.1.   a) Soil water potential at the uneroded subcatchment: 11.10.80   1200 hrs. b) Total potential at
the uneroded subcatchment: 11.10.80   1200 hrs.

control exerted on the magnitude of the storm runoff
response at the uneroded site by the level of the soil water
table prior to the event: if the water table is low there
will be some losses to infiltration; if the water table is
at the soil surface, then all the rainfall must produce
direct surface runoff. At the eroded site, antecedent
conditions may be of less importance, since the water table
remains well below the surface of the ridges at all times,
but continues to remain at the surface of the ephemeral
channels and bare peat areas.

Subsurface runoff processes remain comparatively unimportant except during the latter stages of recession.
Troughs at the uneroded site confirmed that throughflow
does contribute to storm runoff, but only becomes significant in the surface root zone. A number of pipes exist

138

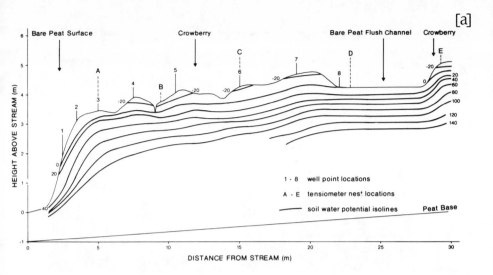

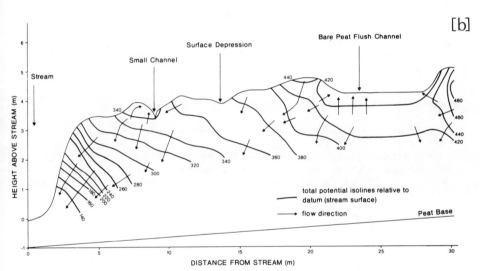

Figure 8.2.   a) Soil water potential at the eroded sub-
    catchment: 21.11.80   1400 hrs.   b) Total potential at
    the eroded subcatchment: 21.11.80   1400 hrs.

at the uneroded site but are of minimal importance in terms
of the quantity of storm runoff they produce.   At the eroded
site, throughflow contributes little to the generation of
storm runoff.   Some pipes are of importance, however, providing a link between the peat flush channels and the main
stream channels.   Such pipes are close to the surface, and
of no more than 1-2 m in length.   Nevertheless, they conduct
large amounts of discharge into the main channel and erode
rapidly, possibly eventually leading to the inception of a
new tributary.

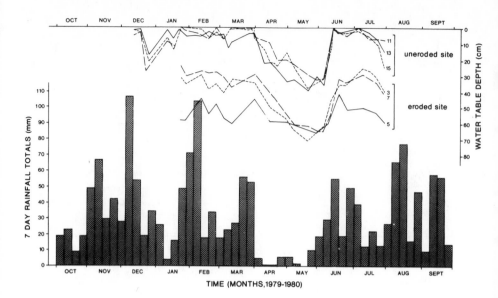

Figure 8.3. Water table levels for selected piezometers in the two subcatchments in relation to weekly rainfall totals.

Storm hydrograph analysis

Relative to most humid temperate catchments, both types of peat moorland catchment produce a larger and more rapid storm runoff response. The low infiltration capacity and permanent saturation of the peat ensures a runoff regime dominated by surface runoff processes. Figures 8.4 and 8.5 show stream discharge hydrographs for the uneroded and eroded subcatchments. The rapidity of the storm runoff response for both catchments is shown by the fact that peak stream discharge occurs very close to the period of maximum rainfall intensity. The 'time to peak' (time delay between the start of the hydrograph rise and the peak discharge) is usually shorter at the eroded subcatchment, except in long duration events when the greater slope of the uneroded site may be important. The runoff percentage at both catchments is large (19.8% at the eroded site; 24.3% at the uneroded site) confirming the widespread occurrence of surface runoff. Peak discharge and runoff volumes, whether expressed in absolute terms or as area-corrected values, are higher at the uneroded site. Regression analysis confirms that the uneroded site produces proportionally more runoff in large events than the eroded site, since the slope coefficient is greater than unity:

Absolute peak discharge $\quad\quad$ A = 0.70 + 1.35B

Area-correct peak discharge $\quad$ A = 0.50 + 1.07B

$\quad\quad\quad\quad\quad\quad$ A = uneroded site

$\quad\quad\quad\quad\quad\quad$ B = eroded site

140

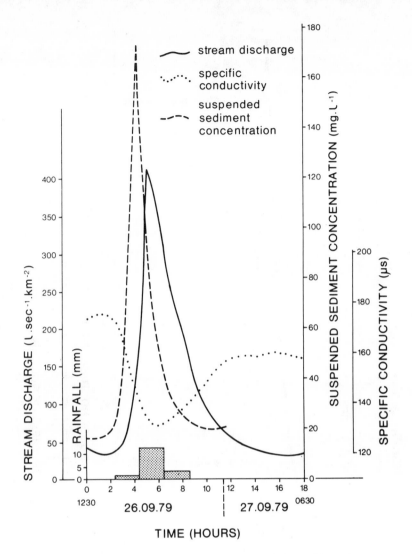

Figure 8.4.   Stream discharge, suspended sediment concentrations, and specific conductance in the uneroded
subcatchment for the storm event of 26.10.79.

This suggests that the contributing area can increase at
the uneroded site during larger events, whereas the source
areas of surface runoff for the eroded catchment remain
relatively more constant.
   A correlation analysis of storm hydrographs for the two
subcatchments confirms that storm rainfall input is the
dominant control of storm runoff at both sites.   Tables
8.2 and 8.3 show that total storm rainfall and maximum
hourly rainfall intensity are highly correlated with peak
discharge, hydrograph rise (peak discharge minus pre-storm
discharge) and storm runoff volume.   By comparison,

141

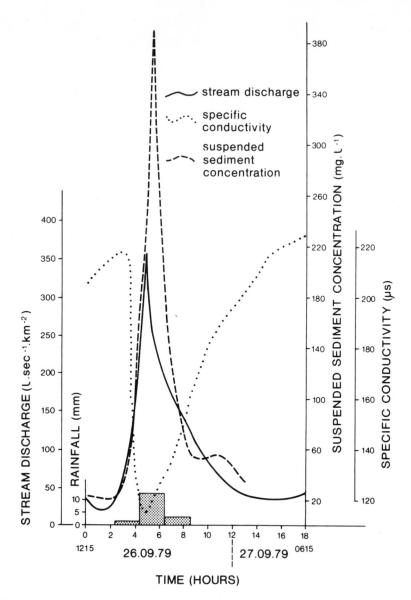

Figure 8.5. Stream discharge, suspended sediment concentrations and specific conductance in the eroded subcatchment for the storm event of 26.10.79.

variables relating to antecedent catchment conditions are generally much more weakly related to storm runoff variables, although antecedent conditions appear to be relatively more important in the uneroded catchment. This may be related to the variable source area for saturation overland flow at the uneroded site, depending on how close the water table is to the soil surface prior to the storm event.

142

Table 8.2. Correlation analysis for storm
hydrographs in the eroded subcatchment

	Peak discharge	Hydrograph rise	Runoff volume	Time to peak
Peak discharge	-	0.9342	0.6973	0.1505
Hydrograph rise	0.9342	-	0.7889	0.0665
Prestorm discharge	0.3960	0.0425	-0.0827	-0.1957
Total storm rainfall	0.7346	0.9345	0.8602	0.4128
Maximum hourly rainfall intensity	0.6344	0.6988	0.4303	-0.0572
A.P.I.	0.3976	0.2739	0.3340	0.1794
Seasonality index	-0.0280	0.0386	-0.2417	-0.1128

——— significant at 0.05 level
═══ significant at 0.01 level
n = 19

Table 8.3. Correlation analysis for storm
hydrographs in the uneroded subcatchment.

	Peak discharge	Hydrograph rise	Runoff volume	Time to peak
Peak discharge	-	0.9524	0.5868	0.0007
Hydrograph rise	0.9524	-	0.6060	0.0148
Prestorm discharge	0.4256	0.2237	0.1871	-0.0495
Total storm rainfall	0.5452	0.6474	0.6068	0.0343
Maximum hourly rainfall intensity	0.5642	0.6830	0.2104	-0.2726
A.P.I.	0.4289	0.3128	0.5445	0.5124
Seasonality index	0.1176	0.0791	0.1626	0.0214

——— significant at 0.05 level
═══ significant at 0.01 level
n = 46

Table 8.4. Multiple regression analysis of factors controlling the magnitude of hydrograph rise in the two subcatchments.

i) Uneroded subcatchment

Variable Added	Multiple R	R^2	R^2 change
Max. hourly intensity	0.6583	0.4333	0.4333
A.P.I.	0.8064	0.6503	0.2170
Time since last event	0.8583	0.7368	0.0865
Seasonality index	0.8666	0.7510	0.0142

ii) Eroded subcatchment

Variable Added	Multiple R	R^2	R^2 change
Total storm rainfall	0.8311	0.6907	0.6907
Time since last event	0.8796	0.7737	0.0830
Seasonality index	0.8923	0.7963	0.0226
Max. hourly intensity	0.9025	0.8145	0.0182

(Additional variables contributed less than 1% to the explained variance.)

Given the more constant source area for surface runoff at the eroded site, antecedent conditions will be of lesser importance to the magnitude of storm runoff. This is confirmed by the multiple regression analyses presented in table 8.4. Storm rainfall is the dominant control of hydrograph rise at both sites, highlighting once again the dominance of surface runoff processes. In the eroded subcatchment, the 'antecedent' variables contribute little to the overall explanation of the hydrograph rise. However, at the uneroded site, the Antecedent Precipitation Index (A.P.I.) contributes 22% to the explanation, confirming the greater importance of antecedent conditions at that site.

Both the observations of hydrological conditions in the catchment and the statistical analyses of storm hydrographs confirm that overland flow processes dominate the storm runoff response from peat moorland catchments. At the uneroded site, overland flow is frequent and widespread on the hillslopes, although the combination of low infiltration capacity and a high water table probably mean that both Hortonian and saturation overland flow are generated. In the eroded subcatchment, the spatial pattern of surface runoff is more complex, being mainly direct runoff from the bare peat areas and from the peat flush channels. The slightly higher ridges, dominated by dense Crowberry vegetation, produce surface runoff only during intense rainfall. There is, therefore, a clear spatial contrast between surface runoff generation in the two subcatchments, which may in turn result in contrasts in erosional processes between the two sites.

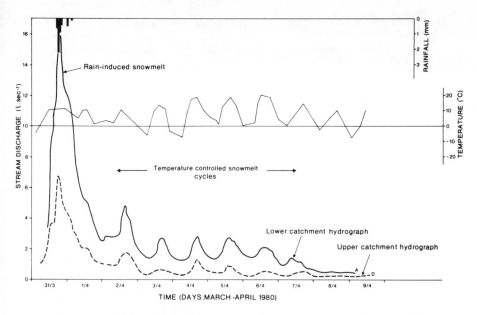

Figure 8.6. Rain-induced and temperature-controlled snowmelt runoff at Shiny Brook.

Snowmelt runoff

Snowmelt runoff from the Pennine peat moorlands is a major
cause of floods in adjacent lowland areas such as the Vale
of York. Despite the interest of Water Authorities in predicting flood response from such events, comparatively
little information is available about the generation of
snowmelt runoff from the peat moorlands. It is clear,
however, that, even if the peat is not frozen, there will
be little infiltration loss from the snowmelt and that the
resultant runoff will be comparatively greater than from
more permeable catchments.

Two types of snowmelt are important in generating storm
runoff at Shiny Brook. In the absence of rainfall, stream
discharge is observed to vary diurnally at times when air
temperatures and an available supply of snow favour significant contributions of thermally-induced snowmelt. Such
a situation is indicated in figure 8.6 where six diurnal
cycles were produced before the snowpack completely melted.
This process may be extremely significant under optimal
circumstances; one diurnal melt cycle produced a peak discharge on 10.4.79 of 93 l s^{-1} in the uneroded subcatchment,
a discharge larger than most 'normal' flood events.

It is, nevertheless, quite clear that rainfall-induced
snowmelt is of much greater importance as a runoff generating process, both in terms of peak discharge and the volume
of runoff produced. The rain-induced flood hydrograph
illustrated in figure 8.6 confirms that a relatively small
rainstorm on a snow-covered peat catchment can produced a
major runoff event. The largest discharges observed at

145

Shiny Brook have occurred during large rainstorms which
followed major snowfalls. Moreover, the high discharges
are often maintained over a period of several hours, in
contrast to the 'peaked' nature of normal storm hydrographs.

THE PRODUCTION OF SUSPENDED SEDIMENT
IN THE TWO SUBCATCHMENTS

The rate of suspended sediment transport involves the
interaction of the sediment production system and the
streamflow system, although both subsystems are causally
related to a common input, rainfall (Gregory and Walling,
1973, p 215). Also, since the suspended sediment is a non-
capacity load, supply conditions are of great importance.
At Shiny Brook, clearly defined differences exist between
the streamflow systems at the two subcatchments, and these
are directly reflected in the production of suspended sediment. At the eroded site, the surface runoff is produced
mainly in the bare peat zones and from the peat flush
channels, both of which provide abundant supplies of sediment. In contrast, at the uneroded site, the hillslope
runoff, though widespread, is generated on a well-vegetated
surface. This suggests that suspended sediment loads of
hillslope runoff are likely to be much greater at the eroded
site. This has been confirmed by observations of the
suspended sediment concentrations of overland flow which
are high in the eroded subcatchment, but very low on the
vegetated hillslopes of the uneroded subcatchment. Channel
bank sources of sediment also differ since the channel at
the uneroded site has incised through the peat layer into
the clay layer below. It is not clear if this is of significance to the production of suspended sediment at the uneroded site, but some of the suspended sediment comprises
mineral rather than totally organic matter as a result.
Channel bank erosion is important in both subcatchments and
undercutting of the banks is commonly observed following
large runoff events at both sites.

Figures 8.4 and 8.5 show typical suspended sediment
hydrographs for the two subcatchments. Peak concentrations
are higher at the eroded subcatchment, which suggests that
the abundant supply of 'hillslope' sediment dominates the
sediment response. The sediment response at the uneroded
site shows clear evidence of exhaustion during the rising
limb of the discharge hydrograph, implying that sediment
sources are relatively limited. In contrast, the suspended
sediment peak at the eroded site occurs after the stream
discharge peak, showing that there is an ample supply of
sediment which is added to by the erosive action of the
storm rainfall and runoff resulting in a continued increase
in sediment supply after the stream discharge peak. This
disparity between the two subcatchments is further illustrated by figure 8.7 which shows the scatter of observations, and the generalised rating curves for the two sites.

The correlation between discharge and suspended sediment
concentration for the eroded site is 0.575 (sample size
= 74), whereas for the uneroded site the correlation is
only 0.199 (sample size = 215). For both sites, simple

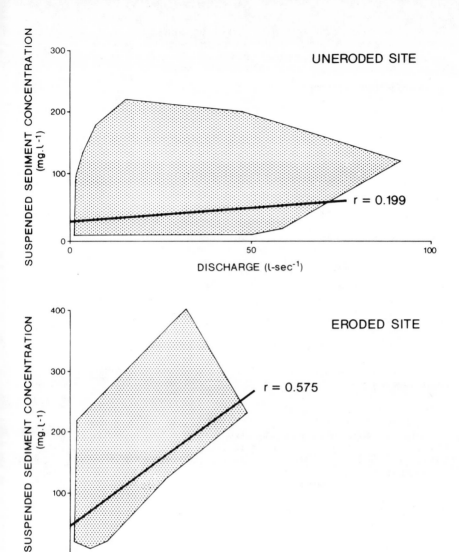

Figure 8.7. Rating curves and data envelopes for suspended
    sediment-discharge relationships for the two subcatch-
    ments.

arithmetic plots produced the highest correlations. The
suspended sediment concentration data on figures 8.4 and
8.5 shows that 'within-storm' exhaustion of sediment supply
occurs at the uneroded site, but not at the eroded site.
Analysis of data from two successive storm events has also
shown that 'between-storm' exhaustion occurs at the uneroded
site, with sediment concentrations being lower during the

Table 8.5. Multiple regression analysis of the controls of peak suspended sediment concentration in the two subcatchments.

i) Uneroded subcatchment

Variable Added	Multiple R	R^2	R^2 change
Max. hourly intensity	0.7404	0.5482	0.5482
A.P.I.	0.7607	0.5787	0.0305

(Number of storms = 48)

ii) Eroded subcatchment

Variable Added	Multiple R	R^2	R^2 change
Total storm rainfall	0.7542	0.5689	0.5689
Max. hourly intensity	0.7881	0.6222	0.0533
Seasonality index	0.8022	0.6435	0.0213
A.P.I.	0.8118	0.6590	0.0155

(Number of storms = 20)

(Additional variables contributed less than 1% to the overall explanation.)

second event despite higher runoff. In the eroded sub-catchment, sediment remains available and concentrations are higher in the second event because of the greater discharges involved. Despite the differences in supply characteristics between the two sites, the dominant control of suspended sediment production is storm rainfall. Table 8.5 presents the results of multiple regression analyses of storm-period peak suspended sediment concentration (within a given storm event) for the two sites. In both cases, storm rainfall input is the most important variable. Neither antecedent conditions nor stream discharge are of any great significance. This implies that suspended sediment is relatively freely available at both sites, although clearly the eroded deep peat zone is the most favourable source of sediment supply. The general level of suspended sediment concentrations (for a given discharge) is quite low at Shiny Brook, compared to other British catchments. However, two points should be noted in this respect: firstly, the density of suspended sediment produced from the peat moorlands is very low, being primarily composed of organic material, so that the volume of material being moved is greater than suggested by the concentration data. Secondly, the very small size of the subcatchments being studied means that overall sediment transport rates per unit area will be relatively high, given the density of the stream network on the peat moorlands.

The results show that suspended sediment concentrations are consistently higher at the eroded site, and are much more closely related to runoff conditions there. At the uneroded site, supply conditions are of much greater importance. This suggests that the differences between the contrasting erosional zones will continue to develop and that progressive growth of the channel network in the eroded interfluve areas is likely.

The transport of solutes and bed load has not been studied in detail at Shiny Brook. The very acid stream-water (below pH 4) and the dominance of organic soils means that solute production in the catchment is a complicated process. In addition, solute input to the catchment, from both dry and wet fallout, is high, given the abundance of nearby sources of air pollution. No measurements of bed load transport have been made, but observations of sedimentation behind weir structures, and in the Wessenden Reservoir (drained in 1980) suggest that rates of bed load transport are high. Most of the bed load is produced in the channels of the 'uneroded' parts of the catchment, since the streams there have incised through the shallower peat into the gritstone substrate.

It is frequently claimed (eg. Shimwell, 1981) that frost and snow are major causes of peat erosion. Observations at Shiny Brook have failed to support this hypothesis. Suspended sediment concentrations during snowmelt runoff, even when rainfall-induced, remain very low; the bare peat surfaces are protected from the action of rain-beat, and snow infill in the channels may also provide some protection there. Frost heave and freeze-thaw processes do produce available sediment. Even so, peak suspended sediment concentrations occur in the late-summer and early-autumn rather than in winter; this implies that desiccation of the surface peat in summer, in combination with intense rainfall events, is the main agent of peat erosion. Snow-melt runoff is probably of much more significance for bed load transport than for the erosion of peat, which seems to be most strongly controlled by storm rainfall inputs.

CONCLUSIONS

i)   The hydrology of blanket peat moorlands is dominated by surface runoff processes. Observations of overland flow, infiltration rates and soil moisture status all show that surface runoff will be produced even during small storms. Hydrograph analysis confirmed that storm rainfall rather than antecedent conditions controls storm runoff.

ii)   Clear spatial differences exist between surface run-off generation on the heavily-eroded deep peat interfluves and on the relatively uneroded, open peat moorlands. The contributing area at the eroded subcatchment is relatively constant, comprising the peat flush channels and bare peat zones. In the uneroded subcatchment, antecedent conditions are more important, and the source areas for surface runoff appear to be variable depending on how close the water table is to the soil surface prior to the storm event.

iii) The differences in surface runoff generation at the
two sites are reflected in contrasts in the production of
suspended sediment. In the eroded subcatchment, the supply
of available sediment is abundant, and suspended sediment
concentrations are consistently higher and better related
to stream discharge than at the uneroded site, where the
sediment supply appears easily exhausted. The cotton grass
vegetation on the uneroded subcatchment hillslopes provides
efficient protection against surface erosion, despite the
large amounts of surface runoff produced. At the eroded
site, both the bare peat and the ephemeral peat flush
channel network provide easily-eroded sources of peat.
Rainfall is the dominant erosive agent, with snowmelt and
frost action appearing to be relatively unimportant.

iv)   The dense channel network of the eroded interfluve
zone will continue to extend and incise into the deep peat.
Network development in the deep peat, once initiated,
occurs 'naturally' - without assistance from overgrazing,
burning or air pollution. By contrast, the peat of the
cotton grass moorland is apparently quite stable; the main
erosion there involves channel incision into the gritstone
below the peat. The high drainage density of the deep peat
interfluve zones will therefore probably increase over
time, and attempts to limit peat erosion should concentrate
on these particular areas, rather than on the whole of the
blanket peat moorlands.

REFERENCES

Bower, M.M., 1961, The distribution of erosion in blanket
    peat bogs in the Pennines, *Transactions of the
    Institute of British Geographers*, 29, 17-31.

Burt, T.P., 1978, Three simple and low-cost instruments for
    the measurement of soil moisture properties,
    *Huddersfield Polytechnic, Department of Geography,
    Occasional Paper*, no.6.

Burt, T.P., 1980, A study of rainfall in the Southern
    Pennines, *Huddersfield Polytechnic, Department of
    Geography, Occasional Paper*, no.8.

Burt, T.P. and Gardiner, A.T., 1982, The permanence of
    stream channel networks in Britain: some further
    comments, *Earth Surface Processes and Landforms*,
    7, 327-332.

Conway, V.M. and Millar, A., 1960, The hydrology of some
    small peat-covered catchments in the Northern
    Pennines, *Journal of the Institute of Water
    Engineers*, 14, 415-424.

Crisp, D.T., 1966, Input and output of minerals for an area
    of Pennine moorland: the importance of precipi-
    tation, drainage, peat erosion and animals, *Journal
    of Applied Ecology*, 3, 327-348.

Crisp, D.T. and Robson, S., 1979, Some effects of discharge upon the transport of animals and peat in a North Pennine headstream, *Journal of Applied Ecology*, 16, 721-736.

Gregory, K.J. and Walling, D.E., 1973, *Drainage basin form and process*, (Arnold, London).

Johnson, R.H., 1957, Observations of the stream patterns of some peat moorlands on the Southern Pennines, *Memoirs of the Manchester Literary and Philosophical Society*, 99, 1-18.

Mosley, M.P., 1972, Gulley systems in blanket peat, Bleaklow, N. Derbyshire, *East Midland Geographer*, 5, 235-244.

Phillips, J., Yalden, D. and Tallis, J.H., 1981, *Peak District Moorland Erosion Study: Phase 1 Report*, Peak Park Joint Planning Board.

Radley, J., 1962, Peat erosion on the high moors of Derbyshire and West Yorkshire, *East Midlands Geographer*, 15, 40-50.

Shimwell, D.M., 1981, Disappearing Peak, *Geographical Magazine*, 53, 684.

Tallis, J.H., 1965, Studies on Southern Pennine peats II: The pattern of erosion, *Journal of Ecology*, 52, 333-344.

Walling, D.E., 1979, Suspended sediment and solute response characteristics of the River Exe, Devon, England, in: *Proceedings of the 5th Guelph Symposium on Geomorphology*, (GeoBooks, Norwich), 169-197.

9 The hydrology and water quality of a drained clay catchment, Cockle Park, Northumberland

Adrian C.Armstrong

INTRODUCTION

Current geomorphological processes in lowland Britain are inextricably intermingled with anthropogenetic effects. In the few thousands of years that man has occupied the land, his effect on its appearance has been increasingly large, and the tempo of this change has accelerated to such an extent that large sections of the landscape are now entirely man-made.

The first, and the most ubiquitous of these anthropogenetic modifications, is the adaption of the land to agricultural production, and so extensive has this process been, that the agricultural landscape is frequently considered by the uninformed to be its 'natural' state. Yet this is not the case, and one of the problems facing the geomorphologist is to distinguish between the natural processes, and those that are man-induced.

One of the first steps in converting much land to agriculture is the artificial drainage of the land. However, the effects of drainage on the hydrological and geomorphological system are by no means clear. The dispute of the 1860's (Bailey Denton, 1861) as to whether drainage increases or decreases runoff is still unresolved, and other questions such as the interaction between artificial drainage and the generation of erosive overland flow, or the leaching of solutes from the soil, have been explored only sketchily.

Consequently, when a long term (10 year) drainage economics experiment was established in 1978 on the University of Newcastle's farm at Cockle Park, Northumberland, with a main aim of studying the economic benefits of drainage, the opportunity was taken to study the discharge of water and nutrients from both drained and undrained areas.

By studying the details of soil water movement in a small area, the results from such an experiment were expected to throw light on the role of artificial drainage in the generation of runoff and pollutant loads. Of particular geomorphological relevance was the question of surface

flows and of high storm peaks, which are discharged from
the experimental area into the receiving stream system.
Further details of the experiment are given in MAFF (1979),
Armstrong (1980a), Armstrong, Carter and Shaw (1980).

ARTIFICIAL DRAINAGE

Field drainage removes free soil water by gravity, and in
doing so, reduces the duration of high water table levels
within the soil, improves soil aeration, and increases soil
strength. It thus both improves the root environment for
the crop, and increases the ease of use of the land by the
farmer.

Water table control is achieved by installing a system
of drains, at depths typically of around 1 m, to provide
both a sink for the water, and a means for transporting the
water away from the site. A considerable body of theory
exists which permits the calculation of the spacing needed
to achieve a given level of water table control (reviewed,
for example, by ILRI, 1973), and these all indicate that
for clay soils with hydraulic conductivities less than
0.01 m day^{-1}, drain spacings of the order of 1 to 2 m are
required to control normal rainfall. Such drain spacings
are clearly both impractical and uneconomic, and a popularly
adopted solution to this apparent impasse is the technique
of mole drainage.

A single tine is pulled through the soil at a depth of
between 500 and 600 mm, and at its base a 'bullet' of
circular cross-section forces the soil to compress and
leave a circular channel behind. The combination of near
vertical cracking in the soil, and the channel left behind,
provides an efficient drainage system that can be installed
at spacings of around 1.5 to 2 m at an economic cost. The
mole channels that are formed are drawn at right angles to,
and through, gravel filled trenches (drains with permeable
backfill) which intercept the flow. Permanent drains at
the base of the trench then transport the water away from
the site. If the soil is suitable, both in terms of its
inherent strength properties, and its water content at the
time of the operation, the mole channels may last for
several years. Claims of 10 years life are not uncommon,
and 5 years is a common useful life span. The suitability
of the soils of this site for mole drainage was however in
dispute, and consequently one of the drainage systems considered in this experiment was mole drainage, with the aim
of establishing the degree of drainage success that could
be achieved.

Present drainage practice in the area does not include
mole drainage, but relies on pipes alone typically at
spacings of 12 m, without either permeable backfill or any
attempt to loosen the subsoil (Armstrong and Smith, 1977).
It has, however, been observed on other clay sites that
drains on their own merely intercept water moving in the
topsoil and do not lower the water table within the soil
(May and Trafford, 1977; Harris, 1977). Further, over a
period of 10 years, the gradual reconsolidation of soil in
the drain trench has been observed to reduce this

interception function to a small fraction of its initial effect (Dennis, 1980). It was thus considered that this local practice was of doubtful value.

SITE DETAILS

The site is located some 6 km north of Morpeth, Northumberland, (Ordnance Survey grid reference NZ188916), approximately 80 m above sea level. The mean annual rainfall for the site is estimated to be 720 mm.

The soil of the site is a clay loam of the Dunkeswick series (Crompton and Matthews, 1970), and is remarkably uniform throughout the experimental area. This soil, which is developed in till derived from Carboniferous material, is common throughout Northeast England. A short profile description, and a particle-size analysis of the subsoil is given in table 9.1. In hydrological terms, the most significant feature of the profile is the sharp transition between the finely structured permeable topsoil, and the massively prismatic subsoil. The mean hydraulic conductivity of the subsoil was measured, using the single auger hole technique, at 0.003 m day^{-1}.

Table 9.1. Soil Profile information

0-250 mm   Very dark greyish brown, very slightly stony, clay loam. Weakly developed medium granular structure. High macroporosity.

250-450 mm   Dark greyish brown and greyish brown, mottled yellowish brown, very slightly stony, clay. Moderately developed medium subangular blocky structure.

450-700 mm   Brown to dark brown, mottled dark grey, very slightly stony clay, with few ferrimanganiferous concretions and dark grey ped faces. Strongly developed coarse prismatic structure, prism faces strongly gleyed.

700-1000 mm   Grey mottled to dark brown clay, with common small fragments of weathered sandstone. Strongly developed coarse prismatic structure, prism faces strongly gleyed.

Mean particle size analysis, 600 mm depth

		content, %
Mineral matter		93.9
Coarse sand	>600 μm	1.7
sand	600-212 μm	5.7
fine sand	212-106 μm	10.7
v. fine sand	106-63 μm	8.5
coarse silt	63-20 μm	13.3
silt	20-2 μm	28.1
clay	<2 μm	35.2

155

EXPERIMENTAL DESIGN

The site offered approximately 3 ha for experimentation, which could be conveniently divided into nine parallel plots, 25 m wide by 100 m long (figure 9.1). The overall design of three drainage treatments with threefold replication was dictated by the requirements for drainage economics experiments, of which this site was just one of a coordinated series. The three drainage treatments considered were:

1) Control, no drainage

2) Drains alone, at 14 m spacing, 900 mm deep, across the slope.

3) Mole drainage. Collector drains with gravel to within 300 mm of the surface laid across the slope at 40 m spacing, 900 mm depth. Above this mole drains drawn up and down the slope, at a depth of 500 mm and a spacing of 1.8 m.

The site was divided into three blocks each of three plots, but the treatments were not randomised separately within each block, and a single randomisation was repeated three times. This design gave the flexibility that any three adjacent plots constituted a complete block, and analysis and instrumentation was not tied to the original block structure.

Prior to installation of the drainage, a ditch approximately 2 m deep was cut downslope of the experimental area. This provided an immediate access point for the interception and monitoring of flows. The plots were isolated using vertical polythene film barriers installed to a depth of at least 1 m, and brought to the surface. An interceptor drain, 1.3 m deep with gravel to the surface, was installed around the experimental area, to prevent the entry of foreign water.

INSTRUMENTATION

For hydrological investigations, only the central block of three plots was intensively instrumented. A sketch plan of the instrument layout is given in figure 9.2.

Flows were measured using special weirs (Talman, 1979) which had a 50 mm high 1/4 'V' notch at the base of two parallel plates 500 mm long with 25 mm separation. Water height over the V plate was recorded using a spring-restrained float, which recorded directly onto a chart, supplemented for part of the period by an electronic data logging system. This design gives good resolution of $0.005 \ l \ s^{-1}$ up to $1.0 \ l \ s^{-1}$, yet permits measurement of flows as high as $7 \ l \ s^{-1}$.

In addition to the drain discharges, surface and plough layer flows were collected at the base of each plot. A drain was installed at a depth of 300 mm, with gravel to the surface. This gravel was kept clear of vegetation, and positive steps were taken to ensure that surface flows were intercepted by this drain, either by cutting small surface

156

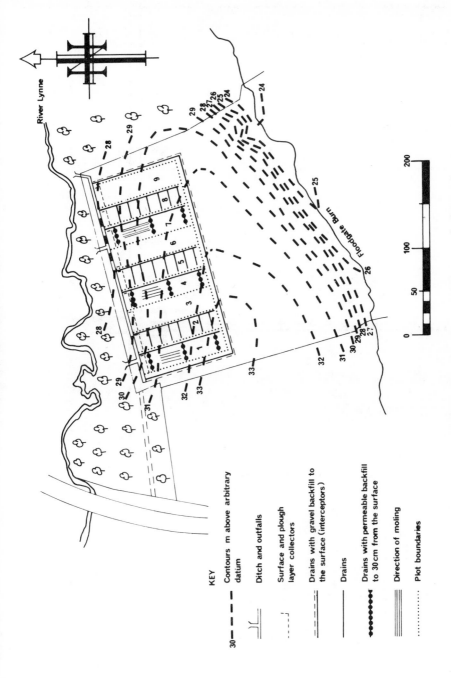

KEY

30 — — —	Contours m above arbitrary datum
	Ditch and outfalls
	Surface and plough layer collectors
	Drains with gravel backfill to the surface (interceptors)
	Drains
●●●●●●	Drains with permeable backfill to 30 cm from the surface
	Direction of moling
⋯⋯⋯	Plot boundaries

Figure 9.1.   Site topography and general layout.

River Lynne

Floodgate Burn

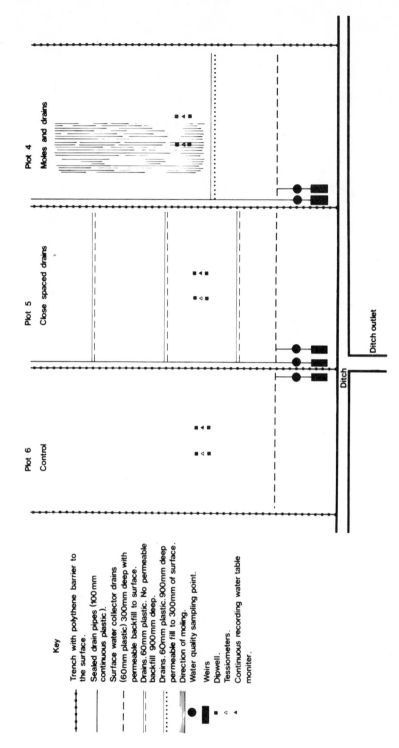

Figure 9.2. Instrumentation layout, plots 4, 5 & 6, for winter 1980-81.

Key

Trench with polythene barrier to the surface.

Sealed drain pipes (100 mm continuous plastic).

Surface water collector drains

Drains. 60mm plastic) 300mm deep with permeable backfill to surface.

Drains. 60mm plastic. No permeable backfill 900mm deep.

Drains. 60mm plastic. 900mm deep permeable fill to 300mm of surface.

Direction of moling.

Water quality sampling point.

Weirs

Dipwell.

Tessiometers.

Continuous recording water table moniter.

Plot 4
Moles and drains

Plot 5
Close spaced drains

Plot 6
Control

Ditch

Ditch outlet

158

grips, and by making sure that the gravel was higher than the surrounding soil.

Water was sampled using a vacuum water sampler situated at the inflow to the sediment control bins associated with each weir. Approximately 1 l of water was sampled every 8 hr and subsequently bulked to give 40 hr samples for analysis. A general picture of the water quality pattern throughout the winter could then be obtained, although no precise details of the variation in water quality over short periods could be observed. The water was analysed using standard ADAS methods for ammonium nitrogen and nitrate nitrogen.

Detailed flow records were obtained for two periods of drainflow, October 1979 to May 1980 and October 1980 to June 1981, and analysed using the trace following system described by Armstrong (1980b), and subsequently archived on computer file.

The depth to the water table was recorded in open auger holes (dipwells). Although the value of dipwell records has been questioned, the use of many replicated wells gives a good impression of the soil water regime as it affects agricultural operations (Armstrong, 1983). For the winter of 1979-80, a total of eight dipwells were installed on each of the nine plots, and read at weekly intervals. Detailed statistical analysis showed there to be no significant differences either between blocks or within plots. Consequently for the winter of 1980-81 the level of instrumentation was reduced to four dipwells on each of the intensively monitored plots (figure 9.2).

These dipwell results were also supported by occasional tensiometer recordings, and some continuous water table recorders. The results from these other instruments support the dipwell results, and are not discussed separately.

AGRICULTURAL OPERATIONS

Details of agricultural operations are kept as an essential component of the economics experiment. A brief summary as it affects the hydrology is given below.

The drainage was installed in June 1978, while the site was in a grass ley. Moles were drawn in August 1978. The site was then ploughed in October 1978 and lay fallow all winter. Spring barley was sown in May 1979. Soil moisture tension data indicated firstly that the autumn cultivation had created an impeding layer which prevented water movement down the profile, and secondly that the mole channels were not functioning. Consequently the moles were redrawn in June 1979, and the site was subsoiled at 300 mm depth after the harvest in October 1979 in order to break up the impeding layer.

The site was ploughed after subsoiling in October 1979, and remained fallow during the winter of 1979-80. A short term grass ley was sown in June 1980, and grazed in September and October. This autumn grazing resulted in damage to the soil surface, termed poaching, which has the effect of reducing infiltration rate of the surface (Kellett, 1978). The site remained in grass during the winter of 1980-81.

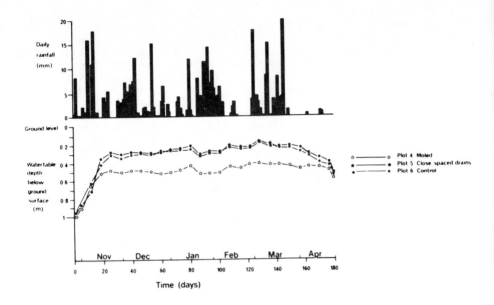

Figure 9.3. Daily rainfall, and soil water regimes of the
site, October 1979 to April 1980.

RESULTS - HYDROLOGY

The results presented here are for two winters, in which
the contrast is between two different land uses. In 1979-80
the site was fallow following an autumn ploughing which had
induced an open structure into the topsoil, whereas in the
winter of 1980-81 the site was in grass and thus had a
closed surface whose infiltration capacity had been further
reduced by poaching damage.

Water tables

The soil water regimes of the site for the two winters are
summarized in figures 9.3 and 9.4, which plot the mean water
tables for each of the three monitored plots, and the daily
rainfall amounts. It can be seen immediately that the mole
drains afford a significant lowering of the water table,
holding it to a value around 500 mm during 1979-80 and
250 mm during 1980-81. By contrast the close spaced drains
offer virtually no benefit in terms of water table control,
being statistically indistinguishable from the undrained
area.
      The higher water tables in the winter of 1980-81 reflect
the surface damage (poaching) of the site caused by cattle
grazing the site late in 1980. One of the consequences of
poaching is the creation of a compacted structureless surface layer (Kellett, 1978) which prevents water movement
down the profile. It is evident that this damage is severe
on both the control and close-spaced drained areas (plots
6 and 5), with the consequence that the water table remains
at or near the surface for long periods. By contrast, the

160

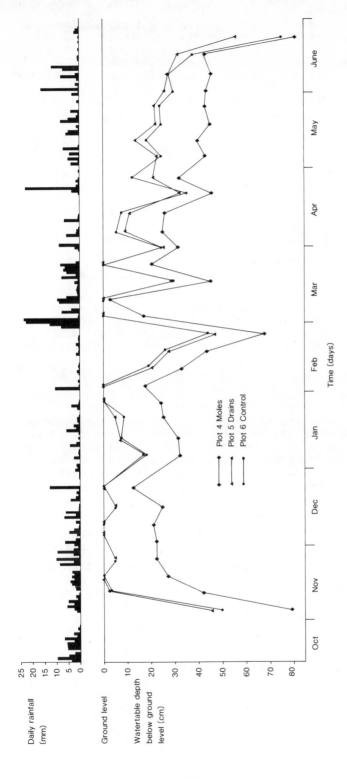

Figure 9.4. Daily rainfall, and soil water regimes of the site, October 1980 – June 1981.

161

Table 9.2. Monthly runoff from the instrumented
plots, 1979-80 expressed as mm runoff

Month	Rainfall	4(surface)	4(drain)	5(surface)	5(drain)	6(surface)
Oct	72.1	0.000	0.378	0.000	0.020	0.003
Nov	73.2	3.78	33.52	7.86	25.79	27.55
Dec	53.9	16.68	58.91	16.09	61.09	54.75
Jan	54.2	7.96	33.76	7.96	33.38	31.70
Feb	84.8	7.00	42.65	7.00	42.42	51.49
March	96.6	12.01	63.16	14.15	65.14	56.93
April	2.6	0.18	0.60	0.32	0.64	1.95
Total	467.4	47.61	232.98	53.38	228.66	224.4
Plot totals		280.59		282.04		224.4

poaching damage is much less severe on the moled area
(Plot 4), which thus retains its ability to control the
water table. The lower degree of poaching in the moled
plot has been confirmed by visual inspection, infiltration
measurement, and detailed analysis of surface micro-
topography.

Water discharge

For each of the instrumented plots (4, 5 and 6) two flows
were measured: combined surface and plough layer flow, and
drainflow (except for the undrained control plot 6 which
had no drainflow component). The overall magnitude of
these components for the winter of 1979-80 is summarised in
table 9.2, and the following points can be made:

i)   there is virtually no difference in total runoff
     between the two drainage treatments.

ii) the control plots lose as much water by surface flow
     and interflow as the two drained plots do by drainflow.

From this it can be seen that the effect of drainage is to
alter the route water takes in leaving the site, not to
greatly affect the overall water balance. Natural drainage
of water through the topsoil is in effect merely diverted
to the drains, either by the trenches of the more closely
spaced drains, or by the fissures that accompany mole
drainage.

This effect is illustrated by figure 9.5, which shows
hydrographs for all 5 measurement points. From this single
storm event, which is typical of the site for the year, the
following points may be noted:

i)   The highest peak is generated by the close spaced drain
     system, and the mole drained plots give only slightly
     lower peaks but virtually identical total volumes. It

162

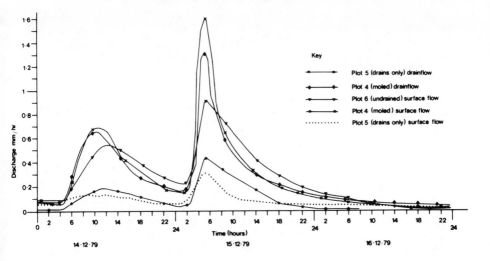

Figure 9.5.   Drainflow hydrographs for 14th-16th December
1979.

is probable that this is a reflection of the closer
spaced permanent drains permitting a more rapid concentration of the flows.

ii) The typical peaked drainflow hydrograph reported on
other sites (Rands, 1973; Trafford and Rycroft, 1973;
Bee *et al.*, 1975; Harris, 1977) is also observed on
this site.

iii) Surface and plough layer flow on the drained plots is
considerably lower than that on the control. The effect
of drainage is thus to redirect the water that is
naturally leaving the site through the plough layer.

Comparable data for the winter 1980/81 are given in table
9.3.   The major change is in the volume of surface runoff.
Because plots 5 and 6 have become virtually impermeable,
they have generated considerable amounts of surface runoff,
whereas on the well drained plot (plot 4) surface runoff
has been virtually eliminated.   This is illustrated by the
pattern of discharge shown by a typical hydrograph, figure
9.6.   Although very similar volumes are discharged by both
sets of drains, the virtual elimination of surface flow on
the moled plot results in a lower total runoff from that
plot.   The mole drained plot thus discharges considerably
less storm runoff than the other plots, since by virtue of
its lower water table, it can accept and store within the
soil a portion of the rainfall input.

This variation in hydrologic response between the plots
is dramatic, and contrasts strongly with the previous year.
The virtual sealing of the surface by poaching results in
an increase in the surface runoff component from plot 5,
from about 20% of the total runoff in 1979-80 to about 50%

163

Table 9.3.   Monthly runoff from the instrumented
plots, 1980-81 expressed as mm runoff

Month	Rainfall	4(surface)	4(drain)	5(surface)	5(drain)	6(surface)
Sept	31.1					
Oct	56.5	0.06	11.45	4.50	6.75	10.43
Nov	78.1	0.82	37.82	6.66	23.41	41.73
Dec	57.6	4.02	50.91	27.83	40.10	63.52
Jan	28.3	0.00	10.26	7.47	5.81	15.20
Feb	38.3	5.35	13.98	6.78	12.05	20.51
Mar	126.4	17.40	91.90	54.06	73.86	125.26
Apr	53.4	7.55	16.76	24.40	11.61	32.20
May	51.0	0.00	3.50	3.54	1.36	3.74
June	50.1	0.00	2.96	1.75	2.18	2.88
Total	570.8	35.186	239.15	137.00	177.13	315.56
Plot totals		274.33		304.13		315.56

Cockle Park 24th-25th March 1981

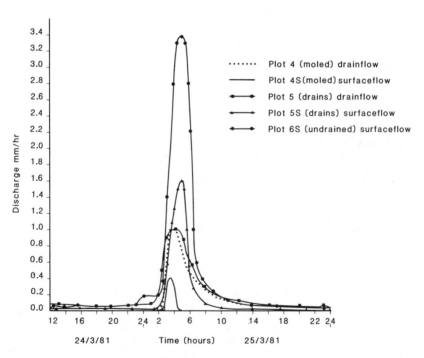

Figure 9.6.   Drainflow hydrographs for 24th-25th March 1981.

in 1980-81. Correspondingly, plot 6 discharges in one very sharp peak, as much runoff from the surface, as the total flow from plot 5.

It can thus be seen that by controlling the surface damage, artificial drainage has very considerably reduced the storm runoff from this site.

Water quality

Data on nitrate and ammonium loads were gathered for the period January to May 1980 and again for November to April 1981. Data for early 1980 were capable of translation to total budgets, and these results are presented in table 9.4.

From these data it can be seen that ammonium concentrations were very low and their contribution to the total budget could be safely ignored. Average values for nitrate-nitrogen concentration of around 15 mg l^{-1} in the drainage waters are similar to those observed at other arable sites in Britain (Cooke 1976). By contrast very low concentrations were observed on the surface flows of the drained plots. The variations in concentration do not, at the sampling frequency used, show any significant variation with discharge, and are approximately constant throughout the period monitored.

The total loads (the product of discharge and concentration expressed as weight loss per ha) reflect the hydrological budgets, in that the bulk of the load is discharged through the drains of the drained plot and through the surface waters of the control. Of significance is the observation that the nitrate load of the drained plots is of the same order of magnitude as that of the control. In view of the fact that these load figures combine all the potential errors in the analysis, it was considered that the differences between the plots could not be considered significant.

A comparable nitrogen budget for 1980-81 is presented in table 9.5. Mean concentrations were very low, usually less than 1 mg l^{-1} for both ammonium and nitrate-nitrogen, and despite considerable runoff volumes the total nitrate losses are less than 1 kg ha^{-1} from each plot. This value is very low, but is to be expected from grassland. In view of the very small size of this contribution, any differences between plots may be safely ignored.

GEOMORPHOLOGICAL IMPLICATIONS

It is difficult to generalize results from a single site, even when it is known that the site is typical of a large area. Nevertheless, with the warning that they are derived from one site only, and require confirmation from other sites, the following general conclusions can be made.

i) the change from grassland to arable has a far greater impact on the water quality of a site than the act of artificial drainage. The loss of nitrogen in the drainage waters is an order of magnitude lower under grassland than under arable.

Table 9.4. Total loads of water quality variables, expressed as kg ha^{-1} loss of nitrogen, and mean concentrations mg l^{-1}, January – May 1980.

Variable		4(surface)	4(drain)	5(surface)	5(drain)	6(surface)
Ammonium Nitrogen	Load	0.45	0.26	0.26	0.04	0.11
Ammonium Nitrogen	mean	0.44	0.11	0.46	0.27	0.77
Nitrate Nitrogen	Load	0.925	17.17	0.896	20.5	16.96
Nitrate Nitrogen	mean	3.23	11.56	2.62	14.63	10.54

Table 9.5. Total loads of water quality variables, expressed as kg ha^{-1} loss of nitrogen, and mean concentrations, mg l^{-1}, 25th November 1980 – 13th April 1981.

Variable		4(surface)	4(drain)	5(surface)	5(drain)	6(surface)
Ammonium Nitrogen	Load	0.002	0.044	0.133	0.056	0.170
Ammonium Nitrogen	mean	0.13	0.73	1.14	0.40	0.88
Nitrate Nitrogen	Load	0.005	0.131	0.257	0.287	10.433
Nitrate Nitrogen	mean	0.37	0.12	3.43	1.87	3.07

ii) The effect of artificial drainage on this site is only to divert flows which occur naturally through and over the topsoil. Only where that drainage increases the storage capacity of the soil by lowering the water table, does it reduce the volume of surface flow and hence of total discharge. Consequently, the incidence of overland flow which is potentially erosive on the site is reduced by effective drainage. Similarly the total volume of storm runoff is decreased by effective storage of water in the soil, which leads to a subsequent decrease in the erosive potential of flows from the site.

Such results show that for this site, the effect of effective artificial drainage is to reduce the potential for erosion, both within the site itself, and downstream. Such a result should not however be expected to apply on a whole basin scale, since there is a correlation between drainage and land use intensity. Consequently it has been observed that in many places there is a positive correlation between drainage rates and both erosive potential and pollutant loads. This is due to the critical position of land drainage as the first step in the transition from low to high intensity land uses (Royal Commission on Environmental Pollution, 1979, p 106). Consequently the observed effect of drainage is frequently the change from undrained grassland, with its low levels of nitrogen losses and its well protected vegetative cover, to an arable system with its higher nitrogen losses, and the deliberate creation of bare soil.

ACKNOWLEDGEMENTS

This study has involved the efforts of many of my colleagues, both in setting up and routine servicing of the experiment. Thanks are particularly due to Cyril Carter, Keith Dignan, and Ken Shaw of the Newcastle office for their help, to ADAS analytical services Newcastle for performing the chemical analyses, and to the University of Newcastle for the use of the land.

REFERENCES

Armstrong, A.C., 1980a, Cockle Park, Northumberland, in: *A Progress report on FDEU sites for 1978*, ed. S.Le Grice (FDEU Technical Report 79/3), 59-73.

Armstrong, A.C., 1980b, A novel system for digitising trace records, *Area*, 12, 123-127.

Armstrong, A.C., 1983, The measurement of watertable levels in structured clay soils by means of open auger holes, *Earth Surface Processes and Landforms*, 8, 183-187.

Armstrong, A.C., Carter, C. and Shaw, K., 1980, The hydrology and water quality of a drained clay catchment, a preliminary report on the Cockle Park drainage experiment, *ADAS Land Drainage Service, Research and Development Report* no.3, (MAFF, London).

Armstrong, A.C. and Smith, Y., 1977, Soil series and drainage practices, *FDEU Technical Bulletin*, 77/5.

Bailey Denton, J., 1861, On the discharge from underdrainage, and its effect on the arterial channels and outfalls of the country, *Proceedings of the Institution of Civil Engineers*, 21, 48-82, and discussion 83-130.

Bee, R., Dennis, C.W. and Farrer, K., 1975, Drayton EHF, Warwickshire, in: *A progress report on FDEU sites for 1974, FDEU Technical Bulletin*, 75/6, 4-14.

Cooke, G.W., 1976, A review of the effects of agriculture on the chemical composition and quality of surface and underground waters, in: *Agriculture and Water Quality*, ed. Dermott, W. and Gasser, J.R., (MAFF Technical Bulletin 32), (HMSO, London), 5-57.

Crompton, A. and Matthews, B., 1970, *Soils of the Leeds District*, (Memoirs of the Soil Survey of England and Wales, Harpenden).

Dennis, C.W., 1980, Drayton EHF, Warwickshire, in: *A Progress report on FDEU sites for 1978*, ed. S.Le Grice (FDEU Technical Report, 79/3).

Harris, G.L., 1977, An analysis of the hydrological data from the Langabeare experiment, *FDEU Technical Bulletin*, 77/4.

ILRI (International Institute for Land Reclamation and Improvement), 1973, *Drainage Principles and Applications*, (Edited from lecture notes on the International Course on land drainage, 4 vols, Wageningen, The Netherlands).

Kellett, A.J., 1978, Poaching of grassland and the role of drainage, *FDEU Technical Bulletin*, 77-1.

MAFF, 1979, *Field Drainage Experimental Unit, Annual Report for 1978*, (MAFF, London).

May, J. and Trafford, B.D., 1977, The analysis of the hydrological data from a drainage experiment on clay land, *FDEU Technical Bulletin*, 77-1.

Rands, J.G., 1973, An analysis of drainflows from FDEU experimental sites, *FDEU Technical Bulletin*, 73-11.

Royal Commission on Environmental Pollution (Chairman Sir Hans Kornbert), 1979, Seventh Report: *Agriculture and Pollution*, (HMSO, London), Cmnd 7644.

Talman, A.J., 1979, Simple flowmeters and water table meters for field experiments, *FDEU Technical Report*, 79-1.

Trafford, B.D. and Rycroft, D.W., 1973, Observations on the soil-water regimes in a drained clay soil, *Journal of Soil Science*, 24, 380-91.

10　Rapid subsurface flow and streamflow

solute losses in a mixed evergreen forest,

New Zealand

M.P. Mosley and L.K. Rowe

INTRODUCTION

There has recently been much interest in the generation of
streamflow by subsurface water movement, and also in the
geomorphic and denudational work accomplished by subsurface
flow (Beven, 1981; Burt et al., 1981; Mosley, 1979, and
references therein). In particular, there has been a
growing realisation that, in natural soils, water movement
tends to be along preferred pathways, often referred to as
macropores (Aubertin, 1971). The hydrologic and denudational implications of such water movement are only now
being examined in any detail (Beven and Germann, 1981;
Mosley, 1982; Scotter, 1978). Freeze's (1974) theoretical
treatment indicated that, under certain circumstances,
rapid subsurface stormflow could be expected, but he did
not consider the effects of preferred pathways in his
calculations. Mosley (1979) observed water moving at
velocities up to 2 cm s^{-1} through root channels and similar
pathways in a forest soil, while Scotter and his co-workers
(Kanchanasut et al., 1978; Scotter, 1978) have experimentally and theoretically demonstrated the preferential
movement of solutes through cylindrical channels and
vertical cracks in the soil.

Data have been gathered during a large-scale, long-term
study of the hydrological implications of different management practices in beech-podocarp-hardwood forest in Tawhai
State Forest, 40 km southeast of Westport, New Zealand,
(Mosley, 1979, 1982; Mosley and Rowe, 1981; O'Loughlin
et al., 1980, 1982). They permit an evaluation of the
importance of subsurface flow and solute loss in an environment that is distinctive for the rapidity of its
hydrological response. The catchments have short (30 m),
steep (30°-35°) slopes and shallow (~ 1 m) soils overlying
impermeable cemented Pleistocene gravels. Subsurface flow
appears both to be unusually rapid and to contribute an
unusually large proportion of storm-period streamflow, as
well as base flow. Normal rainfall is about 2600 mm $year^{-1}$
and runoff is about 1550 mm $year^{-1}$ (Pearce et al., 1976).

The data used herein apply primarily to two small catchments (referred to as catchments 6 and 15, with areas of 1.6 and 2.6 ha respectively) which are covered with undisturbed forest. Other catchments included in the study are managed under a variety of logging methods, burning, and planting practices, and strongly reflect the impact of human disturbance; data for catchment 8 (with an area of 3.8 ha) in its previous undisturbed state are, however, available.

SUBSURFACE FLOW MEASUREMENTS

Subsurface flow has been measured in two circumstances: 1) under storm-period conditions, generated by natural rainfall, and 2) under post-storm conditions, generated by water applied manually. In both cases, measurements were made in pits excavated in the forest soil, down to the cemented gravels. The pits were 2-3 m long (along the contour), with a 1 m long concrete trough constructed in the cemented gravel at the base of the pit in which sub-surface flow, seeping from the face of the pit, was collected. During storm period conditions, flow volumes were measured by tipping buckets connected to an eight-channel event recorder. Under the experimental conditions, flow was collected in a beaker held under the outflow pipe from the trough; the time required for a measured volume of water to flow from the trough was used to compute dis-charges. Under storm conditions, outflow was derived from an undetermined length of ground surface upslope from each pit, although Mosley (1979) concluded that water could move from divide to base of slope during even small storm events. Under the experimental conditions, water was applied 1 m upslope from the pits at a line source (a 1 m length of trough with holes drilled at 5 cm intervals into which water was poured at a constant known rate). Rates of application were within the range of subsurface flow dis-charges observed under natural rainfall, that is, up to 5 l min^{-1} per metre of contour. The experimental runs were therefore intended to determine velocity of water moving downslope from further upslope, rather than the infiltration rate and subsequent rate of movement of precipitation falling immediately upslope of the pits.

SUBSURFACE FLOW UNDER NATURAL RAINFALL

Figure 10.1 shows subsurface flow and streamflow hydrographs during a representative storm, in catchment 8 in its un-disturbed state (figure 10.2). All hydrograph peaks are closely coincident, with lag times (from rainfall centre of mass to runoff peak) of 1.0-1.9 hours. Although the contributing areas to each pit cannot be precisely defined, there is a clear increase in peak discharge and total flow volumes with distance from the catchment boundary, which implies that flow is moving considerable distances through the soil during the hydrograph rise. Subsurface flow in the subwatershed in which the pits are located is more than sufficient to account for the peak discharge measured at

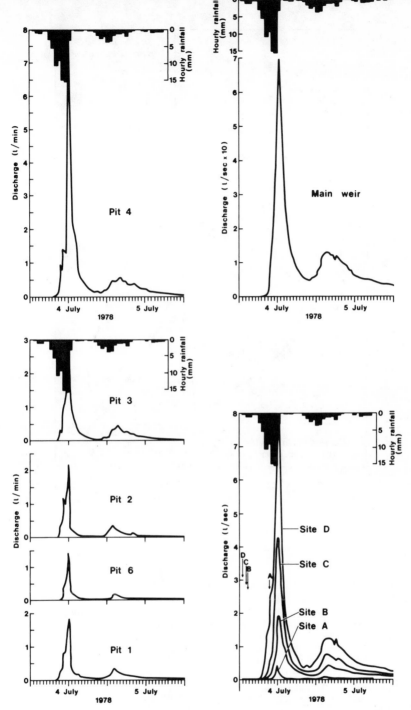

Figure 10.1.   Subsurface flow and streamflow hydrographs
in catchment 8, during a rainstorm in July 1978.
Locations of subsurface flow measurement pits and
streamflow sites are shown in figure 2 (from Mosley (1979)).

171

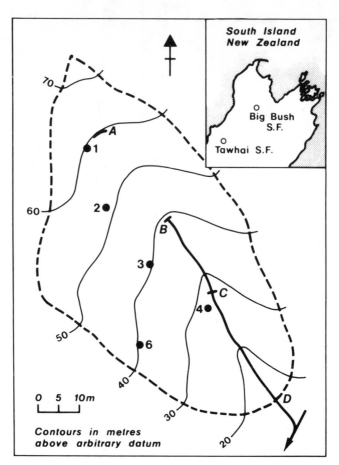

Figure 10.2. Locations of subsurface flow measurement pits and streamflow sites in catchment 8. The main weir is located 150 m downstream from site D.

the main weir. Cumulative runoff curves (Mosley, 1979, figure 9) confirm that subsurface flow during this storm equalled or exceeded streamflow at the main weir.

Cessation of subsurface flow in the subcatchment roughly coincides with cessation of flow at the main weir, although delayed flow volumes from this subcatchment are insufficient to account for delayed flow at the main weir (which is consistent with its storm period specific runoff exceeding that from the catchment as a whole). It is inferred that other side slopes in the catchment contribute proportionately less runoff during storm events, and more during low flow periods. In any case, it is clear that, in this environment, subsurface flow is responsible for most streamflow, both during low flow periods and during storm events (when a certain amount of overland flow may be observed along hillslope depressions and on the limited area of flood plain).

The storm period measurements in catchment 8, together with a number of dye tracing experiments (Mosley, 1979, p 800-1), conclusively demonstrated that subsurface flow through macropores was a reality in the Tawhai Forest environment, and indicated that velocities were up to 2 orders of magnitude greater than velocities through the soil matrix. Experimental work in catchment 6 (together with experiments in catchment 7A - recently clearfelled - and catchment 14 - clearfelled, burned and planted) was undertaken to obtain reliable data on the range of flow velocities through macropores in this type of environment. The data are presented in detail by Mosley (1982); Table 10.1 summarises the mean velocity $\bar{V}$ of subsurface flow under experimental conditions (using time between centres of mass of input and outflow), the maximum flow velocity, V_{max} (using time from start of application of water to start of outflow) and the proportion of the input volume that appeared as outflow, OV/IV. The high velocities of subsurface flow observed in the earlier study (Mosley, 1979) are confirmed, although there is a large amount of variability between experimental sites, in response to the character of the soil, tree root networks, etc.

Table 10.1. Mean values of hydrograph parameters for experiments with input rates of 2500 ml min^{-1} (standard deviations in parentheses).

SITES	$\bar{V}$ (cm s^{-1})	V_{max}(cm s^{-1})	OV/IV	N
Catchment 6, undisturbed	0.30(0.27)	0.46(0.44)	0.29(0.26)	7
Catchment 7A, clearfelled	0.24(0.19)	0.39(0.20)	0.26(0.12)	8
Catchment 14, clearfelled/ burned/planted	0.33(0.22)	0.43(0.10)	0.29(0.12)	8

At three of the experimental sites, storm-period subsurface flow measurements were also made and figure 10.3 shows outflow hydrographs for a storm in which centres of mass of precipitation and outflow were readily defined. Making the assumption that outflow travelled from all parts of the slope and that mean travel distances were 18.5 m and 15.3 m for sites U2 and U3 (slope lengths above of 37 m and 30.5 m), mean subsurface flow velocities of 0.18 and 0.13 cm s^{-1} respectively are indicated (contributing area to site 4 cannot be determined because of contour curvature).

Notwithstanding the variability in subsurface flow velocities at the experimental sites, it appears that slope units behave in such a fashion that water may move rapidly and for long distances during storm events. Since much of the ground surface in the study area is within 30 m of a stream channel, it is possible for all parts of the

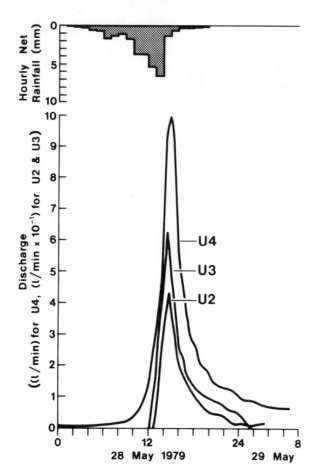

Figure 10.3.   Outflow hydrographs from sites U2-U4 during
    a rainstorm in May 1979.   Precipitation was measured
    beneath the forest canopy (from Mosley (1982)).

catchment to contribute water during a storm event lasting
of the order of 5 hours.   Nevertheless, a large proportion
of the water applied during the experiments was absorbed
into the soil matrix; this was also the case during natural
rainfall (eg. Mosley, 1979, figure 9, indicates that only
about 65% of net rainfall appeared as streamflow; the
remainder would presumably be lost as transpiration.   Data
assembled by Pearce and McKerchar (1979) indicate that,
for the period December 1974-February 1977, runoff was 74%
of net precipitation).

SOLUTE LOADS

Water samples were collected weekly from the streams draining catchments 6 and 15, and analysed for electrical conductivity (C), dissolved reactive phosphate(PO_4), nitrate
nitrogen (NO_3), ammonium nitrogen (NH_4), sodium (Na),

potassium (K), magnesium (Mg) and calcium (Ca) (Mosley and Rowe, 1981). Unfortunately, only a small number of samples have been simultaneously collected at the subsurface flow sites and from the streams, so that it has not been conclusively demonstrated that the chemical characteristics of subsurface flow and streamflow are similar; the data that are available reflect the chemical composition of the concrete subsurface flow collection troughs (that is, elevated concentrations of Ca are observed), but the concentrations of the other solutes are not significantly different from those in streamflow. Because the earlier study of streamflow generation (Mosley, 1979) indicated that both streamflow and baseflow are primarily generated by subsurface flow, it is assumed that the chemical composition of subsurface flow and streamflow are similar, with some dilution effect to be expected from precipitation onto channels and near-channel wetlands during rainfall.

Table 10.2 summarises the water chemistry data for catchments 6 and 15, and for a nearby precipitation collector. For the stream samples the major cations sum to only about 4 mg l^{-1} in most samples, and adjustment for other major constituents (chloride up to about 4 mg l^{-1} and soluble silica up to about 3 mg l^{-1} (Rowe, unpublished data from a limited number of samples)) increases total solute concentrations to only about 11 mg l^{-1}. Values of total dissolved solids determined by evaporation are generally on the order of 20-40 mg l^{-1}, however; the discrepancy is apparently caused by the presence of colloidal organic materials, because evaporated samples all had a brown residue that was not present in the rainwater samples. A poor relationship between total solute concentration, TS, (determined by evaporation) and conductivity is found for the streamflow data, and it is considered that, because of the presence of the colloidal organic material, conductivity is a better index of solute concentration than is TS.

Figure 10.4 shows the variation of conductivity during the data collection period for catchments 6 and 15. There is a slight seasonal influence apparent, with lowest conductivities (hence total solute concentrations) in summer. Analysis of variance of data for 11 additional sampling sites demonstrates that 15% of the variability in conductivity was due to this seasonal effect, and 27% to discharge (Mosley and Rowe, 1981), and this conclusion is expected to apply to catchments 6 and 15. However, the data for the weekly samples apply primarily to low flows; daily information for a nearby site for the period 1978-80 indicates a stronger influence of discharge, with a steady increase in values of conductivity as discharge declines following storms (Mosley and Rowe, 1981, figure 3). Storm period samples have not been collected and analysed for catchments 6 and 15, but some data are available for undisturbed forest catchments at Big Bush State Forest, 100 km northeast of Tawhai State Forest. The forest type is similar, but more open, with rainfall at Big Bush averaging 1500 mm year^{-1} as against 2600 mm year^{-1} at Tawhai. Figure 10.5 shows concentrations of Ca, K, Mg and Na for a storm event at Big Bush. During the storm a concentration effect is apparent for Na, a slight dilution

Table 10.2. Mean ion concentrations (mg l⁻¹) for weekly samples, 1979-80 (standard deviations in parentheses).

SITE	SAMPLE SIZE	PO_4	NO_3-N	NH_4-N	Na	K	Mg	Ca
Raingauge	16	0.03(0.03)	0.02(0.02)	0.13(0.17)	1.26(1.30)	0.19(0.22)	0.15(0.15)	0.21(0.22)
Catchment 6	53	0.04(0.09)	0.04(0.06)	0.01(0.01)	2.71(0.45)	0.28(0.05)	0.29(0.04)	0.58(0.16)
Catchment 15	52	0.05(0.12)	0.05(0.05)	0.01(0.01)	3.09(0.51)	0.35(0.16)	0.33(0.04)	0.69(0.14)

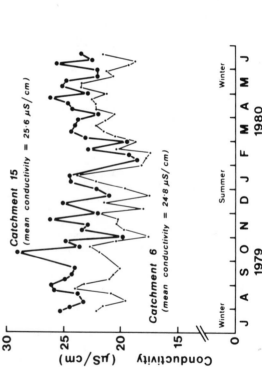

Figure 10.4. Electrical conductivities (corrected to 25°C) of weekly water samples collected at catchments 6 and 15, 1979-80.

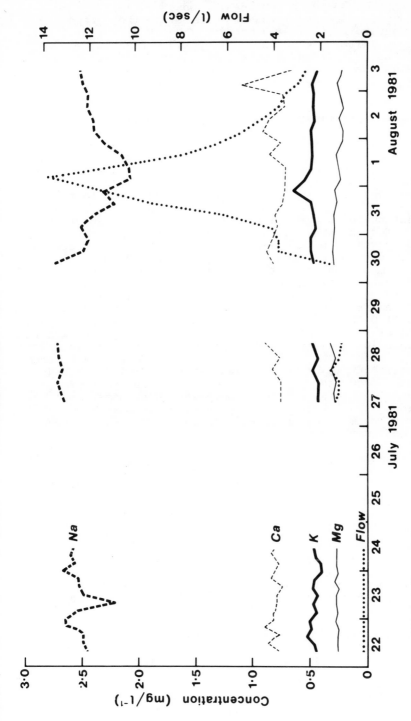

Figure 10.5. Water discharge and concentrations of Ca, K, Mg and Na during a storm event in July/August 1981 at Big Bush State Forest (see inset map in figure 10.2 for location).

effect is apparent for K and no variation caused by streamflow changes could be confidently observed in the data for
Mg and Ca. These patterns are similar to those reported
from Hubbard Brook, New Hampshire, by Likens, et al. (1977)
and from Taita, New Zealand, by Claridge (1980). The net
solute concentration changes little with flow and when
adjusted for ions which were not measured, solute concentrations remain less than 10 mg l^{-1} throughout.

IMPORTANCE OF SOLUTIONAL DENUDATION

Solutional denudation may be estimated from the difference
between solute inputs in precipitation and solute losses in
streamflow, and compared to data for suspended and bed load
sediment transport (O'Loughlin et al., 1980, 1982). The
data presented in table 10.2 require adjustment to take
account of ions for which analyses were not carried out,
so that there is uncertainty about the precise quantities
of solutes in rainfall and streamflow, but fortunately
there proves to be a clear-cut contrast between solute and
particulate losses from the study catchments.

The mean ion concentrations for precipitation listed in
table 10.2 have been summed and adjusted to include chloride
(Cl) and sulphate (SO_4). Chloride and SO_4 levels can be
estimated from the seawater Cl/Na ratio of 1.8 and SO_4/Na
ratio of 0.25 (Hendry and Brezonik, 1980) provided no large
source of air pollution is nearby. Using the mean Na level
in table 10.2, estimated Cl is about 2.3 mg l^{-1} (= 64 μeq l^{-1}).
This is about 40% of the level which can be estimated from
the relationship between Cl levels in precipitation and
distance from the ocean as calculated by Hutton (1976),
160 μeq l^{-1}, using worldwide data including some from
stations in the Wellington region given by Blakemore (1953).
However, as much of the precipitation in the study area
does not come directly from the sea, but from storms moving
more or less parallel to the coast, the estimate is not
unreasonable. Sulphate estimates would be about 0.3 mg l^{-1}.
Using the data in table 10.2 and adjusting for Cl and SO_4
concentrations gives an estimated total solute concentration
for precipitation of about 4.5 mg l^{-1} compared to that for
streamflow of about 11 mg l^{-1}, which also included an
estimate for silica.

These concentrations must again be adjusted for the
difference between precipitation (2600 mm year^{-1}) and
streamflow (1550 mm year^{-1}); it is estimated that solute
inputs in precipitation are about 110 kg ha^{-1} year^{-1}, and
losses in streamflow are about 170 kg ha^{-1} year^{-1}, that is,
there is a net loss of 60 kg ha^{-1} year^{-1} in solution. The
assumption is made that solute concentrations in streamflow
do not vary markedly in response to changes in discharge;
although there is a statistically significant relationship
between concentration and discharge (see above), figure
10.4 indicates that it may be neglected for present
purposes.

In comparison, O'Loughlin et al. (1980) report that
suspended and bed load sediment yields from catchment 6
averaged 113 m^3 km^{-2} year^{-1} during the period December 74-

February 1977, and 33 m^3 km^{-2} year^{-1} during the period January 1977-December 1978 (the latter period being markedly drier than average). Assuming that the sediment (collected in a trap at the catchment weir) had a bulk density of 1600 kg m^{-3} (Guy, 1970, table 4), these yields are equivalent to 1800 and 530 kg ha^{-1} year^{-1} respectively; the average for the period December 1974-December 1978 is 960 kg ha^{-1} year^{-1}. More recently available sediment yield figures (O'Loughlin et al., 1982) for the seven year period December 1974 to December 1981 are, for catchment 6, 1060 kg ha^{-1} year^{-1}, and for catchment 15, 760 kg ha^{-1} year^{-1}. Although inputs of particulate matter to the catchments have not been measured, it is clear that net denudation by solution is an order of magnitude less than by suspended and bed load transport. Moreover, in comparison with solute concentrations which average 11 mg l^{-1} and rarely exceed 20 mg l^{-1}, suspended sediment concentrations range up to over 2000 mg l^{-1} during stormflow (O'Loughlin et al., 1978).

DISCUSSION

The observed rapidity of subsurface flow in the study area has clear implication for denudational processes. Firstly, it would appear that precipitation that moves rapidly from the hillslopes into the channel system via preferred pathways has a limited opportunity to dissolve minerals from the soil mantle or bedrock surface. Secondly, by making a rapid contribution to streamflow, streamflow peaks are increased, and the streams have an enhanced opportunity to erode bed and banks, and to transport eroded sediments. Because these forested catchments have a continuous, thick litter layer covering the mineral soil, and because overland flow occurs only in limited areas, surface erosion is, under natural conditions, of little importance. Most slope erosion appears to be by infrequent mass movements during major storm events (O'Loughlin et al., 1978), which supply sediment to the stream channels. Erosion pin measurements indicate that the most actively eroding sites in catchment 6 and neighbouring catchment 5 are streambanks; presumably, soil creep on the neighbouring slopes renews material eroded from their banks by stream action. However, as shown by studies such as those of Hill (1973) and Hughes (1977), bank erosion occurs primarily during peak flows and its degree is related to stream flow; rapid subsurface flow on the slopes in the study area therefore contributes indirectly to erosion of the channel perimeters, and to removal of sediment supplied by both slow and rapid mass movement processes to the channel system.

REFERENCES

Aubertin, G.M., 1971, Nature and extent of macropores in forest soils and their influence on subsurface water movement, *U.S. Forest Service Research Paper* NE 192.

Beven, K., 1981, Kinematic subsurface stormflow, *Water Resources Research*, 17, 1419-24.

Beven, K. and Germann, P., 1981, Water flow in macropores. II. A combined flow model, *Journal of Soil Science*, 32, 15-29.

Blakemore, L.C., 1953, The chloride content of rainwater collected in the Wellington Province, *New Zealand Journal of Science and Technology B*, 35, 193-197.

Burt, T.P., Crabtree, R.W. and Fielder, N.A., 1981, Patterns of hillslope solutional denudation in relation to the spatial distribution of soil moisture and soil chemistry over a hillslope hollow and spur, Presented to *Annual Conference of the IGU Commission for Field Experiments in Geomorphology*, Exeter, U.K., (see also this volume).

Claridge, G.G.C., 1980, Studies of water quality and quantity at Taita and their possible significance to the Nelson situation, in: *Proceedings, Seminar on Land use in relation to water quality*, Nelson Catchment Board, Nelson, New Zealand, 14-39.

Freeze, R.A., 1974, Streamflow generation, *Reviews of Geophysics and Space Physics*, 12, 627-47.

Guy, H.P., 1970, Fluvial sediment concepts, *U.S. Geological Survey, Techniques of Water-Resources Investigations*, Book 3, Chapter C1.

Hendry, C.D. and Brezonik, P.L., 1980, Chemistry of precipitation at Gainsville, Florida, *Environmental Science and Technology*, 14, 843-849.

Hill, A.R., 1973, Erosion of river banks composed of glacial till near Belfast, Northern Ireland, *Zeitschrift für Geomorphologie*, NF17, 428-42.

Hughes, D.J., 1977, Rates of erosion on meander arcs, in: *River channel changes*, ed. Gregory, K.J., (Wiley, Chichester), 193-206.

Hutton, J.T., 1976, Chloride in rainwater in relation to distance from ocean, *Search*, 7, 207-208.

Kanchanasut, P., Scotter, D.R. and Tillman, R.W., 1978, Preferential solute movement through larger soil voids. II. Experiments with saturated soil, *Australian Journal of Soil Research*, 16, 269-76.

Likens, G.E., Borman, F.H., Pierce, R.S., Eaton, J.S. and Johnson, N.M., 1977, *Biogeochemistry of a forested ecosystem*, (Springer-Verlag, New York).

Mosley, M.P., 1979, Streamflow generation in a forested watershed, New Zealand, *Water Resources Research*, 15, 795-806.

Mosley, M.P., 1982, Subsurface flow velocities through selected forest soils, South Island, New Zealand, *Journal of Hydrology*, 55, 65-92.

Mosley, M.P. and Rowe, L.K., 1981, Low flow water chemistry in forested and pasture catchments, Mawheraiti River, Westland, *New Zealand Journal of Marine and Freshwater Research*, 15, 307-20.

O'Loughlin, C.L., Rowe, L.K. and Pearce, A.J., 1978, Sediment yields from small forested catchments, North Westland-Nelson, New Zealand, *Journal of Hydrology (New Zealand)*, 17, 1-15.

O'Loughlin, C.L., Rowe, L.K. and Pearce, A.J., 1980, Sediment yield and water quality responses to clearfelling of evergreen mixed forests in western New Zealand, *International Association of Hydrological Sciences Publication* no.130, 285-92.

O'Loughlin, C.L., Rowe, L.K. and Pearce, A.J., 1982, Exceptional storm influences on slope erosion and sediment yield in small forest catchments, North Westland, New Zealand. Presented to *Forest Hydrology Symposium, Institution of Engineers Australia*, Melbourne, Australia.

Pearce, A.J. and McKerchar, A.I., 1979, Upstream generation of storm runoff, in: *Physical hydrology - New Zealand experience*, ed. Murray, D.L. and Ackroyd, P. (N.Z. Hydrological Society, Wellington), 165-92.

Scotter, D.R., 1978, Preferential solute movement through larger soil voids. I. Some computations using simple theory, *Australian Journal of Soil Research*, 16, 257-67.

11 Hydrology and solute uptake in hillslope soils on Magnesian Limestone: the Whitwell Wood project

S.T. Trudgill, R.W. Crabtree, A.M. Pickles, K.R.J. Smettem and T.P. Burt

INTRODUCTION

An important step in the study of catchment processes is the evaluation of the links between fluvial processes and soil processes, and particularly the manner in which mobile soil water contributes to the discharge and solute load of rivers. In addition, sideslope solute inputs to a river system constitute chemical denudational outputs from a hillslope system. The work described below has the overall aim of contributing to the understanding of these topics, especially by field measurement and experimentation.

The specific aims of work are to:
1) Assess the routeways and travel times of soil water flow, especially in relation to input characteristics and the influence of soil structure on soil water flow.
2) Assess the nature of solute uptake by mobile soil water in relation to the hydrological processes of (1).
3) Assess the implications of (1) and (2) for (a) temporal variations in stream solute load and (b) solutional denudation on hillslopes.

The work has been undertaken in a lowland, wooded catchment on drift soils overlying a permeable, soluble carbonate bedrock, an environment somewhat neglected in catchment studies, which have tended to focus on upland, headwater systems.

Four main sub-projects have been carried out within the scope of the objectives outlined above.
1) The study of the dissolution of calcium and magnesium from the Magnesian Limestone bedrock and soils - a field and laboratory investigation on Ca:Mg ratios in relation to solid:solvent contact time.
2) A study of soil water residence time and solute uptake. A laboratory study using variable solid:solvent contact times in leaching columns, and a field study of soil water residence times, using fluorescent dyes, in relation to soil solute output. Soil moisture has been monitored using a scanivalve automatic tensiometer system.
3) A study of soil water flow in relation to soil structure. A field and laboratory evaluation of soil water

Table 11.1. Profile characteristics of Elmton soil
           (after Reeve, 1976)

a) PROFILE
    (cm)
    0 - 20/30   Stony loam/clay loam. Strong fine
        A   subangular blocky structure. Abundant
                       fine fibrous roots.
    > 20/30   Weathered bedrock
        R

b) ANALYTICAL DATA

    Horizon   A
    Depth, (cm)   0 - 20
    Sand, 200 μm - 2 mm, (%)   6
          60 - 200 μm, (%)   17
    Silt, 2 - 60 μm, (%)   52
    Clay, < 2 μm, (%)   25
    $CaCO_3$, (%)   10
    Organic carbon (%)   2.5
    pH (water, 1:2.5)   7.7

pathways and preferential flow using tracer-dyes and
tritium, with an assessment of routeways in relation to
structure and antecedent rainfall conditions.
4) A study of hillslope solute sources and solutional
denudation. An assessment of the evolution of solute
characteristics in the precipitation - throughfall/stemflow
- soil water - groundwater - stream water system and of the
distribution of hillslope solutional denudation in relation
to soil type, slope position and solute sources.

                        FIELD SITE

The work has been carried out at Whitwell Wood, 20 km south
east of Sheffield, UK, National Grid Reference SK 523788.
The site supports a 30 year old unfertilised stand of Beech
(*Fagus sylvatica*) planted on Elmton and Aberford soil
series (Reeve, 1976) and a narrow riparian strip of semi-
natural mixed deciduous trees. The soil series are classified as well drained. The Elmton series is a rendzina with
rubbly bedrock at approximately 30 cm depth. The Aberford
series is a brown calcareous earth with bedrock at up to
80 cm depth. The former has a loam texture and the latter
a fine loam or silt loam texture. Both have variable but
small amounts of glacial drift incorporated in the profile.
Soil analytical data are given in tables 11.1 and 11.2.
    The bedrock is Lower Magnesian Limestone, a dolomite
of Permian age, and a fractured aquifer described in other

Table 11.2.   Profile characteristics of Aberford soil
               (after Reeve, 1976)

a) <u>PROFILE</u>
    (cm)

    0 - 20   Clay loam - silty clay loam, moderate
       A   medium subangular blocky structure, common
                    fine fibrous roots.

    20 - 30   Slightly stony clay loam, moderate medium
       B_W subangular blocky, rare large pores,
                    highly calcareous.

    > 30   Magnesian Limestone, subangular or
       R   tabular fragments.

b) | Horizon | A | B_W |
    |---|---|---|
    | Depth, (cm) | 0 - 20 | 20 - 30 |
    | Sand, 200 μm - 2 mm (%) | 7 | 7 |
    | 60 μm - 200 μm (%) | 13 | 15 |
    | Silt, 2 - 60 μm (%) | 56 | 56 |
    | Clay, < 2 μm (%) | 24 | 22 |
    | $CaCO_3$, (%) | 4.2 | 10.8 |
    | Organic carbon (%) | 2.4 | 1.8 |
    | pH (water, 1:2.5) | 7.6 | 7.8 |
    | CEC (meq 100 g) | 31.1 | 23.7 |

areas by Aldrick (1978), Cairney (1972) and Cairney and
Hamill (1977).   The area receives a mean annual rainfall of
620 mm and transpiration plus evaporation normally exceeds
rainfall from April to September.
     Rainfall was monitored using a canopy level natural
siphon recording rain gauge and a check gauge; measurements
were also replicated at ground level at a nearby non-
forested site.   Stemflow and throughfall were also monitored.   Instrumented slope plots (as described below) were
laid out downslope towards the stream and also on flat
interfluve areas.

APPLICATION OF LABORATORY DISSOLUTION EXPERIMENTS TO
FIELD DATA ON CALCIUM AND MAGNESIUM IN RUNOFF WATERS

The work described in this section is an attempt to apply
laboratory experimental results to field data.   The dissolution behaviour of dolomitic bedrock has been studied
both in terms of relative amounts of calcium and magnesium
in solution and also in terms of relative saturation levels
with respect to the minerals calcite and dolomite.   Field
monitoring of calcium to magnesium ratios has been carried
out in conjunction with dye tracing in order to attempt to

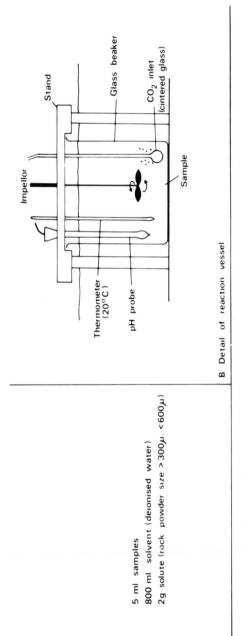

5 ml samples
800 ml solvent (deionised water)
2g solute (rock powder size > 300µ, < 600µl)

B. Detail of reaction vessel

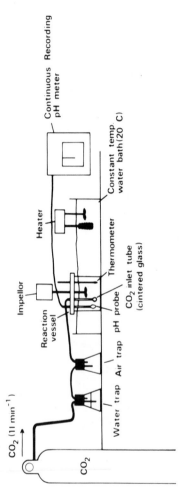

Figure 11.1. Laboratory apparatus for rock powder dissolution experiments.

identify quickflow and baseflow components of stream chemistry.

Trudgill, Smart and Laidlaw (1980) suggested that short residence time components of stream water would have a high calcium:magnesium ratio in response to variations in soil water flow routes into the stream and the behaviour of dolomite during the initial stages of dissolution. This work was a preliminary exercise and further experimental work was required.

Dissolution experiments in the laboratory have been carried out by many workers (including Picknett, 1964; Plummer, Wigley and Parkhurst, 1978; Rauch and White, 1977; Spears, 1976; Weyl, 1958). Of these, most have been concerned with chemical kinetics, often with little or no attempt to directly relate the information to composition of groundwater. Spears (1976) utilised simple experiments to study groundwater composition in the Vale of York and concluded that a valid qualitative interpretation can be taken from results of laboratory experimentation.

Wigley (1973) and Plummer and Mackenzie (1974) performed laboratory experiments on dissolution of calcite and magnesian calcite. From their work the concept of incongruent dissolution of calcium-magnesium carbonates was described. This concept is based on differing rates of solutional release of calcium and magnesium resulting in different rates of approach to saturation levels (Freeze and Cherry, 1979). More recently, Busenberg and Plummer (1981) have performed dissolution experiment using pure dolomite and have described the behaviour of calcium and magnesium ions in solution. These results indicate a more rapid release of calcium into solution.

Laboratory experiments

The initial hypothesis in carrying out laboratory experiments with natural dolomitic bedrock was that the molar calcium to magnesium ratio would progress, during the course of the experiment, from a high value (greater than 1.0) to a lower value (approximately equal to 1.0). The experimental equipment is shown in figure 11.1.

Powdered Magnesian Limestone bedrock was dissolved in deionised water at 20°C. The solution was saturated with CO_2 by bubbling a commercial gas through the solution. This was to keep a constant CO_2 pressure during the course of the experiment. pH was measured using a continuously recording meter. The solution was analysed for Ca^{2+}, Mg^{2+} and alkalinity by removal of 5 ml aliquots at various time intervals, ranging from 15 s to 10 days. Total amounts of a particular species in solution were calculated allowing for removal of samples and loss due to evaporation. Saturation indices and pCO_2 were calculated using the program WATSPEC (Wigley, 1977).

Calcium to magnesium molar ratios plotted against log time for selected experiments are shown in figure 11.2. On examination it can be seen that in general the hypothesis that the Ca:Mg ratio would progress towards unity with time is fulfilled. There is, in several cases, some variation within the overall trend. There are several possible

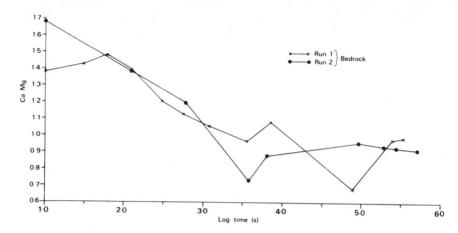

Figure 11.2. Ca:Mg ratios for dissolution of dolomite
bedrock from the field site.

reasons for this. The small sample size makes errors due
to dilution and contamination more likely, atmospheric CO_2
variation may have affected the dissolution characteristics
and, as Busenberg and Plummer (1981) describe, the weathering state of the rock sample is also crucial.

Field experiments

The field site has been described above. Figure 11.3 shows
calcium to magnesium ratios in throughflow drainage waters
and in stream waters over the period of a soil water dye
trace (described in detail below). From the diagram it can
be seen that the Ca:Mg ratio in both stream and throughflow
trough changes in response to input of water from a rainfall event. In both cases the initial Ca:Mg ratio is very
low and on the rapid arrival of dyed input water and
throughflow water the ratio increases. This is followed by
a decline with cessation of rapid throughflow (as indicated
by the dye behaviour).

Discussion

Although the ratio of calcium and magnesium in the field
and laboratory progress in the same direction, the ratios
themselves, and also the concentrations of ions in solution,
differ. The laboratory experimentation system involves
rapid approach to equilibrium whilst the chemical evolution
of the groundwater in the field has taken much longer. No
quantitative correlation can therefore be established in
respect of time scales and solution concentrations but the
evidence of the tracer dyes suggests that the results of
laboratory experimentation can be usefully applied in the
field in a qualitative manner to identify differing hydrochemical flow components.

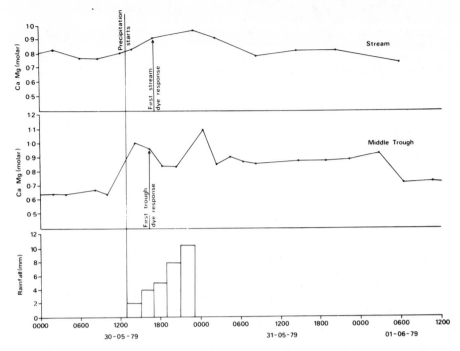

Figure 11.3.   Ca:Mg ratios for a field dye trace.

SOIL WATER RESIDENCE TIME AND SOLUTE UPTAKE

Water which is resident in soil for only a short period of
time has limited opportunity for solute uptake. Only the
most rapidly soluble chemical elements can sustain high
concentrations in solution during quickflow, as appears to
be the case for nitrate, for example (Trudgill, *et al.*,
1981). Soil structural characteristics and rainfall intensity characteristics may combine to encourage a dominance
of rapid flow and the preferential losses of rapidly soluble
material. In soils without preferential structural pathways and under rainfall regimes of low intensity, a more
uniform miscible displacement will occur, with the displacement of soil solutions of concentrations nearer
chemical equilibrium. Laboratory and field experimentation
has been designed to characterize the interactions between
soil solute output and soil water flow rates. Laboratory
column displacement work has been undertaken to vary solid-
solvent contact time and the results of these observations
have been used in conjunction with the results of the
dissolution experiments described above, and field dye
tracing to interpret field soil water solute output
patterns.

Dye tracing

Fluorescent dyes have been used to label infiltrating rain
water, and throughflow troughs have been used to monitor

189

the arrival of dyed water at the stream bank. Three
fluorescent dyes have been used, Rhodamine WT, Lissamine FF
and Amino G Acid (Smart and Laidlaw, 1977). Because of
adsorption, the use of these fluorescent dyes in soils is
limited to the identification of first time arrival of dyed
water. Evaluative work has shown that Lissamine FF behaves
similarly to chloride in column breakthrough in a brown
earth soil but that time to peak concentration may show
marked lags (Smettem and Trudgill, 1983). First times of
arrival for dyes and chloride are, however, similar.

The field site layout is shown in figure 11.4. Soil
water output from throughflow troughs was monitored using
automatic water samplers and the dye emplacement point was
moved progressively upstream as described in Trudgill *et al.*
(1983). A scanivalve automatic tensiometer system was
installed on the slope plot (Burt, 1978b).

Results and discussion

Soil water output was predicted to occur when soil field
capacity was reached (Trudgill, *et al.*, 1983) but this
was found not to predict all cases of dyed water output.
Rather, an intensity model was found to predict soil water
output where rainfall in excess of ped infiltration capacity
of 3.6 mm hr^{-1} led to output occurring (Trudgill *et al.*,
1983; figure 11.5). Dye output was more peaked during low
soil moisture status conditions (high tension). The results
suggest that an even displacement only occurs under conditions of around -20 to +20 cm soil water tension (as
measured at the base of the slope) and that under drier
conditions (-100/-200 cm) and wetter conditions (+40/+50 cm)
a rapid pedal by-passing occurred.

Solute derivation was studied by column leaching, the
sampling of the soil core being achieved by the successive
lowering of a PVC tube over a column of excavated soil.
Breakthrough characteristics were studied using a fraction
collector (figure 11.6) and soil pore characteristics were
studied using a resin impregnation technique and quantimet
analysis of thin section images (see below - figure 11.13).
Dye traces indicated a rapid initial flow, followed by
miscible displacement, with an inverse pattern for solutes
derived from the column (figure 11.7). Repeated column
leachings showed that nitrate displayed a high resistance
to exhaustion (Trudgill, *et al.*, 1981).

The data suggest that a uniform displacement model
(figure 11.8a) and a preferential flow model (figure 11.8b)
may both be applicable to soil water flow. Although the
applicability of these two models has yet to be closely
defined, it would appear that the latter is most applicable
where combinations of intense rainfall and extremes of wet
or dry soils occur (figure 11.9). Water entering a soil
can thus be partitioned into rapid flow and slow flow (or
storage), (figure 11.10), the solute concentration of the
former being dependent upon dissolution kinetics, desorption and some mixing; the solute concentration of the latter
being dependent upon time intervals between displacements.
Data analysis is in progress on the quantification of the
operation of these models. Further work is also aimed at

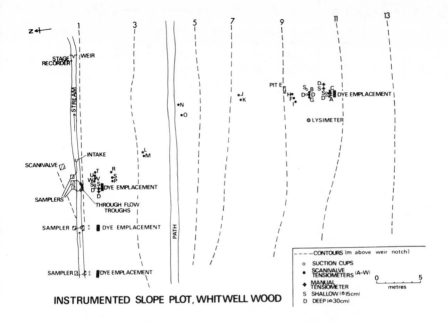

Figure 11.4. Field site layout for the soil water residence time and solute uptake experiments.

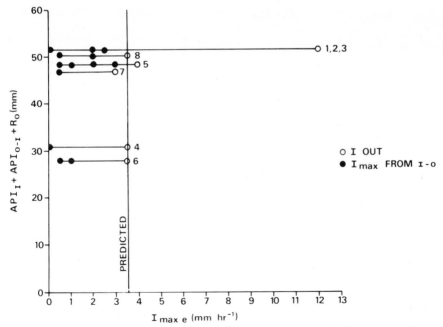

Figure 11.5. Dye output in relation to rainfall intensity and ped infiltration capacity.

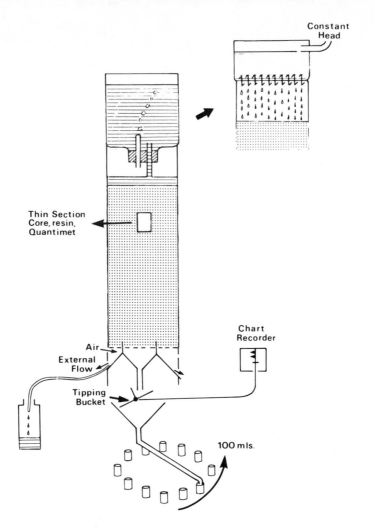

Figure 11.6.   Fraction collector for column leaching
  experiments.

quantifying preferential flow in relation to soil structure
as described below.

  SOIL WATER MOVEMENT IN RELATION TO SOIL STRUCTURE:
  AN ASSESSMENT OF TECHNIQUES FOR MEASURING THE SPATIAL
  DISTRIBUTION AND CONNECTIVITY OF LARGER SOIL VOIDS

This section presents part of an ongoing research project
into soil water movement in relation to soil structure and
input conditions. The project as a whole combines
laboratory experimentation on miscible displacement in
structured soils with field observations of preferential
flow in relation to differing antecedent and input conditions. This work has involved the use of a number of
tracers including Dyes, Cl^-, ^3H_2O and ^{36}Cl.

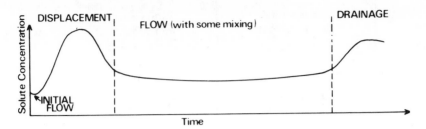

Model solute response for a column run

DISPLACEMENT FLOW (with some mixing) DRAINAGE

Solute Concentration

INITIAL FLOW

Time

Model dye response for a column run

DISPLACEMENT FLOW DRAINAGE

Dye Fluorescence

Time

Figure 11.7. Column solute and dye output.

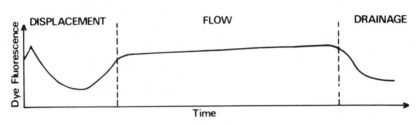

1 Simple displacement model

t_1 t_2 t_3

2 Macropore model

t_1 t_2 t_3

Figure 11.8. Models of soil water movement on the hill-slope.

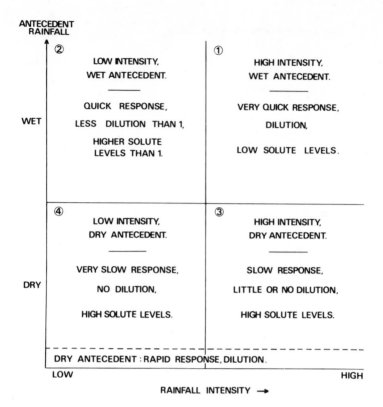

Figure 11.9.   Conditions controlling runoff and solute output.

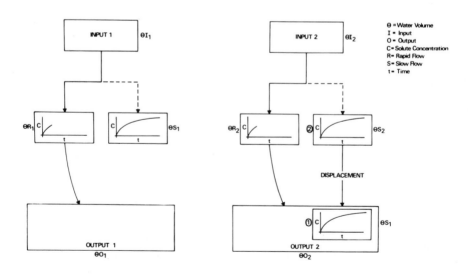

Figure 11.10.   A model of solute output.

The major variables controlling preferential flow are
identified from the field observations and replicated in
further laboratory experiments to characterize:-
1) The preferential flow threshold of a particular soil.
2) The amount of immobile water present under given
conditions.
3) The interaction between fast and slow moving soil
water.
The results of this work are being used to calibrate a
model for preferential flow in structured soils which is to
be tested against data derived from three experimental
lysimeters containing similar soils with differing gross
structures. Central to any quantitative interpretation of
preferential flow and hence a prerequisite for process
modelling is an evaluation of the hydraulic properties
pertaining to the soil under study. In particular, it is
necessary to derive an accurate measure of the spatial
distribution and connectivity of the larger soil voids.
    The difficulties involved in measuring this parameter
will be considered in detail.

Definition of the term 'larger' soil voids

In order to measure the percentage pore space occupied by
the 'larger' soil voids it is necessary to define the pore
size range under study. Use of the general descriptive
term 'macropore' is not sufficient as ambiguities in the
use of this term exist. Aubertin (1971) and Germann and
Beven (1981) for example use the term to imply non-capillary
sized pores, whereas, Anderson and Bouma (1977a), Bouma
et al. (1979) and Bullock and Thomasson (1979) all imply a
lower limit of 60 μm radius thus including all pores
normally containing air at 'field capacity' (arbitrarily,
0.04 bars).
    A practical solution to the problem is to define all
pores greater than 60 μm as 'macropores' and all pores
greater than 1000 μm as non-capillary macropores.
    It will be demonstrated in due course that this dual
division is most useful in analysing the hydraulic properties of a particular soil.

Techniques for the estimation of macroporosity and non-capillary macroporosity

Use of the soil moisture characteristic

Determination of the soil moisture characteristic (drying)
curve by use of tension table and pressure plate apparatus
is well documented (Reeve et al., 1973; Hall et al., 1977).
The curve can be used to calculate equivalent pore size
distributions by capillary theory. Thus, the radius, R, of a
pore holding water at a given tension is defined by;

$$\psi = \frac{2\sigma}{R\rho_w g} \qquad (1)$$

where σ is the surface tension of the air-water interface
       ρ_w is the density of water
       g is the acceleration due to gravity
       ψ is in units of length, equivalent to energy per
            unit weight

195

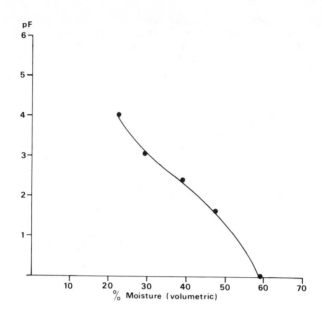

Figure 11.11.   Aberford 'A' horizon moisture characteristic
     curve.

The moisture characteristic for a brown earth soil of the
Aberford series (Reeve, 1976) from a Beech woodland is
shown in figure 11.11.   The curve is constructed from the
mean values of 15 replicate samples each with a surface
area of 44 cm^2 and a depth of 5 cm.   Only the 'A' horizon
is considered, the bulk density of which is 0.96 g cm^3
(mean value).   A pore size distribution derived from
figure 11.11 using capillary theory is shown in figure
11.12.
     Calculation of the pore size distribution in this
manner is subject to a number of errors and limitations:-
1)   The pore size distribution expresses only equivalent
pore sizes and yields no information concerning the shape
and connectivity of the macropores and non-capillary
macropores.
2)   Calculation of pore size classes by dividing the portion
of the curve between saturation and 0.04 bars is subject to
error due to the unknown shape of the curve between these
two points.
3)   The statistical frequency of occurrence of non-capillary
macropores in samples of 44 cm^2 may require a dispropor-
tionately large number of samples for accurate assessment.
This is shown clearly in table 11.3 where the standard
deviation and standard error of the mean shows greatest
variation in the range between saturation and 0.04 bars.
     A solution to the first two limitations is to employ
techniques of thin section image analysis (Quantimet) as
discussed by Jongerius *et al*. (1972a, b), Murphy *et al*.
(1977a, b) and Bullock and Thomasson (1979).   The technique

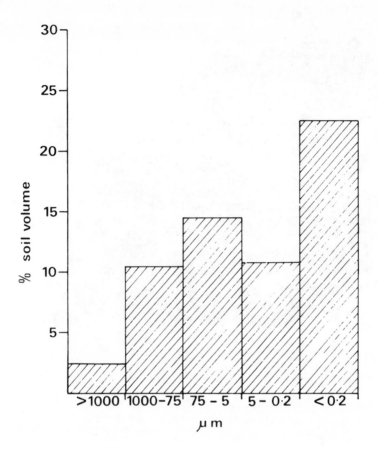

Figure 11.12. Aberford 'A' horizon equivalent pore size distribution.

Table 11.3. Aberford 'A' horizon, means, standard deviations and standard errors over four tension ranges (sample n = 15).

	Moisture Content (θ%)		
Range (bars)	$\bar{x}$	S	$S\bar{x}$
Saturation - 0.04	11.64	7.85	2.10
0.04 - 0.25	6.78	1.68	0.45
0.25 - 1.0	9.85	2.50	0.83
1.0 - 15	7.50	3.0	0.80

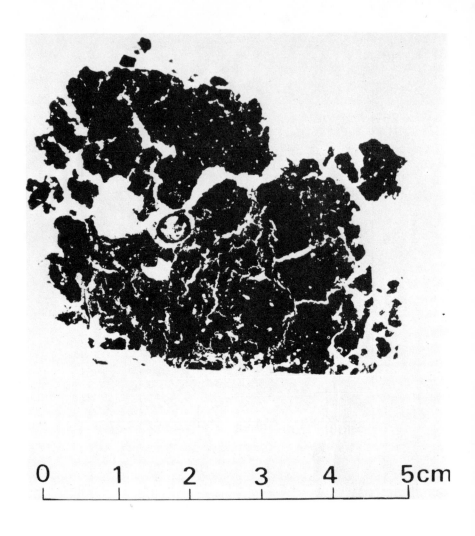

Figure 11.13.   Aberford 'A' horizon quantimet thin section
    image.

involves scanning a thin section of soil under a microscope
or a photograph under an epidiascope connected to a T.V.
scanner and detector module.
    An example of the type of image produced is shown in
figure 11.13 and a pore size distribution calculated from a
portion of this image is shown in figure 11.14. It should
be noted that the lower limit of resolution using this
technique corresponds to a pore radius of approximately
60 μm and therefore, the technique is only suitable for
investigating macropores.

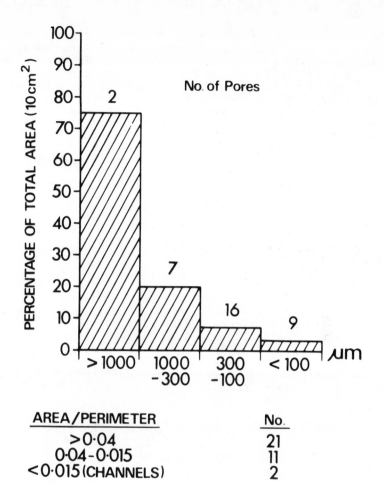

AREA/PERIMETER   No.

AREA/PERIMETER	No.
>0·04	21
0·04-0·015	11
<0·015 (CHANNELS)	2

Figure 11.14.   Pore size distribution derived from
    quantimet.

        In the case presented in figure 11.14 it can be seen
that two non-capillary pores dominate the macroporosity.
Obviously, replicate samples may yield significantly different results and once again a large number of samples
will be required to adequately assess the distribution of
non-capillary macropores.
        To achieve an adequate assessment of non-capillary
macroporosity it is therefore necessary to increase the
space scale of each individual sample in order to obtain a
spatial average that is statistically characteristic of the
soil around a point. An approach to the selection of the
appropriate scale is discussed below.

Spatial sampling for non-capillary macropores

Determination of a representative elementary volume of soil
(REV: Bear, 1972; Beven and Germann, 1981) to include a

statistical average of non-capillary macropores necessitates
the generation of measurement techniques that can be used
on large soil volumes. The technique investigated in the
context of this research has involved the use of infiltrometers. The rationale behind this approach is based on
the fact that non-capillary pores, if occurring within a
sample will exert a dominant control on the magnitude of
the saturated hydraulic conductivity (Anderson and Bouma,
1977a; Bouma and Wosten, 1979). By maintenance of a small
ponded head ($\leq$ 1 cm) the hydraulic conductivity can be
evaluated using the Darcy equation for one dimensional flow:

$$q = Q/At = K \; grad\phi \hspace{3cm} (2)$$

where Q = volume of infiltrated water (cm^3)
       A = cross sectional area (cm^2)
       t = time interval t (hr)
       q = 'steady state' flow of water through the
           profile (cm ha^{-1})
    grad = hydraulic gradient (cm cm^{-1})

The statistical distribution of K for a given field
area can therefore be ascertained by random sampling using
infiltrometers. By increasing the size of the infiltrometer
ring the distribution becomes less extreme (figure 11.15)
and the modal value of K increases. The explanation for
this is that the increasing ring size intercepts non-
capillary macropores with greater frequency and hence begins
to normalise the bi-modal distribution associated with the
use of smaller-sized rings.

Using this technique the REV necessary to include a
spatial average for the non-capillary macropore system at
the field site under study was set at 300 cm^2 (10 cm
radius).

*The problem of hydraulically discontinuous macropores
and non-capillary macropores*

In defining the REV by the infiltrometer technique an
implicit assumption is made that the non-capillary macropore system is hydraulically continuous. In the case of
the Aberford soil discussed in this paper the assumption
has validity as the soil has a mean depth of 25 cm and
displays good connectivity throughout the profile (Smettem
and Trudgill, 1983). However, in the case of particularly deep or cultivated soils the assumption may have
less validity due to discontinuities at the surface or at
depth.

The infiltration approach cannot be used to account for
pores with surface discontinuity and these have to be
assumed non-operational. However, the cumulative infiltration curve can be used to assess the volume of non-
capillary macropores with surface connections terminating
within the profile. The technique is outlined by Collis-
George (1980) and involves separating the main parameters
controlling ponded infiltration by use of the following
equations:-

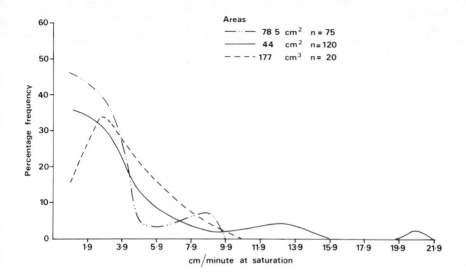

Figure 11.15. Saturated infiltration rates for three sizes of infiltrometer.

$$i(t)L = (i_1^* + i_o) + K^*t \qquad (3)$$

$$i(t)s = i_i + S^*t + K^*t \qquad (4)$$

$$i(t) = i_1 + S^*t \qquad (5)$$

where i(t)L is the solution for long time periods (post 'steady state')

　　　i(t)s is the solution for times up to the establish-
　　　　ment of the steady state

　　　i(t) is a generalised solution for cases where K^*t
　　　　is not established

　　　i_1 = instantaneous component (crack filling)

　　　S^*t = short term sorption component (S = sorptivity
　　　　of Philip, 1957)

　　　K^*t = 'steady state' component

　　$(i_1 + i_o)$ = i axis intercept for t = o in eq (1)

　　　　* denotes experimentally determined parameters

Note the relationship between K^* and q (eq 2);
when grad θ = 1, K^* = K grad θ, K = q

This allows the use of the same data set as previously, provided that the entire curve is available and that K Sat is not excessively high as a result of particularly good non-capillary macropore connectivity.

To test the accuracy of this procedure a number of laboratory simulation experiments were initiated. The results of one such experiment are reported below. For this test, soil passed through a 2 mm sieve was packed into a column (20 cm dia., 30 cm length) containing three rolled gauze cylindrical channels. The channels were open to the

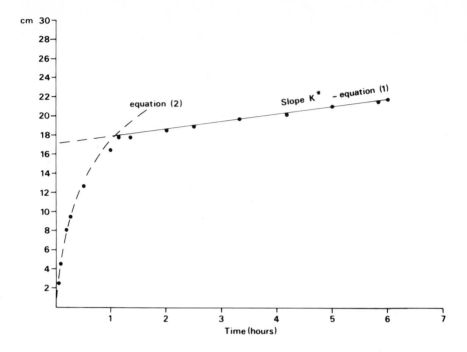

Figure 11.16.   Cumulative infiltration curve for soil
    column with three finite channels.

soil column surface but not connected to the base and had
a total volume of 101.8 cm^3. Water was ponded to a depth
of 1 cm and maintained by use of a constant head device
with 10 cm^3 calibrations. Stop-watches were used to calculate the initial infiltration values. The cumulative curve
to 6 hours is shown in figure 11.16 with slope K^* plotted
(linear regression of 'steady state' phase). To calculate
S^*, values of $i(i_1 + i_0$ eq 3) - K^*t are plotted against
t (hr) and the linear regression is again used (figure
11.17). Projection of the slope S^* to the i axis yields
the division between i_0 and i_1 with i_1 theoretically con-
forming to the volume of the non-capillary macrovoids. The
accuracy of the i_1 calculation depends to a large extent on
the accuracy to which the lower points on the S^* line can
be determined. In this laboratory experiment, i_1 over-
estimated the volume of channel storage by 3% (table 11.4)
which is within the accuracy to which the constant head
device is calibrated. For field use the error limits are
somewhat larger and i_1 should consequently be interpreted
with care. In this case, equations (3) and (4) have been
used rather than the simplified solution (eq 5). Equation
(5) is employed where the correlation coefficient for the
regression of K^* falls below 0.99 (Collis-George, 1980).
For the reported example the correlation coefficient was
0.992.

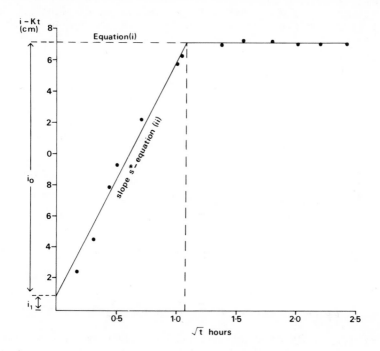

Figure 11.17. Cumulative infiltration curve graphed as $(i - K*t)$ versus $\sqrt{t}$ to obtain S^* and i_1^*.

Table 11.4. Infiltration results from figures 11.16 and 11.17

eq (3) $i(t)L = (0.93 + 16.07) + 0.8$ cm (after 1.12 hrs)

eq (4) $i(t)s = 0.93 + 18.96\ t + 0.8$ cm (to 1.12 hrs)

$\quad\quad i_1 = 0.93$ cm $= 105.56$ cm^3

error against channel volume = +3.7%

error as percent volumetric moisture content = 0.1%

This technique offers the opportunity for assessment of the volume of non-capillary macropores in large volumes of field soil *in situ*, without excessive disturbance and therefore, could have considerable use in catchment studies where the problems of destructive sampling are well known.

To gauge the bulk effective non-capillary macroporosity in
large laboratory soil columns the procedure of Germann and
Beven (1981) can be used. This offers a convenient method
for checking the accuracy of the infiltration approach on
selected columns.

The procedure involves saturating a large soil column
from the base to the surface and subsequently allowing free
drainage, which is collected. The installation of tensiometers allows the final potential profile at the onset of
fully unsaturated conditions to be calculated and hence the
drainage amount can be placed in context of the soil
moisture characteristic for the column. The final water
content (θf) is estimated by;

$$\theta_f = e - \theta_f/V \qquad (6)$$

where   e is the total sample porosity
        Q_f is the final outflow volume (cm^3)
   and V is the sample volume (cm^3)

The total porosity of the sample e, is given by

$$e = 1 - \frac{p}{pr} \qquad (7)$$

where   p is the dry bulk density of the sample
   and pr is the density of the soil material ($g\ cm^{-3}$)

Expression of Q_f as a proportion of θ_f yields the volumetric
content of the freely draining pore system (equivalent to
the non-capillary macroporosity).

Current research indicates that estimates of non-
capillary macroporosity obtained using the infiltration
approach are within 10% of values obtained using the
laboratory column method.

Conclusions

For preliminary studies on a catchment scale, the infiltration approach outlined in this paper has obvious advantages due to the rapidity with which undisturbed sampling
can be undertaken. At this scale, the main use of the infiltration approach is to provide a useful approximation as
to the volume of non-capillary macropores. Such an approach
could find expression in determining areas in which scaling
of infiltration parameters according to the similar media
concept (Warrick *et al.*, 1977) may be inappropriate due to
macropore effects.

For laboratory studies, non-capillary macropores can be
estimated using the approach to Beven and Germann (1981) on
samples of a size determined using the 'steady state' infiltration procedure. Macropores in the size range
60-1000 μm are most suited to determination using thin
section techniques with additional staining procedures
(eg. Bouma *et al.*, 1977) to identify the hydraulically
operational pores. The soil moisture characteristic is

best used in the tension range between 0.04 - 15 bars, where equivalent pore size distributions can be calculated with acceptable confidence limits. For accurate studies the effects of hysterisis, not discussed in this paper, often require careful consideration and where appropriate, the construction of a soil moisture characteristic from both wetting and drying cycles.

HILLSLOPE SOLUTE SOURCES AND SOLUTIONAL DENUDATION

Carson and Kirkby (1972) have modelled the solutional denudation of a temperate limestone hillslope. The potential distribution of solution rates is assumed to be uniform over the slope, but because of increasing evapotranspiration downslope, the model predicts a downslope decrease in solutional denudation which will lead to slope decline. The purpose of this study was to test the hypothesis that spatial variation in solute uptake and solutional denudation, resulting from spatial variation in hillslope water flow, will lead to differential solutional denudation over the hillslope.

Hillslope instrumentation

The hillslope field site, shown in figure 11.18, was instrumented for a one year period from June 1979 to June 1980. The fieldwork programme was designed to monitor hydrological and hydrochemical processes operating on the slope. The geomorphological effect of these processes was then assessed using a bedrock tablet microweight loss technique.
Measurement of stemflow and throughfall was achieved by using 10 throughfall and stemflow gauges. Canopy interception was assessed by comparing the total input to the ground from stemflow and throughfall with the input to the canopy. Canopy input was measured using a canopy level rain gauge and bulk solute collector (Crabtree and Trudgill, 1981).
Soil water samples for chemical analysis were obtained using Soil Moisture Equipment Corporation vacuum soil water samplers, emplaced in two rows up the slope, at depths of 0.17 and 0.27 m. Throughflow troughs and natural lysimeters were also used to obtain soil water samples. Soil water movement was monitored using banks of manual (Burt, 1978a) and automatic 'scanivalve' (Burt, 1978b) tensiometers.
The input of soil water to the stream at the base of the slope was monitored by bracketing the hillslope reach with trapezoidal fibreglass flumes equipped with stage recorders and automatic water samplers. This enabled the input of soil water and solutes from both the instrumented hillslope and the opposite hillslope to be assessed.
A weekly sampling program was used for the collection of precipitation, soil water and stream water samples. A solute budget could not be determined for the slope because the total area of slope contributing to the stream was unknown. However the evolutionary sequence of solute uptake in water moving through the hillslope system does allow spatial variation in solute uptake to be identified. Short term process responses to individual storm events were

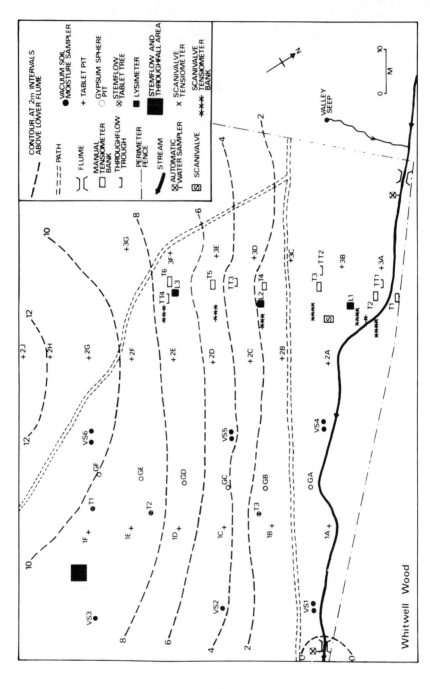

Figure 11.18. Instrumented hillslope section used for the solutional denudation experiments.

monitored using the automatic tensiometers and automatic water samplers.

Previous studies, for example Newson (1970) and Trudgill (1977), have used bedrock tablet micro-weight loss techniques to determine absolute rates of erosion from the solutional weight loss of an accurately weighed rock tablet. Results in terms of an absolute erosion rate are open to question because of the nature of the technique which is based on the emplacement of a fresh piece of rock in a disturbed soil pit. However, the technique is valid for comparison of relative weight loss. In this present study, the micro-weight loss tablets were manufactured from local bedrock material using a standardised procedure, modified from that of Trudgill (1975). The tablets were weighed to ±0.00001 g with a significant weight loss level of ±0.0001 g. The tablets were emplaced in three rows of pits running upslope, as shown in figure 11.18. Each pit contained 10 tablets and the pits were spaced at 10 m intervals upslope. The tablets were emplaced at the soil-bedrock interface for the one year study period. The weight loss of the tablets was assumed to be due to solutional loss by hydrolysis of the carbonate material of the tablet.

Hydrological and hydrochemical process monitoring results

The total precipitation input to the canopy for the 1 year study period was 717 mm. This was partitioned into ground level inputs of 482 mm throughfall (67%), and 90 mm equivalent stemflow (13%). This gives an estimated canopy interception loss of 145 mm (20%). Similar input proportions in beech woodland were found by Delfs (1967) and Nihlgard (1970). Both stemflow and throughfall showed seasonal variations due to the presence or absence of leaves in the canopy (Crabtree, 1981).

The passage of rainfall through a tree canopy was shown by Voight (1960) to cause the leaching of soluble material from the canopy, thereby altering the composition of the incident precipitation. As shown by Mayer and Ulrich (1980) the chemical composition of stemflow and throughfall, in excess of the composition in the incident precipitation, is due to the leaching of impacted aerosols from the atmosphere and not the leaching of substances originating from nutrient recycling. The input of solutes to the slope is shown in table 11.5 as an input loading based on the summation of weekly input loads. Canopy leaching also results in increased acidity of water reaching the ground. The rainfall has a mean pH of 5.8, compared to 5.6 for throughfall and 3.7 for stemflow. Therefore the inputs to the ground surface are of dilute acid solutions. These acid solutions will be neutralised by carbonate dissolution within the soil and bedrock to give rise to solute uptake and solutional denudation.

The inputs to the soil from stemflow and throughfall were considered to be uniform over the slope as a whole. However stemflow represents a point source input of large volumes of water around each stem whereas throughfall provides the input of water to the soil matrix. Dye tracing studies showed that stemflow bypassed the soil matrix by

207

Table 11.5.   Solute inputs to the ground
for the 1 year study period

Solute	Rainfall input to canopy kg m^{-2}	Stemflow kg m^{-2}	Throughfall kg m^{-2}	Total input to ground kg m^{-2}
Ca^{2+}	3.3	1.2	4.2	5.4
Mg^{2+}	4.9	0.8	4.6	5.4
K^+	0.6	0.9	2.1	3.0
Na^+	1.4	0.6	1.8	2.4
HCO_3^-	17.0	0	9.1	9.1
Cl^-	10.9	2.8	9.5	12.3
Si	0.3	0.1	0.4	0.5

flowing along root channels and macropores into the fissured
bedrock without diffusing into the soil.
    The seasonal pattern of soil matrix water behaviour,
resulting from throughfall, showed high soil water tensions
over the slope during summer which were reduced in winter.
However, saturation only occurred at depth in the valley
bottom soils resulting from a rise in groundwater levels.
No throughflow was detected except in the valley bottom in
winter when the throughflow trough pits were flooded by
rising groundwater.   Analysis of soil water potentials
showed that the dominant water movement process over the
slope was vertical percolation to the water table, followed
by lateral displacement flow of groundwater to the stream
by seepage through the valley bottom soils.   This water
movement pattern was also indicated by dye tracing studies.
Over the one year period this seepage accounted for a total
input of 0.7×10^8 l between the two flumes and a total
solute input of:

Ca^{2+}	1214.3 kg
Mg^{2+}	2187.1 kg
K^+	210.5 kg
Na^+	1114.0 kg
HCO_3^-	21860.3 kg
Cl^-	9942.8 kg
Si	248.6 kg

Runoff on the wooded Magnesian Limestone hillslope
can be divided spatially into vertical percolation on the
slope and lateral displacement through the valley bottom.
Also the stemflow and throughfall inputs can be separated.
During storms rainfall produced a rapid displacement of
solute rich groundwater to the stream while the soil matrix
remain unsaturated.   This rapid response would seem to be
due to the stemflow inputs bypassing the soil matrix.
    Assuming a uniform distribution of available solutes or
potential for solution over the slope, soil water solute

Table 11.6. Total major ion concentration changes
upslope in soil water samples at 0.2 m depth

Location	Lysimeters		Suction samplers	
	T.I.C.	pH	T.I.C.	pH
Top of slope	2.68	6.9	3.11	5.5
Midslope	3.59	7.3	7.11	6.4
Bottom of slope	8.08	7.4	9.44	7.4

T.I.C. = sum cations + anions based on annual mean solute
concentrations (mmol l^{-1})

concentrations should be uniform over the slope, except in
the valley bottom which was influenced by alkaline groundwater. However the results from the soil water samplers and
lysimeters, shown in table 11.6, indicated that soil pH,
carbonate content, cation exchange capacity and levels of
readily available and total solute material decreased upslope. This could not be explained by leaching by throughflow or vertical percolation, but was a direct result of
changes in the nature of deposited soil materials on the
slope. These were a calcareous alluvium in the valley
bottom; the Aberford series, a calcareous brown earth formed
from fluvioglacial material on the slope, and the Elmton
series, an acid rendzina, formed from glacial drift, at the
top of the slope. Therefore the spatial variation in soil
water solute uptake was not due to spatial variation in
hydrological processes but due to the spatial variation in
soil chemistry. The geomorphological significance of this
is that as acid water neutralisation, by solute uptake
within the soil, decreases upslope then the potential for
solutional denudation at the soil-bedrock interface will
increase upslope, at least for this particular hillslope.
    The results of the rock tablet emplacement study con-
firmed this prediction that solutional denudation increased
upslope. All three rows of tablets showed a significant
increase in weight loss upslope. The results for the mean
weight loss of each pit are shown in figure 11.19. The
relationship between mean percent weight loss and distance
upslope had the form:

$$\% \text{ weight loss} = 0.13 + 0.006D$$

D = distance upslope (m) Signif. at 99%

$$n = 23, \ 5 = 0.69$$

Problems of spatial autocorrelation limit the statistical
analysis of tablet and soil chemistry data. However general
trends can be indicated by linear regression analysis. The
relationship between mean percentage weight loss and pH
is as follows:

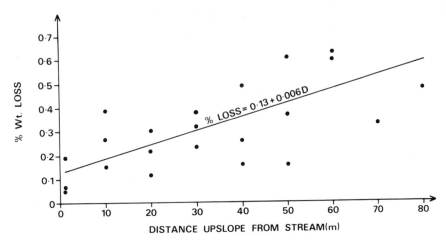

Figure 11.19. The relationship between rock tablet weight
loss and distance upslope.

$$\% \text{ loss} = 1.13 - 0.13 \text{ pH}$$

Significant at 99%   n = 23,   r = -0.76

Both pH and soil carbonate content showed a significant
decrease upslope. However there was no simple relationship
between either pH and carbonate content or mean percent
weight loss and carbonate content (Crabtree, 1981).
    Changes in soil pH appear to be the dominant factor in
controlling solute uptake within the soil and solutional
denudation at the soil-bedrock interface over the hillslope.

Conclusions and discussion

The upslope decrease in soil pH cannot be interpreted in
terms of lateral subsurface flow, because this is virtually
absent on the slope. The soil chemistry changes are related
to the original nature of the three different types of soil
forming parent materials present on the slope. The solutional weathering model proposed by Carson and Kirkby is
therefore inapplicable to the hillslope studied, because,
contrary to the model's assumptions, the potential for
solution at the soil-bedrock interface is not uniform but
increases upslope.
    The soil water solute and micro-weight loss data suggest
that solution rates at the soil-bedrock interface increase
upslope and that solute uptake within the soil decreases
upslope. Continued slope development by solution processes
acting in this manner will ultimately lead to slope decline.
Figure 11.20 contrasts the theoretical pattern of soil
weathering on a limestone hillslope proposed by Curtis,
Courtney and Trudgill (1976), based on the Carson and Kirkby
model, with the actual pattern for the hillslope studied at
Whitwell.

SOIL WEATHERING ON A THEORETICAL LIMESTONE SLOPE FROM CURTIS, COURTNEY AND TRUDGILL (1976)

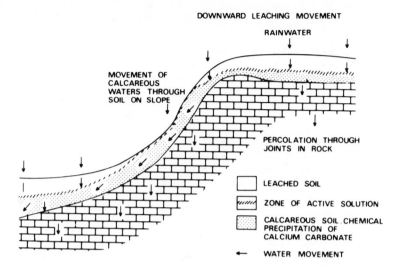

SOIL AND BEDROCK WEATHERING ON THE MAGNESIAN LIMESTONE HILLSLOPE, WHITWELL WOOD

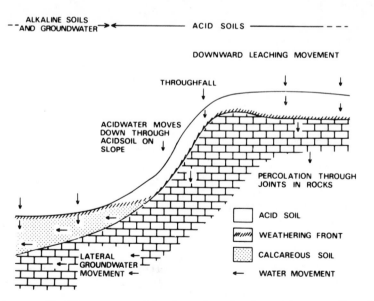

Figure 11.20.   Theoretical and actual patterns of solutional denudation on a limestone hillslope.

The original hypothesis that any spatial variation in solutional denudation results from spatial variation in hillslope runoff must therefore be rejected for the Magnesian Limestone hillslope studied at Whitwell Wood. Spatial variations in solute uptake and solutional denudation over the slope do exist, but were due to spatial variation in soil chemistry and not because of spatial variations in hydrological processes.

The results from this study show that soil chemistry, in particular soil pH, may determine the pattern of solutional denudation over a landscape. A simple generalised solutional denudation weathering model based on soil chemistry can be proposed. Under acid soil conditions, limited solute uptake and neutralisation of acidic input water will occur in the soil resulting in bedrock lowering by dissolution at the soil-bedrock interface. However in base rich alkaline soils, acid water neutralisation will occur within the soil and result in soil solute loss, not bedrock lowering.

ACKNOWLEDGEMENTS

The project was financed by the Natural Environment Research Council under Grant GR3/3459 and Studentships GT4/78/AAPS/46 and GT4/79/AAPS/48. We are grateful to the Forestry Commission for access to the land and continued help, and to Manchester Polytechnic for the loan of equipment. The technical assistance of Marcus Beasant and Cliff Fletcher is gratefully acknowledged. Andy Gardiner and Greg Corscadden assisted with field installations. The following are thanked for their critical appraisal: P. Smart, D. Briggs, R. Cryer, C.D. Curtis, K. Beven and F.M. Courtney.

REFERENCES

Aldrick, R.J., 1978, The hydrogeology of the Magnesian Limestones in Yorkshire between the River Wharfe and the River Aire, *Quarterly Journal of Engineering Geology*, 11, 193-201.

Anderson, J.L. and Bouma, J., 1977, Water movement through pedal soils: 1. Saturated flow, *Journal of the Soil Science Society of America*, 41, 413-418.

Aubertin, G.M., 1971, Nature and extent of macropores in forest soils and their influence on subsurface water movement, *USDA Forest Service Research Paper* No.NE192, Northeast Forest Experimental Station, Upper Darby, Pa.

Bear, J., 1972, *Dynamics of fluids in porous media*, (Elsevier, New York).

Beven, K. and Germann, P., 1981, Water flow in soil macropores II. A combined flow model, *Journal of Soil Science*, 32, 15-29.

Bouma, J., Jongerius, A., Boersma, O., Jager, A. and Schoonderbeek, D., 1977, The function of different types of macropores during saturated flow through four swelling soil horizons, *Journal of the Soil Science Society of America*, 41, 945-950.

Bouma, J., Jongerius, A. and Schoonderbeek, D., 1979, Calculation of saturated hydraulic conductivity of some pedal clay soils using micromorphometric data, *Journal of the Soil Science Society of America*, 43, 261-264.

Bouma, J. and Wosten, J.H.M., 1979, Flow patterns during extended saturated flow in two undisturbed swelling clay soils with different macrostructures, *Journal of the Soil Science Society of America*, 43, 16-22.

Bullock, P. and Thomasson, A.J., 1979, Rothamsted studies of soil structure. II Measurement and characteristics of macroporosity by image analysis and comparison of data from water retention measurement, *Journal of Soil Science*, 30, 391-413.

Burt, T.P., 1978a, Three simple and low cost instruments for the measurement of soil moisture properties, Huddersfield Polytechnic Department of Geography, Occasional Paper No.6.

Burt, T.P., 1978b, An automatic fluid-scanning switch tensiometer system, British Geomorphological Research Group Technical Bulletin No.21.

Busenberg, E. and Plummer, L.N., 1982, Kinetics of dissolution of dolomite in CO_2 - H_2O systems at 1.5 to 65°C and 0 to 1 Atm. pCO_2, *American Journal of Science*, 282, 45-78.

Cairney, T., 1972, Hydrogeological investigation of the Magnesian Limestone of South-East Durham, England, *Journal of Hydrology*, 16, 323-340.

Cairney, T. and Hamill, L., 1972, Interconnection of surface and underground water resources in south-east Durham, *Journal of Hydrology*, 33, 73-86.

Carson, M.A. and Kirkby, M.J., 1972, *Hillslope form and process*, (Cambridge University Press).

Collis-George, N., 1980, A pragmatic method to determine the parameters that characterise ponded infiltration, *Australian Journal of Soil Research*, 18(1), 111-117.

Crabtree, R.W., 1981, Hillslope solute sources and solutional denudation on Magnesian Limestone, Unpublished PhD thesis, University of Sheffield.

Crabtree, R.W. and Trudgill, S.T., 1981, The use of ion-exchange resin in monitoring the calcium, magnesium, sodium and potassium content of rainwater, *Journal of Hydrology*, 53, 361-365.

Curtis, L.F., Courtney, F.M. and Trudgill, S.T., 1976, *Soils of the British Isles*, (Longman, London).

Delfs, J., 1967, Interception and stemflow in stands of Norway Spruce and Beech in West Germany, in: *Forest Hydrology*, ed. Sopper, W.E. and Lull, H.W., (Pergamon, Oxford).

Freeze, R.A. and Cherry, J.A., 1979, *Groundwater*, (Prentice-Hall, New York).

Germann, P. and Beven, K., 1981, Water flow in macropores. I. An experimental approach, *Journal of Soil Science*, 32, 1-13.

Hall, D., Reeve, M.J., Thomasson, A.J. and Wright, V.F., 1977, Water retention, porosity and density of field soils, *Soil Survey Technical Monograph* No.6.

Jongerius, A., Schoonderbeek, O. and Jager, A., 1972a, The application of the Quantimet 720 in soil micro-morphometry, *The Microscope*, 20, 243-254.

Jongerius, A., Schoonderbeek, O., Jager, A. and Kowalinski, S., 1972b, Electro-optical soil porosity investigation by means of Quantimet B equipment, *Geoderma*, 7, 177-198.

Mayer, R. and Ulrich, B., 1980, Input to soil, especially the influence of vegetation in intercepting and modifying inputs - a review, in: *Effects of acid precipitation on terrestrial forest ecosystems*, ed. Hutchinson, T.C. and Havas, M. (Plenum, New York).

Murphy, C.P., Bullock, P. and Turner, R.H., 1977a, The measurement and characterisation of voids in thin soil sections by image analysis. Part I. Principles and techniques, *Journal of Soil Science*, 28, 498-508.

Murphy, C.P., Bullock, P. and Biswell, J., 1977b, The measurement and characterisation of voids in soil thin sections by image analysis. Part II. Applications, *Journal of Soil Science*, 28, 509-518.

Newson, M.D., 1970, Studies in chemical and mechanical erosion by streams in limestone terrains, Unpublished PhD thesis, University of Bristol.

Nihlgard, B., 1970, Precipitation, its chemical composition and effect on soil water in a beech and spruce forest in South Sweden, *Oikos*, 21, 208-217.

Philip, J.R., 1957, Theory of infiltration: 4, Sorptivity and algebraic infiltration equations, *Soil Science*, 83, 345-357.

Picknett, R.G., 1964, A study of calcite solutions at 10°C, *Transactions of the Cave Research Group of Great Britain*, 7(1), 39-62.

Plummer, L.N. and Mackenzie, F.T., 1974, Predicting mineral solubility from rate data: application to dissolution of Magnesian calcites, *American Journal of Science*, 274, 61-83.

Plummer, L.N., Wigley, T.M.L. and Parkhurst, A., 1978, Kinetics of calcite dissolution in CO_2 water systems, *American Journal of Science*, 278, 179.

Rauch, H.W. and White, W.B., 1977, Dissolution kinetics of carbonate rocks. 1. Effects of lithology on dissolution rate, *Water Resources Research*, 13(2), 382-394.

Reeve, M.J., 1976, Soils in Nottinghamshire III: Sheet SK57 (Worksop), *Soil Survey Record*, No.33.

Reeve, M.J., Smith, P.O. and Thomasson, A.J., 1973, The effect of density on water retention properties of field soils, *Journal of Soil Science*, 24, 355-367.

Smart, P.L. and Laidlaw, I.M.S., 1977, An evaluation of some fluorescent dyes for water tracing, *Water Resources Research*, 13. 15-33.

Smettem, K.R.J. and Trudgill, S.T., 1983, An evaluation of some fluorescent and non-fluorescent dyes for use in the identification of water transmission routes in soils, *Journal of Soil Science*, 34, 45-56.

Spears, D.A., 1976, Information on groundwater composition obtained from a laboratory study of sediment-water interaction, *Quarterly Journal of Engineering Geology*, 9, 25-26.

Trudgill, S.T., 1975, Measurement of erosional weight loss of rock tablets, *British Geomorphological Research Group Technical Bulletin*, 17, 13-19.

Trudgill, S.T., Laidlaw, I.M.S. and Smart, P.L., 1980, Soil water residence times and solute uptake on a dolomite bedrock - preliminary results, *Earth Surface Processes*, 5, 91-100.

Trudgill, S.T., Pickles, A.M., Burt, T.P. and Crabtree, R.W., 1981, Nitrate losses in soil drainage waters in relation to water flow rate on a deciduous woodland site, *Journal of Soil Science*, 32(3), 433-441.

Trudgill, S.T., Pickles, A.M., Smettem, K.R.J. and Crabtree, R.W., 1983, Soil water residence time and solute uptake. 1. Dye tracing and rainfall events, *Journal of Hydrology*, 60, 257-279.

Warrick, A.W., Mullen, G.J. and Neilsen, D.R., 1977, Scaling field-measured soil hydraulic properties using a similar media concept, *Water Resources Research*, 13, 355-

Weyl, P.K., 1958, Solution kinetics of calcite, *Journal of Geology*, 66, 163-176.

Wigley, T.M.L., 1973, Incongruent solution of dolomite, *Geochimica et Cosmochimica Acta* 37, 1397.

Wigley, T.M.L., 1977, WATSPEC: A computer program for determining the equilibrium speciation of aqueous solutions, *British Geomorphological Research Group Technical Bulletin*, No.20.

215

PART III

SEDIMENT AND SOLUTE YIELDS

12 Surface and subsurface sources of suspended solids in forested drainage basins in the Keuper region of Luxembourg

A.C. Imeson, M. Vis and J.J.H.M. Duysings

INTRODUCTION

Most forested drainage basins are characterised by low
sediment yields and low rates of erosion. Sediment sources
are concentrated along river channels and where the natural
forest ecosystem has been disturbed by man (eg. Dissmeyer
and Foster, 1981). These generalisations, whilst apparently
valid for most of Luxembourg, are less applicable to the
extensive mixed oak-beech forests found on the Keuper marls
and claystones in the central and southern parts of the
country. In these forests, surface and subsurface sources
of sediment, widely distributed throughout first and second
order drainage basins, give rise to relatively high rates
of soil loss and sediment transport.

Several explanations can be suggested for the unusually
high rates of erosion in the Keuper forests of Luxembourg.
Most forests in temperate regions are found on land marginal
for agriculture, frequently where the soil is coarse
textured, very well drained and inherently infertile. The
Keuper forests occur on land made marginal for agriculture
by the difficult drainage conditions of the heavy-textured
soil. The soils are rich in nutrients and have a high
degree of biological activity. Horton and saturation overland flow are of frequent occurrence. The hydraulic and
chemical characteristics of the Keuper soils result in the
nature and distribution of sediment sources being different
from those normally associated with forests in temperate
regions.

Rates of erosion under forest are usually considered as
indicating a generalised benchmark of erosion (Fowler and
Heady, 1981), or a geologic norm, which can be used as a
standard for assessing the impact of various land use
practices. The relatively high rates of erosion in the
Keuper forests draw attention to the fact that low rates of
erosion reported for forests on marginal coarse-textured
soils may underestimate the benchmark rate of erosion if
extrapolated to formerly forested non-marginal cultivated
land.

In this paper some of the results of a number of field experiments are described which had as their objective locating and evaluating non-channel sediment sources. Laboratory investigations complementary to the field experiments will be described in detail elsewhere. From the laboratory and field measurements, it appeared possible to distinguish between surface and subsurface sources of sediment. This distinction is unusual inasmuch as subsurface sources of particulate inorganic sediment are of relatively infrequent occurrence, requiring specific chemical conditions of the soil.

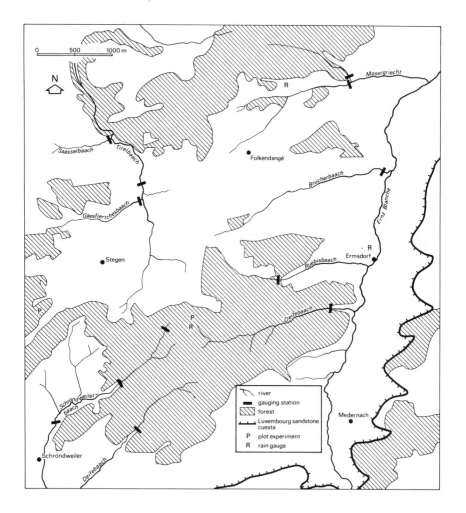

Figure 12.1. The location of gauging stations in the fieldwork area between Larochette and Ettelbrück in central Luxembourg.

THE STUDY AREA

A fieldwork area north of Larochette (figure 12.1) was
chosen where the Keuper outcrops as a clearly defined unit
beneath the cuesta formed by the Luxembourg Sandstone. It
is bounded by the rivers Alzette, Sûre, Ernze Blanche and
Schrondweilerbaach and drained radially by a number of
first and second order streams, most of which were instrumented with weirs and water-level recorders between 1975
and 1980.

The lithology of the fieldwork area is dominated by the
almost horizontally-lying weak shales and marls of the
Steinmergelkeuper which weather into heavy clay soils and
form a gently rolling landscape with a relief of about 40 m.
Along the lower stream courses, the underlying, more sandy
outcrop of Schilfsandstein is an important source of sediment, but it occupies a relatively small area. The lowest
Keuper formation, the Pseudomorphosenkeuper is characterised
by clay soils very similar to those on the Steinmergelkeuper and is difficult to distinguish in the field. The
geology of the Schrondweilerbaach drainage basin, where
most of the measurements described in this paper were made,
is shown in figure 12.2. In this drainage basin, the
forests typically consist of mixed mature stands of oak
(*Quercus robur L*), beech (*Fagus sylvatica L*) and hornbeam
(*Carpinus betulas L*). The gently to moderately sloping
slopes have a microtopography of indistinct shallow depressions and ridges, the former being usually no more than
about 10-15 m wide. During wet weather runoff is generated
from the depressions which become connected by poorly
defined channels. Many of these drain into a network of
drainage ditches superimposed upon the original drainage

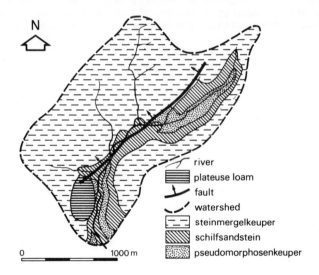

Figure 12.2. The geology of the Schrondweilerbaach catchment in the southeast of the fieldwork area.

221

network. The ditches, dug in the depression of the 1920's, resulted in an increased rate of runoff and in a phase of channel entrenchment. Particularly in their lower reaches, many of the original ditches have developed into gullies (Imeson and Jungerius, 1977).

In general the soil profiles show a similar textural layering throughout most of the region. In almost all cases a silty surface horizon, usually less than 40 cm thick, overlies a heavier textured B horizon which passes gradually into a C horizon usually within 80 cm of the surface. Although the B and C horizons have very low laboratory values of hydraulic conductivity, the transition between the A and B horizons is marked by a coarse strongly developed subangular blocky structure. The individual peds are partially surrounded and separated by interconnected channels and voids which transmit water when the soil becomes saturated. Percolation through the C horizon and deep drainage towards the drainage channels and streams in the Steinmergelkeuper is extremely localised. Even in winter when a perched water table develops above the B horizon over much of the catchment, the headwater streams in the Steinmergelkeuper become dry two or three days after rainfall.

FIELD MEASUREMENTS

Gauging stations, consisting of weirs and water level re-corders, were established at the sites shown in figure 12.1. Along the Schrondweilerbaach, selected for a more intensive study, the discharge from one of the shallow topographic depressions was measured with a tipping bucket gauge and recorder. Discharge from another depression was measured by a specially designed weir in the Laanger Haedboesch. Electrical conductivity measurements were made concurrently with the discharge in water draining the topographic de-pression in the Schrondweilerbaach catchment and recorded on the same chart as the discharge. Downstream, at the gauging station at the forest boundary (figure 12.1), the water level, measured by a pressure sensitive transducer, was additionally recorded together with the turbidity, on a two-channel Rustrak recorder. Turbidity was measured with a Partech suspended solids monitor using a 0-1000 or a 0-100 sensor, calibrated in Jackson turbidity units. It proved necessary to regularly clear and calibrate the turbidity meter and impossible to obtain a reliable turbidity trace unless the instrument was visited about every ten days. Water samples were collected during a number of flood events and also regularly at monthly intervals. Large water samples were taken and the organic material content and clay mineral composition determined in freeze-dried samples. All of the water samples were analysed for major cations and anions.

RESULTS AND DISCUSSION - FIELD OBSERVATIONS

Subsurface sources of sediment

Before this investigation was begun, it was frequently
observed that the water draining the Steinmergelkeuper had
a milky turbid appearance. At the lowest gauging stations,
this was most pronounced a few days after rainfall. The
concentration of suspended solids in the turbid water was
highest in the headwater areas (80-120 mg l^{-1}) and lowest
at the downstream gauging stations (40-60 mg l^{-1}). Water
draining the topographic depressions was also turbid. In
the Schrondweilerbaach, it was also observed that some time
during the passage of the flood peak, the water would rather
suddenly change colour from reddish to greyish brown. The
reddish brown water was assumed to contain sediment from
the river channel banks in the Schilfsandstein, and the
greyish brown water, sediment from the upper soil horizons
of the Steinmergelkeuper.
   During prolonged rainfall visual evidence and observations in piezometers placed in transects across the micro-
topographic depressions and ridges indicated the development of perched water tables above the B or C horizons.
This water could be observed moving downslope through
channels and planes in the upper B horizon even when the
peds themselves remained internally dry. Water also moves
rapidly through large biopores, occasionally emerging under
pressure. The interflow draining the pores and planes and
leaving the topographic depressions is invariably turbid.
The dispersed clay that it contains to some extent enables
it to be traced downstream and a distinction to be made
between:
1) overland flow containing sediment derived from splash
   action on the ground surface,
2) throughflow and return overland flow containing dispersed
   clay,
3) delayed flow free of sediment derived from springs and
   seepages in the Schilfsandstein.
The measurements and samples of throughflow indicate a
widespread occurrence of subsurface dispersion in soils
developed on both the Steinmergel- and Pseudomorphosen-
Keuper. If the relatively coarse surface soil horizon is
a residual accumulation from which dispersed clay has been
removed, it would be logical to assume that subsurface
sediment is supplied from nearly all of the forested parts
of the drainage basins. Nevertheless the intensity of the
dispersion process is likely to vary with soil and drainage
conditions.

Surface sources of sediment

Potential sources of suspended solids occur on the ground
surface where mineral soil is exposed to the erosive action
of raindrops. An earlier investigation (Hazelhoff et al.,
1981) had shown that this area varied from almost zero
between November and February to about 25% in August and
September. Most of the exposed soil results from the
removal of leaves from the surface by the worm *Lumbricus*

terrestris. This removal is most effective in the micro-
topographical depressions which are usually more or less
free of litter between May and October. The generation of
runoff in these depressions provides a waterfilm which enhances the erosive effect of rainfall and provides a
transporting mechanism for mechanically disaggregated and
finely aggregated soil particles.

During rainfall, suspended solids concentrations vary
considerably in the runoff from the depressions but are
usually between 350 and 700 mg 1^{-1}. When rainfall ceases
concentrations drop to between 100 and 120 mg 1^{-1} a short
time later. The variation in suspended solids concentrations seems to correspond to variations in rainfall
intensity. These are sometimes also produced by wind
suddenly reducing the interception storage capacity of the
canopy. The continually varying depth of the water film
produced by rainfall intensity variations has, as might be
expected (Mutchler and Young, 1975), a complex effect on
soil detachment. In the laboratory a rapid increase in
soil detachment accompanying the initiation of a thin water
film was reduced by a factor of between 10 and 100 when the
water layer increased in thickness from 3 to 10 mm. It is
thought that variations in the depth of runoff are also
important in the field.

The microtopographic depressions contribute runoff to
the drainage channels containing sediment entrained as a
result of splash detachment in the depression itself. It
would seem that during runoff as much or more sediment is
splashed in droplets from the flowing water layer to the
surroundingunponded soil, as is gained from splash action
adjacent to the flowing water. Splashed soil particles are
of course concentrated in the depressions during rainfall
events which fail to produce runoff. The sediment entrained
during runoff events is likely to a large extent to consist
of sediment supplied to the depressions during the preceding
weeks by splash transport during non-runoff producing rainfall events, by zoögenic processes and by litter transport
(Imeson and van Zon, 1979; van Zon, 1980). The topographic
depressions, which in effect form partial areas, may be
thought of as containing a store of colluvial material
supplied from all of the areas of bare soil at topographically higher positions. Although the depressions form the
immediate source of sediment during a runoff event, the
sediment is in effect in transit being derived like the
subsurface sediment from sources distributed throughout the
drainage basin.

Sources of suspended solids during flood events

Contributions of sediment from the surface and subsurface
sources mentioned above can sometimes be recognised in the
turbidity and discharge records obtained at the gauging
stations. These records consist of registrations of
turbidity and discharge made at 1 min intervals on the
same chart speed of 1 inch hr^{-1}. Examples of these records
are shown in figures 12.3 and 12.4. Changes in water
colour and sediment size during flood events mean that only
a general relationship between turbidity and suspended

solids concentrations exist. In addition to the two sources
of suspended solids described above, a third general source
formed by the river banks and channel, contributes to the
turbidity. Particularly during large flood events, or the
case of complex hydrographs, all three sources might be
expected to contribute material simultaneously. On certain
occasions hydrological conditions are such that the degree
of overlap in sediment supply from these three different
sources is less and water containing sediment supplied from
the soil profile can be distinguished.

Such an occasion occurred at the forest boundary station
on the Schrondweilerbaach on August 30th 1980, when 11 mm
of rain falling in 15 min produced a clearly defined hydrograph. During the two preceding days about 12 mm of rainfall had been recorded but there had been no runoff and the
river was dry. The effect of the rainfall was to generate
a simple hydrograph with a rather irregular peak and a
double peak in the turbidity trace (figure 12.3). The
first turbidity peak relates to water having a reddish
brown colour and a maximum suspended solids concentration
of about 450 mg l^{-1}. The turbidity began to decrease about
30 min after rainfall ceased. It is assumed that splash
action on the shallow flowing water in the channel and on
the river banks was responsible for the supply of suspended
solids, since the turbid water is rapidly replaced in the
river. The recession constant of the first turbidity peak
(table 12.1) is very similar to that for the quickflow when
separated from the delayed flow. The turbid water is removed so quickly from the catchment that it cannot have
been released from storage outside of the river channel.

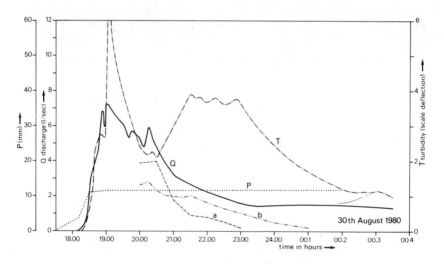

Figure 12.3. The turbidity (T) and discharge (Q) of the
Schrondweilerbaach resulting from the rainfall on the
29th August 1980. (P is the precipitation, a the
quickflow hydrograph and b the discharge of turbid
water from the topographic depressions)

225

Table 12.1. Recession characteristics of turbidity and discharge hydrographs for the Schrondweilerbaach on 30th August 1980 and 7th March 1981.

	α	τ(minutes)	$e^{-\alpha}$
30th August 1980			
1st turbidity peak	-0.0174	57.5	0.983
2nd turbidity peak	-0.0063	158.7	0.994
2nd turbidity peak - background turbidity	-0.0116	86.2	0.988
discharge	-0.0058	172.4	0.994
discharge - base flow	-0.0182	54.9	0.982
7th March 1981			
suspended solids concentration peak first ten minutes	-0.0981	10.195	0.907
suspended solids concentration after first ten minutes	-0.0058	171	0.994
suspended solids concentration (peak - background)	-0.0143	69.8	0.985
discharge	-0.0017	589	0.998
discharge - base flow	-0.003	328	0.997

The second turbidity peak began about 2 hr after the cessation of rainfall and is produced by the arrival in the stream of water having a greyish brown colour caused by the presence of finely dispersed clay. The recession constant of this turbidity peak resembles that of the river hydrograph occurring 5 hr earlier. If it is assumed that the clay producing the second turbidity peak is derived from the subsurface source described above and that this water, when leaving the soil has a clay concentration of between 100 and 120 mg l^{-1}, it is possible to reconstruct the hydrograph of the turbid discharge using a simple mixing model analogous to that used for hydrograph analyses based on river water chemistry (see for example Hall, 1970 and Gregory and Walling, 1973). This is demonstrated in figure 12.3 for the 30th August 1980, using the relationship

$$Q_T = \frac{Q_a\, C_a - Q_b\, C_b}{C_T}$$

where Q_T is the discharge of turbid water, having a turbidity C_T, and C_a and C_b are the turbidity values of respectively the total discharge Q_a and the base flow Q_b. Also shown in figure 12.3 is the quickflow recession curve obtained by the hydrograph separation technique of Singh and Stall (1971). It is apparent that the recession in the second turbidity peak shown in figure 12.3 begins when the quick-

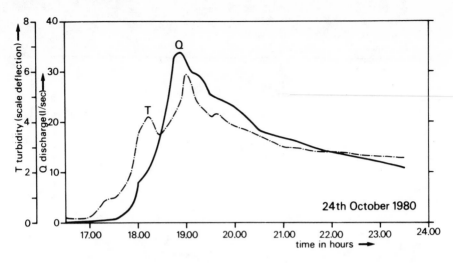

Figure 12.4.   Turbidity and discharge in the Schrondweiler-
      baach on 24th October 1980.

flow contribution ends. However, 3 hr are required for
this turbid water to be replaced by clear water in the
river compared to about 90 min in the case of the first
peak. This difference can probably be attributed on the
one hand to the lower discharge and velocity of the water
in the channel, which means that more time is needed to
replace the turbid water stored in irregular channel pools,
and on the other hand to the location and storage characteristics of the interpedal channels supplying the turbid
water. This interpretation implies that the quickflow runoff component represented in the single runoff peak consists of a channel precipitation component, carrying sediment producing the first turbidity peak, and an interflow
component transporting the dispersed clay. No overland
flow component can be recognised on this occasion, presumably since there was insufficient rainfall.
     At higher discharge levels sediment supplied from the
three general sources mentioned above tends to merge but
frequently a double peak in the turbidity trace can still
be recognised (figure 12.4). In general, the higher the
discharge, the shorter the time separating the two peaks.
At higher discharge levels the double peaks are related to
the sediment source areas in a different way. This is well
illustrated by the runoff and suspended solids concentrations resulting from the short but intense storm on
March 7th 1981 (figure 12.5) when 20 mm of rainfall were
recorded between 16.00 and 17.00 hrs.
     The first peak in the suspended solids concentration
occurred about an hour after rainfall had begun and like
the peak described for August began to decline 30 min after
rainfall had ended. The sediment had a reddish colour
characteristic of the Schilfsandstein and it was observed
that much of the sediment was the result of small scale
slumping and mudflows. Suddenly at 20.05 hrs, the river

227

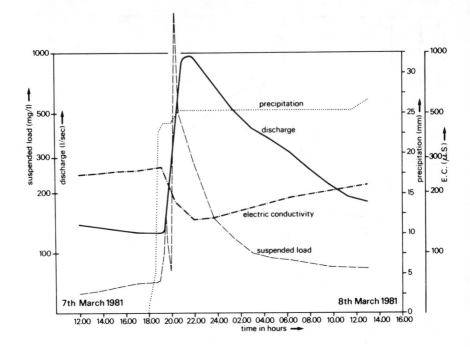

Figure 12.5.   The discharge, suspended solids concentration
and electrical conductivity of the Schrondweilerbaach
following the rainfall on 7th March 1981.

water became browner and the suspended solids concentrations
began to increase for a second time.   The maximum concentration was reached within 30 min and the concentrations
declined to about 20% of this maximum value when the peak
discharge was attained about 80 min later.   The dark brown
sediment is thought to be derived mainly from the river and
drainage channels in the Steinmergelkeuper region of the
drainage basins.   Most of the runoff contributing to the
peak is supplied from the topographic depressions in the
Steinmergelkeuper area.   From the suspended solids concentrations and colour of the sediment water suspension it is
difficult to determine exactly when sediment supplied from
subsurface sources begins to reach the channel due to contributions from other sources.   By 03.00 hrs on the 8th
March however, the river has a grey-brown appearance caused
by the presence of dispersed clay and it is clear that the
sediment load is dominated by the subsurface source.

Forested and cultivated drainage basins

The subsurface sources of suspended solids are especially
important in forested drainage basins.   Partly to illustrate
this, but also to place the measurements from the forested
Schrondweilerbaach in perspective the hydrographs and
turbidity traces recorded on the Keiwelsbaach and

228

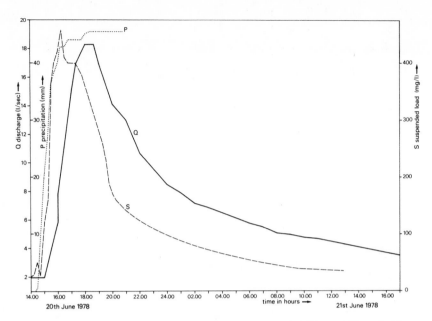

Figure 12.6. The discharge (Q) and turbidity (S) of the
Keiwelsbaach on 20th and 21st June 1978.

Mosergriecht in the north of the fieldwork area are shown
for the 20th June 1978 in figures 12.6 and 12.7.   The
Keiwelsbaach is the northern tributary of the Mosergriecht
(figure 12.1) and drains a completely forested catchment
93 ha in area.   The upper Mosergriecht catchment in contrast is mainly agricultural (273 ha).   The yearly specific
sediment discharge of the Mosergriecht, measured between
1978 and 1980, was usually about four times that of the
Keiwelsbaach.   The yearly runoff coefficients of both rivers
vary between 30 and 45%.
     On the 20th June 1978, the turbidity of both rivers was
being recorded when 46 mm of rain fell in less than 2 hr
during a storm with a recurrence interval of about 1 year.
The turbidity traces shown in figures 12.6 and 12.7 were
calibrated in concentrations of suspended solids by means
of samples collected at regular intervals during the flood
events.   In spite of the large amount of rainfall, the runoff coefficients for both drainage basins were very low
(table 12.2).   About fifty times as much suspended solids
were discharged from the Mosergriecht as from the forested
Keiwelsbaach.
     On both the Mosergriecht and Keiwelsbaach, the dis-
charge began to increase at the same time but the peak
discharge of the forested Keiwelsbaach was recorded almost
an hour later than on the Mosergriecht.   This and the low
maximum discharge of the Keiwelsbaach (table 12.2) demonstrates the contrasting soil conditions in these two catchments.   The turbidity of the Keiwelsbaach began to rise
20 min earlier in the Keiwelsbaach than in the Mosergriecht

229

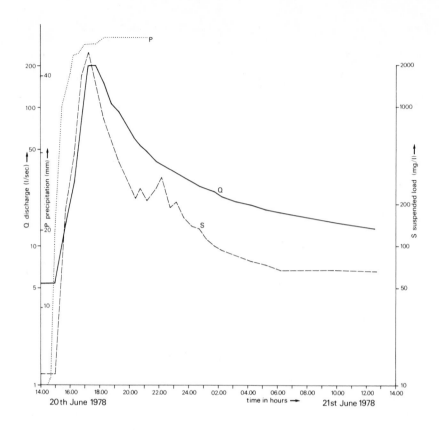

Figure 12.7　The discharge and turbidity of the Mosergreicht
on 20th and 21st June 1978.

and reached a peak value more than 2 hr ahead of the runoff
peak and an hour earlier than turbidity in the Mosergriecht.
On the Keiwelsbaach, the concentration of suspended solids
had declined to almost one half of its maximum value when
the discharge began to decline.

The fact that on the Keiwelsbaach high suspended solids
concentrations decline almost directly when rainfall
ceases, and reach maximum values while the discharge is
still relatively low points to the importance of splash
action, in the river channel and on the channel banks, in
supplying suspended solids to the river.　At low discharge
levels on the rising stage large areas exist in the relatively wide channel bed where the water depth is very
shallow and splash action during high intensity rainfall
is effective in entraining sediment.　As the river rises
and occupies the entire channel bed, splash action in the
channel itself will become much less effective.　Another
factor promoting high suspended solids concentrations during
the rising hydrograph stage is the large amount of aggregated soil material supplied by the process of soil fall
to the channel bed during dry periods (Imeson and Jungerius,
1977).

230

Table 12.2. Runoff and sediment transport during the floods of 20th June 1978 on the Keiwelsbaach (forested, 93 ha) and Mosergriecht (cultivated, 237 ha)

	Keiwelsbaach		Mosergriecht	
	discharge	sediment	discharge	sediment
time of commencement of hydrograph or turbidigraph rise	15.10	14.40	15.10	15.00
peak discharge ($1 \ s^{-1}$) or turbidity ($mg \ l^{-1}$)	18.5	440	210	2500
time of peak discharge or concentration	18.00	16.15	17.25	17.20
volume of water (m^3) and sediment discharge (kg)	677	42	3150	2300
runoff coefficient (%) and proportion of sediment transported as dispersed clay after 20.00 hrs (%)	1.5	36	3	8

Most of the runoff of the Keiwelsbaach during the hydrograph rise can be explained by channel precipitation, but this is not the case for the discharge during the peak and falling stages. Whilst most of the drainage basin can clearly retain the 46 mm of precipitation and overland does not occur, the relatively late arrival of the runoff peak compared to the Mosergriecht and the dispersed clay it contains suggest that much of the runoff has moved as interflow through the soil before entering the topographic depressions or drainage channels on its way to the main river. From 20.00 hrs onwards it is thought that virtually all of the suspended solids being transported by the river are derived from subsurface sources. At about this time an inflexion is also observed in the turbidity trace. The suspended solids transported in dispersion after 20.00 hrs amount to 36% of the total amount of sediment transported by the flood.

The Mosergriecht produces more runoff than the Keiwelsbaach (table 12.2) probably because of differences in interception and because the relatively permeable Ap soil horizons have poorer structural characteristics and are much thinner than the equivalent A horizons or colluvium in the forests. Perched water tables will develop more rapidly at topographically low lying positions than under forest. The relatively poor structure of the soil is thought to be a major factor in explaining the much higher suspended solids concentrations in the runoff. From a great number of aggregate stability determinations it was found that on average 74% of the aggregates 2-4 mm in size were water stable under forest as compared to 89% under pasture and only 13% for cultivated farmland. Soil particles are much more easily detached and transported

231

from aggregates under farmland than under forest. The amount of sediment transported as dispersed clay after 20.00 hrs forms only 8% of the total amount of suspended solids transported by the floodwave. The extent to which this material is derived from the cultivated soil or the forested parts of the Mosergriecht drainage basin is not known.

CONCLUSION

In the forested drainage basins in the Keuper region of Luxembourg, subsurface sources of suspended solids contribute material to the rivers during periods of hydrograph recession. This transport is important but its exact quantitative significance still has to be determined. This is difficult in the main stream channels where the sub-surface sediments can be recognised only after the passage of the main flood peak. If the dispersed clay in the rivers is derived from the soil, other particle-size fractions might be expected to be residually enriched. Silt-rich surficial horizons occur throughout the forest areas and it is tempting to relate these to the loss of the clay. However, this cannot be done until the removal of dispersed clay is quantified and until other hypotheses relating to the genesis of the silty surficial horizons have been tested. The pedological significance of clay transport is such that future research is being concentrated on quantifying the output of surface and subsurface sediment from the topographic depressions where the transport of suspended solids is not influenced by channel processes.
    The results described above also indicate the difficulty of establishing rating curves for very small drainage basins where suspended solids are supplied to the river by a variety of processes which are out of phase with the runoff (Imeson, 1977). Walling (1977) points out that the most important cause of scatter in the relationship between sediment concentration and water discharge are factors related to the dynamics of erosion and sediment discharge. In the case of the forested catchments in the Keuper region of Luxembourg, sediment supply is a function of dispersion as well as of mechanical erosion. Particularly during the periods of hydrograph recession the dynamics of the dispersion process (Imeson and Verstraten, 1981) is a factor which should be considered in investigating water discharge and suspended solids concentration relationships.

ACKNOWLEDGEMENTS

We wish to thank P.D. Jungerius, J.M. Verstraten, F.J.P.M. Kwaad and M. Hendriks for their useful discussion of this work. Mrs. M.C.G. Keijzer-v.d.Lubbe is thanked for preparing the manuscript and Mr. A.J. van Geel for preparing the illustrations.
    This work was supported by a University of Amsterdam research grant.

REFERENCES

Dissmeyer, G.E. and Foster, G.R., 1981, Estimating the cover-management factor (C) in the Universal Soil Loss Equation for forest conditions, *Journal of Soil and Water Conservation*, 36, 235-240.

Fowler, J.M. and Heady, E.O., 1981, Suspended sediment production on undisturbed forest land, *Journal of Soil and Water Conservation*, 36, 47-53.

Gregory, K.J. and Walling, D.E., 1977, *Drainage basin form and process*, (Arnold, London).

Hall, F.P., 1970, Dissolved solids - discharge relationships. 1. Mixing models, *Water Resources Research*, 6, 845-850.

Hazelhoff, L., van Hooff, P., Imeson, A.C. and Kwaad, F.J.P.M., 1982, The exposure of forest soil to erosion by earthworms, *Earth Surface Processes and Landforms*, 6, 235-250.

Imeson, A.C., 1977, Splash erosion, animal activity and sediment supply in a small forested Luxembourg catchment, *Earth Surface Processes*, 2, 153-160.

Imeson, A.C. and Jungerius, P.D., 1977, The widening of valley incisions by soil fall in a forested Keuper area, Luxembourg, *Earth Surface Processes*, 2, 141-152.

Imeson, A.C. and Verstraten, J.M., 1981, Suspended solids concentrations and river water chemistry, *Earth Surface Processes and Landforms*, 6, 251-263.

Imeson, A.C. and van Zon, H.J.M., 1979, Erosion processes in small forested catchments in Luxembourg, in: *Geographical approaches to fluvial processes*, ed. Pitty, A.F., (GeoBooks Norwich), 93-107.

Mutchler, C.J. and Young, R.A., 1975, Soil detachment by raindrops, in: *Present and perspective techniques for predicting sediment yield and sources*, Agricultural Research Service, (USDA) Publication ARS-S-40, 113-117.

Singh, K.P. and Stall, J.B., 1971, Derivation of base flow recession curves and parameters, *Water Resources Research*, 7, 292-303.

Walling, D.E., 1977, Limitations of the rating curve technique for estimating suspended sediment loads, with particular reference to British rivers, *International Association of Hydrological Sciences Publication* no. 122, 34-48.

Zon, H.J.M. van, 1980, The transport of leaves and sediment over a forest floor: A case study in the Grand Duchy of Luxembourg, *Catena*, 7, 97-110.

13 Sources of variation of soil erodibility

in wooded drainage basins in Luxembourg

P.D. Jungerius and H.J.M. van Zon

INTRODUCTION

Studies of erosional processes in natural or semi-natural environments are relatively few, and have been concerned mainly with sediment yield to stream channels. Substantially more research has been done in agricultural environments, but here too it has most often been focused on sediment production and yield more than on variability between catchments and the causes of the observed differences. Variations in erosion within a catchment have received least attention. Consequently in some countries like the USA there is a substantial body of information about the amount of sediment reaching streams, though it is generally not known quantitatively where the sediment comes from (Glymph, 1975). This is not surprising in view of the physical impossibility of measuring erosion at every point in the drainage basin.

On the other hand much research has been devoted to the processes of erosion, such as gullying, bank erosion, splash erosion, etc. However, the results of these studies are not usually placed in the natural framework in which the processes operate, i.e. translated into the sediment movements within and sediment production of any catchment. This is also not surprising since sediment yield is a function of a number of (possibly interacting) processes, which in their turn are dependent on a wealth of different variables.

It is, however, clear that a catchment should be more to a geomorphologist than a sediment-delivering black box. To explain differences in sediment yield and benefit from the knowledge built up from studying the output of different catchments, it is necessary to consider erosional processes not as isolated occurrences; they must be placed in their natural context to understand their relative importance. This is only possible if the effects of erosional processes are integrated in such a way that their combined effect becomes clear. This type of integration is basic to ecological studies, and this ecological approach is of great importance in geomorphology.

The approach above described has been adopted for the studies of the forest ecosystems in Luxembourg carried out by the University of Amsterdam during the last decade. The long-term aim of these studies is to assess the factors important for sediment production and to ascertain the relations between these factors and explain variations in sediment production between and within catchments, in order to be able to predict sediment yield. In short-term experimental studies, parts of this general problem have been tackled in a sort of trial and error method. This paper forms an example of such an experimental study.

Earlier studies in the Grand-Duchy have already shown that in the woods of Luxembourg, which can be regarded to be in almost a natural state, flora and fauna are of prime importance to erosion. For example, the efficacy of splash erosion is mainly determined by the presence or absence of a protective litter cover, and furthermore, the fall height and drop-size distribution are greatly influenced by vegetational characteristics (Hazelhoff et al., 1981). For the Luxembourg example, it can be stated that the areas where the forest floor is exposed to the impact of water drops falling from the canopy are the most important sediment-producing areas. The occurrence of exposed soil is in its turn a function mainly of faunal activity. Most of the exposures of bare soil are caused by the consumption of leaves by earthworms (Lumbricus terrestris). Consequently the spatial distribution of bare patches depends on the ecological requirements of the earthworms. It appears, for example, that clay soils have up to 80 times more unprotected surface than sandy soils. If it is accepted that patches of bare soil form the sediment producing areas of wooded slopes, the next question to be answered is whether there are differences between the erodibility of these patches in different catchments and/or within catchments.

Because not every bare patch can be investigated, a sampling strategy must be developed which suits this specific purpose. In catchments with an irregular distribution of ecological conditions, it would seem appropriate to adopt a stratified sampling procedure. In the main, the choice made from the differentiating criteria results in a subdivision into landscape units. The criteria are then mostly based on parent rock, soils, relief and vegetation. On the basis of these discriminating variables, the catchments in Luxembourg could be subdivided into three clearly defined landscape units which are described below. In section 4 the extent to which these landscape units represent areas that are homogeneous in terms of soil erodibility will be investigated.

CHARACTERISTICS OF THE LANDSCAPE UNITS

The investigated catchments border the Lias escarpment in the southern part of Luxembourg. They have been subdivided into three units, each with a specific combination of ecological and geomorphological characteristics. They have been named after the underlying rocks.

Landscape units on Lias sandstone

This landscape unit occupies the highest parts of the
catchments: the plateaux and the upper parts of the escarpment slopes of the Lias cuesta. The underlying sandstone
decomposes to a sandy substratum. Forests here consist
mainly of tall specimens of beech (*Fagus sylvatica*) with
little undergrowth. The beech produces leaves which are
not very palatable to earthworms (Satchell, 1967). Perhaps
because of this, and because of other factors such as the
sandy texture of the soil and its low pH, the earthworm
population is low. Evidence of digging by larger soil
fauna is present in places. Faunal activities in combination with the influence of tree roots has produced a bioturbation layer of 50 to 70 cm depth. The homogenization
counteracts the differentiation in soil horizons, and soil
profile development is accordingly rather weak. Only the
thin Ah horizon at the surface is well developed. Nevertheless, the compact horizon below the bioturbation layer has
an increased clay percentage, presumably due to illuviation.

Landscape units on Lias marls

Occasional steps in the escarpment slope are underlain by
the Psilonoten or Lias marls which have been converted at
the surface to a clayey weathering produce. Apart from
beech, the forests in this area consist of hornbeam
(*Carpinus betulus*) and oak (*Quercus robur*). The soil profile is characterized by an upper zone of bioturbation
which is only 15 to 25 cm thick and is often bounded at its
base by a concentration of tree roots. The soil texture
in this zone is more silty and less clayey than in the
subjacent zone (the silt : clay ratio increases from .7 in
the C horizon to 2.1 in the Ah horizon). This is perhaps
partly due to the well known process of downward leaching,
but partly also to dispersion by splash of soil material
brought to the surface by moles, with subsequent removal of
the clay fraction by overland flow.
     In spite of homogenization, there is a clear concen-
tration of organic matter in the thin Ah horizon. This
accounts for the difference in organic matter content of
material brought from the less organic subsoil to the
surface by moles, and the material excreted at the surface
by worms of Allolobophora species, which live in the upper
few cm of the soil. Burrows abandoned by moles are filled
with humic material from the surface (krotovinas) resulting
in a very irregular and discontinuous distribution pattern
of organic material in the surface soil.

Landscape units on Keuper marls

The area underlain by the Keuper marls occupies the lowest
part of the catchments. The ecological conditions resemble
those of the Lias marls in several respects but there are
important differences, presumably connected with the somewhat higher silt content of the weathered Keuper marls.
The composition of the forest is not very different, but
the soils seem to be more poorly drained than on the Lias

marls. Profile development with the characteristic bioturbation layer is comparable with that on the Lias marls
(silt : clay ratio increases from 1.2 in the C horizon to
2.1 in the Ah horizon), but there is a much more even
distribution of organic matter in the surface soil. This
is presumably due to the dense population of *Lumbricus
terrestris* in these soils. The eating habits of these
worms involve pulling large amounts of leaves from the
litter layer into the soil. The occurrence of *Lumbricus*
is highly irregular, which is as yet unexplained (Hazelhoff
et al., 1981).

The distribution pattern of soil texture and organic
matter content is further complicated by slow downward
creep of the bioturbation zone on slopes. Also permeability
of the soils varies from place to place due to cracking,
formation of biopores and differences in structure. Many
routine soil profile descriptions mention two types of
structure conjointly in the Ah horizon: moderate or even '
strong granular (or crumb) and weak subangular blocky.

Table 13.1a. Response to rainfall experiments (n=5)

		Lias Sandstone	Lias Marls	Keuper Marls	average
a	splash erosion $(mg\ dm^{-2}.min^{-1})$	55.9	62.5	58.2	58.6
b	time to ponding (min)	2.8	1.4	.9	1.7

Table 13.1b. Analysis of variance for response
to rainfall experiments

	source of variation	SS	% of total variation	DF	MS	F	sign. level
a splash erosion	landscape units	97.3	.7	2	48.7	.04	.96
	residual	13889.6	99.3	11	1262.7		
	total	13986.9	100.0				
b time to ponding	landscape units	9.9	30.0	2	4.9	2.36	.14
	residual	23.0	70.0	11	2.1		
	total	32.9	100.0				

SOURCES OF VARIATION IN SOIL ERODIBILITY

When experiments were carried out in the different landscape units to study the response of the bare areas to
splash, it appeared that variation in this response within
the landscape units, was not significantly different from
variations between units (table 13.1). So it must be
assumed that the average sensitivity to splash erosion is
independent of the subdivision into landscape units.
Response in respect to time to ponding was more consistent
with the character of the landscape units, sandy soils
being more permeable than soils derived from marls. Apart
from the relationship with overland flow, this feature
controls the occurrence of a water film, the depth of which
in turn determines the effect of splash (Mutchler and Young,
1975). However, the variation of time to ponding within
the landscape units was also found to be large (table 13.1a).
    Not surprisingly, there was no significant difference
between the landscape units with regard to the Universal
Soil Loss Equation's soil erodibility factor K (table 13.2).
Only about one third of the total variation found by
analysis of variance was explained by the landscape factor.
    It had to be concluded that the variables - parent rock,
soil, relief and vegetation - were not sufficient to differentiate areas with respect to soil erodibility. It was
therefore decided to pay more attention to variations of
soil erodibility within the bare areas. The surface of the
exposed soil is modified by two groups of animals: moles,
including other burrowing rodents, and worms (Imeson and
van Zon, 1979). Although the occurrence of these animals
is largely dependent on the ecological conditions of the
recognized landscape units, it was hypothesized that they

Table 13.2a. USLE soil erodibility factor K (n=5)

	Lias Sandstone	Lias Marls	Keuper Marls	average
factor K	.078	.113	.184	.125

Table 13.2b. Analysis of variance for factor K

source of variation	SS	% of total variation	DF	MS	F	level of significance
landscape units	.03	35.35	2	.015	3.28	.07
residual	.05	64.65	12	.004		
total	.08					

Table 13.3a.   Soil-texture properties related to
soil erodibility (n=5)

			Lias Sandstone	Lias Marls	Keuper Marls	row average
a	average clay content	exposed surface (cf. worm casts)	8.3	32.1	35.3	25.2
		surface horizon	7.8	28.5	33.7	23.3
		subsurface hor. (cf. mole hills)	4.9	29.2	39.1	24.4
		column average	7.0	29.9	36.0	24.3
b	average silt + very fine sand content (%)	exposed surface (cf. worm casts)	30.0	58.2	58.1	48.8
		surface horizon	14.7	54.3	58.3	42.4
		subsurface hor. (cf. mole hills)	13.4	55.3	53.6	40.8
		column average	19.4	55.9	56.7	44.0
c	average organic matter content (%)	exposed surface (cf. worm casts)	7.2	22.4	11.0	13.5
		surface horizon	9.1	14.9	9.5	11.2
		subsurf. horizon (cf. mole hills)	2.7	6.7	4.3	4.6
		column average	6.4	14.7	8.3	9.8

are able to modify soil erodibility to the extent that
differences between the landscape units are appreciably
reduced.

The investigations were carried out in two steps. The
first step implied examining in more detail the variation
in the main parameters of the USLE factor K: clay (<2 μm),
silt and very fine sand (2 to 100 μm), and organic matter
content in soil materials that could be related to the
activity of the soil fauna. The results are shown in
tables 13.3 and 13.4. These determinations require sample
bulking, and it is known (Imeson and Jungerius, 1976) that
the strength of the soil may vary from aggregate to
aggregate. In a second step, therefore, individual soil
aggregates collected from specific sites of faunal activity
were subjected to a dispersion test. To investigate the
influence of season upon erodibility, this test was carried
out twice, on aggregates sampled in springtime when the
bare areas are increasing rapidly, and at the end of the
summer when they reach their maximum size. The results are
shown in table 13.5.

Table 13.3b. Analysis of variance for
soil texture properties

	source of variation	SS	% of total variation	DF	MS	F	sign. level
a clay content	landscape units	7033.9	61.1	2	61.1	29.25	<.001
	soil(-fauna)	27.4	.2	2	13.7	.11	.90
	interaction	119.5	1.0	4	1.0	.25	.91
	residual	4328.2	37.6	36	37.6		
	total	11509.0	100.0				
b silt + fine sand content	landscape units	13622.7	70.4	2	6811.3	61.35	<.001
	soil(-fauna)	534.8	2.8	2	267.4	2.02	.15
	interaction	425.9	2.2	4	106.5	.80	.53
	residual	4774.9	24.7	36	132.6		
	total	19358.2	100.0				
c organic matter content	landscape units	567.1	29.0	2	283.5	18.83	<.001
	soil(-fauna)	646.3	33.1	2	323.2	21.50	<.001
	interaction	197.4	10.1	4	49.4	3.28	.022
	residual	542.2	27.8	36	15.1		
	total	1953.0	100.0				

Two-way analysis of variance was used to compare the
two sources of variation in soil erodibility (landscape
units and faunal activity). Table 13.3 shows the data of
the main parameters of the USLE factor K for three specific
domains of faunal activity. The upper row for each parameter represents material sampled where the forest floor
was transformed by earth worm activity and exposed to
splash erosion, the second row represents the Ah horizon,
and the third row the zone immediately below the Ah horizon,
from which moles derive the material for their mounds. In
table 13.5, with the results of the dispersion test, the
latter entry has been replaced by material from the actual
mole hills. The so-called first approximation of the
factor K (Wischmeier and Smith, 1978), calculated from
grain-size and organic matter content, is given in table
13.4 because the corrections for structure and permeability
are not possible for some of the materials.

Tables 13.3 and 13.4 show a clear and significant difference between the landscape units as far as soil properties related to erodibility are concerned. The clay and
silt content are higher on marls than on sandstone, whereas
the organic matter content is higher on Lias marls than in
soils of the other landscape units. Only organic matter is

241

Table 13.4a. USLE soil erodibility factor K
(first approximation, n=5)

Landscape units	Lias Sandstone	Lias Marls	Keuper Marls	average
faunal activity				
exposed surface (cf. worm casts)	.089	.036	.109	.073
surface horizon	.040	.052	.099	.090
subsurface horizon (cf. mole hills)	.089	.182	.210	.139
average	.078	.063	.160	

Table 13.4b. Analysis of variance for USLE factor K

source of variation	SS	% of total variation	DF	MS	F	sign. level
landscape units	.04	10.8	2	.02	3.27	.05
faunal activity	.08	22.2	2	.04	6.71	.003
interaction	.02	6.5	4	.01	.99	.43
residual	.22	60.5	36	.01		
total	.37	100.0				

Table 13.5a. Dispersion (n=5)

		Lias Sandstone	Lias Marls	Keuper Marls	average
a March 1980	worm casts	.0	1.4	1.8	1.1
	surface horizon	.1	1.6	1.9	1.2
	mole hills	.9	3.5	3.0	2.4
	average	.3	2.1	2.2	1.6
b Sept. 1980	worm casts	.3	1.2	.9	.8
	surface horizon	.0	.9	1.0	.6
	mole hills	1.0	2.6	1.8	1.8
	average	.4	1.6	1.2	1.1

Table 13.5b. Analysis of variance for dispersion

	source of variation	SS	% of total variation	DF	MS	F	sign. level
a March 1980	landscape units	33.97	37.5	2	16.98	16.53	<.001
	soil fauna	17.22	19.0	2	8.61	8.38	.001
	interaction	2.33	2.6	4	.58	.57	.69
	residual	37.00	40.9	36	1.03		
	total	90.52	100.0				
b Sept. 1980	landscape units	10.18	20.7	2	5.09	7.16	.002
	soil fauna	11.87	24.2	2	5.93	8.35	.001
	interaction	1.52	3.1	4	.38	.54	.71
	residual	25.58	52.1	36	.71		
	total	49.14	100.0				

significantly different in the three zones of faunal activity. Interaction between the two sources of variation is low (table 13.3b). The residual variation, i.e. variation observed within the samples and not accounted for by the two main factors, is not particularly high, but increases to more than 60% when the grainsize and organic matter parameters are multiplied for the computation of factor K (table 13.4b). In the latter case it is particularly interesting to note that differences are better explained by zones of faunal activity than by landscape units. This is because the stabilizing effect of the higher clay content of the marls is offset by the opposite effect of their equally higher silt content. It explains why the landscape units cannot be differentiated on the base of the factor K (table 13.2).

Dispersion is also significantly different in the three landscape units (table 13.4) but equally pronounced is the difference between the zones of faunal activity, especially at the end of the summer. Dispersion of sandy soils is higher than that of soils derived from marls, and dispersion of mole hills is higher than that of material reworked by worms. Interaction is again low, but the portion of unexplained variation is very high.

DISCUSSION

Sediment output is determined by the efficiency with which different erosion processes operate and influence each other. In our case if it is attempted to integrate the effects of the different erosion processes, it appears that

this should be done at different levels. Each level represents a source of variation in soil erodibility and in
bareness of the soil surface. The first level is defined
by the spatial differentiation in a catchment; there appear
to exist landscape units in which similar processes operate
with similar intensity. These landscape units are specified
on the basis of the variables parent material and soil,
vegetation and relief.

Although erodibility and consequently the amount of
erosion are certainly to a large extent determined by the
aforementioned variables, in our case it appeared that
within a landscape unit, erodibility is, at least for a
significant part, modified by faunal activity; this represents another level of integration.

It appears that the differences between the landscape
units are smoothed by the faunal activities: worms and
moles reduce by homogenization differences in erodibility
between landscape units. In this context it is interesting
to note that moles enhance erodibility by bringing material
to the surface which is poor in organic matter, while worms
reduce erodibility by improving the structure and enlarging
the organic matter content. So conflicting tendencies are
present; for example the Keuper soils, in which bioturbation is most pronounced, have the lowest range of textures
and therefore the smallest difference in the erodibility of
the various surface materials.

For reasons which are little understood there is
apparently much variability in ecological conditions, even
within the landscape units, and this results in a highly
irregular pattern of sediment producing areas (bare
patches), and of erodibility. Also there is a regular
seasonal fluctuation in the size of the exposed areas and
in the activity of the animals during the year.

In general, erodibility is lowest at the end of the
summer when the patches of bare soil reach their maximum
extent. The influence of faunal activity is at that time
so large that it overrules the differences which must exist
between the landscape units: the integration level of faunal
activity is then more important than that of the landscape
unit.

APPENDIX: METHODS

Samples were taken on randomly chosen upland sites in level
terrain. Soil profiles were described and sampled at 15
sites, five for each of the three landscape units.
Laboratory investigations included grain-size distribution,
organic matter content by dichromate oxidation (organic
carbon x 1.73), susceptibility to splash erosion and time
to ponding. Splash was generated by artificial rain falling
from 4 m height at a rate of 60 mm hr^{-1} for 5 min. This
has a one-year recurrence interval. Samples for testing
the dispersion in relation to the soil fauna were taken
twice, in March 1980 and September 1980. A modified
version of the dispersion test proposed by Jungerius
(1982) was used. The degree of dispersion is expressed
on a scale from 0 to 8. The soil erodibility factor K was
computed with the equation given by Wischmeier and Smith

(1978). The organic matter content of forest soils is often outside the range for which the equation was developed. In such cases a rather arbitrarily chosen limit of 11% was used to avoid negative or otherwise nonsensical values of K.

For the analysis of variance, use was made of the SPSS library (Nie *et al*., 1975). The method introduced by Shapiro and Wilk (1965) was applied for testing normality. Serious departure from normality (test statistic in the critical region when the level of significance is chosen as .01) was observed in only three out of the 65 samples.

Bartlett's chi-square statistic was used to test the homogeneity of variance (Snedecor and Cochran, 1967). In a number of tables the inequality of the variances would not permit the application of the analysis of variance. This is mainly because the standard deviation varies directly as the mean. In these cases logarithmic transformations were used to stabilize the variances. Analysis of variance on the log-transformed data produced slightly different F-values but the relative importance of the various sources of variation as apparent from the tables in this paper was not affected.

REFERENCES

Glymph, L.M., 1975, Evolving emphases in sediment yield predictions, in: *Present and prospective technology for predicting sediment yield and sources,* Agricultural Research Service, (USDA) Publication, ARS-S-40, 1-4.

Hazelhoff, L., Hooff, P. van, Imeson, A.C., and Kwaad, F.J.P.M., 1981, The exposure of forest soil to erosion by earthworms, *Earth Surface Processes and Landforms*, 6, 235-250.

Imeson, A.C. and Jungerius, P.D., 1976, Aggregate stability and colluviation in the Luxembourg Ardennes; an experimental and micromorphological study, *Earth Surface Processes*, 1, 259-271.

Imeson, A.C. and Zon, H. van, 1979, Erosion processes in small forested catchments in Luxembourg, in: *Geographical Approaches to Fluvial Processes*, ed. Pitty, A.F., (GeoBooks, Norwich).

Jungerius, P.D., 1982, A rapid method for the determination of soil dispersion, and its application to soil erosion problems in the Rif Mountains, Morocco, *Studia Geomorphologica Carpatho-Balcanica,* 15, 31-38.

Mutchler, C.K. and Young, R.A., 1975, Soil detachment by raindrops, in: *Present and prospective technology for predicting sediment yield and sources,* Agricultural Research Service, (USDA) Publication, ARS-S-40, 113-117.

Nie, N.H., Holl, C.H., Jenkins, J.G., Steinbrenner, K. and Bent, D.H., 1975, *Statistical Package for the Social Sciences*, (McGraw-Hill, New York), 398-405.

Satchell, J.E., 1967, Lumbricidae, in: *Soil Biology*, ed. Burger, A. and Raw, F., (Academic Press, London), 259-322.

Shapiro, S.S. and Wilk, M.B., 1965, An analysis of variance test for normality, *Biometrika*, 52, 591-611.

Snedecor, G.W. and Cochran, W.G., 1967, *Statistical Methods*, (Iowa State University Press), 6th ed.

Wischmeier, W.H. and Smith, D.D., 1978, Predicting rainfall erosion losses: A guide to conservation planning, *U.S. Department of Agriculture, Agricultural Handbook* no. 537.

14 Microerosion processes and sediment mobilization in a road-bank gully catchment in central Oklahoma

Martin J. Haigh

INTRODUCTION

Attempts to transform geomorphology into a precise science tend to be foiled by the fact that while landform morphology appears relatively simple, the landforming process is extremely complex. This situation is the bane of both the field and the laboratory geomorphologist. In nature, it is very unusual to discover a geomorphic system which is simple in both structure and history and whose evolution can be totally resolved in terms of the operation of an unequivocably measurable suite of present-day processes. In the laboratory, it is quickly discovered that there are few geomorphic systems which do not resist being scaled down to a size and timescale appropriate to analysis. As the laboratory worker retreats further from nature, so the uncertainty involved in accurately recreating simple forms from a complex of process interactions magnifies, and explanation becomes lost amidst a plethora of equifinal solutions.

This study represents part of a search for an intermediate scale of investigation. The author is dedicated to the study of landforms which are small, simple, and active enough to be treated as long-running laboratory experiments, yet large enough, and 'natural' enough, to be considered cousins of real-world geomorphological systems. The landforms which seem to fit the specifications of a natural laboratory most nearly are those which are created by mankind's engineering activities such as mine dumps and road-banks.

Road-banks are landforms which have not yet received much attention in the textbooks of geomorphology. However, they are among the most common and widespread of all man-made landforms. It is calculated that the USA has 1 km of roadway for every 1 km² of land. On average, each kilometre of roadway is flanked by a little more than 1 ha of road-bank. Nationwide, 5.8 million km of roadway are surrounded by 6.9 million ha of road-bank (Mowbray, 1969; Disecker and Richardson, 1961).

In much of the western and southwestern United States, Oklahoma included, road-bank erosion is a conspicuous problem. However, devegetated and gullied road-banks are more than just unsightly. Studies of soil loss from road-banks in Georgia have documented sediment yields ranging from 230 to more than 500 t ha^{-1} year^{-1}. Such high sediment yields can contribute significantly to the pollution of the surrounding landscape through the silting of streams, drainage ditches, culverts, ponds and reservoirs (Disecker and Richardson, 1961, 1962; Disecker and McGinnis, 1967; Scheidt, 1967; Parizek, 1971; de Belle, 1971). Road-bank gullies can damage road structures and surfaces as well as neighbouring fields. The development of deep hollows in road margins is a traffic hazard that can cost lives. Further, the drainage from roads, road-banks and road-bank gullies can significantly affect the character of the streams to which they act as tributaries (Hayden, 1982). Road-banks and road-bank gully catchments, then, are subjects worthy of study in their own right.

This experiment, however, forms part of a sequence which began with studies of the erosion of slope and gully systems on free-standing colliery spoil mounds. Later studies have attempted to examine the role of time, climate, and basal stream control on similar landforms. The distinguishing feature of this study is a complication of slope structure. The slope is not composed of purely homogeneous materials but contains three distinct layers including a rock outcrop.

The road-bank selected for study is typical of the smaller road-banks of Oklahoma. The experiment may be regarded as a case study of a single road-bank and its included gully catchment. Its dimensions are: relief - 3 m, ground surface length - 8.5 m, and mean slope angle - 27° (51% grade).

THE SITE

This experiment is situated in Cleveland County of central Oklahoma (Sec 19, T.9.N. R.1.E). The road-bank is adjacent to a paved road and appears to have been created around 1965 as a part of the engineering works associated with the construction of the Lake Thunderbird Reservoir in the Little River valley.

The road-bank is incised into the soil and bedrock of an area of gently sloping (3-5°), savanna, upland. It abuts an area of semi-natural woodland dominated by low growing oak species. However, the canopy of this scrubby woodland is far from complete and there is a dense, herbaceous, ground layer. This is dominated by Little and Big Bluestem grasses (50% composition: dry weight) with seven other grass species well represented. Forbs account for just 10% of the understorey biomass, and best represented is the perennial Sunflower (Bourlier et al., 1975).

The soil which caps this central Oklahoma roadcut is a loamy fine sand called the Darsil, a member of the Stephenville-Darsil-Windhorst Complex. The type member of the series, the Stephenville, is a moderately deep (500-1000 mm) Udic Haplostalf. The Darsil, which may be considered a beheaded Stephenville, is much more shallow (< 480 mm) and is classified as an Udic Ustochrept (Bourlier *et al.*, 1975, 1976; Bourlier, 1979, Mayhugh, 1977).

The physical properties of the Darsil may be summarised as follows: its permeability ranges upwards from 50-500 mm hr^{-1} and seepage rates tend to be high. Available water capacity ranges from 0.7-1.6 cm cm^{-1}. The Plasticity Index runs up to about 7 whilst the Liquid Limit is less than 26. The shrink-swell potential is described as low. The soil erodibility factor from the Universal Soil Loss Equation (K) is estimated as 0.2 (Bourlier *et al.*, 1975).

The Darsil loamy fine sand outcrops over the upper 2 m of the road-bank surface, and the A horizon over the upper 0.7 m. Beneath the subsoil, bedrock outcrops over a distance of some 0.1 m on the interfluve and 0.4 m in the floor of the gully channel. Bedrock in this area is the Garber Sandstone, a poorly cemented, fairly coarse-grained, red sandstone, and is the ultimate source of all the sediments involved by this study (Wood and Burton, 1968).

Downslope of the bedrock outcrop and underlying some 4 m of the road-bank surface, from the bedrock outcrop to the upper limit of the slope foot alluvial and colluvial deposits, are the deposits of the debris slope. The physical properties of these deposits differ markedly from those of the overlying Darsil soils. Permeability is low, 20 mm hr^{-1} or less. Further, both the Plasticity Index (30 plus) and the Liquid Limit (40 plus) are very much higher. These deposits have a very well-developed shrink-swell capacity and sport well-developed desiccation crack polygons. The soil erodibility factor of the Universal Soil Loss Equation (K) is estimated as 0.24.

CLIMATE

In the old-style climatic classification of Koppen, central Oklahoma is described as humid subtropical (Cfa). The average rainfall in Cleveland County is around 875 mm $year^{-1}$ (figure 14.1). However, there is a pronounced seasonality. The heaviest and greatest rainfalls are associated with the convectional storms of the later spring months, especially May, which commonly records more than 150 mm of precipitation. By contrast, the winter months, November, December, January, are normally rather dry and monthly precipitation rarely reaches 50 mm (figure 14.2). The character of the rains ensures that they tend to be relatively erosive. Wischmeier and Smith (1965) estimate the rainfall erosion factor of the Universal Soil Loss Equation (R) as 250. They calculate that the months November through February contribute less than 20% of the year's erosive rains. By contrast the months April through July contribute 55%, whilst the months August through October each contribute about 10%.

The average July temperature in central Oklahoma, 27°C, ensures that the summer evaporation of rainfall is high. Annual lake-level evaporation losses are estimated as 1500 mm $year^{-1}$. This affects the rate of runoff generation

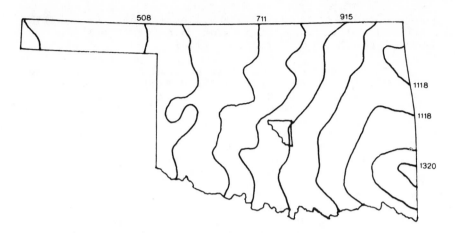

Figure 14.1.   The State of Oklahoma showing Cleveland
County and mean annual precipitation (mm).

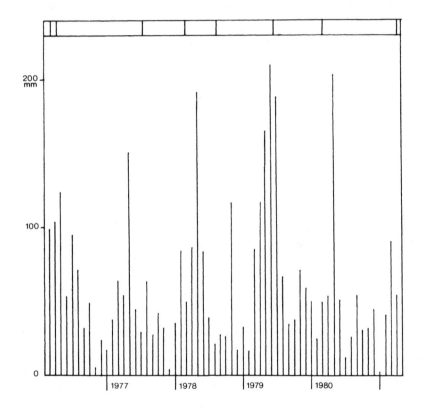

Figure 14.2.   Monthly precipitation totals for Norman,
Oklahoma, during the experiment, and the timing of
ground-loss data collection.

which is only 130 mm year^{-1} (Oklahoma Water Resources
Board, 1975). Winter temperatures are also high. The
January average at Oklahoma City is 3°C. However, in spite
of this there is still significant freeze-thaw activity.
Climatic records from Oklahoma City indicate some 65
freeze-thaw cycles during each year of the investigation.
However, the freezing wave does not penetrate far into the
soil. Even at a depth of 100 mm, extreme low temperatures
in January or February are no lower than 0°C.

TECHNIQUES

Changes in the elevation of the road-bank surface are
monitored by means of erosion pins (Haigh, 1977). Three
profiles, including one located in the central gully
channel and two on the gully catchment interfluves, have
been monumented at 1 m intervals with 600 mm by 15 mm mild
steel rods. Changes in the exposure of these pins are
recorded to the nearest 1 mm and are accurate to about 1 mm.
    The erosion pins were established in late March 1976.
Data were collected in mid-April 1976, late July 1976,
early August 1977, mid-March 1978, early September 1978,
late June 1979, mid-March 1980, and early April 1981
(figure 14.2). The total period of observation was 5 years.
    In the middle of the experiment, it was decided to try
and supplement these erosion pin data by more detailed
measurements of changes in the micro-morphology of the
ground surface. It was observed that the lower debris
slope of the road-bank was covered with a dense network of
desiccation cracks and that many of these cracks were
integrated into a system of tiny channels or microrills
(Haigh, 1978). An attempt was therefore made to try and
discover the magnitude of the contribution of microrill
erosion to the general erosion of the whole slope.
    The device employed for this aspect of the study was a
scaled down version of the 'micro-topographic profile gage'
(Curtis and Cole, 1972), more familiarly known as the
carpenter's contour gauge. This consists of a frame holding
150 sliding rods, each 1 mm in diameter, bound side by side,
which when in contact with the ground surface provide an
accurate representation of its form. For the purposes of
this study, the contour gauge was fitted with two attach-
ments. First, a spirit level was added to the frame to
ensure that all measurements could be referred to the
horizontal. Secondly, a notch was made in the frame so
that the device could be placed in register with an erosion
pin.
    Records are collected as follows: first, the device is
lodged next to one of the interfluve erosion pins; second,
the frame is oriented at right-angles to the line of
steepest slope; third, the bars in the frame are levelled
and the device lowered until it rests on the highest point
of the ground surface and against the adjacent erosion pin;
fourth, the frame is levelled; fifth, each bar is adjusted
downwards until it rests on the soil surface; and finally,
the device is removed and the disposition of the sliding
bars traced into a record book. After some experimentation,

251

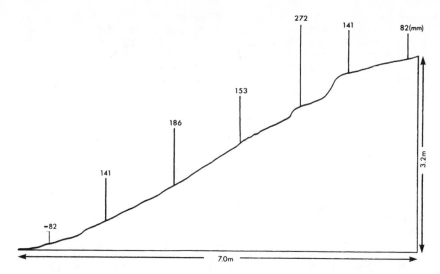

Figure 14.3. Morphology of the experimental road-bank:
profile based on 150 mm unit-length slope measurements;
vertical lines represent erosion pin sites and measured
ground retreat, March 1976 - April 1981.

effective data collection was initiated in September 1979.
Subsequent collections were undertaken in March 1980 and
April 1981.

The data produced by the repeated measurement of the
erosion pins and collection of contour gauge configurations
were supplemented by detailed measurements of the catchment's morphology. Profiles of the interfluve slopes and
gully channel were recorded by means of a miniature slope
pantometer. Slope angles were recorded by 150 mm unit-
lengths.

Soil samples were collected at sites adjacent to each
row of erosion pins. These samples were subjected to
particle-size fractionation by the wet-sieve method.

RESULTS AND DISCUSSION

The five year results from the non-gully slope sections have
been represented at their appropriate locations on a true-
scale slope profile reconstructed from the 150 mm unit-
length slope angle measurements (figure 14.3). At first
glance, these results seem typical of those recorded on
young, artificial slopes elsewhere (Haigh, 1979). They
indicate that there are two zones of especially intense
erosion, one on the lower part of the upper convexity, and
the other on the upper end of the lower concavity.
Experience suggests that these two zones of intense erosion
will tend to migrate towards the mid-slope as the profile
ages.

The pattern of ground retreat across this slope is,
however, affected by the complications of the road-bank's

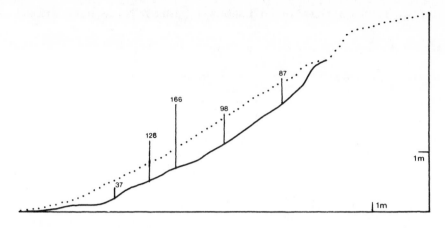

Figure 14.4. Morphology of the gully channel: vertical
    lines represent erosion pin sites and measured ground
    retreat: March 1976 - April 1981.

structure. Thus, while it is usual for the upper zone of
intense erosion to be stronger than the lower zone, it is
obvious that the massive retreat at the upper convexity
can be related to the outcrop of sandstone immediately
downslope. It is probable that this caprock is protecting
the slope below and effectively reducing the rates of
ground loss on the lower slopes. It is further possible
that the rate of slope retreat above the rock outcrop is
being enhanced by basal sapping. Observations during rainstorm conditions indicated that the rock layer acts as a
barrier to throughflow and encourages seepage on its upper
surface.
    Results from the gully channel are recorded on a true-
scale representation of the gully long profile (figure
14.4). This channel is best developed on the lower debris
slope, but its upper section is floored by the bedrock outcrop. Beyond the rock outcrop the channel deepens progressively towards the start of the interfluve's lower concavity where it abruptly shallows. Peak incision is recorded at the point where the existing channel begins to
shallow. However, it is worth noting that this zone of
peak incision may, in fact, be the only point on the entire
gully-long profile where that gully is actually getting
deeper. Elsewhere, retreat records from the gully channel
are greatly exceeded by retreat records from the surrounding
non-gully slopes. On average, the depth of gully incision
decreased by 17 mm year^{-1} during the course of the experiment. This gully is therefore healing at a rate which
could lead to its complete elimination in approximately
25 years and within 45 years of its date of origin.

Comparison of predicted and measured soil
losses from nongully slopes

At present, it is usual to estimate soil losses by employing
the Universal Soil Loss Equation (USLE) (Wischmeier and

Smith, 1965; Mitchell and Bubenzer, 1980). The USLE states that:

$$A = (2.24) \text{ RKLSVM}$$

where: A is soil loss in tonnes per hectare (t/ha),
R is a measure of rainfall erosivity,
K is a measure of soil erodibility,
LS is a topographic factor which compensates for the effects of slope angle and slope length,
and VM includes measures of the inhibitory effects of vegetation and soil surface properties.

The values of the USLE which apply in this study are:

R = 250 (from the charts of Wischmeier and Smith (1965)).
K = 0.22 (calculated from the Agriculture Research Service nomograph (USDA, 1975) for the debris slope deposits, supplemented by published Soil Conservation Service K values for Darsil soil (Bourlier et al., 1975) and scaled to the volume of ground loss recorded on each soil area).
and LS = 9.0 (calculated for the length and mean angle of the road-bank above the zone of visible deposition, from published equations and nomographs (Mitchell and Bubenzer, 1980)).

Given these values, the USLE estimate of soil loss from this experimental road-bank's nongully slopes would be 1310 t ha^{-1} year^{-1}. This, of course, is very high. However, research elsewhere indicates that it may be a reasonable estimate for soil losses from the original disturbed surface of the newly created road-bank (cf. Brown et al., 1979).

It is recommended that the value of VM should be set to unity for slopes which are devoid of a vegetative cover and which have been disturbed, for example by up-and-down slope cultivation (Marsh, 1978). Clearly, the experimental road-bank has not suffered major disturbance since its creation in the mid-1960s. Equally, while the soil contains insufficient coarse fragments to develop an erosion-inhibiting stone layer, its upper layers are both more coarse and liable to protection by rainsplash compaction and the development of a soil crust. Further, while the road-bank supports no vegetation and its soils contain very little organic matter except at the soil crest, the area is slightly overshadowed by a peripheral shrub at the slope head.

The actual total for ground retreat can be estimated by averaging the retreat records from the nongully slope erosion pins above the zone of visible deposition. The results of this calculation indicate that the mean annual ground loss during the course of this experiment was 32.5 mm year^{-1}. The particle specific gravity for the materials included in this slope is 2.6. However, even at the soil surface, there are well developed soil pores and it is unlikely that the bulk density of the eroding soil ever exceeds 1.6 gm cc^{-1}. If this figure is employed to estimate soil loss, total annual soil mobilization on the experimental road bank is 520 t ha^{-1} year^{-1}. This confirms that the value of the VM factor for this site should be about 0.4.

Comparison of nongully slope ground loss totals with climatic parameters

An attempt has been made to correlate three climatic variables: total precipitation, volume of precipitation over 12.5 mm day^{-1}, and number of freeze-thaw cycles, with the averages of annual sediment mobilization on the nongully slopes. Curiously, the best of these variables for predicting erosion is the total precipitation ($r=0.98$), with total precipitation over 12.5 mm day^{-1} a close second ($r=0.96$). The correlation between the number of freeze-thaw cycles and average annual ground loss is poor ($r=0.01$). However, if one correlates the freeze-thaw variable with the residuals left by the correlation with total precipitation, the result is very much better ($r=0.61$). For practical purposes, however, it is hard to find a better or more simple predictor for annual ground loss in mm than:

$$E = 0.29 \ (P - 90)$$

where: E is ground loss (mm)
           P is annual precipitation (mm)

in the case of these results.

Microrilling and microerosion processes

The results of repeated remeasurement of ground surface detail with the modified carpenter's contour gauge are represented at true scale in figures 14.5-14.7. The graphs describe the changing configuration of the ground surface adjacent to the erosion pins on the nongully slope for three occasions during the last 18 months of this study. Each profile has been corrected to the horizontal during data collection. The separation of the profiles is determined by changes in the exposure of the erosion pin located on the immediate right of each profile. Each profile is thus identified with a particular row of erosion pins. Row 0 is at the slope foot. Row 1 is 1 m upslope and so on. Each profile is a cross-profile gathered at right angles to the line of steepest slope.

Microrilling is a feature of the debris slope beneath the bedrock outcrop. The profiles of rows 1-3 have been collected on this debris slope. Row 0 is a record of deposition at the slope foot. Rows 4 and 5 are located on the eroding subsoils of the Darsil Loamy Fine Sand. The impact of microrill activity can be demonstrated by a comparison of the sinuosities of microrilled and non-microrilled profiles. Sinuosity, the ratio of ground-surface to through-the-air distance, is far greater for the microrilled profiles. The mean value is 1.16 (S.D.=0.09) compared to 1.03 (S.D.=0.04) for the non-microrilled profiles. A t-test indicates that the difference is highly significant ($p>0.99$).

255

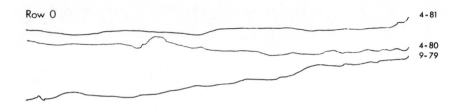

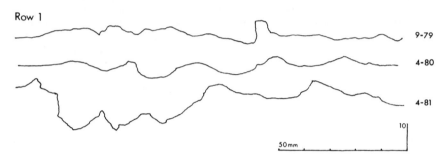

Figure 14.5. True-scale cross profiles of the road-bank
lower slope surfaces.

These observations may be supported by the fact that
there is no significant change in total sinuosity during
the course of the experiment.

A similar result obtains if the results of another
method for the comparison of the microprofiles are considered. This second approach is founded in the concept of
relative relief. Differences between the highest and
lowest points in each 10 mm increment of each 150 mm section have been scored. Averages and their standard deviations have been calculated for each row profile and each
date of observation (table 14.1). The data confirm that
the three microrilled sections have a much greater local
relief than the profiles measured on the soil and wash
deposits and that this local relief is subject to much
greater variation during the course of the experiment. The
mean local relief per cm on the three microrilled sections
was 3.67 mm compared with 1.95 mm on the sections without
microrills. The mean profile standard deviation during the
course of the experiment was 1.86 mm for the microrilled
profiles but only 0.90 mm for the non-microrilled profiles.
Clearly, the microrilled profiles are significantly more
rugged and, more importantly, significantly more active
than those which do not feature microrill channels.

Microrills are small, ephemeral, lines of concentrated
wash which seem to form on soils which are prone to desiccation cracking, and which are not routinely subjected to
sheet-floods, numerous cycles of wetting and drying, or to

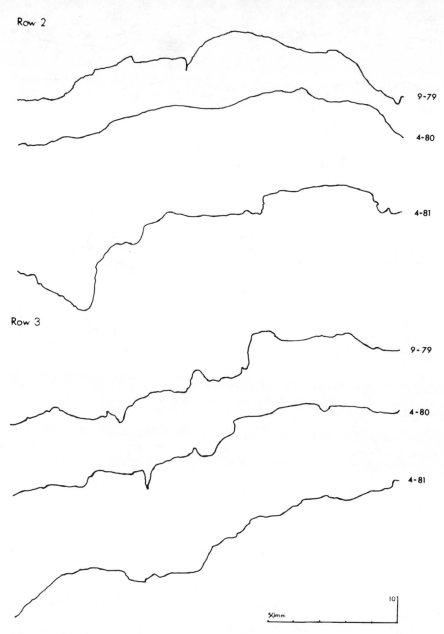

Row 2

9-79

4-80

4-81

Row 3

9-79

4-80

4-81

10
50mm

Figure 14.6. True-scale cross profiles of the road-bank
    middle slope surfaces.

frequent frost action (cf: Engelen, 1973; Haigh, 1978).
They are intimately associated with the soil crust and with
the desiccation crack which is customarily found in the
floor of their channel. Their dimensions, both width and
depth, can be expressed in terms of a few millimetres.
Consequently, they are quickly obliterated by natural

Row 4

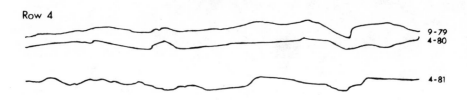

9-79
4-80

4-81

Row 5

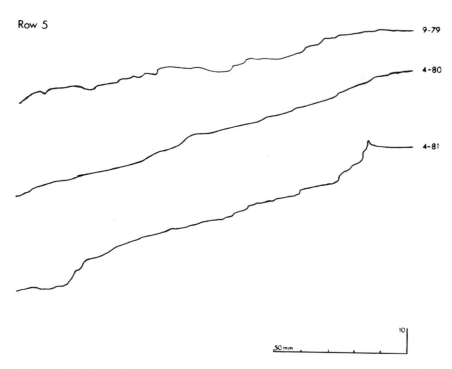

9.79

4-80

4-81

10
50mm

Figure 14.7. True-scale cross profiles of the road-bank
upper slope surfaces.

movements at the soil surface, especially freeze-thaw or
soil rehydration. They develop best in dry, frost-free
climates, subjected to short, intense, rainstorms.
     In Oklahoma, one might expect the development of micro-
rills to have a pronounced seasonal component. Whilst
admitting that microrill development, like soil crust
strength, is to some extent a function of the soil moisture
content and the duration of the previous rainfall event, it
might be expected that microrills would be least evident
at the end of the winter frost period and best developed
at the end of the long, hot, rainy summer. Unfortunately,
the results of this experiment do not confirm this hypothesis. The profiles measured in September 1979 are more
dissected than those of April 1980, but less dissected than

Table 14.1. Mean relative relief per centimetre ($mm\ cm^{-1}$) and profile standard deviation for cross-slope microprofile data (figures 14.5 - 14.7).

	9/1979	4/1980	4/1981
Row 0	1.35 (0.71)	1.31 (1.10)	0.75 (0.54)
Row 1	2.24 (2.10)	2.30 (1.66)	4.73 (3.12)
Row 2	4.06 (2.63)	2.95 (2.17)	5.08 (4.69)
Row 3	3.87 (4.04)	3.80 (2.55)	3.95 (3.13)
Row 4	1.81 (1.19)	1.36 (0.65)	1.66 (0.78)
Row 5	2.37 (1.45)	3.19 (1.04)	3.78 (2.63)

Values in brackets are profile standard deviations

those of April 1981. A more sustained and more intense pattern of data collection would be required to elucidate the dominant controls of microrill development. The profile measurements undertaken in this study are too separate in both time and space to yield any clear idea of the character of the continuity of ground surface changes.

Detailed examination of the character of the cross-sections, however, reveals a little more concerning the patterns of sediment redistribution in this catchment. Row 0 documents 18 months of debris accumulation at the slope foot during a period when colluvial deposition was more active than alluvial deposition. The alluvial fan which rises towards the right and which is indented by tiny distributary channels is eliminated by sediments originating from the nongully slopes during the course of observation. Row 1 shows the effects of microrilling and a pattern of microrilling which increases through the experiment. The differential lowering effected by microrill activity across this section amounts to an average ground loss of 12 mm $year^{-1}$. The profile also shows the way in which the larger gravel fragments can retard ground retreat. The small square peak on the profile for 9-79 is a small soil pillar capped by a 5 mm wide stone. The remains of this soil pillar survive on the profile for 4-80. The jagged peaks to the left of the profile for 4-80 and at several locations on the profile for 4-81 have a similar cause.

The most dramatic instance of the development of a microrill channel is to be found on Row 2, to the left of profile 4-81. There is also evidence of repeated micro-channel development at the base of the erosion pin to the right of the section. Row 3 shows an example of the lateral downslope migration of a small microrill channel and of the sharp v-notch of a well developed and persistent desiccation crack.

The two sections recorded on the Darsil soils show relatively little in the way of microrelief. There is a

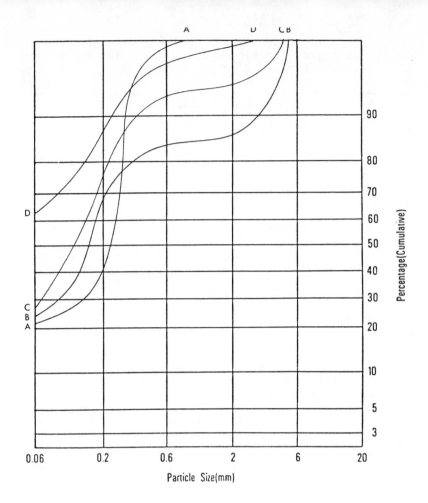

Figure 14.8. Particle-size characteristics of the Darsil
A-horizon (A), the Darsil B horizon (B), the wash
deposit (C) and the debris slope deposit (D).

small persistent channel towards the right of Row 4 and the
evolution of a spectacular platform to the right of Row 5,
profile 4-81, which is caused by a trapped piece of wood.

Particle size fractionation

Representative soil samples have been collected for four
zones on the experimental road-bank and have been subjected
to particle-size fractionation by the wet sieve method.
The four soil types sampled are the Darsil A-horizon, the
Darsil B-horizon, the debris slope soil, and the slope foot
wash. Two samples were collected for each soil and the
averaged results are portrayed as figure 14.8. The results
highlight the cardinal reason for the differences in the
surface morphology of the different soils. The silt-clay
fraction of the Darsil A-horizon is 22%, that of the Darsil

B-horizon 24-26%, and that of the wash deposits 28-29%, but that of the debris slope deposits is 55-72%. In geo-technical terms, the Darsil soil is a very fine sand, as is the wash deposit, but the debris slope (D) soil is a clay. The textural characteristics of the wash deposit (C), which is derived from the other three, most closely re-sembles the Darsil subsoil.

Comparison of the particle-size characteristics of the mobilized soil with those of the deposits accumulating at the road-bank foot

The particle-size characteristics of the mobilized, (ie. eroded) soil can be reconstructed by scaling the soil loss record at each erosion pin by the particle-size parameters of the local, eroding, soil. For want of a better hypo-thesis, it might be suggested that this particle-size distribution should be preserved in the slope-wash and alluvial fan deposits accumulating at the road-bank foot. However, direct measurement of the growth of these slope foot deposits indicate that only 10% of the total volume of sediment mobilized is, in fact, retained in this micro-catchment, and there is therefore plenty of scope for the selective removal of the more erodible soil components.

Since it is the largest soil particles which are least readily mobilized, and since there is little scope for mechanical disintegration in a catchment whose longest flow-line scarcely exceeds 10 m, one might expect that the coarse fragments should represent an enhanced proportion of the retained slope foot deposits. In fact, this is not the case. The gravel fraction elements are retained in pre-cisely the same proportion that they are mobilized (table 14.2). The pattern of sediment retention is curiously bimodal. Relatively small proportions of coarse and coarse-medium (2 mm - 0.4 mm) sized particles are retained and the same is true of particles smaller than 0.06 mm, the silt/clay fraction. On the other hand, a relatively large

Table 14.2. Comparison of the particle-size characteristics of mobilized sediment with the proportion retained by deposition at the road-bank foot.

Fraction	2-6 mm	0.6-2 mm	0.4-0.6 mm	0.2-0.4 mm	0.1-0.2 mm	0.06-0.1 mm	<0.06 mm
Eroded Soil	4.6%	0.8%	1.1%	17.4%	22.4%	6.7%	46.9%
Retained by Deposition	0.4%	0.05%	0.04%	1.7%	3.4%	1.3%	3.0%
Relative * Retention	9.4%	6.3%	3.7%	9.7%	15.3%	19.7%	6.3%

* overall, 10% of the eroded soil is retained in the catchment by deposition.

proportion of particles in the fine to very fine sand size range (0.2-0.06 mm) are retained in the wash deposits.

CONCLUSIONS

This study examines the pattern of morphogenesis on a simple artificial slope and included gully system. It forms part of a series of studies of simple, man-made, landforms. The distinguishing feature of this investigation is that the slope is not composed of homogeneous materials but has three layers including a rock outcrop.

The pattern of ground loss measured on the nongully slopes is similar to that recorded on similar, young, artificial slopes elsewhere. There are two zones of especially intense erosion, one at the base of the upper convexity and the other at the head of the lower concavity. The unusually high differences in the magnitude of these erosion peaks is related to the rock outcrop which, partly by concentrating seepage, is encouraging retreat upslope and which, by resisting erosion itself, is protecting the lower slopes from erosion.

The gully channel is best developed on the deposits below the bedrock outcrop. Peak incision is associated with the lower break of slope. However, on average, the depth of the gully declined through the course of this experiment at a rate which could lead to its complete elimination in approximately 25 years, that is about 45 years after the creation of the slope.

Soil losses predicted by the Universal Soil Loss Equation are confirmed by empirical measurement. The mean annual ground loss from the area of this catchment above the zone of visible deposition is 32.5 mm year^{-1}, which if one assumes a soil bulk density of 1.6 gm cc^{-1}, converts to a soil loss of 520 t ha year^{-1}. It is suggested that the combined effects of the rock outcrop, 10 to 15 years without surface disturbance, and slight overshading by marginal vegetation, produces a VM (cf. CP) factor of 0.4.

Annual average ground loss from the nongully slopes shows a strong correlation with total annual precipitation ($r=0.98$) and with precipitation received in daily totals of more than 12.5 mm ($r=0.96$). There is no direct correlation between ground loss and the number of frosts. However, number of frost days correlates well with the residuals from the ground loss - precipitation correlation ($r=0.61$).

Microprofiles of 150 mm cross-slope sections of the nongully slopes have been collected during the last 18 months of this study. These data indicate that desiccation crack-related microrills are persistent features on certain sections of the slope and that their development contributes significantly to the surface roughness of affected areas.

Ten percent of the sediment mobilized by erosion in this catchment is retained by deposition. The particle-size characteristics of the eroded soils have been reconstructed from the particle-size characteristics of the soils undergoing erosion, scaled by their records of ground loss. The proportions of the particle-size fractions contained by the eroded soil have been compared with those retained in the catchment. Relatively small proportions of particles larger than 0.4 mm or smaller than 0.06 mm are retained. Relatively high proportions of particles in the size range 0.2-0.06 mm are retained.

ACKNOWLEDGEMENTS

Thanks go to the many individuals who have assisted in the field work connected with this project, especially: G. Rydout (University of Arizona), W. Wallace (University of Chicago), G. Latz (University of Tokyo), M. Smith (University of Oklahoma), B. Bourlier (Soil Conservation Service) and H. MacLean.

REFERENCES

Agricultural Research Service, US Department of Agriculture, 1975, *Control of water pollution from cropland*, I, (Report ARS-H-5-1).

Belle, G. de, 1971, Roadside erosion and resource implications in Prince Edward Island, Canada, *Department of Energy, Mines and Resources, Policy Coordination Branch, Geographical Paper*, no.48.

Bourlier, R., 1979, Personal Communication.

Bourlier, R. *et al.*, 1975/1976, Soil Survey Interpretation 1: Stephenville-Darsil-Windhorst Complex, 3 to 8% Slopes and Darsil Laboratory Data, Oklahoma State University, Department of Agronomy, Soil Morphology, Genesis, and Classification Lab. (Unpublished)

Brown, W.M. (III), Hines, W.G., Rickert, D.A. and Beach, G.L., 1979, A synoptic approach for analyzing erosion as a guide to land-use planning: river-quality assessment of the Willamette Basin, Oregon, *United States Geological Survey Circular* 715-L.

Curtis, W.R. and Cole, W.D., 1972, Micro-topographic profile gage, *Agricultural Engineering*, 53, 17.

Disecker, E.G. and McGinnis, J.T., 1967, Evaluation of climatic, slope, and site factors on erosion from unprotected roadbanks, *Transactions American Society of Agricultural Engineers*, 10, 9-11,14.

Disecker, E.G. and Richardson, E.C., 1961, Roadside sediment production and control, *Transactions American Society of Agricultural Engineers*, 4, 62-64,68.

Disecker, E.G. and Richardson, E.C., 1962, Erosion rates and control methods on highway cuts, *Transactions American Society of Agricultural Engineers*, 5, 153-155.

Engelen, G.N., 1973, Runoff processes and slope development in badlands, National Monument, South Dakota, *Journal of Hydrology*, 18, 55-79.

Haigh, M.J., 1977, Use of erosion pins in the study of
    slope evolution, *British Geomorphological Research
    Group Technical Bulletin* no.18, 20-47.

Haigh, M.J., 1978, Microrills and desiccation cracks: some
    observations, *Zeitschrift für Geomorphologie*,
    N.F.22, 457-461.

Haigh, M.J., 1979, Ground retreat and slope evolution on
    plateau-type colliery spoil mounds at Blaenavon,
    Gwent, *Transactions Institute of British Geo-
    graphers*, N.S.4, 321-328.

Marsh, W.M., 1978, *Environmental analysis for land use and
    site planning*, (McGraw Hill, New York).

Mayhugh, R., 1977, *Soil Survey of Pottawatomie County,
    Oklahoma*, United States Department of Agriculture,
    Soil Conservation Service and Oklahoma Agricultural
    Experiment Station report.

Mitchell, J.K. and Bubenzer, G.D., 1980, Soil loss esti-
    mation, in: *Soil Erosion*, ed. Kirkby, M.J. and
    Morgan, R.P.C., (Wiley, Chichester) 17-62.

Mowbray, A.Q., 1969, *Road to ruin* (Lippincott, Philadelphia).

Oklahoma Water Resources Board, 1975, *Oklahoma Comprehen-
    sive Water Plan: Phase 1, Section 2.*

Parizek, R.R., 1971, Impact of highways on the hydrogeologic
    environment, in: *Environmental geomorphology*, ed.
    Coates, D.R., (Binghamton, State University of New
    York, Geomorphology Symposium 1), 151-199.

Scheidt, M.E., 1967, Environmental effects of highways,
    Journal of the Sanitary Engineering Division, ASCE,
    93, 17-25.

Utah Water Resources Laboratory, 1976, Erosion control
    during highway construction, *Manual of erosion
    control principles and practices*, N.C.H.R.P. Project
    16-3.

Wischmeier, W.H. and Smith, D.D., 1965, Predicting rainfall
    erosion losses from cropland east of the Rocky
    Mountains, *United States Department of Agriculture,
    Agricultural Handbook* no.282.

Wood, P.R. and Burton, L.C., 1968, Ground-water resources
    in Cleveland and Oklahoma Counties, Oklahoma,
    Oklahoma Geological Survey Circular, 71.

15 Water and sediment dynamics of the Homerka catchment

W. Froehlich and J. Slupik

INTRODUCTION

The lack of a satisfactory method of extending research results has proved a considerable problem and a major shortcoming of catchment investigations (Rodda, 1976). The extrapolation of research results obtained from a small plot to the whole drainage basin by multiplication of data, and the interpolation of results obtained from a stream gauging station to a small unit area (eg. 1 ha or 1 km^2) by division of data have been questioned (eg. Ketcheson et al., 1973; Yair et al., 1978), because both assume homogeneity in what will in general be a naturally heterogeneous drainage basin (Amerman, 1965). This assumption cannot be reconciled with the existence of variable source areas of runoff production (eg. Dunne and Black, 1970; Walling, 1971). Correct extrapolation necessitates a knowledge of the mechanics of water circulation and sediment transport within a catchment, ie. a knowledge of runoff production on the slopes and its routing through the stream channel (figure 15.1). Evidence for this is provided by investigations undertaken in the Homerka catchment (figures 15.2, 15.3).

STUDY AREA

The Homerka stream drains an area of 18 km^2, situated in the flysch Beskidy Mountains (400-1000 m a.s.l.), which form part of the Carpathians. The catchment of the Homerka stream can be subdivided into an upper and a lower part (Niedziałkowska, 1981). The upper part, underlain by more resistant flysch series, has more permeable skeletal soils, steeper slopes, and supports a *Fagetum Carpathicum* forest community. The lower part of the catchment, underlain by less resistant flysch series, has less permeable silty-clayey soils, which are used for agriculture. Various environmental conditions in these two units reflect their contrasts in water storage, potential evaporation and precipitation input. Hence, two types of water circulation

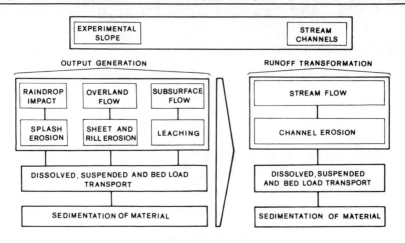

Figure 15.1. Water circulation and sediment transport
investigations in the Homerka catchment.

on the slopes have been distinguished (Słupik, 1981). The
'slower' one in the zones with foothill relief, and the
'quicker' one in the zones with montane relief. Mean
annual precipitation ranges from 700 mm near the outlet
of the basin to more than 1000 mm in the headwater parts
of the catchment (Niedźwiedź, 1981).
     Measurements of water discharge and sediment transport
have been made both at the foot of the slopes and in the
stream channel at gauging sites representing sub-catchments
of different sizes (Froehlich, 1979; Froehlich and Słupik,
1979). Investigations of the slope and channel subsystems
have focussed on the recognition of spatial variability of
both water circulation and sediment transport processes in
a small catchment and thereby on elucidating the source
areas of runoff and sediment production (Słupik, 1981;
Froehlich, 1982).

RESEARCH RESULTS

In terms of runoff production, the major role is played by
the unmetalled roads*, and the natural erosional hollows
from which water and sediment are supplied by concentrated
flow (figures 15.3, 15.4). Inter-channel areas, ie. areas
which are not drained in a linear way, produce only a few
per cent of the water and sediment input into the stream
channel during high water discharges (figures 15.5, 15.6),
but this proportion is an order of magnitude greater at low
flows (Froehlich and Słupik, 1980a). The unprepared roads
increase the drainage density and since the permeability
of the road surface is low, their response to rainfall is

* In the Flysch Carpathians the unmetalled roads include simple paths,
  cart tracks in the fields and woods, lumber tracks and sunken roads.

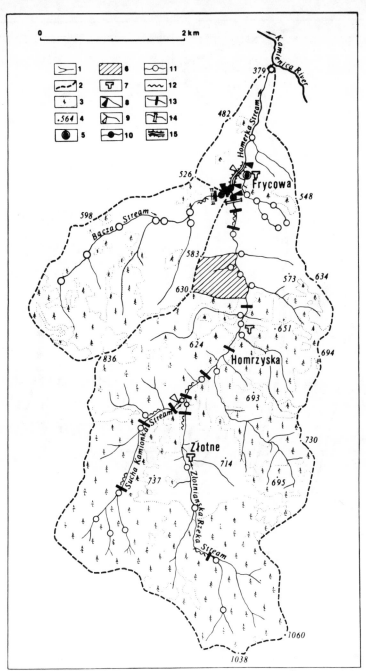

Figure 15.2. The Homerka drainage basin (after Froehlich,
1979). 1. Stream network, 2. Water divides, 3. Forest,
4. Altitude, 5. Research Station of the Department of
Geomorphology and Hydrology of the Polish Academy of
Sciences, 6. The experimental slope, 7. Rain gauges,
8. Analogue water level recorders, 9. Water level posts,
10. Permanent stream sampling stations, 11. Periodic
stream sampling stations, 12. Measuring points for bed
load transport, 13. Drop structure dams, 14. Sites for
measuring channel erosion rates, 15. Experimental
stream channel segment.

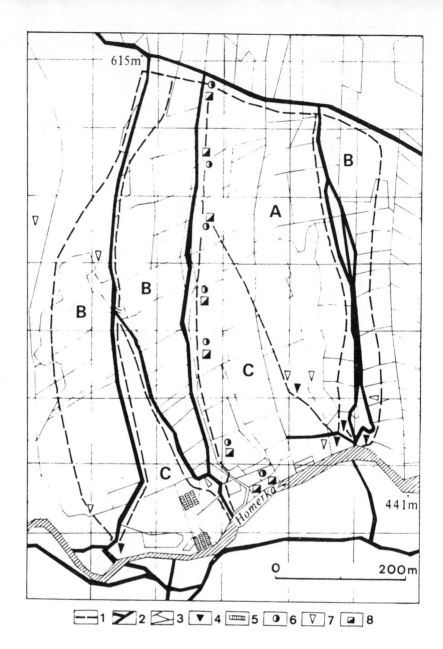

Figure 15.3.   The experimental slope in the Homerka catch-
    ment.   1. Water divides; A) watershed of the Holocene
    incision, B) sunken roads watersheds, C) watersheds of
    the inter-channel areas.   2. Unmetalled roads, 3. Field
    boundaries, 4. Points for measuring concentrated flow
    and taking water samples, 5. Containers for measuring
    sheet flow and taking water samples, 6. Wells for
    measuring water level changes and taking water samples,
    7. Points for measuring water discharge of springs and
    taking water samples, 8. Points for measuring soil
    moisture.

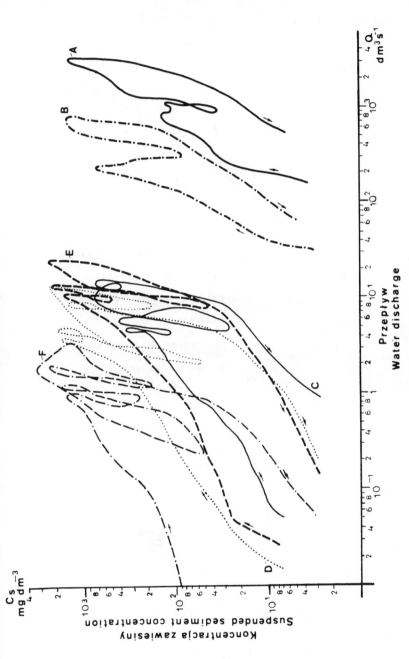

Figure 15.4. The relationship between water discharge (Q) and suspended sediment concentration (Cs) during high water flow, May 1978 (based on Froehlich, 1982). A. Homerka stream, B. Bacza stream, C. Holocene incision; D, E, F. Sunken roads.

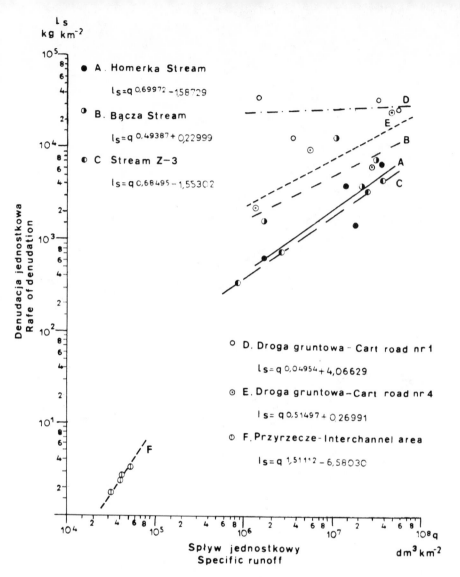

Figure 15.5. The relationship between specific runoff (q)
    and denudation rate (ls) in catchments of different
    size; including the Homerka and Bacza streams, and the
    watersheds of sunken roads and the Holocene incision
    (based on Froehlich, 1982).

rapid. The abundance of poorly consolidated materials
capable of being transported is responsible for the high
suspended sediment concentrations (figures 15.4, 15.6),
which are further increased by the impact of raindrops
(Froehlich and Słupik, 1980b).
    The concave portion of the experimental slope is drained
by a natural erosional incision with a permanently wet zone
at its outlet (figure 15.3). This feature therefore
responds quickly to rainfall, but the sediment concentrations

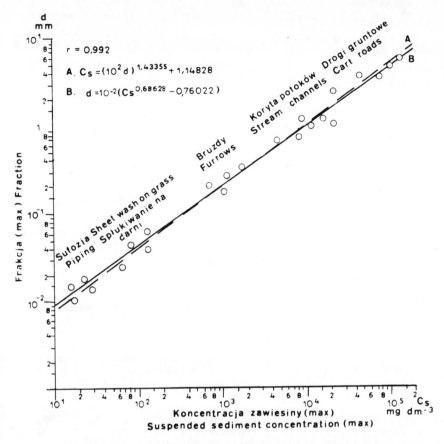

Figure 15.6. The relationship between maximum values of
    suspended sediment concentration (Cs) and maximum
    diameter of mineral particles transported in suspension
    (d) in the Homerka stream and on the experimental slope
    (based on Froehlich, 1982).

are much lower than those associated with the sunken roads
(figures 15.4, 15.6) because of the permanent grass cover.
    The inter-channel areas on the footslopes and on the
valley floor support a high density grass cover and are
natural 'traps' for both water and particulate material
derived from the fields by sheet flow. Thus the above
areas tend to store water at high discharges and to favour
deposition. The natural erosional incisions, the unmetalled
roads and the inter-channel areas without linear drainage
constitute the three potential areas contributing streamflow within the Carpathian drainage basins (Słupik, 1981).
    The intensity of streamflow and sediment transport
processes changes as the drainage basin increases in size
from source to mouth (eg. Chorley, 1969; Thornes, 1980), in
response to changes in the geographical environment which
determine both the nature and intensity of processes of

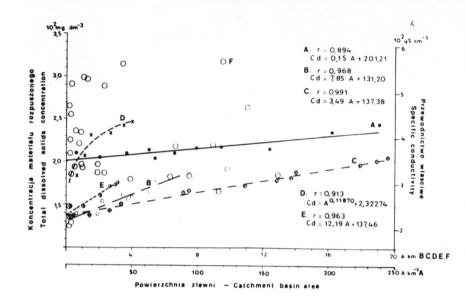

Figure 15.7.   The relationship between drainage basin
    area (A) and dissolved solids concentration in sub-
    catchments of different sizes (based on Froehlich,
    1982).   A. Kamienica Nawojowska stream, B. Kryściów
    stream, C. Homerka stream, D. Bacza stream, E. Sucha
    Kamionka stream, F. Small tributaries of the Homerka
    and Bacza streams.

water circulation and sediment transport. Arnett (1979)
has shown that in a relatively small area of only 900 km^2,
yields of particulate and dissolved material are highly
variable.   This variation increases as the scale of investigation becomes more detailed.
    The above factors control the decrease in specific
runoff as the drainage basin increases in area (Froehlich
and Słupik, 1980a).   This is also reflected in the spatial
variability of dissolved solid concentrations (figure 15.7),
and of denudation rates per unit area (figure 15.8) at low
flows.   During high water discharges a similar statistical
relationship is valid for suspended sediment and bed load
transport (Froehlich, 1982).   Water discharges and sediment
loads increase in a manner directly proportional to the
increase in drainage basin area, with a linear relationship (figure 15.8).   Conversely, streamflow rate (specific
runoff) and denudation rate decrease with increasing
drainage basin area, ie. down the trunk stream.   The model
demonstrating the spatial variability of both values is
best approximated by a hyperbolic function (figure 15.8).
In this model the values of streamflow and denudation rate
determine the precise relationship between the response of
a sub-catchment and that of the whole drainage basin
(figure 15.9).   They also enable spatial variations in
water circulation and sediment transport processes to be
presented in the form of statistical relationships.   The

272

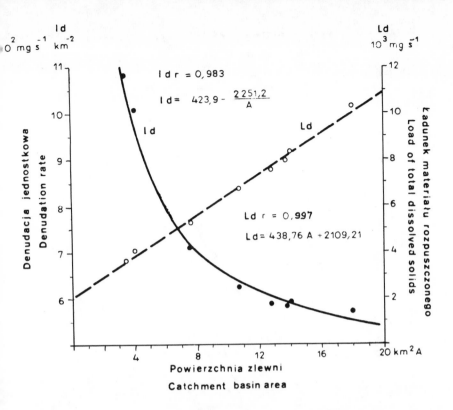

Figure 15.8. The relationship between drainage area increase down the trunk stream (A) and load of total dissolved solids (Ld) and denudation rate (ld) (based on Froehlich, 1982).

correct manner of extrapolating research results is through application of this model.

CONCLUSIONS

1) The above principles concerning the increase in stream-flow discharge and loads, and the decrease in both stream-flow and denudation rates (per unit area) as the drainage basin increases in area, (ie. down the trunk stream) apply to the catchments of streams that cross the boundaries of climatic belts and other environmental conditions at right angles (figure 15.7, catchments A,B,C,E). These are 'symmetric' drainage basins. The model does not apply to the catchments of streams that cross the different vertical zones at oblique angles (figure 15.7, catchment D). These are 'asymmetric' drainage basins. Similarly, events varying in timing (eg. the aspect-controlled progress of snowmelt in the mountains) and in spatial extent (eg. storms covering an area smaller than the overall drainage basin) complicate this pattern of spatial variability found in the Homerka catchment. Such cases cannot form the basis for extrapolation.

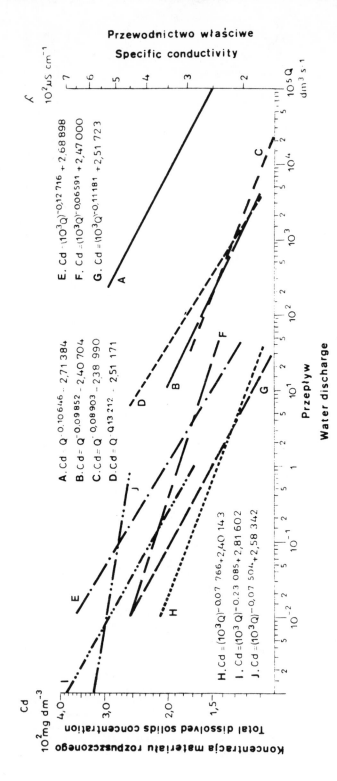

Figure 15.9. The relationship between water discharge (Q) and dissolved solids concentration (Cd) in sub-catchments of different size (based on Froehlich, 1982). A. Kamienica Nawojowska stream, B. Kryściów stream, C. Homerka stream, D. Bacza stream. E. Holocene incision, F, G, H. Sunken roads, I. Spring on the experimental slope, J. Spring in valley bottom of the Bacza stream.

Within the figure:

Przewodnictwo właściwe
Specific conductivity

$10^2 \mu S \; cm^{-1}$

$10^5 Q$
$dm^3 \, s^{-1}$

Przepływ
Water discharge

A. $Cd = Q^{-0.10646} \cdot 2,71384$
B. $Cd = Q^{-0.09852} - 2,40704$
C. $Cd = Q^{-0.08903} - 2,38990$
D. $Cd = Q^{0.13212} - 2,51171$

E. $Cd = (10^3 Q)^{-0.12716} + 2,68898$
F. $Cd = (10^3 Q)^{-0.06591} + 2,47000$
G. $Cd = (10^3 Q)^{-0.11181} + 2,51723$

H. $Cd = (10^3 Q)^{-0.07766} + 2,40143$
I. $Cd = (10^3 Q)^{-0.23085} + 2,81602$
J. $Cd = (10^3 Q)^{-0.07504} + 2,58342$

Cd
$10^2 mg \; dm^{-3}$

Total dissolved solids concentration
Koncentracja materiału rozpuszczonego

2) In experimental catchments, the extrapolation of research results is made difficult by the lack of a model
describing spatial variability of water circulation and
sediment transport processes. Such a model was found by
examining both an experimental slope and the sub-catchments
in the Homerka drainage basin. This model demonstrates the
quantitative representativeness of the study area. Recognition of the spatial variability of water circulation and
sediment transport makes it possible to review and change
(if necessary) the location of the network of measurement
points necessary for the extrapolation of research results.

REFERENCES

Amerman, C.B., 1965, The use of unit source watershed data
    for runoff prediction, *Water Resources Research*, 1,
    499-507.

Arnett, R.R., 1979, The use of differing scales to identify
    factors controlling denudation rates, in: *Geo-
    graphical approaches to fluvial processes*, ed.
    Pitty, A.F., (GeoBooks, Norwich), 127-147.

Chorley, R.J., 1969, The drainage basin as the fundamental
    geomorphic unit, in: *Water Earth and Man*, ed.
    Chorley, R.J., (Methuen, London), 77-99.

Dunne, T. and Black, R.D., 1970, An experimental investi-
    gation of runoff-production in permeable soils,
    Water Resources Research, 6, 478-490.

Froehlich, W., 1979, Characteristics of the catchment
    basins of the Kamienica Nawojowska and its tributary
    stream Homerka. Programme of permanent investi-
    gations of fluvial processes, in: *Excursion guide-
    book field meeting of the IGU Commission on Field
    Experiments in Geomorphology*, (University Press,
    Wrocław), 127-141.

Froehlich, W., 1982, Mechanizm transportu fluwialnego i
    dostawy zwietrzelin do koryta w górskiej zlewni
    fliszowej (The mechanism of fluvial transport and
    waste supply into the stream channel in a moun-
    tainous flysch catchment), *Prace Geograficzne
    Polska Akademia Nauk*, 143, 1-144.

Froehlich, W. and Słupik, J., 1979, Methods and results of
    slope experimentation (in the Homerka catchment
    basin), in: *Excursion guide-book field meeting of
    the IGU Commission on Field Experiments in Geo-
    morphology*, (University Press, Wrocław), 142-152.

Froehlich, W. and Słupik, J., 1980a, The pattern of the
    areal variability of the runoff and dissolved
    material during the summer drought in flysch
    drainage basins, *Quaestiones Geographicae*, 6, 11-34.

Froehlich, W. and Słupik, J., 1980b, Importance of splash
    in erosion process within a small flysch catchment
    basin, *Studia Geomorphologica Carpatho-Balcanica*,
    14, 77-112.

Ketcheson, J.W., Dickinson, W.T. and Chisholm, P.S., 1973, Potential contributions of sediment from agricultural land, in: *Fluvial processes and sedimentation* (Proceedings of Hydrology Symposium held at University of Alberta, Edmonton, 1973), 184-191.

Niedziałkowska, E., 1981, Rzeźba terenu (Relief), in: Warunki naturalne zlewni Homerki i jej otoczenia (Natural conditions of the Homerka catchment basin and its surrounding), *Dokumentacja Geograficzna Polska Akademia Nauk*, 3, 13-21.

Niedźwiedź, T., 1981, Klimat (Climate), in: Warunki naturalne zlewni Homerki i jij otoczenia (Natural conditions of the Homerka catchment basin and its surrounding), *Dokumentacja Geograficzna Polska Akademia Nauk*, 3, 22-42.

Rodda, J.C., 1976, Basin studies, in: *Facets of Hydrology*, ed. Rodda, J.C., (Wiley, London), 257-297.

Słupik, J., 1981, Rola stoku w kształtowaniu odpływu w Karpatach fliszowych (Role of slope in generation of runoff in the Flysch Carpathians), *Prace Geograficzne Polska Akademia Nauk*, 142, 1-98.

Thornes, J.B., 1980, Erosional processes of running water and their spatial and temporal controls; a theoretical viewpoint, in: *Soil Erosion*, ed. Kirkby, M.J. and Morgan, R.P.C., (Wiley, London), 129-182.

Walling, D.E., 1971, Streamflow from instrumented catchments in south-east Devon, *Exeter Essays in Geography*, (University of Exeter, Exeter), 55-81.

Yair, A., Sharon, D. and Lavee, H., 1978, An instrumented watershed for the study of partial area contribution of runoff in the arid zone, *Zeitschrift für Geomorphologie, Supplement Band*, 19, 71-82.

16 Sources of sediment and channel changes

in small catchments of Romania's hilly regions

D. Balteanu, G. Mihaiu, N. Negut and L. Caplescu

INTRODUCTION

The hilly regions of Romania cover one third of the country and supply 41% of the average annual runoff and the maximum suspended sediment loads. Values of mean annual suspended sediment transport are highest in the Curvature Subcarpathians (> 25 t ha^{-1}year^{-1}) but high sediment yields occur in all the other hilly regions (5-25 t ha^{-1}year^{-1}).

Our investigations focused on three small catchments situated in the southeast Subcarpathians (Curvature Subcarpathians), which are formed of Neogene sedimentary rocks and affected by positive neotectonic movements and frequent earthquakes. The region shows clear evidence of valley deepening, which in turn promotes the transfer of material from the slopes to the channels. Torrential rainfall is common from April until October. Over large areas the slopes are deforested. The small catchments are affected by active geomorphological processes which are related to the underlying lithology and the current land use (Balteanu and Taloescu, 1979).

Contemporary processes in the three catchments have been investigated by using repeated aerial and terrestrial photographs taken over a period of 10-17 years. The photogrammetric evidence has been correlated with topographic surveys, measurements of material accumulation behind dams and periodic measurements of reference marks located on the slopes. In this way quantitative information was obtained concerning the following parameters: the rate of gully growth, the increase in gully surface area, the volume of material displaced, the volume of material eroded from different zones of the catchments, and the rate of filling of reservoirs. The choice of these parameters took into account their direct application in the design of control measures.

The Draghici basin has an area of 4.87 km^2 and is developed
on unconsolidated piedmont sediments in which marls, clays,
sandy clay and loess deposits are dominant. The area is
used as agricultural land (arable and pasture) and acacia
trees have been planted along the valley bottom.

Erosion control measures including the building of 16,
3-5m high, dams and of slope erosion control devices were
carried out during 1971 - 1974. Successive measurements
performed on terrestrial and aerial photographs as well as
geomorphological mapping undertaken in the years 1960, 1971,
1972 and 1977 have provided evidence on the sources of sediment and the development of the channels and slopes before
and after the introduction of improvement works. (Mihai
et al, 1979).

During the period studied, 127 772 m^3 of material were
eroded from the basin, corresponding to a volume of 15.43 m^3
ha^{-1}year^{-1}. This erosion was unevenly distributed over the
years and between the different sectors of the main valley
(table 16.1).

The period 1960-1971, that preceded the inception of
conservation works in the basin, was characterized by low
rainfall totals (the total amount of precipitation falling
between April and October averaged 364 mm). Nevertheless,
the long profiles and cross profiles shown in figures 16.1
and 16.2 emphasize that the valley deepened and the slope
profiles changed correspondingly, a trend exhibited by most
catchments in Romania's Subcarpathians and piedmont hills.
The valley deepened by 0.05 - 0.10 m year^{-2}. Most of the
material eroded originated from the channel and from the
lower sector of the slopes. Gullies whose banks are affected
by mass movements, were the main source of channel bed
material. During the 11 year period, a total volume of
36 884 m^3 of material was transported from the basin, corresponding to an annual average of 3 353 m^3 or 6.88 m^3 ha^{-1}
year^{-1}. During that same period the area affected by active
geomorphic processes increased at an average rate of 1.3 m^2
year^{-1}, and the rate of increase of the valley volume was
0.9% year^{-1}.

In contrast to the preceding period, the years 1971-1977
experienced high rainfall totals (950 mm of precipitation
between April and October), with one year (1975) featuring
extreme values. Under these conditions geomorphological processes were intensified, despite the erosion control works.
Thus in the 7 years analysed, 90 888 m^3 were eroded from the
basin, producing an average annual rate of 12 985 m^3, i.e.
26.65 m^3 ha^{-1}year^{-1}. Sixty two per cent of this quantity
was retained by the dams. The reservoirs were almost completely filled during the first two years after completion
(table 16.2). The material came largely from extensive landslides, which filled the channel in places.

Landslides and mudflows were periodically reactivated
every year. Their development was also assisted by the formation of temporary lakes behind the dams, which produced
conditions of excessive moisture at the base of the slopes.

NUMBER OF SECTION	NUMBER OF CROSS PROFILES	LENGTHS (m)	SEDIMENT YIELD (m³)					MEAN SEDIMENT YIELD (m³ yr⁻¹)				
			1960-1971	1971 - 1977				1960-1971	1971 - 1977			
				1971-1972	1972-1973	1973-1977	TOTAL 1971-1977		1971-1972	1972-1973	1973-1977	TOTAL 1971-1977
I a	P_{51}, P_{50}	235	2 805	2 490	1 647	2 472	6 609	225	2 490	1 647	618	1 101
II a	P_{49}	103	2 060	-721	-515	-	-1 236	187	-721	-515	-	-206
I b	P_{44}, P_{43}, P_{42}	260	11 960	6 110	5 824	8 736	20 670	1 087	6 110	5 824	2 184	3 445
II b+	P_{41}, P_{40}, P_{48},											
IIIa	P_{39}, P_{38}	574	-6 464	6 762	5 990	-	-772	588	-6 762	5 990	-	-129
IV	P_{37}, P_{36}, P_{33}	250	2 170	12 613	1 464	2 196	16 273	197	12 613	1 464	549	2 712
V	P_{32}	205	2 455	3 895	1 845	-	5 740	223	3 895	-	-	957
VI	P_{31}	25	1 525	175	175	-975	-620	139	175	175	-244	-103
VII	P_{30}, P_{29}, P_{28}, P_{27}, P_{26}, P_{25}, P_{24}, P_{23}	930	4 985	-108	16 290	24 432	40 614	435	-108	16 290	6 108	6 769
VIII	P_{22}, P_{21}, P_{20}, P_{18}	240	2 460	-1 840	5 450	-	3 610	224	-1 840	5 450	-	602
TOTAL		2 822	36 894	15 852	38 170	36 861	90 888	3 353	15 852	38 170	9 215	15 148

Table 16.1   Rates of erosion of the Drăghici valley 1960 – 1977.

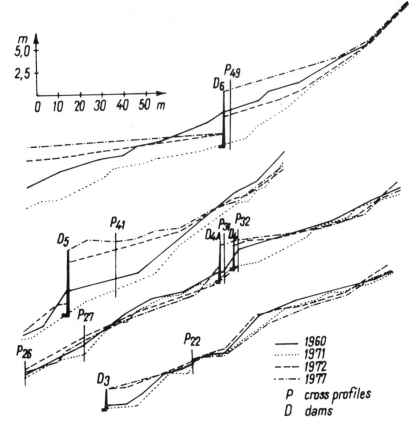

Figure 16.1   Draghici valley.   Changes of the long-profile
between 1960 and 1977.

The filling of the reservoirs decreased the slope of the
long profile and the water flow tended to meander, thus
eroding the base of the slopes.   In this way the cross-
sectional area of the valley increased at an average annual
rate 5-50 times higher than the one characterising the
period before improvement works had begun.   Considerable temporal variability in the quantity of eroded material was
apparent during this time: 90.5 m^3 $ha^{-1}year^{-1}$ in 1971-1972;
23.3 m^3 $ha^{-1}year^{-1}$ in 1972-1973; 18.9 m^3 $ha^{-1}year^{-1}$ in 1974-
1977.   However the quantity of sediment tended to decrease
as erosion control works were started.   This tendency was
particularly evident over the years 1973-1977 when the acacia
and poplar trees planted to stabilise the channel bed and
banks began to grow.

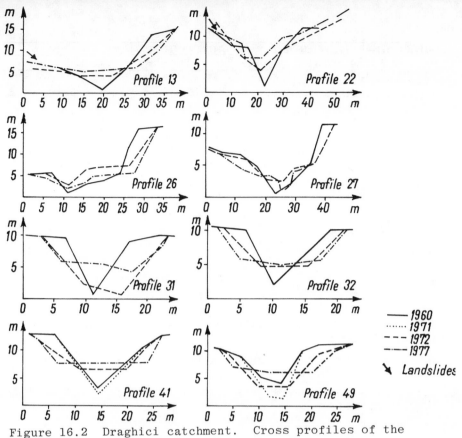

Figure 16.2   Draghici catchment.   Cross profiles of the
main valley.

THE TATAR BASIN

For comparative purposes, the response of a small catchment
(1960-1976) under similar conditions, but not subject to
conservation measures was studied.   The Tatar basin occupies
0.75· km² and is underlain by an alternation of sands, argillaceous sands, clays and sandy marls.   It is totally deforested, and is used as pasture land.   In 1960 80% of the
catchment was covered by stable colluvial deposits (old
landslides) dissected by a 5.5 km long gully network (Mihai
and Negut, 1981; Negut *et al*, 1980).

The observations made on this basin over a 16 year period
showed that the 441 440 m³ of sediment eroded corresponded
to an average of 367.86 m³ ha⁻¹year⁻¹ (table 16.3).   In this
study we again focused our research on the sources of the
material and the rate of operation of the geomorphological
processes.   The findings revealed the following:

(i)   In wet years large landslides and mudflows occurred,
filling the channel with sediment.   Sliding was intensified
in the middle and upper reaches of the basin producing an
average annual supply rate of 38.7 m³ and 21.8 m³ respectively, for every linear metre of channel.   Once in the chan-

NUMBER OF SECTION	NUMBER OF CROSS PROFILES	LENGTHS (m)	1971 – 1972 m^3	%ˣ	1972 – 1973 m^3	%ˣ	1974 – 1977 m^3	%ˣ	TOTAL m^3	%ˣ
I a	P_{51}, P_{50}	235	-	-	-	-	-	-	-	-
II a	P_{49}	103	8 532	79	2 088	20	-	-	10 440	99
I b	P_{44}, P_{43}, P_{42}	260	-	-	-	-	-	-	-	-
II b+ III a	P_{41}, P_{40}, P_{48}, P_{39}, P_{38}	570	26 843	46	6 697	11	-	-	35 540	57
IV	P_{37}, P_{36}, P_{33}	250	-	-	-	-	-	-	-	-
V	P_{32}	205	6 400	57	1 595	14	-	-	7 995	71
VI	P_{31}	25	-	-	-	-	1 540	62	1 540	62
VII	P_{30}, P_{29}, P_{28}, P_{27}, P_{26}, P_{25}, P_{24}, P_{23}	930	-	-	-	-	-	-	-	-
VIII	P_{22}, P_{21}, P_{20}, P_{18}	240	2 500	3	965	1	-	-	3 465	4
TOTAL		2 822	44 095	27	11 345	7	1 540	62	56 980	35

ˣ % of the original volume of the valley upstream of the dam.

Table 16.2   Sedimentation in reservoirs behind dams in the Drăghici catchment 1971 – 1979.

Table 16.3 Rates of erosion of the Tatar gully (1960–1977)

Name of the gully	Number of section	Volume of the gully 1960 (m³)	Mean sediment yield (m³year⁻¹)	Mean sediment yield/length 1976 (m³ m⁻¹year⁻¹)	Mean sediment yield/surface of the gully 1976 (m³ ha⁻¹year⁻¹)
Tataru	1	29 862	8 242	21,8	1 303
	2	12 117	2 234	12,3	676
	3	4 332	9 757	28,7	1 927
	4	5 698	2 091	11,0	1 231
	5	1 263	680	5,1	1 152
	Total	53 272	23 004	20,3	1 355
Tributary 1	1	10 317	784	2,5	516
	2	23 720	2 546	15,9	1 010
	Total	34 037	3 330	7,0	824
Tributary 2	1	7 950	230	2,1	478
	2	2 114	764	5,7	991
	3	5 327	262	1,6	515
	Total	15 391	1 256	3,0	714

nel some of the material was transported further by mudflows.

ii) In relatively dry years the non-consolidated material transported to the channel was removed by gullying. In
this way there is a permanent alternation between years
dominated by landslides and mudflows and years of gully erosion when sediments are discharged from the basin (figures
16.3 and 16.4).

iii) In the upper part of the catchment there was strong
regressive erosion accompanied by landslides and piping processes. The gully head retreated at an average rate of
5.4 m year^{-1} in the case of the main gully (86.4 m over the
16 years) and 3.7-1.3 m year^{-1} in the case of secondary gullies (table 16.4).

iv) In our 16 year observation period the area affected
by active sliding and gullying increased by an average of
0.9-1 ha year^{-1} from 7.4 ha in 1960 to 15.06 ha in 1976.
(table 16.5).

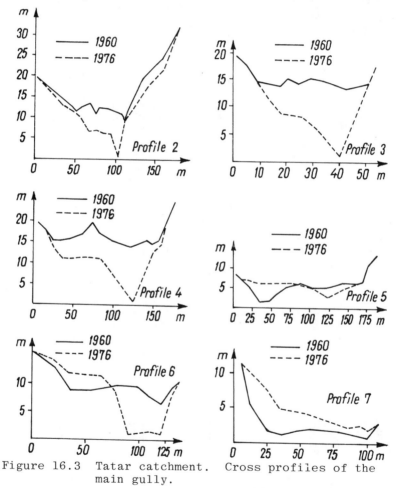

Figure 16.3 Tatar catchment. Cross profiles of the
            main gully.

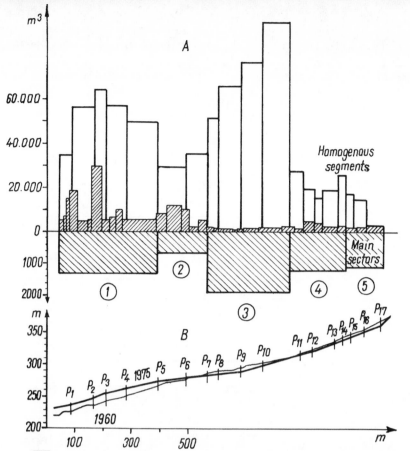

$1 \begin{smallmatrix} a \\ b \end{smallmatrix}$ ▨▢ The gully volume (m³) in (a) 1960, and in (b) 1976.

2 ▨ Ratio between mean gully volume/gully surface area in 1976 (m^3 ha^{-1} $year^{-1}$).

3 ▥ Cross profiles.

Figure 16.4   Tătar catchment: A) variation in gully volume, and in ratio between average volume and active area, 1960 - 1976; B) the long profile in 1960 and 1976.

Figure 16.4   Tătar catchment: C) stereoreturned orthogonal image.

Table 16.4   Tatar   Catchment. Gully head retreat 1960-1976

Name of the gully	Number of section	Length (m)		Gully head retreat (m year^{-1})
		1960	1976	
Tataru	1	378	378	-
	2	182	182	-
	3	252	252	-
	4	190	190	-
	5	46	132	5,4
	Total	1 048	1 134	5,4
Tributary 1	1	312	312	-
	2	100	160	3,7
	Total	412	472	3,7
Tributary 2	1	96	96	-
	2	108	130	1,4
	3	164	186	1,3
	Total	368	412	2,7

THE PORCAREATA BASIN

The Porcareata basin covers 1.3 km^2 and is developed though-
out on gravels and Quaternary sands. The main valley is a
gully with 10-30 m high vertical banks, and the channel is
filled with gravel and sands. In 1970 the lower sector of
the gully was stabilised and five dams (2.5-4 m high and
30-40 m wide) were built. Since the gravels underlying the
basin are pervious, only one or two floods are recorded each
year and these occur when precipitation exceeds 30 mm in
24 hours. Storms with smaller totals or with low intensities
infiltrate or are largely intercepted by the forest, since
the catchment is wholly afforested.
    In the period 1970-1974 seven floods eroded a total of
2 100 m^3 of sediment at an average rate of 550 m^3 year^{-1}
(4.23 m^3 ha^{-1}year^{-1}). An extreme flood caused by a storm
rainfall of 177 mm in 24 hours filled four reservoirs with
sediment, and deposited a layer of sediment 3.5 m thick in
the fifth reservoir. In the space of one day 14 895 m^3 of
sediment was discharged from the basin (114.6 m^3 ha^{-1}).
Sediment is transferred from the slopes to the channels by
isolated sand and gravel falls and by gravel avalanches
caused by freeze-thaw cycles and by drying and moistening
of the material. The rate of steep slope retreat was monitored by positioning wire nets at the base of the slope.
Slopes with a south and east orientation were found to

Table 16.5   Rates of gully surface changes

Name	Number of section	Surface 1960 (m^2)	Rate of gully surface increase ($m^2 year^{-1}$)	Rate of gully surface increase/ length in 1976 ($m^2\ m^{-1} year^{-1}$)
Tataru	1	16 853	2 900	7,7
	2	7 313	1 607	8,8
	3	5 952	2 793	11,1
	4	10 066	432	2,3
	5	1 662	265	2,0
	Total	41 846	7 997	7,1
Tributary 1	1	9 092	182	0,6
	2	12 530	793	5,0
	Total	21 622	975	2,1
Tributary 2	1	3 516	81	0,7
	2	2 286	339	2,5
	3	4 760	20	0,1
	Total	10 562	440	1,0

supply 3-4 times more material than slopes with a north or
west aspect. Once in the channel, the sediment is transported once or twice a year by floods, although the extreme
events are of greatest signifance. The sediment discharge
capacity of the channels is smaller than the quantity of
material supplied by the slopes.

CONCLUSIONS

In all· the basins studied, soil erosion rates are very high
and relate to a young region with marked endogenic and exogenic mobility of the relief. In all the catchments a great
variability of geomorphological processes in time and space
was found. Lithological contrasts are responsible for major
variations on the mechanisms of channel and slope material
movement. Temporal variation in annual precipitation cause
very significant differences in the annual rate of soil
erosion.
    The investigations emphasize the value of photogrammetric
techniques for geomorphological experiments in young regions
with a marked mobility of relief.

REFERENCES

Bălteanu, D., 1976. Some investigations on the present-day mass movements in the Buzău Subcarpathians. *Revue Roumanie de Géologie, Géophysique et Géographie, série de Géographie,* 20, 53-61.

Bălteanu, D. and Taloescu, I., 1978. Asupra evoluţiei ravenelor. Exemplificări din dealurile şi podişurile de la exteriorul Carpaţilor (On gully evolution in the outer-Carpathian hills and tablelands). *Studii şi Cercetări de Geografie,* 25, 43-54.

Mihai, G. and Negut, N., 1981. Observaţii preliminare privind evoluţia ravenelor formate pe alternanţe de orizonturi permeabile şi impermeabile (Some preliminary results on gully evolution in alternating permeable and impermeable strata). *Studii şi Cercetări de Geografie,* 28, 114-117.

Mihai, G., Taloescu, I. and Negut, N., 1979. Influenţa lucrărilor transversale asupra evoluţiei ravenelor formate pe alternanţe de orizonturi permeabile şi impermeabile (The influence of dams on gullies formed in alternating permeable and impermeable strata). *Buletinul Informativ al Academiei de Stiinte Agricole şi Silvice,* 8, 103-116.

Negut, N., Taloescu, I., Mihai, G. and Săvulescu, C., 1980. A photogrammetric solution for the detailed study of erosion processes and landslides. *14th Congress, International Society for Photogrammetry, Hamburg, Commission V.*

17 Landsliding, slope development and sediment yield in a temperate environment: Northeast Romania

Ionita Ichim, Virgil Surdeanu and Nicolae Radoane

BACKGROUND

Romania is traversed by latitude 45 degrees north and longitude 25 degrees east (figure 17.1), and has a temperate-continental position. The Carpathian Mountains provide altitudinal zonation from steppe to alpine domains. Erosion rates have been documented by evaluating the sediment yields from individual drainage basins and by studying specific processes of slope erosion. Analysis has focussed on the seasonal cyclic character of both measures of erosion rate; the dynamics of slopes with landslides or gullies (basic forms of linear erosion); and the identification of several types of temperate slope development.

DATA COLLECTION

Sediment yield data have been obtained from measurements of suspended sediment load on 200 rivers draining catchments of more than 150 km^2 and with 15 years or more of record. Slope erosion data are based on 15 years of fieldwork on gullies and experimental plots in the forest steppe of the Moldavian Plateau, and 6 years of fieldwork in the forest zone of the East Carpathians. The dynamics of landslides have been investigated from 1973 to 1980 in nine small catchments (between 0.86 and 2.27 km^2 in area) in the Bistriţa valley, by repeated topographic surveys and photographs, and by measurements of the deformation of tubes inserted in the landslides. In each case, precipitation, temperature and soil humidity have been registered, and certain physical indices were determined for the landslide deposits.

INTERPRETATION OF RESULTS

Sediment yields are greatest in Spring (figure 17.1). The precise date varies with altitude, and ranges from March in steppe, forest-steppe and hill-forest zones, through April, in mountain forests below 1000 m, to May in mountains

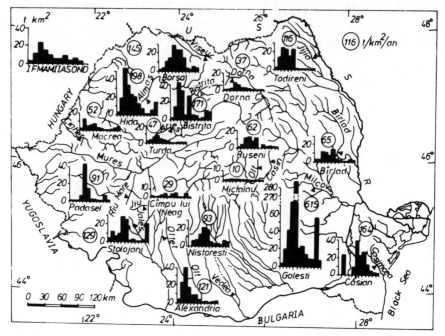

Figure 17.1. Variations in monthly sediment yield (t km^{-2})
    for 17 selected stations in Romania.

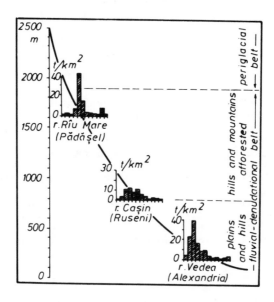

Figure 17.2. Examples showing the effect of altitude on
    monthly sediment yield in Romania.

above 1000-1200 m (figure 17.2). In the plains and hills, the critical season for slope (wash) erosion is from May to the beginning of July, and this cannot explain the March peak in sediment yield. Instead, we believe that this maximum reflects not intense slope erosion, but the prolonged continuity of discharge in gullies, provided by snow melt and soil thaw reinforced by the Spring period of frontal rains. Furthermore, thawing greatly increases susceptibility to erosion, as has been clearly shown by experimental data from agricultural fields (table 17.1). Although the processes of slope erosion are more effective during the critical season, intense rainstorms usually cover small areas. This means that only locally is much sediment carried into the rivers, and that the effect is not spread over large drainage basins as with the Spring erosion.

The average annual sediment yield for drainage basins larger than 150 km^2 varies from 15 to 4000 t km^{-2}. In the mountain forest zone, slope erosion is usually less than sediment yield, except where mass movements are important. This difference may reflect the importance of river beds as sources of sediment. The tendency here is to form <u>basal slopes convex in profile</u>.

Conversely, in hill regions sediment yield is much less than slope erosion: at the scale of a Strahler 6th order basin, the ratio is 1/18 to 1/20. Hence there is a tendency to form <u>concave footslopes of accumulation</u> in these hill regions. Both of these tendencies can be considered typical for temperate-continental climatic conditions, in the absence of disturbing factors such as landslides.

Extensive regions in Romania, especially in the Subcarpathian region, have annual sediment yields in excess of 1000 t km^{-2}. This stems from intense deforestation in the 19th and 20th centuries. Since 1944 a major program of reforestation and erosion control measures has been implemented. Without neglecting the negative effect of deforestation, recent research leads to the conclusion that <u>landslides are the main factor in increasing sediment yield</u>. Their distribution is governed by lithologic and local conditions, rather than by climate, and they may therefore be considered as an azonal influence on the variation of sediment yield.

Two examples serve to illustrate the effect of landslides on slope dynamics and the transfer of material to rivers. Both are from the Flysch mountain region of the Bistriţa valley. Figures 17.3 and 17.4 indicate the main characteristics of the materials.

In the Buba basin, the upper part of the slope material was displaced, from 1973 to 1980, at speeds of up to 19 m yr^{-1} (figure 17.3). This amounts to at least 40 or 60 m^3 yr^{-1} per metre of slope width. The most important displacements occurred in the middle of the slope profile. In the lower part the speed reached almost 3 m yr^{-1} which, considering the thickness of material moving, is equivalent to at least 9 m^3 m^{-1} yr^{-1}. Thus it is clear that in the absence of a river (or waves) periodically removing material from the slope foot, the latter will be aggraded by deposition of landslide debris. This is a frequent mode of blocking of small valleys by landslide deposits.

Table 17.1. Coefficients of runoff (percentage of precipitation) and soil erosion (t ha^{-1}) for torrential rains on six days in 1969, on slopes steeper than 20%, (Based on Popa, 1977)

Date 1969	COEFFICIENT OF RUNOFF, %					SOIL EROSION, t ha^{-1}					Soil moisture status before rain
	Grass 1st yr.	Grass 2nd yr.	Wheat	Maize	Peas	Grass 1st yr.	Grass 2nd yr.	Wheat	Maize	Peas	
15.II	41	37	94	81	86	8.3	7.4	48.1	26.2	25.2	frozen
17.II	35	29	96	97	93	7.5	8.1	59.4	35.0	34.7	frozen
27.III	40	5	75	68	65	9.1	7.9	18.0	14.1	15.1	wet
8.VI	27	12	38	23	55	10.2	0.2	8.1	5.2	25.9	wet
1.VII	40	1	42	50	79	14.3	0.5	12.4	15.1	38.2	wet
12.VII	35	14	27	39	78	30.4	4.1	26.4	31.5	149.7	v. wet

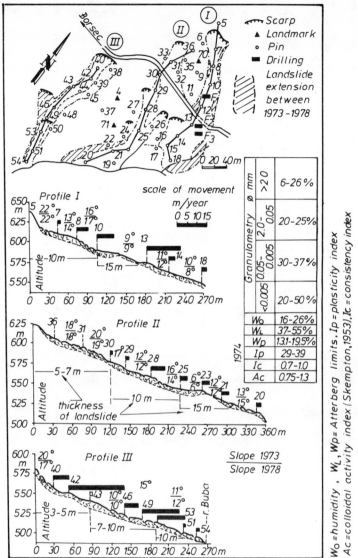

Figure 17.3. Landslides at Buba: the River Buba is at the base of the slope.

On the slope of Huiduman, during the same 1973-80 period, mean rates of movement did not exceed 4 m yr^{-1}. The greatest average volumetric rate of movement was 35 to 40 m^3 m^{-1} yr^{-1}, with an evacuation of 25 m^3 m^{-1} yr^{-1} at the base. (This more rapid evacuation was achieved by coastal erosion on the Izvoru Muntelui Reservoir at the slope foot.) In both these examples, modifications due to landsliding are visible both in the shape and in the gradient of the slopes (figure 17.3 and 17.4). The rate of landsliding is directly influenced by seasonal variations in

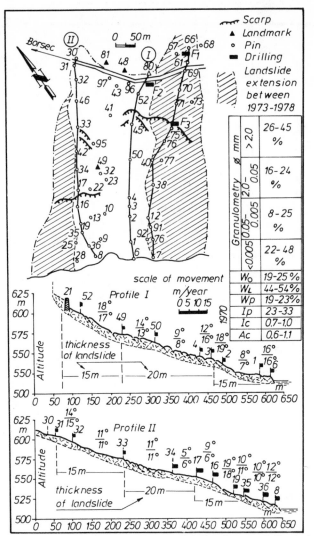

Figure 17.4. The landslide at Huiduman: the Izvoru Muntelui Reservoir is at the base of the slope.

precipitation: greater movements occur in the warm season (figures 17.5 and 17.6).

In the mountain forest zone, on slopes without important mass movements, river sediment comes mainly from gullies. Measurements on experimental plots undertaken between 1976 and 1980 showed that diffuse erosion of the slope surface provided only 1% of the sediment removed from the 18 km[2] Pingaraţi basin. In contrast, gullying plays a very important role. Thus it is concluded that the slopes are being dissected by the continued development of gullies.

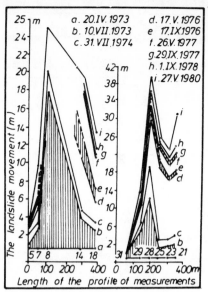

Figure 17.5. Surface displacement as a function of distance downslope, Buba landslide, 1973-78 (cumulative).

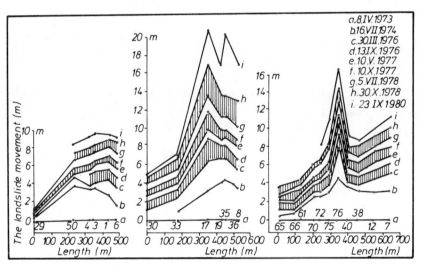

Figure 17.6. Surface displacement as a function of distance downslope, Huiduman landslide, 1973-78 (cumulative).

During the cold season (December-April, with freeze-thaw) the channel bed is aggraded by material detached from the channel sides by gelifraction. During spring and the warm season, when snowmelt starts and runoff increases, the channels are enlarged (figure 17.7). Hence channel development is essentially restricted to the warm season, when erosion exceeds accumulation. On average, 0.35 m³ yr⁻¹ is removed per metre of gully length, so that a 70 m long gully yields 25 m³ yr⁻¹.

295

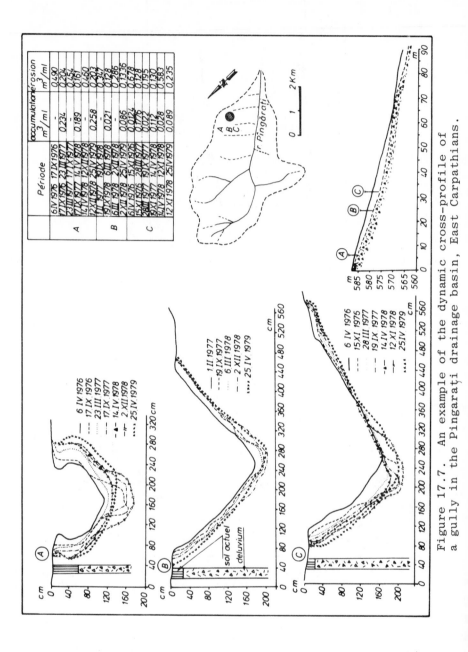

Figure 17.7. An example of the dynamic cross-profile of
a gully in the Pingaraţi drainage basin, East Carpathians.

296

In the Pingaraţi river basin, similar observations give rates up to $0.80 \text{ m}^3 \text{ m}^{-1} \text{ yr}^{-1}$, but the average is between 0.05 and $0.20 \text{ m}^3 \text{ m}^{-1} \text{ yr}^{-1}$.

CONCLUSIONS

In the temperate zone of Romania, landslides are azonal processes of slope development, and introduce a perturbing factor into the variation of sediment yield. At least two zonal types of slope development can be identified for temperate-continental climates - slopes with convex profiles at the base, in first-order valleys of the forest zone (belt) of the mountain region; and slopes with concave profiles at the base resulting from the development of accumulation surfaces (glacis), in hill regions of the forest and forest steppe zones.

(Translated by I.S. Evans, from the French.)

REFERENCES

Ichim, I., 1980, Present tendencies in the dynamics of Romania's relief, *I.G.U., Commission on Field Experiments in Geomorphology, Meeting of Commission, Kyoto, August.*

Popa, A., 1977, Cercetări privind erofuinia si măsurile de combatire a croziunii pe terenuril agricole div Podisul Moldovenesc, MAIA, ASAS, Bucuresti.

Radoane, N., 1980, Contributii la cunoasterea unor procese torenţiale din bazinul rîului Pîngaraţi, în perioada 1976-1979, *St. cerc., geol. geofiz., geografie., seria Geografie*, XXVII, 53-63, (Bucuresti).

Skempton, A.W., 1953, Soil mechanics in relation to geology, *Proceedings Yorkshire Geological Society*, 29, 33-62.

Surdeanu, V., 1980, Récherches expérimentales de terrain sur les glissements, *Studia Geomorphologica Carpatho-Balcanica*, 15, (in press).

18 Development of field techniques for assessment of river erosion and deposition in mid-Wales, UK

Graham J. Leeks

INTRODUCTION

Many costly river training and bank protection works bear witness to man's frequent failure to successfully manage river sediment systems. Problems in the effective application of existing theories, laboratory results and localised empirical studies to real world rivers are repeatedly stressed by those burdened by management responsibility.

The Institute of Hydrology is carrying out a five year programme of research to generate and analyse a data base on river erosion and deposition in mid-Wales, using techniques which could be easily applied in regional studies elsewhere in Britain. The mid-Wales landscape can be divided into Upland extending up to 600 m, which is dominated by plateaux incised by steep glacially modified valleys with peat and podsol soils, and Piedmont characterised by floodplain development and partial storage of river sediments with mainly brown earth soils. Rock types are chiefly Ordovician and Silurian shales and mudstones, and these are the main constituents of river gravels and cobbles derived both from solid rock exposures and from fluvio-glacial and peri-glacial landforms. The project is designed to be of relevance to land use practice and particularly to riparian farmers.

The research effort is split into Piedmont/Upland areas on an environmental basis, and into local reach/regional network studies on the basis of scale. One aim of the project has been to extend as far as possible across the range between these idealised extremes in order to produce results applicable to real world erosion and deposition problems. There is arguably a need to extend from 'research catchment scales' (eg. of 0.1 km^2 - 10 km^2) up to 'practical catchment scales' (of 10 km^2 upwards). Following localised process studies, there is a need to study gross processes which can be related to gross form. To this end the study includes large areas of the five main catchments of mid-Wales; namely, the Severn (580 km^2), Wye (1280 km^2), Ystwyth (170 km^2), Rheidol (182 km^2) and Dyfi (471 km^2).

The 'whole river approach' is justified as a means of investigating the hypothesis that upland land use changes such as afforestation and pasture improvement (which involve changing the vegetation cover and extending drainage nets) will lead to significant changes down river, where river regulation and lowland farming practices may also have significant influences. The general principle that the dynamics of a river sediment system at a particular cross-section may relate to both local and up-river factors is often ignored by riparian owners in dealing with perceived erosion and/or deposition problems at-a-site. In the paper some of the instrumentation and techniques developed over the last two years are described and the uses of data obtained within the context of the project aims are briefly reviewed.

TOPOGRAPHIC SURVEY

Reconnaissance work has led to the selection of 35 sample sites of significant erosion and/or deposition (see figure 18.1). These sites form the regional network. The geomorphological features which are being monitored include shoals, low river banks (0.1 m to 3.0 m high) and, with most difficulty, high river cliffs (eg. 3-20 m high). The topographic survey of these features is carried out by use of Infra-red Electronic Distance Measurement (E.D.M.) equipment, and a theodolite. Data on orientation, angle of slope and distance, can be input directly into a solid state memory. This is then connected directly to a microcomputer to provide survey plans, sections and a variety of derived indices including bed and water surface gradients. The accuracy of measurements is ± 5 mm + 5 ppm over distances of up to 1000 m using one reflector. The E.D.M. allows more rapid and accurate survey than is possible with tapes, levels or optical range finders. It is also remote from the bank and therefore does not cause erosion, as could be the case if erosion pins were used. In the case of high river cliffs, composed of vertical or ·near vertical loosely consolidated fluvioglacial material, distance is measured from a fixed point to a prism reflector set on a pole placed flat against the cliff face. Accurate relocation of the instrument is necessary for repeated survey and for this purpose steel rods are pushed flush into the ground (to reduce disturbance by sheep, cattle and humans and to prevent damage to agricultural implements). These are relocated with a metal detector.

For profiles of lower river banks a pro-forma device is used, with distance to the bank face measured perpendicular to a straight beam which is inclined down the bank from a fixed point on the bank top.

Shoals are surveyed using the E.D.M. and theodolite to give longitudinal, cross-sectional and planform topographic data.

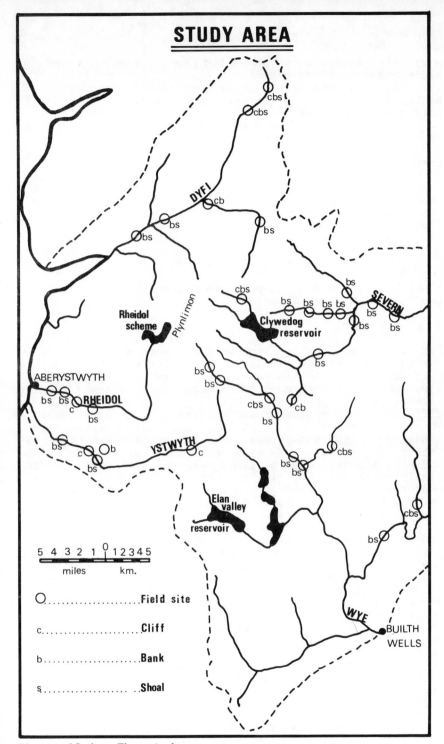

Figure 18.1. The study area.

SEDIMENT SAMPLING

A variety of techniques are being used to sample river
sediments. Sampling schemes which are longitudinally and
laterally representative, sufficient in size, and of
hydraulic significance are not easy to design. At river
cliffs and banks core samples are taken at the free face,
and bulk samples of debris are collected at the cliff base
for size analysis. In the case of shoals, systematic grid
point and bulk samples (both surface and immediately subsurface) are collected. In addition a large number of
35 mm photographs are taken to provide an additional data
source as described by Adams (1979). Comparative tests of
these sampling methods are being undertaken. A summary of
the sampling techniques is provided by table 18.1 and
figure 18.2).

SEDIMENT TRANSPORT

Studies of sediment transport are concentrated upon suspended and bed load movement.

Suspended load

Sampling strategies for suspended sediment monitoring have
varied in accordance with instrumentation and the magnitude
of flow events. A summary of these strategies is shown in
figure 18.3. It has been pointed out several times in past
literature (eg. Leopold, Wolman and Miller, 1964) that
different samplers can produce different concentrations.
There is therefore a need to inter-calibrate measurement
techniques. For the purposes of this project, US DH depth-
integrating samplers are used as a standard, as they are
the most widely accepted. They have the advantage of being
highly portable, and allow collection of samples representative of a whole cross-section. They can be used
quickly at a large number of sites during high flows or on
routine weekly runs, and at a smaller number of sites
during lower flow conditions, bearing in mind the need to
provide calibration with other instrumentation at certain
sites. The main problem in the use of US DH samplers is
the need for laboratory analysis of sample concentrations
which can be time-consuming. The record provided of variation in suspended sediment discharge through time is also
discontinuous. Automatic vacuum samplers can reduce the
labour input in the field. The 'Northants' sampler has
been found useful for collecting bottle samples at frequencies ranging from 2.5 min. to 8 hours and can be automatically triggered during high flow events. It is particularly useful in upland catchments where response-times
to heavy rain are sometimes less than 30 min. and it is
therefore difficult to deploy other instruments sufficiently
rapidly. However, the need for time-consuming laboratory
filtration remains.
     An alternative to collection of bulk samples is the use
of turbidity monitors. Portable instruments based on
nephelometric or absorptiometric principles can be used as

Table 18.1. Field and laboratory sampling and analysis of sediment from survey sites.

Sampling site	No. of samples	No. of Pebbles per sample	Method of Sampling				Sieve and sieve equivalent	Photo-graphic
			Random	Stratified	Systematic	Total Pop.		
SHOAL								
(a) Surface m² at Principal Point	1	(i)10 (ii)Bulk	(i)✓			(ii)✓	(i)✓	✓
(b) Near Surface m² at Principal Point	1	Bulk	✓			✓	✓	✓
(c) Down-stream, Up-stream, Bankwise and Streamwise from Principal Point	10	10	✓		✓		✓	✓
(d) Channel bed	1	100	✓				✓	
(e) Opposite bank	1	Bulk		✓			✓	
RIVER CLIFF								
(f) Free face	1	Bulk		✓			✓	
(g) Debris Slope	1	Bulk		✓			✓	
(h) Channel Bed	1	100	✓				✓	
BANK								
(i) Free face	1	Bulk		✓			✓	
(j) Debris slope	1	Bulk		✓			✓	
(k) Channel Bed	1	100	✓				✓	

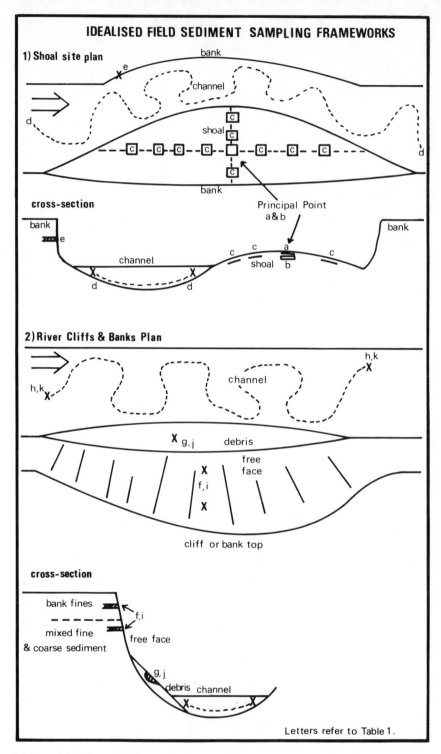

Figure 18.2.   Idealised field sediment sampling frameworks.

SUSPENDED LOAD MONITORING – SAMPLING STRATEGIES

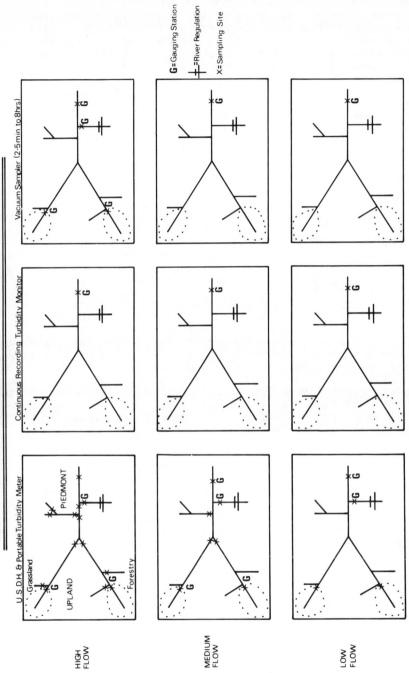

Figure 18.3.   Sampling strategies for monitoring suspended sediment loads.

305

a faster alternative to US DH samplers for flood monitoring.
Coupled with a recorder, they have proved of value at downstream sites where discharge gauging stations are also
available. They do not, however, provide a concentration
representative of the whole cross-section and a correction
factor based on values obtained using depth-integrating
US DH samplers is considered useful. The eventual aim of
this work is to derive estimates of suspended sediment
yield by use of rating curve methods for which there is at
present apparently no alternative when dealing with a large
number of sample sites without automatic samplers or continuous monitoring equipment. The main sub-catchments
contributing suspended load during floods can therefore be
identified.

Bed load

Work at the Institute of Hydrology upland experimental
catchments at Plynlimon has shown that bed load yields from
traps are much higher in catchments subjected to dense open-
ditching, as is often carried out as part of afforestation
and pasture land improvement (Newson, 1980). This suggests
that the sediment input into the downstream Piedmont may
also be increased as a result of land use change. Unfortunately, sediment trapping becomes more difficult on
the larger rivers of the Piedmont zone and a practical
limit is reached for bed load traps. Additional data are
provided by repeated topographical surveys of hollows in
the river beds created by farmers excavating for gravel.
These hollows can be regarded as bed load traps of varying
efficiency and co-operation from several farmers has
permitted calculation of minimum bed load transport rates.
An example is given from the Afon Llwyd, a tributary of the
River Severn, in table 18.2 and figure 18.4. However, bed
load data are not comparable in number, consistency or
extent with those for suspended load.

SEDIMENT TRACING

The deficiencies in the bed load data mentioned above are,
to some extent, filled by the development of magnetic bed
load tracing. The Institute of Hydrology, in close association with F. Oldfield and B. Arkell of the University of
Liverpool, UK, have carried out three bed load tracing
experiments in the Wye and Severn catchments with the
objectives of defining threshold conditions for bed load
movement, distance travelled and volume of bed load movement (Arkell et al., 1983). Natural bed material has
been taken from the rivers and heated to temperatures of
1100°C to enhance magnetic susceptibility. It is then
replaced in the stream and the spatial distribution of
magnetically enhanced particles is measured before and
after high flows. Advantages of the technique include
preservation of the original density, size, shape and
texture of bed material, and the ease of detection with
cheap commercially available metal detectors or specially
designed susceptibility bridges, with which numerical

306

TABLE 18.2 Transects of Afon Llwyd Shoal taken after the excavation of ~200t of gravel, August 1979.

PERIOD	FLOODS date	peak Q ($m^3 s^{-1}$)		TRANSECT 1	2	3	4	5	6	TOTAL WEIGHT LOSS OR GAIN *
13.09.79-17.10.79	26.09.79	11.0	Erosion	0.15	-	2.0	0.85	1.7	2.5	+14.7t
			Deposition	0.725	7.9	1.4	0.23	0.2	-	
			NET / VOL	d 0.575 / 3.45	d 7.9 / 31.6	e 0.6 / 2.4	e 0.62 / 2.48	e 1.5 / 12.0	e 2.5 / 10.0	
17.10.79-20.11.79	03.11.79 / 05.11.79	12.0 / 30.0	Erosion	0.05	3.4	2.8	-	1.5	-	+13.7t
			Deposition	0.35	-	2.4	7.5	-	0.75	
			NET / VOL	d 0.30 / 1.80	e 3.4 / 13.6	e 0.4 / 1.6	d 7.5 / 30.0	e 1.5 / 12.0	d 0.75 / 3.0	
20.11.79-20.12.79	29.11.79 / 5.12.79	10.5 / 90.0	Erosion	0.05	0.4	0.1	0.25	0.3	0.2	+144.7t
			Deposition	0.45	1.9	4.6	6.5	3.8	0.45	
			NET / VOL	d 0.40 / 2.40	d 1.5 / 6.0	d 4.5 / 18.0	d 6.25 / 25.0	d 3.5 / 28.0	d 0.25 / 1.0	
20.12.79-24.04.80	26.12.79 / 27.12.79 / 05.02.80 / 08.02.80	10.0 / 24.5 / 16.0 / 16.0	Erosion	0.90	-	1.075	3.675	1.55	1.1	-32.8t
			Deposition	-	4.9	1.4	-	-	-	
			NET / VOL	e 0.90 / 5.40	d 4.9 / 19.6	d 0.325 / 1.3	e 3.675 / 14.7	e 1.55 / 12.4	e 1.1 / 6.6	
24.04.80-27.08.80			Erosion	-	5.9	-	-	-	0.2	+34.0t
			Deposition	1.15	-	1.6	2.5	2.1	0.8	
			NET / VOL	d 1.15 / 6.90	e 5.9 / 23.6	d 1.6 / 6.4	d 2.5 / 10.0	d 2.1 / 16.8	d 0.6 / 2.4	
27.08.80-31.10.80	6.10.80 / 26.10.80	90.0 / 10.0	Erosion	0.725	0.175	0.225	4.1	0.175	-	-11.9t
			Deposition	-	0.05	0.175	-	1.6	0.9	
			NET / VOL	e 0.725 / 4.35	e 0.17 / 0.68	e 0.05 / 0.2	e 4.1 / 16.4	d 1.425 / 11.4	d 0.9 / 3.6	

$$+152.4t = 19.05t\ km^{-2}\ yr^{-1}$$

NET = Net erosion or deposition along transect (m^2)
VOL = Total volume of erosion or deposition in reach of channel (m^3)
d = deposition   e = erosion

*assuming a density conversion factor of $1.8tm^{-3}$

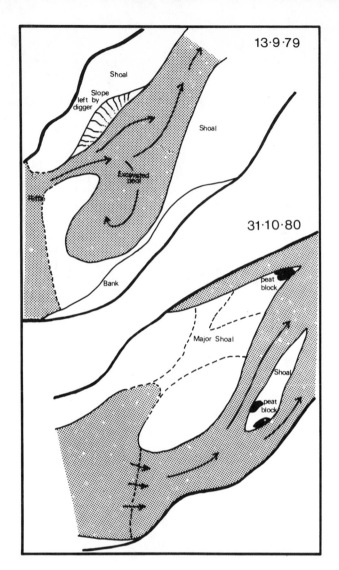

Figure 18.4.   The bed load monitoring site on the Afon
    Llwyd.

values can be assigned to the magnetic properties of
individual particles or areas of the stream bed.
    It was found possible to trace magnetised sediment up
to 500 m downstream of the injection site and the method
is proving particularly successful in tracing the movement
of different size particles across shoal features in the
downstream zone.

CONCLUDING COMMENTS

The emphasis of this paper has been on rapid techniques which can be widely deployed with a minimal labour input. Many public authorities (eg. Highway authorities) whose responsibilities occasionally have to include interest in river channel stability, appear to show little perception of the dynamics of the river sediment system until a critical failure occurs, which leads to large expenditure on corrective measures. It is hoped that some of the suite of techniques outlined will encourage greater interest in routine monitoring of the river sediment system, which is often ignored at great cost.

ACKNOWLEDGEMENT

This paper is based on work carried out by the Institute of Hydrology under contract to the Ministry of Agriculture, Fisheries and Food.

REFERENCES

Adams, J., 1979, Gravel size analysis from photographs. *Journal of the Hydraulics Division,* ASCE, 104, 1247-1255.

Arkell, B., Leeks, G.J.L., Newson, M.D. and Oldfield, F., 1983, Trapping and tracing: Some recent observations of supply and transport of coarse sediment from upland Wales, in: *Modern and Ancient Fluvial Systems,* ed. Collinson, J.D. and Lewin, J., Spec. Publs. Int. Ass. Sediment., 6, (Blackwell, Oxford), 107-119.

Leopold, L.B., Wolman, M.G. and Miller, J.P., 1964, *Fluvial processes in geomorphology,* (Freeman, San Francisco).

Newson, M.D., 1980, The erosion of drainage ditches and its effect on bed load yields in mid-Wales: Reconnaissance case studies, *Earth Surface Processes,* 5, 275-290.

Newson, M.D. and Leeks, G.J.L., 1980, *Annual Report to Ministry of Agriculture, Fisheries and Food, I.H. Project 73, Erosion and Deposition by rivers,* 42pp.

19 Suspended sediment properties

and their geomorphological significance

D.E. Walling and P. Kane

INTRODUCTION

Measurements of the suspended sediment loads of rivers
occupy an important position in the study of fluvial geomorphology, both at the scale of the small catchment and
the larger river basin. Such measurements can be traced
back to 1846 on the Mississippi (Nordin, personal communication), to 1863 on the Seine and Marne in France (Tixeront,
1974) and to 1874 on the Nile (Ball, 1939). Since these
early beginnings, measuring activity has expanded greatly
and has been paralleled by improvement and standardisation
of basic sampling equipment (eg. FIASP, 1963) and by the
development of automatic sampling apparatus and continuous
monitoring equipment (eg. IAHS, 1981). A large body of
information relating to the magnitude of suspended sediment
loads now exists, and this has provided an improved knowledge of both the suspended sediment loads transported by
the major rivers of the world and spatial and temporal
variations in the suspended sediment yields of small
drainage basins. Such data have proved to be of considerable value to the geomorphologist in, for example, docu-
menting rates of erosion and their control by climatic and
physiographic conditions (eg. Dunne, 1979), establishing
the magnitude and frequency characteristics of geomorphological processes (eg. Webb and Walling, 1982), studying
the relationship between erosion and sediment yield at the
basin scale (eg. Meade, 1982) and assessing the significance
of sediment loads to channel morphology (eg. Jansen, 1979).
   In more recent years, concern for non-point pollution
and contaminant transport within the fluvial system has
directed attention to suspended sediment-associated transport of nutrients and contaminants (eg. Shear and Watson,
1977; Allan, 1979; Knisel, 1980). This has in turn highlighted the importance of the particulate component in the
total transport of certain elements such as phosphorus and
heavy metals (eg. Förstner and Wittman, 1981) and the
significance of sediment properties in influencing sediment-
associated transport (eg. Ongley, 1982). Properties such
as particle-size, mineralogy, surface chemistry and organic

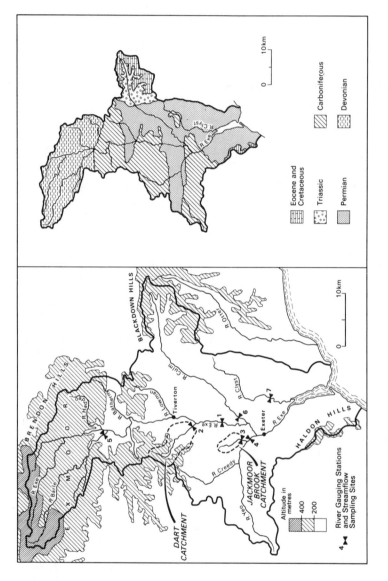

Figure 19.1. The study catchment showing the location of the suspended sediment sampling sites (cf table 19.1), and the generalised geology of the area.

312

matter content all exert important controls on the composition of the particulate load and the ability of the sediment to adsorb and release individual contaminants (eg.
Walling and Kane, 1982; Peart and Walling, 1982; Ongley
et al., 1981). This aspect of sediment loadings constitutes
a new dimension in the study and analysis of suspended
sediment transport. Attention is no longer restricted to
the magnitude of suspended sediment yields, the composition
and properties or quality of the sediment must also be
considered. To meet these demands, and more particularly
to provide bulk samples of suspended sediment for subsequent
analysis, new sampling techniques are now being developed
(eg. Ongley et al., 1981).

It is suggested that these new developments in the
study of suspended sediment transport have important implications for the geomorphologist and that he can profitably
broaden his own perspective to include the properties of
the sediment as well as the magnitude of the transported
loads. The potential significance of studies of sediment
properties to interpreting the basin sediment system, and
therefore to catchment studies, will be illustrated in this
contribution which presents the results of work undertaken
by the authors in the Exe basin, Devon, UK during the past
few years.

THE EXE BASIN INVESTIGATION

The Exe basin investigation represents an ongoing study of
suspended sediment and solute transport within a 1500 km^2
area of Devon, UK (figure 19.1) (cf. Walling, 1978; Walling
and Webb, 1981; Webb and Walling, 1982). Initially attention focussed on magnitude aspects of suspended sediment
generation, conveyance and yield, but more recently the
study has been broadened to include the properties of the
sediment transported by the River Exe and its various
tributaries. More traditional monitoring of suspended
sediment loads at a number of gauging sites (figure 19.1,
table 19.1) using manual sampling and continuous recording
turbidity meters has been complemented by investigation of
the nature of the particulate material.

Table 19.1.   The measuring sites.

River (site number figure 19.1)	Catchment area (km^2)	Estimated mean annual suspended sediment yield (t km^{-2} year^{-1})
Exe at Thorverton (1)	601.0	25
Creedy at Cowley (4)	262.0	50
Dart at Bickleigh (2)	46.0	70
Clyst at Clyst Honiton (7)	98.2	50
Culm at Rewe (6)	273.0	40
Jackmoor Brook (3)	9.8	50

Field and laboratory techniques

Two approaches have been used for collecting samples of
suspended sediment for laboratory analysis of sediment geochemistry. The first involved the collection of bulk
(>100 l) storm water samples in polyethylene containers
using a submersible pump suspended from a bridge. Sediment
was separated from the samples by continuous flow centri-
fugation in the laboratory and the sediment recovered was
freeze-dried for storage prior to analysis. The centrifuge
units have been shown to provide highly efficient recovery
of suspended sediment >0.45 μm and the bulk samples typically provided between 30 and 250 g of sediment depending
on the ambient suspended sediment concentration. The second
method made use of the 400 ml water samples collected by
manual and automatic sampling equipment from the various
measuring stations. The sediment was recovered from these
samples by sedimentation and decantation and was subsequently freeze-dried. Yields ranged between 0.1 and 1.0 g
per bottle.
    Use of the former method was essential to obtain suffi-
cient sediment for a large number of disparate laboratory
analyses, and bulk water samples were collected from all
sites over a wide range of flood conditions. The second
method possessed the advantage of automatic sample collection at hourly intervals at sites where such equipment was
installed, but was restricted in its usefulness by the
small quantities of sediment involved. These restricted
the scope of subsequent laboratory analysis.
    Laboratory analysis of sediment samples involved
standard laboratory procedures which included particle-size
analysis using Coulter Counter and Sedigraph apparatus, and
determination of major element concentrations by X-ray
fluorescence (XRF), clay mineralogy by X-ray diffraction
(XRD), C and N concentrations by pyrolysis/thermal conductivity, P by chemical fractionation/UV spectroscopy, Fe by
chemical fractionation/atomic absorption spectroscopy and
^{137}Cs by gamma spectroscopy.

Sediment composition

Background information on the major element and normative
composition, clay mineralogy, and typical particle-size
characteristics of suspended sediment collected from six
sampling sites is presented in tables 19.2, 19.3, 19.4 and
19.5. In terms of normative composition, the phyllosilicate, quartz, iron oxide and organic matter components
account for >99% of the total weight. Quartz is predominantly silt-sized and is rare in the clay (<2 μm) fraction.
The total phyllosilicate component is not directly equivalent to the total clay content, because some mica occurs
in the silt fraction. The particle-size data relate to the
dispersed mineral fraction and demonstrate that fine-
grained sediment predominates, with more than 75% of the
load being composed of particles <10 μm. The data on major
element composition contained in table 19.4 represents
average values for samples collected over a wide range of
flow conditions and reflect the dominance of the clay

Table 19.2. Normative composition of suspended sediment

| River | Composition in % of freeze-dried sediment | | | | | |
| | Phyllosilicates | | | Quartz | Iron oxides | Organic matter |
	Mica	Others	Total			
Exe at Thorverton	32.2	25.0	57.2	27.2	3.0	12.6
Creedy at Cowley	31.1	30.4	61.5	24.9	4.0	9.6
Dart at Bickleigh	25.8	25.4	51.2	34.4	3.0	11.4
Clyst at Clyst Honiton	45.7	20.0	65.7	21.3	4.5	8.5
Culm at Rewe	37.2	18.8	56.0	24.1	4.1	15.8
Jackmoor Brook	38.1	24.6	62.7	25.2	3.7	8.4

Table 19.3. Mean particle-size composition of suspended sediment

| River | Particle size distribution (%)* | | | |
	<0.5 μm	0.5-2 μm	2-10 μm	10-62 μm
Exe at Thorverton	19	22	38	21
Creedy at Cowley	29	24	33	14
Dart at Bickleigh	26	19	31	24
Clyst at Clyst Honiton	39	33	22	6
Culm at Rewe	32	29	29	10
Jackmoor Brook	54	26	15	5

* Particle size data relate to the chemically dispersed mineral fraction which has an effective upper limit of 62 μm.

mineral fraction and its chemical uniformity. Thus most of the Al, K and Mg is attributable to the clay, whilst the Si is allocated between the clay and quartz components. Some 60% of the iron is 'free' iron with between 20% and 30% being organically bound or in an inorganic 'amorphous' form, as is most of the Mn. The Ti content is attributable to fine-grained rutile and anatase intimately associated with mica. Within the clay fraction, four major clay mineral groups have been identified. Mica is the dominant group in all the rivers, though a mixed-layer mica-vermiculite component is present in substantial but variable amounts. Kaolinite and chlorite are present in small quantities.

The significance and value of studying spatial and temporal variations in these and other properties can be illustrated by considering, firstly, the particle-size characteristics of fluvial suspended sediment and particularly 'in stream' behaviour; secondly, the relationship of the nature of the material leaving the basin to that of

Table 19.4. Major element composition of
suspended sediment

River		Concentration in % of freeze-dried sediment							
		Fe	Mn	Ti	Ca[1]	K[1]	Si	Al	Mg[1]
Exe	mean	3.61	0.20	0.46	0.25	2.14	24.74	7.62	2.25
	c.v.[2]	7	21	5	35	6	3	7	33
Dart	mean	3.14	0.16	0.47	0.23	1.71	26.41	6.75	1.97
	c.v.	10	10	4	19	5	4	5	18
Creedy	mean	3.78	0.17	0.47	0.32	2.06	24.62	8.09	2.63
	c.v.	8	23	3	17	8	2	8	14
Clyst	mean	3.81	0.15	0.46	0.49	3.03	23.58	8.65	3.86
	c.v.	3	27	4	12	6	2	5	12
Culm	mean	3.67	0.14	0.37	0.66	2.47	21.53	7.38	5.34
	c.v.	8	42	9	11	13	8	11	17
Jackmoor Brook	mean	3.93	0.15	0.44	0.44	2.53	24.52	8.26	3.72
	c.v.	10	35	10	13	11	6	14	12

1 Includes exchangeable cations.   2 Coefficient of variation (%).

Table 19.5. Semi-quantitative clay mineralogy
of suspended sediment

River		Major clay mineral groups (%):			
		Mica	Mixed layer mica-vermiculite	Kaolinite	Chlorite
Exe	mean	64	24	9	3
	c.v.	22	70	26	53
Dart	mean	52	36	7	5
	c.v.	33	55	14	42
Creedy	mean	54	30	13	3
	c.v.	24	56	28	71
Clyst	mean	56	34	10	0
	c.v.	22	30	14	-
Culm	mean	57	36	5	2
	c.v.	17	22	7	51
Jackmoor Brook	mean	61	30	6	3
	c.v.	17	44	19	65

the source material for erosion within the basin; thirdly
the potential for 'fingerprinting' sediment sources; and
finally, the use of studies of sediment properties in
elucidating the processes operating within the basin sediment system.

PARTICLE-SIZE BEHAVIOUR

The particle-size characteristics of suspended sediment are
clearly important in influencing the sediment delivery and
transport processes operating within a drainage basin. It
can be demonstrated theoretically that settling velocity,
and hence the potential for deposition, are directly related to particle diameter and Williams (1975) has suggested
that the sediment delivery ratio for the 432 km^2 basin of
Elm Creek, Texas would fall from 37% to 21% if the median
particle-size (d_{50}) of transported sediment increased from
1.0 μm to 10 μm, and to 6.1% if this was further increased
to 100 μm. Models for routing eroded sediment through the
drainage basin system must inevitably take account of this
important property.
      Information on the particle-size characteristics of
suspended sediment is frequently provided in data compil-
ations, but detailed study of this apparently simple
property reveals many potential inconsistencies. More
specifically, the results obtained from particle-size
analysis may depend heavily upon the sample preparation
and the method of analysis employed. Furthermore, very
considerable differences may exist between the particle-
size distribution obtained from laboratory analysis and
that which actually exists in the river under natural conditions. The latter has been termed the 'effective'
particle size distribution by Ongley et al. (1981).

The 'effective' particle-size distribution

Information on the 'effective' or in-stream particle-size
distribution of suspended sediment is frequently of more
relevance to the geomorphologist than data relating to the
ultimate particle-size of the chemically dispersed mineral
fraction. To meet these requirements, it is necessary to
produce particle-size data which most closely reflect the
in-stream conditions. The authors have experimented with
a number of methods of achieving this objective and suggest
that Coulter Counter analysis of untreated samples collected
using a traditional suspended sediment sampler provides an
effective approach.
      Figure 19.2 presents data for six of the study sites
which compare the typical result of a traditional particle-
size analysis of the chemically dispersed mineral fraction
with such Coulter Counter analyses. Marked contrasts exist
between the two size distributions for all sites and there
is clear evidence of aggregation under in-stream conditions.
Particles less than 1.0 μm in diameter are essentially
absent from the 'effective' particle-size distribution and
there is almost an order of magnitude difference between
the median particle-size of the two distributions. The

317

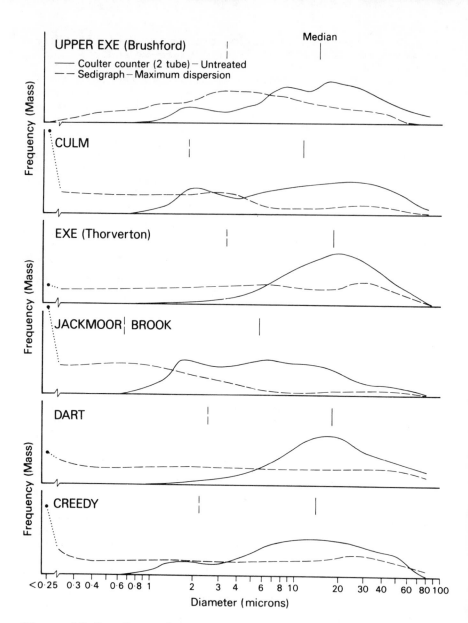

Figure 19.2.   Comparisons of the particle-size distribution
     of dispersed mineral sediment (Sedigraph) with an
     approximation of the 'effective' in-stream particle-
     size distribution obtained using a Coulter Counter.

form of the 'effective' size distributions suggest that
aggregates of between 10 μm and 20 μm diameter are generally
dominant and that particles <1.0 μm in diameter are an
important constituent of such aggregates.   Differences in
the form of the 'effective' particle-size distributions
between the individual sampling sites can be explained in

318

terms of contrasts in the upstream geology and of the influence of variable organic matter contents on the aggregation processes. Thus the relatively low median diameter evidenced by the 'effective' particle-size distribution for the Jackmoor Brook can be related to the high incidence of clay-sized mineral particles associated with soils developed on the Permian geology of the catchment and to the relatively low organic matter content of the sediments (cf. tables 19.2 and 19.3). Conversely, the maximum median diameter and the high degree of aggregation exhibited by sediment collected from the River Exe at Thorverton may be explained in terms of the high proportion of organic matter found in samples from this site (cf. table 19.2).

The major contrasts between the 'effective' particle-size distributions and those associated with more traditional analyses of the chemically dispersed mineral fraction evident in figure 19.2 clearly demonstrate the importance of a knowledge of the former to the geomorphologist. The existence of nearly an order of magnitude difference in the median particle size (d_{50}) of the two distributions emphasises the need to consider aggregation mechanisms in any attempt to model particle-size effects in suspended sediment delivery and transport. The importance of soil aggregates has become increasingly recognised in the study of soil loss processes (eg. Young, 1980) but much less is known about their significance in the routing of eroded sediment through the basin to the catchment outlet.

SEDIMENT PROPERTIES AND CATCHMENT DENUDATION

Measurements of sediment yield at catchment outlets (t year $^{-1}$) are frequently converted to estimates of the denudation rate in the upstream area (mm year $^{-1}$). Problems associated with using a short period of record to estimate long-term rates, with selecting the most appropriate density conversion factor, and with assuming that denudation processes operate uniformly over the drainage basin are now well recognised. Furthermore, several workers have pointed to the need to distinguish the organic and mineral components of the measured yield (eg. Arnett, 1978). However, little attention has been given to the selectivity of the erosion and delivery processes and the relationship between the properties of the material leaving the basin and those of the material mantling its slopes. The selectivity of soil erosion processes has been well documented by plot and laboratory studies of soil loss (eg. Lal, 1976; Young and Onstad, 1978) and it is clear that a preferential loss of both fine-grained material and low specific gravity organic material generally occurs. Furthermore the sediment delivery ratio concept (eg. Robinson, 1977) implies that a significant, if not large, proportion of the eroded material will be deposited as it is routed through the catchment to the basin outlet (cf. Meade, 1982). Again it is likely that this deposition will be preferential and that it will primarily involve the coarser size fractions. Significant contrasts may therefore exist between the

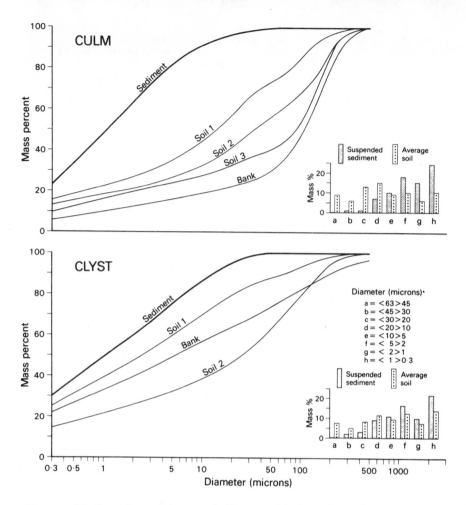

Figure 19.3. Comparison of the particle-size characteristics of suspended sediment transported by the Rivers
Culm and Clyst with those of typical soils and river
bank material within their catchments. All data refer
to the chemically dispersed mineral fraction.

properties of the material leaving the basin and of those
characterising its surface. The magnitude of the contrast
will reflect both the degree of selective erosion and the
efficiency of the sediment delivery system with its associated potential for redistribution of sediment within the
basin. Both are important facets of the denudation process.

Figure 19.3 provides a comparison between the particle-
size characteristics of suspended sediment transported by
the Rivers Culm and Clyst (sites 6 and 7, figure 19.1) and
those of typical soils and river bank material within their
catchments. In this case, the particle-size data refer to
the chemically dispersed mineral fraction, since an absolute
comparison of material properties is required. In both
cases the suspended sediment load is composed of material

Table 19.6. Estimates of the average maximum
percentage of mineral soil that can contribute to
the suspended sediment yield.

	Catchment				
	Exe	Creedy	Culm	Clyst	Dart
No. of soils	32	38	51	39	29
Average maximum potential contribution to sediment yield	55	63	50	63	71

Table 19.7. Enrichment factors for specific particle
size classes of suspended sediment with respect to
average soil <63 μm.

Catchment	Particle size classes in microns*							
	Depletion				Enrichment			
	<63>45	<45>30	<30>20	<20>10	<10>5	<5>2	<2>1	<1>0.3
Culm	<0.01	0.15	0.09	0.48	1.12	1.78	2.49	2.36
Clyst	<0.01	0.40	0.39	0.79	1.14	1.31	1.34	1.5

* Mineral fraction only.

<63 μm in diameter, whereas the soils and bank material
contain a considerable proportion of coarser particles.
If 63 μm is accepted as the upper limit of the particle
size fraction capable of being transported as suspended
sediment, it is possible to calculate the proportion of the
superficial material that can contribute to the suspended
sediment yield. Table 19.6 presents these values for the
catchment areas above five of the sampling sites, and
provides a range between 71% for the River Dart and 50% for
the River Culm.
     More detailed comparison of the particle-size distri-
butions of suspended sediment and the <63 μm fraction of
the superficial material can provide information on the
selectivity of the erosion and delivery processes. The
insets to figure 19.4 provide a visual comparison of the
proportions of various particle-size classes in suspended
sediment and 'average' soil for the Rivers Culm and Clyst
and table 19.7 expresses the depletion or enrichment of the
suspended sediment in terms of the ratio of the proportion
of a particular size class in the sediment to that in the
soil. With both rivers there is clear evidence of depletion of the coarse size fraction and enrichment of the
fine fraction in the sediment, with depletion occurring
for particle sizes >10 μm and enrichment for particle sizes
<10 μm. A well defined relationship exists between the
precise degree of depletion or enrichment and the magnitude
of the particle size grouping. Maximum depletion occurs

within the size range 45-63 µm and maximum enrichment is evident for particles <2 µm in diameter.

The contrasts apparent between the two catchments, in that sediment from the River Culm exhibits a greater degree of depletion of coarse material and greater enrichment of particles <10 µm, can be explained partly in terms of the coarser nature of soils in the Culm basin and partly in terms of contrasts in the efficiency of sediment delivery between the two basins. The sampling site on the River Culm is downstream of a 10 km reach of well developed low lying floodplain which is frequently inundated during times of high flow. This floodplain area must act as a sediment sink and will reduce the delivery efficiency of the basin. A reduction in delivery efficiency is likely to be reflected by enrichment of fines in the suspended sediment. Preferential sedimentation of coarser size fractions in sediment sink areas will reduce the concentration of dense primary minerals and other independent mineral grains whilst similar sized but low density aggregates of clay and organic matter with low settling velocities will tend to remain in suspension.

If it is assumed that there is little or no loss of the clay (<2 µm) fraction during the processes of erosion and delivery, and this seems reasonable in view of the expected inverse relationship between the efficiency of these processes and particle-size, the ratio of the proportion of clay in the suspended sediment (C_{sed}) to that in the <63 µm fraction of the soil (C_{soil}) provides an efficiency index for the overall erosion and delivery system operating in a basin viz:

$$\text{Efficiency Index (\%)} = \left[C_{soil} \, (\%)/C_{sed} \, (\%) \right] \times 100$$

Values of this index for the Rivers Culm and Clyst are 35% and 59% respectively. The contrasts between the two basins can again be related to variations in the composition of the <63 µm fraction of their soils and in their sediment delivery dynamics.

In addition to the particle-size characteristics of the chemically dispersed mineral fraction, a comparison of the characteristics of the suspended sediment transported from a basin with those of the soils mantling its slopes can usefully consider other properties. Table 19.8 presents data illustrating the degree of enrichment in organic carbon (C), nitrogen (N), total phosphorus (P) and total 'free' iron (Fe) of suspended sediment from the Rivers Exe and Dart and the Jackmoor Brook, when compared to the upper horizons (0-6 cm) of typical soils under arable cultivation and pasture within their respective catchments. Again, considerable enrichment is evident, with values rising to in excess of 4.0 times for C and N levels in sediment from the River Dart. Contrasts occur both between the catchments and between the four constituents and these can be related to differences in soil composition and in the efficacy of the erosion and delivery processes. Since the Fe content of the soils is closely related to the inorganic (clay) fraction it is interesting to compare its enrichment response to that of C and N which reflects the behaviour

322

Table 19.8.   Enrichment of suspended sediment in total
C, N, P and 'free' iron with respect to average soils.

Catchment	Land Use	Enrichment Factor			
		C	N	P	Fe
Jackmoor Brook	Arable	1.85	2.00	1.20	1.23
	Pasture (slopes)	1.24	1.49	1.17	1.00
Dart	Arable	4.47	4.00	1.83	1.20
	Pasture (slopes)	1.25	1.28	1.00	1.15
Exe	Arable	3.29	3.09	1.28	1.41
	Pasture (slopes)	1.54	1.55	1.07	1.21

of the organic fraction.   Whereas there is relatively
little difference between the degree of enrichment of Fe
and C and N in sediment collected from the Jackmoor Brook,
significant contrasts are apparent for the Rivers Dart and
Exe.   This would suggest that most of the organic fraction
associated with sediment from the Jackmoor Brook is well
humified and intimately associated with the clay fraction,
whereas a considerable proportion of the organic matter
transported by the Rivers Exe and Dart is in a 'free' form
and not clay-associated.   This conclusion is substantiated
by other analyses and the contrast may in turn reflect
important differences in sediment sources between the
Jackmoor Brook and the other two basins.   In this context,
it can be noted that arable land use is dominant within the
Jackmoor Brook catchment, whereas it is extremely limited
within the other two drainage basins.

FINGERPRINTING SEDIMENT SOURCES

The need to break down the spatial lumping inherent in the
conversion of measurements of sediment yield to an equivalent depth of surface lowering over the entire catchment
has prompted many attempts to determine the relative
importance and location of specific sediment sources (eg.
Imeson, 1974; Lehre, 1981).   These attempts have traditionally involved the monitoring of small plots or slope
segments, the isolation of individual micro-catchments and
the use of erosion pins.   More recently, however, a number
of workers have suggested that it is possible to 'fingerprint' potential sediment sources using various physical
and chemical properties of the material involved.   The
properties of the sediment transported from the basin can
then be studied with a view to assessing the likely importance of the 'fingerprinted' source areas.   Thus Klages
and Hsieh (1975) have used mineralogical indicators to
distinguish sediment generated within various sub-basins of
the Gallatin River in Montana, USA, and the relative
importance of slope and channel sources within individual

drainage basins have been assessed using clay mineralogy (Wall and Wilding, 1976), magnetic measurements (Walling et al., 1979) and sediment colour (Grimshaw and Lewin, 1980).

This application of information on sediment properties would seem to have considerable potential in geomorphological studies, but it is not without problems. In many situations it will not prove possible to differentiate conclusively sediment from different sources such as catchment slopes under different land use, and river banks. In the study described by Wall and Wilding (1976) these two basic sources were clearly distinguished, since the slopes were developed on a cover of glacial drift, whereas the stream channels were incised into the underlying bedrock, but in most studies similar parent material will be involved.

Table 19.9 provides a comparison of several properties of the suspended sediment transported from three of the study basins with those of potential source areas. These include surface soils from slopes under arable cultivation and pasture and from floodplain pasture, and channel bank material. Several of the properties clearly distinguish the different sources and would seem potentially valuable as 'fingerprints'. However, no clear picture of the relative importance of the various sources emerges. With most of the properties, the levels associated with the suspended sediment are considerably in excess of those characterising the sources and it is clear that the enrichment mechanisms discussed previously exert a major control on the precise relationship between the properties of sediment and source material. Although it seems realistic to argue that channel banks may be effectively excluded as a potential source, in view of the extremely low values of the various properties associated with this material, it is impossible to distinguish clearly the relative importance of the remaining three sources. If definite results are required it is clearly necessary to select properties or measures which are effectively uninfluenced by enrichment or depletion mechanisms or on which the influence of such mechanisms can be readily interpreted.

Elemental ratios provide an example of the former, since they are much less susceptible to enrichment effects. However, the choice of ratio is important, because there are two major components in suspended sediment, namely the organic and inorganic fractions, and, despite the incidence of aggregation, these may have different sediment delivery characteristics. It seems desirable, therefore, to choose ratios which are derived exclusively from one fraction or the other. In table 19.10 a number of these ratios are introduced to complement the data presented in table 19.9. Of these ratios, the greatest interpretational confidence may be placed on that for C/N. Although the C/P and organic/inorganic P ratios might theoretically be supposed to be unsuitable, the inorganic P is so closely associated with the organic fraction that these data can be used with some confidence. In contrast, the iron ratio values will clearly be influenced by both fractions and are thus of more limited use for fingerprinting. Using the ratio data available, it becomes apparent that, based on the

Table 19.9. Comparison between suspended sediment
and source area material for selected properties

| | Total % | | | Iron % | | | |
	C	N	P	Fed[1]	Feox[2]	χ[3]	ΣC[4]
JACKMOOR BROOK							
Suspended sediment	4.84	0.58	0.147	2.48	0.74	7.60	28
Soils							
Arable (slopes)	2.61	0.29	0.123	2.01	0.39	4.03	14
Pasture (slopes)	3.91	0.39	0.126	2.47	0.54	9.15	13
Floodplain pasture	3.91	0.28	0.104	2.25	0.52	6.01	12
RIVER DART							
Suspended sediment	6.61	0.64	0.148	2.02	1.06	3.88	24
Soils							
Arable (slopes)	1.48	0.16	0.081	1.68	0.27	4.45	10
Pasture (slopes)	5.29	0.50	0.159	1.75	0.57	4.63	16
Alluvial pasture very poorly represented in this catchment							
RIVER EXE							
Suspended sediment	7.11	0.68	0.153	2.07	0.95	4.33	28
Soils							
Arable (slopes)	2.16	0.22	0.120	1.47	0.37	6.77	16
Pasture (slopes)	4.63	0.44	0.144	1.71	0.53	5.70	14
Floodplain pasture	4.50	0.45	0.105	1.87	0.56	-	8
BANKS	<1.0	<0.1	<0.05	-	-	~2	1-2

1. Total 'free' iron (dithionite extraction).

2. Organic and inorganic 'amorphous' iron (oxalate extraction).

3. Magnetic susceptibility $m^3 g^{-1}$

4. Exchangeable cations meq $100 g^{-1}$

composition of its organic component, the dominant source
of suspended sediment transported by the Jackmoor Brook is
ploughed fields. For the other catchments, the existence
of fresher organic matter (higher C/N values) suggests that
pasture fields are a significant source. Although the
ratio data are not themselves conclusive in this context,
channel banks can probably be excluded as a significant
source in both catchments because of the extremely low
levels of organic matter documented in table 19.9.

Caesium-137 originating as fallout from atmospheric
nuclear weapons testing has now been successfully used by
a number of workers to study soil erosion and accumulation

Table 19.10. Comparative elemental ratio
data for three catchments

| | Elemental ratio | | OP/IP[1] | Feox/Fed[2] |
	C/N	C/P		
JACKMOOR BROOK				
Suspended sediment	8.4	32	0.51	0.30
Soils				
Arable (slopes)	9.0	24	0.41	0.18
Pasture (slopes)	10.0	33	0.50	0.28
Floodplain pasture	13.0	38	0.87	0.26
DART				
Suspended sediment	10.5	46	0.61	0.53
Soils				
Arable (slopes)	9.1	18	0.58	0.16
Pasture (slopes)	10.4	33	0.85	0.31
Alluvial pasture very poorly represented in this catchment.				
EXE				
Suspended sediment	10.4	46	0.61	0.47
Soils				
Arable (slopes)	9.8	17	0.37	0.23
Pasture (slopes)	10.3	34	0.72	0.31
Floodplain pasture	10.0	45	1.50	0.35
BANKS	7-9	5-30	0.05-0.70	<0.20

1. Ratio of organic to inorganic phosphorus

2. Ratio of organic and inorganic 'amorphous' to total 'free' iron

Table 19.11. Comparison between the ^{137}Cs activity
of source area materials and suspended sediment

| | ^{137}Cs activity (pCi g^{-1}) | | | | | |
| | Pasture | Arable | Bank | Suspended sediment | | |
				Exe	Dart	J. Brook
Observed (average)	0.37	0.12	0.02	0.30	0.13	0.23
Corrected*	1.55	0.23	0.054	-	-	-

* Corrected for particle size effects at the 10 μm level.

within the landscape (eg. McHenry and Ritchie, 1977; McCallan *et al.*, 1980; Campbell *et al.*, 1982). It would seem that this isotope also has considerable potential for use as a 'fingerprinting' parameter. [137]Cs levels in surface material vary considerably according to precipitation levels and land use practices. Maximum concentrations are generally found in the surface layers of soils under permanent pasture or woodland, whereas levels are considerably reduced on cultivated land due to mixing, and are very low in channel bank material due to its limited exposure to precipitation (table 19.11). [137]Cs is strongly adsorbed onto the clay mineral fraction of surface soils, and studies undertaken by the authors suggest that it is possible to produce meaningful estimates of the enrichment of ^{137}Cs to be expected as a result of the particle-size enrichment discussed earlier. Table 19.11 therefore provides a corrected value of ^{137}Cs concentration for each potential source based on a comparison of the particle-size characteristics of source material and those of the suspended sediment. It is clear from table 19.11 that the [137]Cs 'fingerprinting' evidence is again consistent with arable fields providing the major sediment source in the Jackmoor Brook catchment. In the case of the River Exe, the higher levels of ^{137}Cs found in the sediment imply that pasture areas are a significant contributor, whereas the lower levels reported for the River Dart suggest that bank sources are of some significance in this drainage basin.

[137]Cs data provide the best 'fingerprinting' parameter currently employed by the authors for the inorganic fraction of suspended sediment in mineralogically uniform catchments, but it is apparent that results from a number of organic and inorganic 'fingerprints' will probably be necessary to provide conclusive results on the relative importance of particular sediment sources in the study area. One 'fingerprint' parameter may point to two possible sources, whereas an additional parameter may effectively exclude one of these. In the case of the Jackmoor Brook, current results point convincingly to arable fields as the dominant sediment source. Further work is required to produce definitive statements for the Rivers Exe and Dart, but certain tentative conclusions may nevertheless be advanced on the basis of evidence obtained to date. With the River Exe, this evidence points to a minimal contribution from channel banks and to pasture areas as the dominant source, which is in turn consistent with the low proportion of arable land use within this drainage basin. In the Dart basin, there is some evidence that pasture areas provide the major source, but the ^{137}Cs data suggest that channel erosion also provides a significant contribution.

More work is required to fully assess the potential of elemental ratios, ^{137}Cs content and similar sediment properties in fingerprinting sediment sources, but this general approach would seem to offer considerable scope in geomorphological investigations.

Inductive reasoning based on interpretation of spatial and temporal variations in the output of suspended sediment from drainage basins (eg. Guy, 1964; Temple and Sundborg, 1972; Dunne, 1979) has provided an important contribution to the understanding of the processes involved. Such reasoning has hitherto involved magnitude characteristics of sediment response, including storm-period concentrations and loads, and the sediment concentration/discharge relationship, but scope clearly exists for including quality considerations in any interpretation. The potential for using sediment properties as 'fingerprints' has been discussed in the previous section and it has, for example, been shown that differences in the ^{137}Cs content of sediment or its C/N ratio could be used to infer sediment source contrasts between a number of drainage basins. At a more detailed level, studies of the relationship between sediment properties and discharge and of intra- and inter-storm variations in the behaviour of specific sediment properties would seem to offer very considerable potential for elucidating processes of erosion and sediment yield.

Figure 19.4 presents plots of the relationships between values of several sediment properties and storm discharge for the River Exe at Thorverton (site 1) and the River Dart (site 2). In the case of the River Exe, values of organic carbon content and C/N ratio are considered. A clear inverse trend is apparent in the organic carbon relationship, indicating that the organic fraction of the suspended sediment decreases at times of maximum flow. This could reflect an exhaustion effect, consequent upon the preferential mobilisation of organic particles during the early stages of a storm runoff event, but the positive trend in the relationship between the C/N ratio and discharge points to a more complex situation. The increase in the C/N ratio with increasing discharge suggests that the decrease in the magnitude of the organic fraction is coupled with a change in the nature of this fraction. One possible explanation for this pattern could invoke the increased erosive energy available, and the expansion of the area contributing to storm runoff and therefore to sediment generation, at such times. The former could account for an increasing proportion of relatively high specific gravity mineral material (cf. lower density organic material) and the latter could be expected to make available organic matter of a different character within the newly accessed areas.

Similar inverse relationships between the organic carbon content of suspended sediment and the magnitude of storm discharge are exhibited by other rivers in the study area (eg. the River Dart, figure 19.4). These may again be accounted for in terms of exhaustion of readily available organic material, and increased erosive energy, during times of maximum flow. The tendency for the C/N ratio to increase with increasing discharge magnitude is, however, less clearly apparent at other sites. With the River Dart this general positive relationship is only poorly developed and it is non-existent in the case of the Jackmoor Brook.

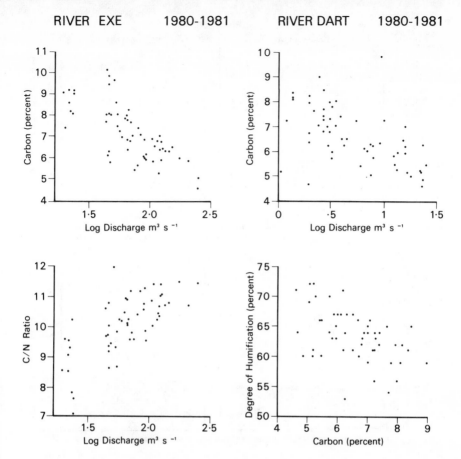

RIVER EXE   1980-1981   RIVER DART   1980-1981

Figure 19.4. Relationships between several suspended sedi-
    ment properties and storm discharge for the River Exe
    at Thorverton and the River Dart.

This contrast between the three sites further substantiates
the differences in sediment sources and sediment generation
mechanisms suggested by tables 19.10 and 19.11. Further
study is required to establish the precise significance of
these relationships, and the behaviour of other sediment
properties must also be explored. For example, figure 19.4
illustrates an inverse relationship between the organic
carbon content and the degree of humification of suspended
sediment collected from the River Dart. This points to
further complexities in catchment behaviour.
    Examples of typical information provided by monitoring
the behaviour of sediment properties during individual
storm runoff hydrographs are provided in figure 19.5.
Without discussing these in detail, it is clear that very
considerable variations in the carbon and nitrogen content
of the sediment and its C/N ratio occur during these events.
These trends must in turn reflect temporal variations in
the operation of sediment generation and conveyance

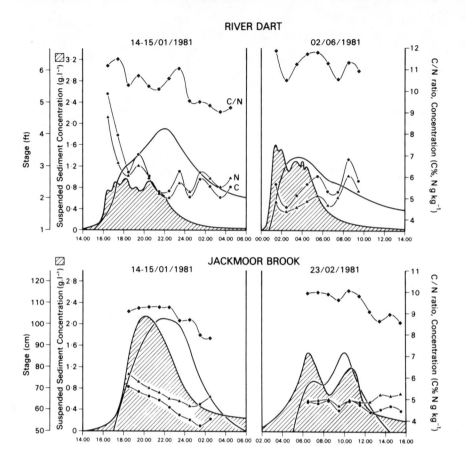

Figure 19.5. Inter- and intra-storm variations in the C
and N content and the C/N ratio of suspended sediment
transported by the River Dart and the Jackmoor Brook.

processes. For example, the storm of 14-15 January, 1981
in the River Dart and Jackmoor Brook basins exhibits declining C and N concentrations and C/N ratios throughout
the event, which are consistent with a progressive
exhaustion of readily available topsoil-derived organic
material. These changes could have important implications
for the operation of sediment transport processes. Furthermore, significant inter-storm contrasts in the behaviour
of these properties are apparent for both basins from a
comparison of the two storms illustrated. In the second
event, C and N concentrations show little or no evidence
of decline during the storm for the Jackmoor Brook and an
overall increase is apparent for the River Dart. These
inter-storm contrasts must in turn reflect differences in
the nature or magnitude of the processes operating or in
their spatial distribution. Other sediment properties
could be expected to demonstrate equally complex behaviour
in terms of inter- and intra-storm variations and again it

will probably be necessary to study the behaviour of a
suite of properties in order to piece together a meaningful
understanding of the erosion and conveyance properties
operating in these basins.

THE PROSPECT

Recent interest in the properties of suspended sediment has
stemmed primarily from investigations of biogeochemical
cycling and pollutant transport through the fluvial system.
It is suggested, however, that the new perspective afforded
by such studies if also of relevance to the geomorphologist
and that he should recognise the potential for including
sediment quality studies in catchment experiments. This
will inevitably necessitate the development of new monitoring techniques and the full potential will not be realised
until our understanding of the factors controlling the
behaviour of sediment properties has improved. Moreover,
more work is required to determine those properties which
offer the greatest potential for geomorphological research.
In some contexts the choice is clear but in others such as
'fingerprinting' only a small proportion of the wide range
of possibilities have as yet been tried.

ACKNOWLEDGEMENT

The authors gratefully acknowledge the support of the
Natural Environment Research Council in providing a research
grant for work on suspended sediment properties in local
rivers.

REFERENCES

Allan, R.J., 1979, Sediment-related fluvial transmission of
    contaminants: some advances by 1979, *Inland Waters
    Directorate, Environment Canada, Scientific Series*,
    No.107.

Arnett, R.R., 1978, Regional disparities in the denudation
    rate of organic sediments, *Zeitschrift für Geo-
    morphologie*, Supplementband 29, 169-179.

Ball, J., 1939, *Contributions to the geography of Egypt*,
    (Government Press, Cairo).

Campbell, B.L., Loughran, R.J. and Elliott, G.L., 1982,
    Caesium-137 as an indicator of geomorphic processes
    in a drainage basin system, *Australian Geographical
    Studies*, 20, 49-64.

Dunne, T., 1979, Sediment yield and land use in tropical
    catchments, *Journal of Hydrology*, 42, 281-300.

FIASP, 1963, *Determination of fluvial sediment discharge*,
    Federal Inter-Agency Sedimentation Project Report
    No.14.

Förstner, U. and Wittmann, G.T.W., 1981, *Metal pollution in the aquatic environment*, (Springer Verlag, New York).

Grimshaw, D.L. and Lewin, J., 1980, Source identification for suspended sediments, *Journal of Hydrology*, 47, 151-162.

Guy, H.P., 1964, An analysis of some storm-period variables affecting stream sediment transport, *US Geological Survey Professional Paper*, 462-E.

IAHS, 1981, *Erosion and sediment transport measurement, Proceedings of the Florence Symposium*, IAHS Publication No.133.

Imeson, A.C., 1974, The origin of sediment in a moorland catchment with special reference to the role of vegetation, in: *Fluvial processes in instrumented watersheds*, ed. Gregory, K.J. and Walling, D.E., Institute of British Geographers Special Publication No.6, 69-72.

Jansen, P.P., 1979, *Principles of river engineering*, (Pitman, London).

Klages, M.G. and Hsieh, Y.P., 1975, Suspended solids carried by the Gallatin River of Southwestern Montana: II Using mineralogy for inferring sources, *Journal of Environmental Quality*, 4, 68-73.

Knisel, W.G. (ed.), 1980, *CREAMS: A field-scale model for chemicals runoff and erosion from agricultural management systems*, USDA Conservation Research Report No.26.

Lal, R., 1976, *Soil erosion problems on an alfisol in Western Nigeria and their control*, IITA Monograph No.1, (Ibadan, Nigeria).

Lehre, A., 1981, Sediment budget of a small California Coast Range drainage basin near San Francisco, in: *Erosion and sediment transport in Pacific rim steeplands, Proceedings of the Christchurch Symposium*, IAHS Publication No. 132, 123-139.

McCallan, M.E., O'Leary, B.M. and Rose, C.W., 1980, Redistribution of Caesium-137 by erosion and deposition on an Australian soil, *Australian Journal of Soil Research*, 18, 119-28.

McHenry, J.R. and Ritchie, J.C., 1977, Estimation of field erosion losses from fallout Caesium-137 measurements, in: *Erosion and solid matter transport in inland waters, Proceedings of the Paris Symposium*, July 1977, IAHS Publication No.122, 26-33.

Meade, R.H., 1982, Sources, sinks, and storage of river sediment in the Atlantic drainage of the United States, *Journal of Geology*, 90, 235-252.

Ongley, E.D., 1982, Influence of season, source and distance on physical and chemical properties of suspended sediment, in: *Recent developments in the explanation and prediction of erosion and sediment yield, Proceedings of the Exeter Symposium*, IAHS Publication No.137, 371-383.

Ongley, E.D., Bynoe, M.C. and Percival, J.B., 1981, Physical and chemical characteristics of suspended solids, Wilton Creek, Ontario, *Canadian Journal of Earth Science*, 18, 1365-1379.

Peart, M.R. and Walling, D.E., 1982, Particle size characteristics of fluvial suspended sediment, in: *Recent developments in the explanation and prediction of erosion and sediment yield, Proceedings of the Exeter Symposium*, IAHS Publication No.137, 397-407.

Robinson, A.R., 1977, Relationships between soil erosion and sediment delivery, in: *Erosion and solid matter transport in inland waters, Proceedings of the Paris Symposium, July 1977*, IAHS Publication No.122, 159-167.

Shear, H. and Watson, A.E.P. (eds.), 1977, *The fluvial transport of sediment-associated nutrients and contaminants*, (International Joint Commission, Windsor, Ontario, Canada).

Temple, P.H. and Sundborg, A., 1972, The Rufiji River, Tanzania, hydrology and sediment transport, *Geographiska Annaler* (A), 54, 345-368.

Tixeront, J., 1974, La couverture végétale et les debits solides des cours d'eau dans le bassin de la Seine, in: *Effects of man on the interface of the hydrological cycle with the physical environment, Proceedings of the Paris Symposium*, IAHS Publication No.113, 28-35.

Wall, G.J. and Wilding, L.P., 1976, Mineralogy and related parameters of fluvial suspended sediments in Northwestern Ohio, *Journal of Environmental Quality*, 5, 168-173.

Walling, D.E., 1978, Suspended sediment and solute response characteristics of the River Exe, Devon, England, in: *Research in fluvial systems*, ed. Davidson-Arnott, R. and Nickling, W., (Geobooks, Norwich), 169-197.

Walling, D.E. and Kane, P., 1982, Temporal variation of suspended sediment properties, in: *Recent developments in the explanation and prediction of erosion and sediment yield, Proceedings of the Exeter Symposium*, IAHS Publication No.137, 409-419.

Walling, D.E., Peart, M.R., Oldfield, F. and Thompson, R., 1979, Suspended sediment sources identified by magnetic measurements, *Nature*, 281, 110-113.

Walling, D.E. and Webb, B.W., 1981, The spatial dimension in the interpretation of stream solute behaviour, *Journal of Hydrology*, 47, 129-149.

Webb, B.W. and Walling, D.E., 1982, The magnitude and
    frequency characteristics of fluvial transport in
    a Devon drainage basin and some geomorphological
    implications, *Catena*, 9, 9-23.

Williams, J.R., 1975, Sediment routing from agricultural
    watersheds, *Water Resources Bulletin*, 11, 965-974.

Young, R.A., 1980, Characteristics of eroded sediment,
    *Transactions of the American Society of Agricultural
    Engineers*, 23, 1139-1142 and 1146.

Young, R.A. and Onstad, C.A., 1978, Characteristics of rill
    and interrill eroded soil, *Transactions of the
    American Society of Agricultural Engineers*, 21,
    1126-1130.

20 Dynamics of water chemistry in hardwood and pine ecosystems

Wayne T. Swank and W.T. Scott Swank

INTRODUCTION

Historically, theories and interpretations of landscape development have emphasized fluvial erosion. Less attention has been given to denudation by solution or chemical weathering except where bedrock is highly soluble. However, several recent forest studies in some sections of the eastern United States have shown that present day denudation is strongly dominated by chemical solution processes with a solutional to mechanical weathering ratio of about 5 to 1 (Cleaves et al., 1970; Likens et al., 1977).

Fyfe (1981) has pointed out the need for team effort and interfacing of disciplines to solve urgent geochemical problems of the next decade. An interdisciplinary research effort at the Coweeta Hydrologic Laboratory in the southern Appalachian Mountains of the United States (Monk et al., 1977) provides a unique opportunity to examine the role of solutional weathering in land forming processes. The co-operative research includes scientists trained in geology, hydrology, botany, microbiology, entomology, forestry, and system analysis. We are investigating biogeochemical cycling in forests and the effects of natural and man-induced disturbances on nutrient cycles and forest productivity. The approach is to explain whole system behaviour as revealed in catchment nutrient and water budgets, by studying internal ecosystem processes.

At Coweeta, solution export is an important form of fluvial denudation for relatively mature hardwood forests (table 20.1). The major cation export is in solution form rather than extractable cations in sediment transport, while the reverse situation is apparent for nitrogen. In terms of gross export of mass, dissolved substances amount to 215 kg ha^{-1} year^{-1} compared to total sediment losses of 258 kg ha^{-1} year^{-1}. However, export data represent the net production of processes and it is known that the internal rate of nutrient transfers is usually much greater than export (Swank and Waide, 1980). Furthermore, the mechanisms, patterns and amounts of nutrient transfer may differ between forest types which, in turn, could produce

Table 20.1. Annual dissolved and sediment exports of selected nutrients from a 60 ha mature hardwood covered catchment at Coweeta Hydrologic Laboratory.

Export form	Total	Organic matter	Constituent[1]				
			N	Ca	K	Mg	Na
	-------- kg ha^{-1} year^{-1} ----------						
Sediment (bedload and suspended)[2]	258	6.16	1.80	.16	.006	.01	.003
Dissolved[2]	215	15.0 (DOC)	.08	9.68	5.63	4.38	11.02

[1] Sediment cations extracted using double acid extraction procedure ($0.025N$ H_2SO_4 + $0.05N$ HCl).

[2] Averages are based on an annual flow of 1185 mm and sediment sampling during 1974-1976; DOC for 1979-1980

differential weathering rates. This paper summarizes one aspect of the research at Coweeta - the alteration of water chemistry during its passage through the ecosystem. Specifically, we compare seasonal concentrations of selected cations and anions in precipitation, throughfall, litter and soil water, and streamflow for three forest types in the region.

Except for Douglas-fir (*Pseudotsuga menziesii* (Mirb.) Franco) ecosystems of the Pacific Northwest (Sollins *et al.*, 1980; Cole and Johnson, 1977), there is little published information on the alteration of water chemistry through successive compartments of a given ecosystem. Rather, most attention has been directed towards chemical alteration by compartments such as the tree canopy (Tarrant *et al.*, 1968; Miller *et al.*, 1976; Henderson *et al.*, 1977), forest floor (Buldgen and Remacle, 1981; McColl, 1972) or soil (McColl and Cole, 1968; Johnson and Cole, 1976).

SITE DESCRIPTION

The 2184 ha Coweeta Hydrologic Laboratory is located in the Nantahala Mountain Range of western North Carolina, within the Blue Ridge Physiographic Province. The topography of Coweeta is steep and elevations range from 686 to 1600 m with average side slopes of 50%. The climate of the region is characterized by cool summers, mild winters, and abundant rainfall in all seasons. Average annual precipitation varies from 1700 mm at lower elevations to 2500 mm on the upper slopes. There are 69 km of first-, second-, and third-order perennial streams within the basin. The drainage pattern is dendritic, stream channels are generally V-shaped, and drainage density is uniform at 3.15 km km^{-2}. The range of monthly streamflow is rather narrow, but discharge is highest and most variable during February and

March and lowest and most stable during late summer and early fall. Quickflow (or direct runoff) for Coweeta watersheds varies between 7 and 22% of the total annual runoff, and there is no overland flow on undisturbed catchments. The indigenous vegetation is comprised of uneven-aged, mixed hardwoods. The overstory is dominated by oak (*Quercus*) and hickory (*Carya*) species, red maple (*Acer rubrum* L.), and yellow poplar (*Liriodendron tulipifera* L.), and the understory has an abundance of dogwood (*Cornus florida* L.), mountain laurel (*Kalmia latifolia* L.), and rhododendron (*Rhododendron maximum* L.). More complete descriptions of the Coweeta site and research history are given by Johnson and Swank (1973) and Swank and Douglass (1977).

The regolith within the Coweeta basin is deeply weathered and averages about 7 m in depth. Soils generally occur within two orders - fully developed Ultisols and immature Inceptisols. The underlying bedrock is termed the Coweeta Group and has been mapped and described by Hatcher (1974, 1979, 1980). The group consists of a series of metasedimentary and possibly metaigneous rocks which overlie the older rocks of the Tallulah Falls Formation of Precambrian origin. This formation is predominately biotite paragneiss and biotite schist occurring in medium to thick layers. Biotite paragneiss is coarse grained and medium to medium dark gray; contains quartz, plagioclase, biotite, muscovite, and garnet. The biotite schist is coarse grained and medium to dark gray; contains biotite, orthoclase, quartz, garnet, and sillimanite. Within the formations are interlayers of pelitic schist, metasandstone to metagraywacke, and minor calc-silicate quartzite. The minimum thickness of the formation in this area is about 1000 m; a more complete description is given by Hatcher (1980).

SPECIFIC STUDY SITES AND METHODS

Data were derived from three forest stands in three separate studies conducted over the past 10 years. The mixed hardwood forest and white pine (*Pinus strobus* L.) plantation are on catchments at Coweeta. The pitch pine (*Pinus rigida* Mill.) forest is about 10 km north of the Coweeta basin. Table 20.2 summarizes site characteristics.

The mixed hardwood stand is uneven-aged, with some trees more than 200 years old. The overstory is dominated by a variety of oak and hickory species, red maple, and yellow poplar and the understory has an abundance of dogwood, mountain laurel, and rhododendron. The pitch pine stand is typical of early forest succession on more xeric sites in the region; that is, a pine overstory with a mixture of hardwoods in the understory. The white pine forest is on a catchment that was clearcut in 1940. Hardwood sprouts were cut annually for a long period and the area was planted to pine in 1955. Information given in this paper was collected when the plantation was 13 years old (Best and Monk, 1976). Crown closure was complete in all stands and basal area ranged from 15 to 30 m^2 ha^{-1}. Because the stands differed greatly in size, the number of throughfall and lysimeter collectors also varied widely. Since decomposition of the

337

Table 20.2. Summary of site and study characteristics

Characteristic	Forest Type		
	Quercus-Carya	Pinus rigida	Pinus strobus
Age (years)	Uneven-aged (200+ years)	45	13
Soil type	Tusquitee and Chandler loam (Humic Hapludult and Typic Dystrochrept)	Evard sandy loam (Typic Hapludult)	Same as Quercus-Carya
Basal area (m² ha⁻¹)	26	30	15
Number trees ha⁻¹	3032	850	1670
Number throughfall collectors	16	6	20
Number lysimeter collectors	32	6	8

litter layer is most rapid in the hardwood stand (Cromack and Monk, 1976), its forest floor is thinnest. The litter layer is intermediate under white pine (5 cm) and thickest in pitch pine (10 cm).

Bulk precipitation chemistry has been routinely measured within the Coweeta basin since 1969 (Swank and Douglass, 1977). Samples were collected in clearings with a 45° sky view from the gauging site. Throughfall was measured with troughs measuring 15 x 200 cm and connected to 20 l polyethylene carboys. Litter and soil water solutions were sampled with lysimeters; the majority of data presented in this paper is for collections taken with porous plate or porous cup ceramic tension lysimeters (Cole et al., 1961; Hansen and Harris, 1975). Flow proportional samplers were used to sample stream water. All collectors were serviced on either a storm event or weekly basis. Samples were composited by volume each month for chemical analyses in the hardwood and white pine studies. Sample collection and analysis were based on a storm event basis for the pitch pine study. Because the pitch pine study included adequate samples, a statistical analysis of growing and dormant season data was applied only to that data set. A t-test was applied to differences between mean concentrations of nutrients in water passing through successive compartments. Water samples were analyzed with an autoanalyzer and atomic absorption spectrophotometer.

Table 20.3. Comparison of H ion concentrations in
hydrologic compartments for combined dormant and
growing seasons for mixed hardwood and pitch pine
forests (analysis of variance, *p < .05; **p < .01).

Compartment	H^+ Concentration (μeq 1^{-1})	pH
A. Pitch Pine		
Precipitation	98.2	4.01
Throughfall	52.9*	4.28
Litter water	5.6**	5.26
Soil water (25 cm)	17.4**	4.76
B. Mixed Hardwoods		
Precipitation	17.1	4.77
Throughfall	5.1*	5.30
Litter water	4.6	5.34
Soil water (25 cm)	1.0*	6.01
Stream water	0.2*	6.82

RESULTS AND DISCUSSION

The hydrogen ion is important in driving chemical weathering
reactions and also is active in exchange processes within
soil and vegetation (Ulrich et al., 1980; Eaton et al.,
1973).   Table 20.3 shows average annual hydrogen ion concentrations for the pitch pine and oak-hickory forests.
Both forest types show a significant depletion of hydrogen
as water passes through the canopy. As water moves through
the forest floor of the pitch pine stand, concentrations
are reduced tenfold or almost 50 μeq 1^{-1}.   Between the soil
surface and 25 cm depth, there is significant release of
hydrogen which, as will be shown later, is associated with
sulphate movement. The major depletion of hydrogen in
hardwoods occurred in the canopy but the process continued
as water passed through successive ecosystem compartments.
The net change in concentrations between precipitation and
streamflow is equivalent to about 2 pH units for hardwoods.
     Other ionic data are summarized for the hardwood forest
in table 20.4 for the dormant (November-April) and growing
(May-October) seasons.   Concentration changes for most ions
were greater in the growing season than in the dormant
season.   This result was expected because foliage is
present and biological activity is substantially greater
during the growing season.   In both seasons, inorganic
forms of nitrogen were somewhat depleted within the canopy
and concentrations were greatly reduced in the litter layer,
probably due to utilization by micro-organisms.   Stream
water concentrations were near detection limits.

Table 20.4. Average weighted concentrations of selected cations and anions in hydrologic comparments of a mature, mixed hardwood forest for selected time periods.

| Compartment | Mean Concentration (mg 1^{-1}) | | | | | | |
	NO_3-N	NH_4-N	K	Na	Ca	Mg	SO_4
A. Dormant Season							
Precipitation	.12	.07	.04	.17	.24	.03	1.31
Throughfall	.07	.03	.96	.18	.50	.21	2.43
Litter water	.01	.02	2.39	.30	1.06	.59	4.56
Soil water (25 cm)	.01	.01	1.02	.49	.66	.68	3.10
Stream water	.001	.002	.48	.90	.78	.35	.55
B. Growing Season							
Precipitation	.10	.07	.11	.10	.22	.04	1.36
Throughfall	.05	.18	2.94	.19	1.09	.44	4.96
Litter water	.03	.06	3.20	.29	1.55	.74	5.51
Soil water (25 cm)	.02	.07	1.54	.37	1.07	.62	3.21
Stream water	.004	.003	.53	1.00	.89	.40	.45

Concentrations of all cations increased in throughfall, with K showing more than a twentyfold increase. Elevated levels of cations were also observed in water passing through the litter. At the 25 cm depth, cation concentrations were either depleted or remained relatively unchanged. Between the 25 cm depth and discharge from the catchment, K and Mg concentrations were halved, Na was doubled, and Ca appeared unchanged.

The canopy contributes large amounts of SO_4 in throughfall, and the source is primarily impacted aerosols which are not collected in precipitation samples (Lindberg and Harriss, 1981). In our studies, sulphate was further enriched as water passed through the forest floor, but there was some depletion at 25 cm and substantial depletion between this level and the stream. Other research at Coweeta has shown that this soil has a large SO_4 adsorption capacity (Johnson et al., 1980). Furthermore, recent unpublished studies indicate that about 30% of the inorganic SO_4 in soil is metabolized by microbes to ester- and carbon-bonded sulphur.

The dynamics of cation and anion movement in the pitch pine ecosystem differs somewhat from the hardwood results (table 20.5). In the dormant season there was substantial leaching of NO_3 from the pine canopy while in the growing season there was little difference in concentrations of precipitation and throughfall and, in fact, there may have been depletion as in hardwoods. During the growing season,

Table 20.5. Average concentrations of selected cations and anions in hydrologic compartments of a 45-year-old pitch pine forest for the dormant and growing seasons

Compartment	Mean Concentration (mg l^{-1})						
	NO_3-N	NH_4-N	K	Na	Ca	Mg	SO_4
A. Dormant Season							
Precipitation	.34	.23	.24	.42	.43	.06	3.00
Throughfall	.82*	.10	1.42*	.36	2.02*	.91*	6.57*
Litter water	.01**	.004*	1.04*	.83*	2.13	1.04*	5.59
Soil water (25 cm)	.08	.02	5.17**	.91	2.30	1.34*	11.88*
B. Growing Season							
Precipitation	.17	.03	.16	.10	.26	.04	4.96
Throughfall	.13	.09	3.73*	.18**	1.71**	.54**	8.22*
Litter water	.04*	.02*	2.31	.61**	2.33	1.24**	8.89
Soil water (25 cm)	.01*	.01	3.22	.45*	2.10	.98	8.33

the hardwood understory of the pine stand was in leaf. If we assume that NO_3 was also leached from the pine canopy in the growing season, then it appears that the hardwood understory is an important consideration in NO_3 dynamics. This behavior may reflect the lack of epiphytic lichens and microflora in the pine canopy which are known to utilize nutrients and alter throughfall chemistry in other tree canopies (Lang *et al.*, 1976; Pike, 1978) and/or absorption by the understory (Yarie, 1980). As with hardwoods, the greatest NO_3 depletion was in the litter compartment. The pattern of SO_4 movement in pitch pine also differed from that in hardwoods. Concentrations in throughfall were almost double precipitation values; there was little change as water passed through the forest floor, but within the next 25 cm, SO_4 concentrations doubled during the dormant season but no concentration changes occurred during the growing season. Reasons for this behavior are unknown but may be due to sulphate transforming microbial populations, or high rates of sulfolipid mineralization. Microbial regulation seems plausible because activity would be greater in the warm temperatures of the growing season.

The pattern of K concentration in litter and soil water beneath pitch pine was opposite that beneath hardwoods but Na, Ca, and Mg generally followed the same trend in both forest types. The magnitude of cation enrichment appeared to be greater for pitch pine in litter and soil compartments during both seasons.

Less information is available for the white pine stand but data in table 20.6 again show that changes in

341

Table 20.6. Average volume weighted concentrations of selected cations in hydrologic compartments of a 13 year-old white pine plantation for selected time periods.[1]

Compartments	Mean Concentrations (mg l^{-1})			
	K	Na	Ca	Mg
A. Dormant Season				
Precipitation	.08	.28	.26	.03
Throughfall	.71	.56	.38	.08
Litter water	1.53	.69	.93	.23
Soil water (25 cm)	1.98	1.25	1.25	.21
Stream water	.34	.60	.39	.17
B. Growing Season				
Precipitation	.36	.44	.20	.04
Throughfall	2.79	.45	.55	.21
Litter water	4.69	.96	2.54	.51
Soil water (25 cm)	2.56	1.28	2.56	.51
Stream water	.62	.93	.62	.21

[1] Taken from Best (1971)

concentrations of most constituents were greatest in the growing season. Patterns were similar to the hardwoods, with enrichment in throughfall and litter water and depletion or little change at 25 cm. However, in all cases, ions were depleted between 25 cm and the stream. Compared to hardwoods, this greater depletion is attributed partially to white pine's higher growth rate (Swank and Schreuder, 1973), and substantially greater evapotranspiration from pine, which reduced water movement to the stream (Swank and Douglass, 1974).

CONCLUSIONS

Taken collectively, these data allow some general statements or inferences about solution alteration as water moves through these temperate forest ecosystems. The largest change in concentration of most ions occurs during the growing season, when plant uptake and other biological processes are most active. Patterns of cation changes through compartments are generally similar for the three types, but there are a few exceptions. Patterns appear to be most different for NO_3 and SO_4, which are mediated by microbial transformations.

The forest canopy, litter, and upper 25 cm of soil are major compartments of ion exchange. Hydrogen ions delivered in precipitation are important in leaching cations from these compartments. Washdown of aerosols impacted in the canopy is significant for some ions. Foliar absorption and utilization by microflora also play a role in altering water chemistry, as do decomposition and microbial transformations in the forest floor. As water moves through the soil, plant uptake, ionic exchange on mineral and organic sites, microbial transformations, and weathering all interact to further alter solution chemistry. The influence of evaporative processes on water chemistry is probably most important in the soil in association with transpiration.

It is clear that the presence or absence of forest vegetation strongly affects the chemical composition of water which eventually contacts the underlying saprolite and bedrock. Furthermore, it appears that solution chemistry also varies with the type of forest vegetation. These alterations in water chemistry could influence solubility and oxidation-reduction rates of minerals as well as the rate of chemical weathering. Since solutional denudation is an important erosional process, the consideration of present and past plant cover in association with biological processes could be important in interpretations of land forms.

ACKNOWLEDGEMENTS

This research was supported in part by the Southeastern Forest Experiment Station and in part by the National Science Foundation Grant DEB 7904537. The senior author gratefully acknowledges the sponsorship provided by the International Geographical Union Commission on Field Experiments in Geomorphology which facilitated his participation in the meeting. We also thank Mr. James E. Douglass, Dr. Gray S. Henderson, and Dr. Bruce L. Haines for their technical review of the manuscript.

REFERENCES

Best, G.R., 1971, Potassium, sodium, calcium and magnesium flux in a mature hardwood forest watershed and an eastern white pine forest watershed at Coweeta, Master of Science Thesis, University of Georgia, Athens.

Best, G.R. and Monk, C.D., 1976, Cation flux in hardwood and white pine watersheds, in: *Mineral cycling in Southeastern ecosystems*, ed Howell, F.G., Gentry, J.B. and Smith, M.H., (ERDA Symposium Series, Conf-740513), 847-861.

Buldgen, P. and Remacle, J., 1981, Influence of environmental factors upon the leaching of cations from undisturbed microcosms of beech and spruce litters, *Soil Biology and Biochemistry*, 13, 143-147.

Cleaves, E.T., Godfrey, A.E. and Bricker, O.P., 1970, Geochemical balance of a small watershed and its geomorphic implications, *Geological Society of America Bulletin*, 81, 3015-3032.

Cole, D.W., Gessel, S.P. and Held, E.E., 1961, Tension lysimeter studies of ion and moisture movement in glacial till and coral atoll soil, *Soil Science Society of America Proceedings*, 25, 321-325.

Cole, D.W. and Johnson, D.W., 1977, Atmospheric sulfate additions and cation leaching in a Douglas-fir ecosystem, *Water Resources Research*, 13, 313-317.

Cromack, K., Jr. and Monk, C.D., 1976, Litter production, decomposition, and nutrient cycling in a mixed hardwood watershed and a white pine watershed, in: *Mineral cycling in Southeastern ecosystems*, ed. Howell, F.G., Gentry, J.B. and Smith, M.H., (ERDA Symposium Series, Conf-740513), 609-624.

Eaton, J.S., Likens, G.E. and Bormann, F.H., 1973, Throughfall and stemflow chemistry in a northern hardwood forest, *Journal of Ecology*, 61, 495-508.

Fyfe, W.S., 1981, The environmental crisis: quantifying geosphere interactions, *Science*, 213, 105-110.

Hansen, E.A. and Harris, A.R., 1975, Validity of soil-water samples collected with porous ceramic cups, *Soil Science Society of America Proceedings*, 39, 528-536.

Hatcher, R.D., Jr., 1974, *Introduction to the tectonic history of northeast Georgia*, (Georgia Geological Survey Guidebook 13-A).

Hatcher, R.D., Jr., 1979, The Coweeta Group and Coweeta syncline: major features of the North Carolina-Georgia Blue Ridge, *Southeastern Geology*, 21, 17-29.

Hatcher, R.D., Jr., 1980, *Geologic map of the Coweeta Hydrologic Laboratory, Prentiss Quadrangle, North Carolina*, (State of North Carolina, Department of Natural Resources and Community Development, prepared in cooperation with Tennessee Valley Authority).

Henderson, G.S., Harris, W.F., Todd, D.E. and Grizzard, T., 1977, Quantity and chemistry of throughfall as influenced by forest type and season, *Journal of Ecology*, 65, 365-374.

Johnson, D.W. and Cole, D.W., 1976, Sulfate mobility in an outwash soil in western Washington, in: *Proceedings of the first international symposium on acid precipitation and the forest ecosystem*, (General Technical Report NE-23, Forest Service, Northeastern Forest Experiment Station, Broomall, Pennsylvania), 827-836.

Johnson, D.W., Hornbeck, J.W., Kelly, J.M., Swank, W.T. and Todd, D.E., Jr., 1980, Regional patterns of soil sulfate accumulation: relevance to ecosystem sulfur budgets, in: *Atmospheric sulfur deposition: environmental impact and health effects*, ed. Shriner, D.S., Richmond, C.R. and Lindberg, S.E., Proceedings of the second life sciences symposium, (Ann Arbor Science Publications, Inc., Ann Arbor, Michigan), 507-520.

Johnson, P.L. and Swank, W.T., 1973, Studies of cation budgets in the southern Appalachians on four experimental watersheds with contrasting vegetation, *Ecology*, 54, 70-80.

Lang, G.E., Reiners, W.A. and Heier, R.H., 1976, Potential alteration of precipitation chemistry by epiphytic lichens, *Oecologia*, 25, 229-241.

Likens, G.E., Bormann, F.H., Pierce, R.S., Eaton, J.S. and Johnson, N.M., 1977, *Biogeochemistry of a forested ecosystem*, (Springer-Verlag, New York).

Lindberg, S.E. and Harriss, R.C., 1981, The role of atmospheric deposition in an eastern United States deciduous forest, *Water, Air, and Soil Pollution*, 16, 13-31.

McColl, J.G., 1972, Dynamics of ion transport during moisture flow from a Douglas-fir forest floor, *Soil Science Society of America Proceedings*, 36, 674-688.

McColl, J.G. and Cole, D.W., 1968, A mechanism of cation transport in the forest soil, *Northwest Science*, 42, 134-140.

Miller, H.G., Cooper, J.M. and Miller, J.D., 1976, Effects of nitrogen supply on nutrients in litter fall and crown leaching in a stand of Corsican pine, *Journal of Applied Ecology*, 13, 233-248.

Monk, C.D., Crossley, D.A., Jr., Todd, R.L., Swank, W.T., Waide, J.B. and Webster, J.R., 1977, An overview of nutrient cycling research at Coweeta Hydrologic Laboratory, in: *Watershed research in eastern North America: a workshop to compare results*, Vol. I, ed. Correll, D.L., (Smithsonian Institute, Edgewater, Maryland), 35-50.

Pike, L.H., 1978, The importance of epiphytic lichens in mineral cycling, *The Bryologist*, 81, 247-257.

Sollins, P., Grier, C.C., McCorison, F.M., Cromack, K., Jr., Fogel, R. and Fredriksen, R.L., 1980, The internal element cycles of an old-growth Douglas-fir ecosystem in western Oregon, *Ecological Monographs*, 50, 261-285.

Swank, W.T. and Douglass, J.E., 1974, Streamflow greatly reduced by converting deciduous hardwood stands to pine, *Science*, 185, 857-859.

Swank, W.T. and Douglass, J.E., 1977, Nutrient budgets for undisturbed and manipulated hardwood forest eco-systems in the mountains of North Carolina, in: *Watershed research in eastern North America: a workshop to compare results*, Vol. I, ed. Correll, D.L., (Smithsonian Institute, Edgewater, Maryland), 343-363.

Swank, W.T. and Schreuder, H.T., 1973, Temporal changes in biomass, surface area, and net production for a *Pinus strobus* L. forest, in: *IUFRO biomass studies, International Union of Forest Research Organizations, working party on the mensuration of the forest biomass*, (College of Life Sciences and Agriculture, University of Maine at Orono), 171-182.

Swank, W.T. and Waide, J.B., 1980, Interpretation of nutrient cycling research in a management context: evaluating potential effects of alternative management strategies on site productivity, in: *Forests: fresh perspectives from ecosystem analysis*, Proceedings of the 40th annual biology colloquium, (Oregon State University Press, Corvallis), 137-158.

Tarrant, R.F., Bollen, W.B. and Chen, C.S., 1968, *Nutrient cycling by throughfall and stemflow precipitation in three coastal Oregon forest types*, (Research paper PNW-54, Forest Service, Pacific Northwest Forest and Range Experiment Station, Portland, Oregon).

Ulrich, B., Mayer, R. and Khanna, P.K., 1980, Chemical changes due to acid precipitation in a loess-derived soil in central Europe, *Soil Science*, 130, 193-199.

Yarie, J., 1980, The role of understory vegetation in the nutrient cycle of forested ecosystems in the mountain hemlock biogeoclimatic zone, *Ecology*, 61, 1498-1514.

21 Variable solute sources and hydrologic pathways in a coastal subalpine environment

T.M. Gallie and H. Olav Slaymaker

INTRODUCTION

This paper describes sources and pathways of solutes within a 2 ha, subalpine watershed in southwestern British Columbia. Solute flowpaths in this environment are variable and the nature of this variability has important implications for hydrochemical sampling strategies.

STUDY AREA

The study site is situated on the crest of a ridge adjacent to the Lillooet River, between 1800 m and 1900 m above sea level. The area lies in the Pacific Ranges of the Coast Mountains, in southwestern British Columbia. The mesoclimate is cold, perhumid. Annual precipitation exceeds 1600 mm and most of this falls as snow. Continuous snow cover persists for 7 to 9 months per year.

The local bedrock is a convoluted association of steeply dipping, well jointed metasediments and metaplutonics that have suffered moderate hydrothermal alteration. These late Mesozoic units are part of a 25 km^2 roof pendant exposed within the dioritic Coast Mountains pluton.

Local surficial materials include at least one late Pleistocene basal till which is predominantly dioritic. This stony till is overlain by several fine textured Holocene units of low bulk density. Pyroclastic ejecta from the Mt. Mazama (6600 BP) and Bridge River (?) (2600 BP) volcanic events occur in local sediment traps interspersed with 10 to 30 cm thick organic deposits. The organic units are decayed remnants of Hypsithermal Hemlock forests. The ash and organic deposits are rich in aeolian silts winnowed from local talus and perhaps from late Neoglacial moraines approximately 3 to 4 km to the west.

The basal till is buried by colluvium in geomorphically active sites. Coarse talus occurs beneath glacially oversteepened cliffs. Stone-banked terraces are active below bedrock slabs, particularly on north-facing slopes. Well washed, sandy gravel fans form on gentle foot slopes.

The soil landscape is areally dominated by acidic Brunisols (Inceptisols) and Cumulic Regosols (Entisols). Brunisols are weakly developed soils, in this case having low base status. Cumulic Regosols are disturbed soils with poorly developed horizons (Canadian Soil Survey Committee, 1978). Within this 2 ha watershed, however, pedons range from Ferrohumic Podzols (Spodosols) beneath well drained tree islands to Orthic RegoGleysols (Aquents) in poorly drained lowland sites.

The vegetation is equally heterogeneous. The watershed is located near the upper altitudinal limit of mountain parkland in the local alpine-subalpine ecotone. Five vegetation associations informally describe the major communities:

> tree islands
> heath communities
> dryland sedge
> wetland sedge
> sparsely vegetated, active sites

There is strong covariance among surficial materials, soil groups, and vegetation associations. Contiguous soil-vegetation complexes are mapped in figure 21.1 and, as will be shown, are important surrogates for infiltrability (table 21.1).

HYDROLOGY OF ENVIRONMENTAL SYSTEMS

Runoff is rapid and substantial. Most stream discharge occurs from May to July as snowmelt, but rainstorms and rain-on-snow events generate several significant runoff events in the autumn. Peak discharge intensities exceed 0.62 m^3 s^{-1} km^{-2}, and daily snowmelt-runoff ratios approach unity. Clearly, event contributing areas can be large and dynamic in this watershed (Dunne et al., 1975). These large contributing areas occur because water does not move uniformly through the soil but moves along preferred pathways.

Precipitation-excess overland flow (Horton, 1933; Freeze, 1974) is generated on bedrock outcrops and in sedge hollows because these systems have low infiltrability. Turfy, organic-rich epipedons in sedge hollows exhibit strong water-repellency and ponding is necessary to induce significant infiltration. The causes of the water-repellency have not been isolated but it is suspected that this is due to accumulations of hydrophobic organic compounds near the soil surface (Barrett, 1981).

Saturation overland flow (Dunne, 1978) is generated in heath, debris and wetland complexes where concavities force subsurface flowlines to converge. Soil moisture storage and transmission capacity are exceeded as profiles become saturated from bottom to top.

Preferred subsurface pathways occur as vertical fingers, horizontal, coarse-textured, subsurface channels or pipes, and thin, transient saturated lenses. Fingers are most common in soil-vegetation complexes with surface water-repellency and transient surface detention storage (sedge

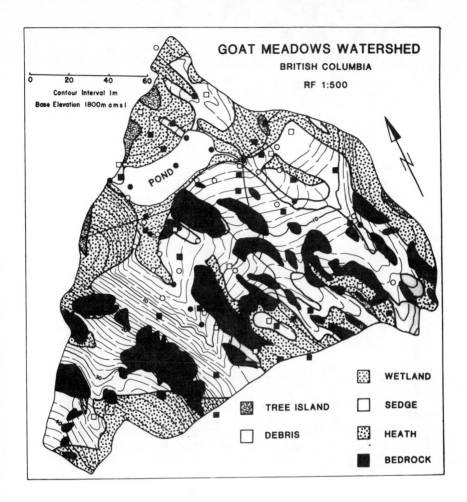

Figure 21.1. Map of Goat Meadows Watershed. Solute
sampling sites are indicated by symbols (streamwater,
closed circles; overland flow, open squares; saturated
soil water, open circles; unsaturated soil water,
closed squares).

and heath), but fingering also appears to be important
beneath tree islands.

Subsurface channels have been observed only in active
debris sites along contacts between till and colluvium.
They are of coarser texture than the surrounding matrix
and, based on current surface processes, we speculate that
the pipes are former surface channels that have been buried
by mass movements. Because of their coarse texture and
their protracted discharge following runoff events, we
infer that they are activated by saturated zones that rise
to intersect the channel. Indeed, multiple saturated zones
can be observed in heath and sedge soil pit walls during
and following sustained events. They occur along textural

Table 21.1. Soil-vegetation complexes characterized according to slope position, general stability, and infiltrability. Numbers in parentheses indicate areal occurrence in the watershed (4% occupied by pond).

	STABLE, WELL DEVELOPED SOIL HORIZONS. WELL VEGETATED. ASH, LOESS AND ORGANIC MATTER OVER TILL		UNSTABLE OR POORLY DEVELOPED SOIL HORIZONS. POORLY VEGETATED. COLLUVIUM OVER BEDROCK OR TILL	
	COMPLEX	INFILTRABILITY	COMPLEX	INFILTRABILITY
Interfluves	TREE ISLANDS: (2%) Rarely if ever saturated. Fingering important.	HIGH	BEDROCK OUTCROPS: (16%) No cover. Overland flow.	LOW
Midslopes	HEATH COMMUNITIES: (30%) Fingering. Subsurface saturation at textural contact.		ACTIVE OR RECENTLY ACTIVE DEBRIS: (33%) Fingering. Subsurface channels activated by rising saturation zones.	
Footslopes	SEDGE HOLLOWS: (12%) Dense, turfy root mats. Hydrophobic. Overland flow and fingering.	LOW	SATURATED WETLANDS: (3%) Usually saturated.	HIGH

contacts, particularly loess over till beneath the stable heath communities.

Finally, irregular soil horizon boundaries and thick organic matter and sesquioxide accumulations along large roots and cobbles, suggest that soil macropores may be conducting disproportionate quantities of infiltrating surface water. This observation is consistent with work by Chamberlain (1972), Cheng et al., (1975), and DeVries and Chow(1978) for lower altitude forest soils in British Columbia.

Taken together, these phenomena point to a hierarchy of preferred pathways for event runoff. The primary pathways, or contributing areas, for event runoff are the stream channel network, adjacent wetlands, sedge percolines and bedrock outcrops, which become extensions of the channel network during prolonged events. Subordinate to this is a subsurface network of saturated vertical fingers, subsurface channels and transient saturated zones in heath and debris sites which carry relatively rapid event runoff to the primary contributing areas. Subordinate to this is a network of high connectivity soil channels (macropores) at all sites which allows some water to bypass the soil matrix as it moves towards a local pathway. Each soil-vegetation complex (table 21.1) has a different combination of these preferred hydrologic pathways.

WATER CHEMISTRY

Soil-vegetation complexes were expected to show significant differences in water chemistry because of their different hydrology and pedochemical properties. Specifically,

> i) variations in hydraulic properties should lead to variations in soil-water contact time between environments, affecting effective reaction rates and

> ii) variations in corrosive potential between environments should lead to variations in actual reaction rates.

For this reason the sampling design was based on the areal coverage of the individual soil-vegetation complexes as well as the major flow components. The total number of samples taken over 12 months of field work was 645. These included (figure 21.1):

> i) snowmelt and bulk precipitation

> ii) stream chemistry measured throughout the channel network (closed circles)

> iii) persistent overland flow and surface detention storage sites (open squares)

> iv) saturated subsurface waters (open circles) in piezometers, throughflow troughs, and at seeps

> v) unsaturated subsurface waters (closed squares) collected using vacuum samplers.

Results from this sampling program showed small but significant differences in soil solution chemistry between soil-vegetation complexes. However, these differences were minor in comparison with the difference discussed below

Unexpectedly, there is a dramatic difference in Ca/Na molar ratios within the data set as a whole. Using this criterion, only two major populations result (figure 21.2). One population is predominantly sodic and consists of capillary soil water (water held under tension) withdrawn under low suction (45 cm Hg or about 1 bar). The other population is relatively richer in calcium, becoming more calcic as total dissolved solids concentrations increase. These samples have a variety of origins but are predominantly overland flow, saturated zone soil water, and streamflow.

Several experiments were initiated to clarify the nature of these two populations. First, small hillslope plots were isolated. Saturated zone flows were sampled with throughflow gutters, and unsaturated zone flows with vacuum samplers. Secondly, undisturbed soil monoliths were leached with distilled water in the laboratory. Finally, water soluble salt extractions were analysed from samples of each horizon of every soil group in the watershed.

Water from mineral soil horizons, whether from saturated or unsaturated zones, was strongly sodic in all experiments. The Ca/Na molar ratios clustered around 0.25. This occurs largely because soil water reacts with acidic volcanic ash, loess and till which are the soil parent materials.

In contrast, spring waters from the metamorphic bedrock have Ca/Na molar ratios greater than 8.0 and their concentrations range up to 800 mmoles m^{-3} Ca^{2+}. Bedrock is petrologically heterogeneous but calcic silicate assemblages are abundant in all units and are likely Ca^{2+} sources. Groundwater flow nets indicate a theoretical flow pattern of discharge to the wetlands adjacent to the pond and recharge beneath the pond floor (figure 21.1). The wetlands are, in fact, the predominant source areas of concentrated calcic surface water in the catchment. Furthermore, pond level fluctuations during July and August registered net losses of 14 mm day^{-1} to groundwater. These observations are consistent with the hypothesis that saturated zone soil water is contaminated by calcic groundwater.

Thus calcic bedrock groundwater and sodic mineral soil water are two distinct ionic sources. Precipitation and snowmelt comprise a third source. This meteoric source is very dilute and only slightly sodic. Variable mixing of ions from bedrock, soil and atmospheric sources produces the calcium-rich concentration pattern of free water in figure 21.2.

Dilute free water samples, such as overland flow and transient surface detention storage, are the least calcic free water samples, because they are only slightly altered meteoric waters. They are a significant fraction of the data set because 30% of the watershed is composed of bedrock outcrops, sedge and wetlands. These are discontinuously impermeable surfaces during all but the most sustained events. This limits soil-water contact and

352

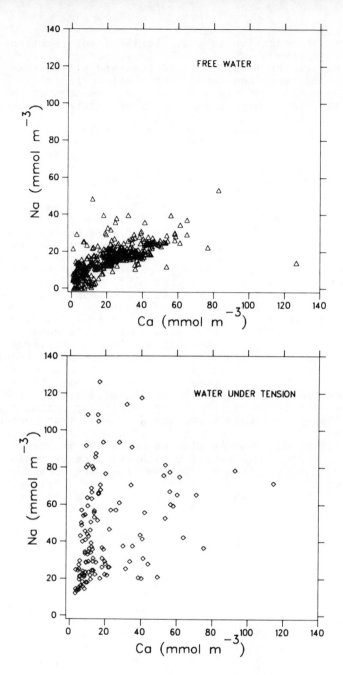

Figure 21.2.   Field data contrasting molar calcium and
   sodium concentrations of capillary soil water (water
   under tension) and gravity or free water. Capillary
   soil water samples with higher calcium content are
   from humic horizons or from the capillary fringe above
   a calcic saturated zone.

reaction concentration levels. Furthermore, reactions at the soil surface occur within humic epipedons, which are relatively richer in all soluble ions and silica than mineral horizons. Leachates from humic horizons are generally sodic, but can have molar ratios as high as 1:1 Ca/Na, presumably reflecting an abundance of loosely bound ions in decaying plant litter.

The more concentrated free water samples are deficient in sodium relative to capillary soil water (figure 21.2) because capillary soil water contributions to free water are either limited or else highly diluted. In all field and laboratory experiments soil effluent is uncharacter-istically dilute relative to water held under tension. This may occur because soil water flow is not uniform.

Soil pores within a given matrix are neither uniform in size nor in connectivity. When soil is unsaturated, small pores are filled with water and conduct most of the flow because the largest pores are filled mostly with air. When soil is saturated, large pores become filled with water and control water flow. Thus, flow volumes in saturated zones can be several orders of magnitude greater than in unsaturated zones. In both conditions there exist a number of 'dead end' pores which have poor communication with regions of flow. This leads to variable contributing volumes within a single monolith. DeSmedt and Wierrenga (1979) refer to contributing and non-contributing fractions as 'active' and 'inactive' zones. The active zone is characterized by mobile pore water, relatively short water residence times and dispersive mixing of solutes. The inactive zone is characterized by relatively immobile pore water, longer residence times and diffusion transport of ions toward the active zone along concentration gradients.

Similar processes may also be operative at larger scales in preferred subsurface pathways. Water residence times in fingers, pipes and lenses are short because saturated soils are hydraulically efficient. Primary soil-water reactions are thus limited. Additional solute contributions from nearby unsaturated zones should be minimal because ions must migrate toward the saturated zone by diffusion or by unsaturated flow, which are relatively inefficient mass transfer processes.

SUMMARY

Much water appears to bypass the soil matrix as overland flow or moves through the soil along discrete, saturated, subsurface flowpaths. These preferred pathways limit primary soil and water reactions as well as ionic inputs from adjacent, relatively inactive zones. Soil effluent thus contains a dilute chemical signature from the un-saturated zone regardless of the pedon of origin. Ground-water on the other hand is relatively concentrated and only small additions in low lying sites produce calcic stream-flow from a watershed areally dominated by sodic soils. This occurs because groundwater discharge zones coincide with the saturated wetlands and channels which are the primary contributing areas in the watershed.

Three fluctuating solute sinks result from this mixing
pattern; unsaturated, transient saturated, and permanently
saturated soil water zones. Unsaturated zones are relatively inactive and contain concentrated capillary soil-
water. Transient saturated zones are active and conduct
dilute mixtures of meteoric and capillary soil water during
and following events. Permanently saturated zones are
reservoirs of intermediate concentration, receiving additions from all sources within the watershed.

DISCUSSION

We conclude that the sources and pathways of solute discharge from these slopes are highly variable. In this
case, solute variability at individual sites is as high as
solute variability between sites. Mixing and transport
rates are time variant because flowpaths become active
under different event intensities and antecedent moisture
conditions. Details of these processes are controlled by
the hydraulic properties and topography of the solum.
Thus, soil-vegetation complexes are rational and convenient
sampling units for hydrologic process studies.
There is a general decrease in solute variability from
slope interfluves to the basin outlet. Solute variability
at channel sites is less than at slope sites. Tributary
confluences and permanently saturated wetlands exhibit the
most stable water chemistry. This trend appears to reflect
the spatial integration of flowpaths and sources as they
converge on the channel network. Mixing becomes increasingly more complete as water moves towards the basin outlet
while individual flow components become less distinct below
each pathway intersection.
If this conceptual model has general validity, it
implies that the information content of channel-based hydrochemical sampling programmes is limited. Channel sites in
the Goat Meadows watershed are useful for discriminating
gross flow components and variations in component solute
loads. Unfortunately, these channel sites are not useful
by themselves for interpreting mixing and transport processes on adjacent slopes, primarily because these phenomena
are time-variant, perhaps non-linear, and the chemical
signals from each source become indistinct with increasing
distance from the source. Upon reaching the channel network, sources are so well integrated (mixed) as to be
reduced to only two major components of flow:

> i) concentrated, long residence time
> water from saturated near-channel soil
> bodies and groundwater discharge zones and

> ii) dilute, short residence time water
> from direct channel precipitation, over-
> land flow, and quick event flow from a
> variety of subsurface pathways.

The former comprise indirect event flow during storm
sequences and sustain streamflow between storms. The
latter comprise direct event flow during and immediately

following the peak of each event. Discharge hydrographs can be separated into these two components using mass budget, load mixing equations (Steele, 1968; Pinder and Jones, 1969; Hall, 1970, 1971; Pilgrim et al., 1979). As stated earlier, near-channel chemistry is relatively stable, while precipitation and overland flow are sufficiently dilute as to be distinctive. One can substitute these ionic concentrations to produce approximate solutions to the definition of direct and indirect runoff quantities. However, if more physically based hydrograph separation is desired, the choice of component concentrations becomes more complex.

For example, event effluent from these soils is as much a product of dispersive mixing in active pathways as it is a product of primary pedogenic reaction kinetics. Much soil water is not shunted (replaced by fresh water) during storms so that uniform mixing in the soil reservoir appears to occur rarely. Situations arise where volumes of relatively inactive soil pores contain solutions near equilibrium concentration, while pore water in adjacent active zones exhibit near infinite dilution. Indeed, at this stage of the research it is difficult to characterise relatively static capillary soil water and groundwater concentrations. One can confidently identify their molar ratios; soil waters have Ca/Na < 0.5 while groundwaters have Ca/Na > 8.0. However, their Ca, Na concentrations range from 10 to 800 mmoles m^{-3}. Processes that may explain this variability in soils have already been discussed. Perhaps analogous phenomena occur in groundwater networks such as these, where flowpaths are short, fracture patterns may not be homogeneous, and mixing may not be complete.

In short, detailed sampling programmes within soil units or environmental systems having similar hydrologic properties appear necessary to understand the variability of hydrologic processes on slopes. The high variability of water chemistry at slope sites occurs because transport and mixing processes are time and space variant. Channel environments show much less variability due to spatial integration associated with flowpath convergence, but the information content of these signals is correspondingly reduced. It is possible that the standard assumptions of simple flow components and static ionic sources used in most hydrochemical hydrograph separations to date may be in part an artifact of inadequate spatial sampling designs.

REFERENCES

Barrett, G., 1981, Streamflow generation in the Coast Mountains of British Columbia, Unpublished MSc. Thesis, University of British Columbia, Vancouver B.C.

Canadian Soil Survey Committee, 1978, The system of soil classification for Canada, (Agric. Canada, Queens Printer, Ottawa).

Chamberlain, T.W., 1972, Interflow in the mountainous
    forest soils of coastal British Columbia, in:
    Mountain Geomorphology, ed. Slaymaker, H.O. and
    McPherson, M.J., B.C. Geographical Series No. 14,
    (Tantalus Research Ltd., Vancouver), 121-127.

Cheng, J.D., Black, T.A. and Willington, R.P., 1975, The
    generation of stormflow from small forested water-
    sheds in Coast Mountains of southwestern British
    Columbia, in: *N.R.C. Hydrology Symposium Pro-
    ceedings, Winnipeg*, 542-551.

DeSmedt, F. and Wierrenga, P.J., 1979, A generalized solu-
    tion for solute flow in soils with mobile and im-
    mobile water, *Water Resources Research*, 15, 1137-
    1141.

DeVries, J. and Chow, T.L., 1978, Hydrologic behavior of a
    forested mountain soil in coastal British Columbia,
    Water Resources Research, 14, 935-942.

Dunne, T., 1978, Field studies of hillslope flow processes,
    in: *Hillslope Hydrology*, ed. Kirkby, M.J., (Wiley-
    Interscience, New York), 227-293.

Dunne, T., Moore, T.R. and Taylor, C.H., 1975, Recognition
    and prediction of runoff-producing zones in humid
    regions,

Freeze, R.A., 1974, Streamflow generation, *Reviews of
    Geophysics and Space Physics*, 12, 627-647.

Hall, F.R., 1970, Dissolved solids-discharge relationships:
    1 Mixing Models, *Water Resources Research*, 6, 845-
    850.

Hall, F.R., 1971, Dissolved solids-discharge relationships:
    2 Application to field data, *Water Resources
    Research*, 7, 591-601.

Horton, R.E., 1933, The role of infiltration in the hydro-
    logical cycle, *Transactions of the American Geo-
    physical Union*, 14, 446-460.

Pilgrim, D.H., Huff, D.D., and Steele, T.D., 1979, Use of
    specific conductance and contact time relations for
    separating flow components in storm runoff, *Water
    Resources Research*, 15, 329-339.

Pinder, G.F. and Jones, J.F., 1969, Determination of the
    groundwater component of peak discharge from the
    chemistry of total runoff, *Water Resources Research*,
    5, 438-445.

Steele, T.D., 1968, Seasonal variation in chemical quality
    of surface water in the Pescadero Creek watershed,
    San Mateo County, California, Unpublished PhD
    dissertation, Stanford University.

.

22 Some implications of small catchment

solute studies for geomorphological research

I.D.L. Foster and I.C. Grieve

INTRODUCTION

Chemical weathering has been identified as a significant
component in landscape evolution (Garrels and Mackenzie,
1971), and studies on non-limestone areas are now relatively common in the geomorphological literature. Quantification of the chemical processes involved can be undertaken by a number of methods, eg. comparison of the
weathered rock and its unweathered equivalent (eg. Loughnan,
1969), comparison of unweathered rock with drainage waters
to identify relative mobilities (eg. Anderson and Hawkes,
1958) or on the basis of theoretical calculations (Garrels
and Christ, 1965).
      Investigations using small drainage basins permit the
testing of theoretical calculations and allow calculation
of budgets, provided external inputs to the solute system
can be identified (Likens *et al.*, 1977). These inputs are
primarily derived from the atmosphere in natural systems,
but in rural areas artificial fertilisers, pesticides and
animal slurries must also be identified. In addition,
analysis of ecosystem dynamics can be achieved by such an
approach and the impact of nutrient cycling and pedological
processes on rates of chemical denudation established.
      In this investigation, an attempt is made to establish
links between geochemical, ecological and pedological
processes operating in a small drainage basin, utilising
the budget approach, in order to estimate rates of solutional denudation and the major mechanisms involved.

THE FIELD AREA

The study was carried out in a small(95.16 ha) second order
drainage basin in North Warwickshire (figure 22.1A and B).
The catchment lies at approximately 130 m OD, and the area
has a relative relief of approximately 40 m with an average
channel slope of 0.046. The catchment is underlain by a
variable thickness of boulder clay overlying the Keele Beds
of the Upper Coal Measures. These latter comprise dominant

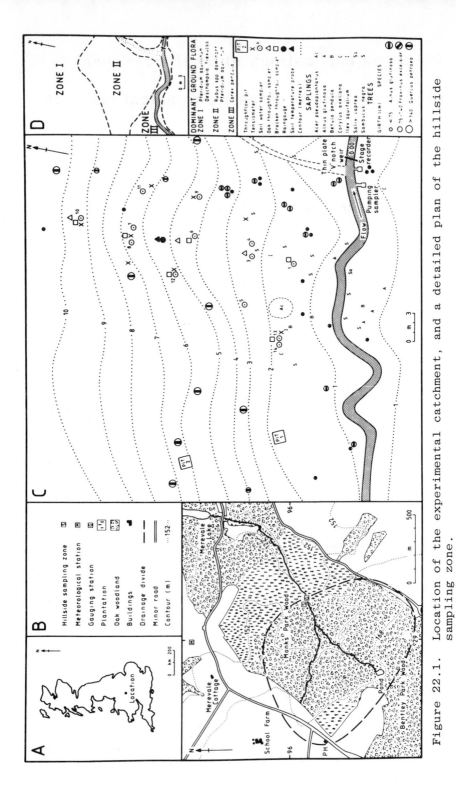

Figure 22.1. Location of the experimental catchment, and a detailed plan of the hillside sampling zone.

360

clay shales, some of which are calcareous, and subsidiary
sandstone outcrops. Soils of the majority of the area are
cambic stagnogleys of the Bardsey series, with gleyic brown
earths of the Melbourne series on the narrow sandstone
bands (W.A.D. Whitfield, Pers. Comm.). 78.4% of the total
area is covered by mature deciduous woodland and small
areas of grassland (14.1%) and coniferous plantation (7.5%)
occupy an area mainly confined to the margins of the
drainage divide in the south-west. The deciduous woodland
comprises *Quercus petraea* with an understorey of *Pteridium
aquilinum* and *Deschampsia flexuosa*, although *Alnus glutinosa*
with a more diverse ground flora dominates the narrow flood
plain.

Climatically, the area may be classified as humid
temperate with a cold season (Gregory, 1976). Rainfall
records for Elmdon (some 18.5 km to the south-west), available since 1945, indicate a mean annual precipitation for
the region of 681 mm and low seasonal variation (February -
46 mm, minimum; August - 71 mm, maximum). The coefficient
of variation of annual precipitation lies somewhere between
14 and 16% (Gregory, 1955). Precipitation for the period
under detailed investigation, 1978-1980, was some 9% higher
than the long term average, although on-site data indicate
a reduction in rainfall catch of around 110 mm in comparison
with the Elmdon records.

FIELD AND LABORATORY ANALYSIS

The data discussed in this paper relate to a period of
almost 2 years from December 1978 to September 1980 with
samples collected at 2 weekly intervals up to September
1979 and weekly thereafter. Inputs of precipitation were
monitored in an open field at the main meteorological
station (figure 22.1B). Volume was measured with a Casella
tipping bucket raingauge and samples for chemical analysis
were collected using 20 cm diameter polythene funnels
plugged with glass wool to prevent contamination.

Throughfall and soil water were monitored in a small
hillslope hollow adjacent to the main stream. This was
fairly representative of the main soil and vegetation zones
identified in the catchment (figure 22.1C and D). Throughfall was collected above and below the bracken canopy at
three and six sites respectively using 30 cm diameter polythene funnels plugged with glass wool.

Samples of soil water were collected from the B-horizon
of the soil using vacuum lysimeters (Parizek and Lane,
1970) at a network of seven sites on the hillslope, and
from the A and B horizons at two sites on the flood plain
and one site at the junction between the flood plain and
slope. In an adjacent hollow two throughflow pits were
constructed and water samples collected from the base of
the H, A and B horizons. Volumes were measured in the
field and samples were also taken for chemical analysis
(figure 22.1C).

Samples were collected from the main stream adjacent to
the hillslope segment by hand on each visit to the site,
and at 8 hourly intervals between visits during the second

361

Table 22.1. Chemical composition (mgl^{-1}) of selected water samples (1979-1980)

SAMPLE	Ca^{2+}		Mg^{2+}		Na^{+}		K^{+}		NH$_4^{+}$		NO$_3^{-}$		NO$_2^{-}$	
	$\bar{x}$	SD	$\bar{x}$	SD	$\bar{x}$	SD	$\bar{x}$	SD	$\bar{x}$	SD	$\bar{x}$	SD	$\bar{x}$	SD
A. Precipitation and throughfall														
Bulk Precipitation	2.78	4.22	0.76	1.17	7.18	13.89	1.83	2.17	1.86	1.70	4.66	3.71	0.10	0.22
Oak Throughfall	4.12	4.15	1.43	1.72	6.92	7.86	9.46	17.03	0.80	0.74	3.61	2.98	0.12	0.26
Bracken Throughfall	4.13	3.73	1.51	1.70	7.10	7.09	11.60	15.51	0.88	0.99	3.66	3.41	0.10	0.40
B. Hillslope Samples (soil water)														
S10	0.77	1.49	1.56	0.55	7.18	3.28	14.60	3.60	0.05	0.10	2.57	1.94	0.05	0.18
S8	1.86	1.50	2.44	1.98	46.02	41.40	13.79	9.99	0.08	0.12	12.66	13.49	0.02	0.03
S6	1.66	1.66	2.07	1.24	4.15	2.19	20.64	3.93	0.07	0.12	4.42	3.41	0.05	0.18
C. Hillslope Samples (throughflow)														
T1A	6.23	4.21	3.01	1.27	10.93	5.98	18.39	8.80	-	-	8.07	5.13	-	-
T1B	1.88	0.93	2.87	1.51	16.81	8.14	16.28	6.14	-	-	5.11	3.03	-	-
T1C	2.78	2.23	4.32	2.50	14.74	7.56	14.90	7.56	-	-	4.91	3.33	-	-
T2A	3.56	2.98	2.50	1.10	15.11	8.43	24.19	13.27	-	-	3.53	2.99	-	-
T2B	3.96	3.11	2.75	1.12	11.47	4.32	23.52	9.38	-	-	3.56	2.13	-	-
T2C	2.48	1.28	3.43	1.40	14.18	6.62	20.66	8.60	-	-	3.12	5.00	-	-
D. Flood plain samples (soil water)														
S4	18.87	5.16	11.78	3.60	28.24	9.80	6.17	1.70	0.12	0.22	45.27	39.25	0.03	0.06
S3	31.26	8.58	20.93	3.89	17.52	2.85	6.17	1.74	0.11	0.16	3.90	4.92	0.06	0.19
S13	15.17	6.86	21.71	3.31	14.36	2.11	2.20	0.72	0.09	0.26	1.99	1.93	0.06	0.18
S1	35.22	9.73	23.03	4.94	16.27	3.09	3.80	1.25	0.34	0.98	34.84	48.36	0.04	0.08
E. River Water Samples														
ST1	43.85	11.78	10.15	2.47	16.54	3.09	3.04	1.79	0.08	0.15	4.16	3.13	0.03	0.06

(Table 22.1. cont.)

SAMPLE	Cl⁻		HCO₃⁻		CO₃²⁻		SO₄²⁻		pH		SC[1]	
	$\bar{x}$	SD	$\bar{x}$	SD	$\bar{x}$	SD	$\bar{x}$	SD	$\bar{x}$	SD	$\bar{x}$	SD
A. Precipitation and throughfall												
Bulk Precipitation	11.41	20.14	7.74	6.51	0.0	0.0	8.05	10.00	5.24	0.99	83.91	73.16
Oak Throughfall	9.77	8.59	5.90	7.46	0.0	0.0	12.84	11.20	4.58	0.82	134.14	68.79
Bracken Throughfall	11.71	9.16	13.24	15.65	0.0	0.0	12.35	11.88	4.75	1.04	155.68	93.65
B. Hillslope Samples (soil water)												
ST0	19.75	6.29	0.06	0.33	0.0	0.0	25.77	17.08	4.06	0.25	231.66	18.06
S8	18.00	8.01	0.0	0.0	0.0	0.0	29.61	10.99	3.97	0.36	412.63	218.42
S6	11.53	6.65	0.44	1.63	0.0	0.0	20.74	14.80	4.21	0.45	178.67	24.95
C. Hillslope Samples (throughflow)												
T1A	12.00	6.64	–	–	–	–	–	–	4.65	0.47	257.56	83.20
T1B	16.47	8.20	–	–	–	–	–	–	4.40	0.41	253.50	70.21
T1C	19.18	9.65	–	–	–	–	–	–	4.89	0.72	254.68	53.97
T2A	18.34	9.45	–	–	–	–	–	–	4.32	0.46	349.04	145.44
T2B	22.40	12.29	–	–	–	–	–	–	4.08	0.32	358.88	132.94
T2C	16.61	7.44	–	–	–	–	–	–	3.99	0.64	341.15	95.19
D. Flood plain samples (soil water)												
S4	8.34	3.36	26.11	8.07	0.0	0.0	29.99	9.56	6.02	0.43	444.76	106.77
S3	28.99	6.04	220.65	65.77	0.0	0.0	24.59	15.09	6.73	0.24	494.94	64.15
S13	40.07	7.86	37.05	30.22	0.0	0.0	31.67	15.12	5.68	0.30	392.33	48.97
S1	28.35	6.26	57.35	15.80	0.0	0.0	41.39	17.78	6.22	0.37	547.5	62.45
E. River Water Samples												
ST1	31.81	8.92	170.40	45.77	4.53	7.73	31.33	9.86	7.50	0.23	523.19	93.08

[1] in μ cm⁻¹

year using an automatic vacuum sampler. Flow was recorded continuously on a Munro vertical drum stage recorder linked to a compound v-notch thin-plate weir.

All water samples were filtered through Whatman GF/C filters prior to analysis. The relative proportion of organic and inorganic material in suspended sediment was determined from filter paper residues on a limited number of river water samples, using wet oxidation techniques similar to those described by Maciolek (1963).

Ca^{2+} and Mg^{2+} concentrations in water samples were determined by atomic absorption spectrophotometry, Na^+ and K^+ by flame photometry, NO_3^- and Cl^- by autoanalyser (colorimetry), and pH and specific conductance (SC) were also measured. In the second year, HCO_3^-, CO_3^{2-}, (titration), SO_4^{2-}, NO_2^- (autoanalyser), NH_4^+ (colorimetry) and dissolved organic matter (DOM: wet oxidation and titration) were also determined on most water samples. Some analyses of Mn^{2+}, Fe^{3+}, Al^{3+} and Si^0 (atomic absorption spectrophotometry) were also carried out.

Specific conductance was used as an indirect measure of total dissolved solids (TDS) for all samples but was not found to be suitable for prediction of individual ionic concentrations (cf. Foster *et al.*, 1981).

AVERAGE CONCENTRATIONS AND FLOW VOLUMES

Data representing the typical chemical composition of water samples collected at various sites within the catchment are presented in table 22.1. Significant downslope trends are evident for the soil water samples, with most ions increasing in concentration towards the stream. Throughflow chemistry demonstrates a similar pattern to soil water data obtained from vacuum lysimeters at the slope foot. Differences between the sites are further emphasised by comparison of Fe^{3+}, Al^{3+}, Mn^{2+} and Si^0 concentrations, where the first three ions dominate slope samples but are only present at low levels in river water. Silica levels are, however, considerably less variable (table 22.2).

Total runoff for the second year of study was 198.4 mm, from an input of 635 mm. By subtraction, this indicates that the majority of moisture loss from this environment is by evapotranspiration, and the loss of 436.6 mm is comparable with other estimates of evaporation losses for the area. MAFF (1967) gives potential transpiration for Warwickshire as 491.7 mm per annum and actual evapotranspiration losses published by the Water Data Unit (1978) are in the range 385-455 mm for the Upper Trent to 428-508 mm for the Upper Nene.

BIOGEOCHEMICAL INTERACTIONS

The significance of small drainage basin studies for analysing biological, hydrological and geochemical processes has received increased attention in recent years, with studies by Likens *et al.*, (1977), Verstraten (1977),

Table 22.2. Comparison of solute concentrations for occasionally determined ions in solution (mg l^{-1})

(1979-1980)

	Site	Fe^{3+}	Al^{3+}	Mn^{2+}	Si^0
A.	Slope soil water samplers	1.00	2.52	2.81	9.1
B.	Flood plain soil water samplers	0.05	0.18	0.42	5.4
C.	Stream	0.0	0.45	0.03	6.8

Waylen (1979), Webb and Walling (1980) and Reid et al. (1981). These studies clearly demonstrate the applications in terms of identifying both rates of chemical weathering and the mechanisms and mineralogical changes involved (cf. Garrels and Mackenzie, 1971).

Preliminary geochemical analyses usually employ tri-linear diagrams (Piper, 1944), which enable changes in the relative chemical composition of water samples to be compared, with units expressed in terms of chemical equivalance. Figure 22.2, based on annual mean water and soil chemical data, clearly demonstrates significant alterations in chemical compositions through the hillslope system. Such large scale modifications represent a combination of pedological, geochemical and nutrient cycling processes operating within the system. Cation inputs are dominated by Na^+ and K^+, with similar compositions in throughfall. Some alteration, notably a decrease in Mg^{2+} and Ca^{2+}, occurs on the slope head sites, with slight increases in Mg^{2+} downslope. Concentrations of Ca^{2+} and Mg^{2+} on the soil exchange complex were high relative to concentrations in solution. This reflects the strong adsorption of these ions and partly explains the relative reduction in Ca^{2+} and Mg^{2+} between precipitation and hillslope soil water. The most dramatic change occurs in the floodplain site which, to some extent, interacts hydrologically with the river. Here, dramatic increases in the amount of Ca^{2+} and Mg^{2+} in solution are recorded. River water samples demonstrate a much higher level of Ca^{2+}, which suggests non-equilibrium conditions with the flood plain. This is thought to reflect the weathering of calcareous shale outcrops, which occur over limited areas of the basin.

Anion inputs are dominated by NO_3^-, Cl^- and SO_4^{2-}, and the slope sites reflect a similar chemical composition. In contrast, flood plain sites and the main stream demonstrate marked increases in HCO_3^- in terms of total chemical composition, again reflecting the input from weathering of carbonate rocks.

At the present time, soil mineralogy and bedrock composition data are unavailable and detailed weathering reactions cannot, therefore, be established.

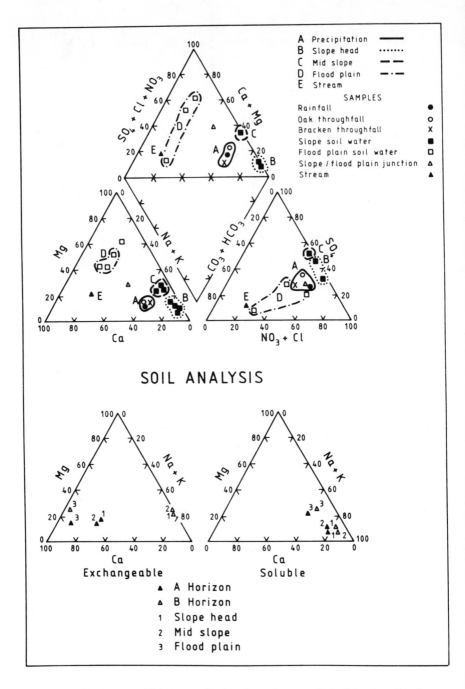

Figure 22.2.   Trilinear plot of water and soil chemical
analyses.

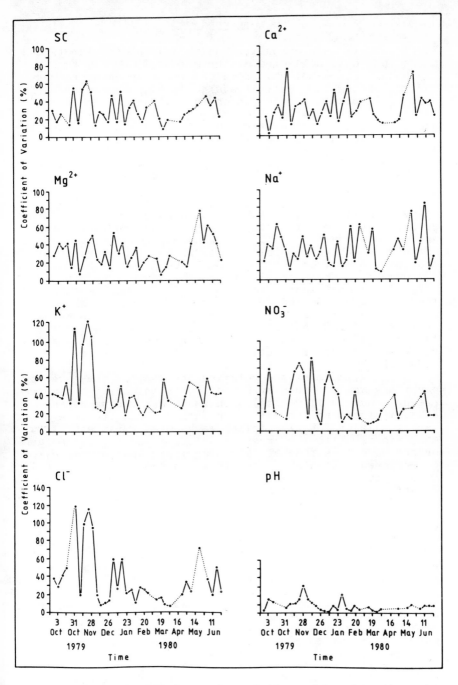

Figure 22.3. Coefficient of variation of bracken throughfall samples.

Input budgets were calculated on the basis of volumetric and solute data collected at the meteorological site and in throughfall samplers. Some caution must be expressed for the throughfall estimates since a high degree of variability in chemical composition is indicated, with coefficients of variation exceeding 50% for the six gauges used to collect bracken throughfall (figure 22.3). Average concentration and volumetric data were used to calculate budgets.

Calculation of outputs in river waters presents a far more complex problem since the use of rating curves for calculation of sediment and solute loadings may produce significant errors, (eg. Walling, 1978; Foster, 1980). However, malfunction of the vacuum sampler for long periods precluded the use of instantaneous solute and flow data, and rating curves for all ions and suspended sediment were established where possible (table 22.3). No significant models were derived for DOM, SO_4^{2-}, CO_3^{2-}, NO_2^- and NH_4^+, although with the exception of SO_4^{2-}, these ions represent only minor constituents in runoff. SO_4^{2-} budgets were therefore approximated on the basis of volume-weighted average concentrations. Seasonal rating equations were applicable for SC and Cl^- data, and all other rating relationships were based on a single analysis. All relationships were of power-function form with no evidence of models higher than first order.

Some estimate of the relative accuracy of these rating equations was obtained by comparing rating curve estimates (using 6 hourly flow data) with loads derived using the 8 hourly flow and concentration data derived from the vacuum sampler for 21 sampling weeks over a range of flow conditions. With the exception of Ca^{2+} and NO_3^-, average errors were less than 5%, although individual weekly errors were somewhat higher on occasions. Annual rates of chemical denudation based on these analyses are considered sufficiently accurate, but some caution must be expressed in relation to budgets calculated at shorter time periods.

Throughput budgets were not calculated, since the volume of water collected in the throughflow pits could not be assigned to a definite catchment area. However, volume-weighted annual mean concentrations were calculated for the base of the B horizon. Comparison of these with volume-weighted input and output concentrations gives an indication of the net effect of the soil system, assuming that water percolating below this depth was subject to negligible evaporation and plant uptake losses (table 22.4). The right-hand columns show the ratio of concentration in each output to input concentration. If the net effect of passage through the soil and groundwater zones is simply concentration due to evaporation, the output ratios should be approximately 3.2 (635 mm input divided by 198.4 mm output). From table 22.4, Na^+, K^+, NO_3^- and H^+ concentrations in soil water are clearly adequate to account for total stream losses, but Ca^{2+}, Mg^{2+}, and Cl^- must be augmented by weathering in the ground water zone. This latter appears to act as a sink for K^+ and H^+. K^+ may be adsorbed on

Table 22.3. Rating curve analysis for calculation of river loadings

A. Single Season Ratings

Component	Intercept (a)	Slope (b)	Sample size (n)	Correlation coefficient (r)	Coefficient of Determination (r^2, %)	Significance level
Ca^{2+}	1.599	-0.096	638	0.46	21.5	.001
Mg^{2+}	1.068	-0.130	638	0.74	54.7	.001
Na^+	1.208	-0.050	638	0.40	16.3	.001
K^+	0.482	-0.107	638	0.50	25.1	.001
NO_3^-	0.145	0.303	638	0.37	13.3	.001
HCO_3^-	2.318	-0.213	44	0.77	59.3	.001
Suspended solids	0.195	0.708	87	0.43	18.2	.001

B. Seasonal Rating Equations

(a) Summer (April - September)

		Slope (b)	Sample size (n)	Correlation coefficient (r)	Coefficient of Determination (r^2, %)	Significance level
SC	2.784	-0.123	353	0.78	61.1	.001
Cl$^-$	1.530	-0.054	353	0.34	11.7	.001

(b) Winter (October - March)

		Slope (b)	Sample size (n)	Correlation coefficient (r)	Coefficient of Determination (r^2, %)	Significance level
SC	2.697	-0.060	285	0.57	32.2	.001
Cl$^-$	1.460	0.025	285	0.22	4.8	.001

[1] Model form $C = aQ^b$
where C = concentration
       Q = discharge

[2] No significant models were derived for Dissolved organic matter, SO_4^{2-}, CO_3^{2-}, NO_2^- and NH_4^{2+}.

Table 22.4. Effects of passage of water
through the soil and ground water zones

	Concentration[1] (mg l^{-1})			Ratio to concentration in precipitation input	
	Input[2]	Soil output[3]	Stream output	Soil output	Stream output
Ca	1.49	2.52	30.44	1.70	20.43
Mg	0.38	3.96	8.21	10.42	21.61
Na	2.78	11.07	14.00	3.98	5.04
K	1.02	13.15	2.87	13.28	2.81
H	0.028	0.132	0.00	4.70	0.00
NO$_3$	3.40	4.85	3.43	1.43	1.01
Cl	6.25	16.3	30.25	2.63	4.84

(1) volume-weighted annual mean

(2) bulk precipitation

(3) base of B horizon, average of 2 pits

exchange sites in the clay-rich C horizon, which contains
moderate amounts of micaceous clays (Whitfield, pers. comm.)
with a high buffering capacity for K^+ (Russell, 1973). The
loss of H^+ may be explained by weathering in the ground
water zone. The effect of the soil system is to increase
the concentration of Mg^{2+} and K^+ beyond the increase expected for simple concentration by evaporation, but Ca^{2+}
and NO_3^- are decreased in concentration, thus reflecting a
net gain to the soil by atmospheric input.

The biosphere not only alters the relative chemical
composition of bulk precipitation as previously discussed,
but also has a selective impact on solute loadings (figure
22.4). Of the cations, K^+ is relatively enriched as a
result of passage through the oak and bracken canopy, but
diminishes in runoff, which indicates the impact of nutrient
cycling on certain chemical elements. In contrast, Ca^{2+},
Mg^{2+} and Na^+ show little relative change. NO_3 plus NO_2
presents a similar picture to K^+ which further emphasises
the importance of nutrient cycling.

The overall input/output balance is presented in table
22.6, where cations and anions balance fairly closely
indicating no gross errors. With the exception of K^+,
NH_4^+, H^+ and NO_3^-, all other species are a net loss to the
system. Cation losses are dominated by Ca^{2+} and anion
losses by HCO_3^-.

Chemical losses dominate in this forested environment,
with an output of 467.52 kg ha^{-1} year^{-1} (table 22.5).
Suspended material, of which approximately 13% is organic,
constitutes less than 4% of the total output and DOM losses
are of the order of 1%. Although absolute rates of

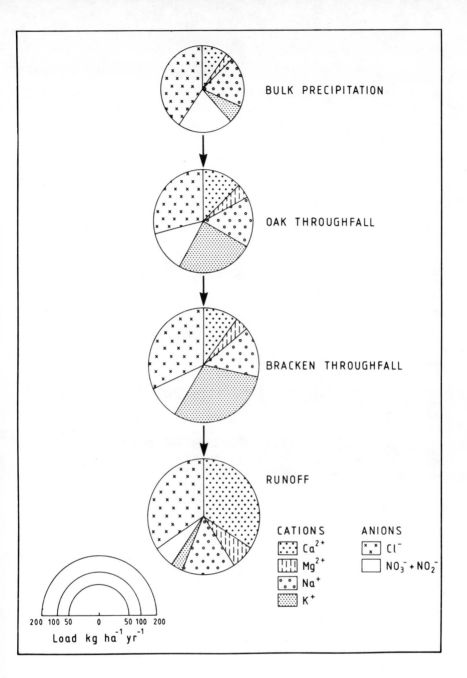

BULK PRECIPITATION

OAK THROUGHFALL

BRACKEN THROUGHFALL

RUNOFF

CATIONS

- :::: Ca^{2+}
- ⊞ Mg^{2+}
- ∘∘ Na^+
- ▒ K^+

ANIONS

- ×× Cl^-
- ☐ $NO_3^- + NO_2^-$

200 100 50   0   50 100 200
Load kg ha^{-1} yr^{-1}

Figure 22.4. Annual solute loadings in precipitation, throughfall and runoff.

371

Table 22.5. Inputs and outputs of major dissolved constituents for the year 1979-1980

A. Total Loadings (kg ha^{-1} year^{-1})

Source	Ca^{2+}	Mg^{2+}	Na$^+$	K$^+$	NH$_4^+$	H$^+$	NO$_3^-$	Cl$^-$	HCO$_3^-$	SO$_4^{2-}$
Bulk Precipitation	9.43	2.40	17.66	6.49	11.80	0.18	21.61	39.67	44.40	22.17
Runoff	60.89	16.43	27.95	5.74	0.00	0.00	6.86	60.50	236.00	53.15
Balance	-51.46	-14.03	-10.29	+0.75	+11.80	+0.18	+14.75	-20.83	-191.60	-30.98

B. Relative Loadings (equiv. ha^{-1} year^{-1} 10^3)

Source	Ca^{2+}	Mg^{2+}	Na$^+$	K$^+$	NH$_4^+$	H$^+$	NO$_3^-$	Cl$^-$	HCO$_3^-$	SO$_4^{2-}$
Bulk Precipitation	0.47	0.20	0.77	0.17	0.44	0.18	0.35	1.12	0.73	0.46
Runoff	3.04	1.35	1.22	0.15	0.00	0.00	0.11	1.71	2.87	1.11
Balance	-2.57	-1.15	-0.45	+0.02	+0.44	+0.18	+0.24	-0.59	-3.05	-0.65

C. Ion Balance

	Σ +	Σ -
Input	2.23	2.66
Output	5.76	5.80

D. Annual output in streamflow (kg ha^{-1} year^{-1})

i) Dissolved
    DOM   7.22
    TDS   467.52

ii) Particulate
    Inorganic 15.53
    Organic   2.35
    Total   17.88

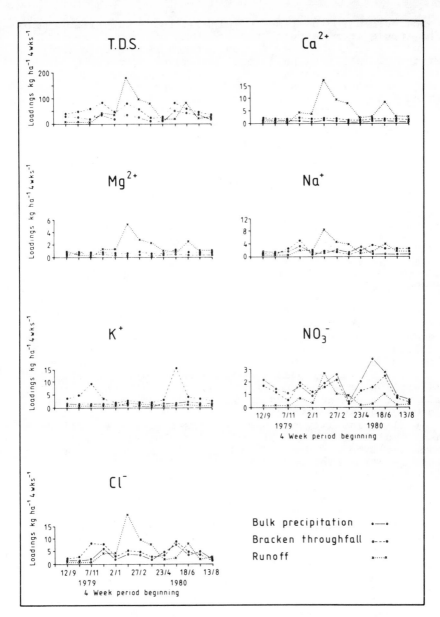

Figure 22.5. Solute loadings in precipitation, throughfall and runoff (4 weekly).

denudation are much higher than those presented for a forested watershed by Likens *et al.* (1977), the relative proportions are similar.

The distribution of inputs, throughputs and outputs for the year 1979-80 are shown in figure 22.5. Maximum outputs were associated with high runoff in the winter period,

which were also the times of highest input. Subtraction of
TDS inputs from outputs suggests that maximum losses still
occur during the winter period, with a net loss to the
system of 291.71 kg ha^{-1} year^{-1} of which over 50% occurred
in the month of February. K$^+$ and, to a lesser extent Na$^+$
have two major periods of input from leaching of the vegetation canopy; during the autumn period of biomass reduction
and in the early regrowth period in May and June. Fluctuations in these and other inputs are rapidly buffered by
the soil/ground water subsystem and are not represented in
the runoff loadings.

DENUDATION RATES

It is common practice to present figures for chemical denudation in terms of surface lowering using Bubnoff units
(Young, 1974), which are equivalent to volumetric losses
of m^3 km^{-2} year^{-1} or rates of lowering in mm year^{-1} x 10^3.
Total solutional outputs obtained from budget calculations
require an assumed particle density for volume conversion
(2650 kg m^{-3} was assumed for this study) and it is also
necessary to subtract the non-denudational component; in
this case a bulk precipitation input. With the exception
of the Hubbard Brook experiment (Likens et al., 1977), most
records reported in the literature are less than 4-5 years
in length. Interpretation of data presented here and in
other studies must therefore be approached with caution,
when extrapolating over long periods of time. Furthermore,
assumptions must be made regarding the bulk density of the
material assuming no reduction resulting from solutional
losses, and such data do not provide information on the
spatial variability of these processes over the catchment
as a whole. However, calculation of denudation rates
does permit comparison with other environments, although
this is often hindered by the presence of different
lithologies, vegetation assemblages and varied climatic
conditions. Estimates in other areas of non-carbonate
rocks usually range between 0.5 and 80 m^3 km^2 year^{-1} (table
22.6). Slowest rates of solution are associated with
crystalline rocks, with highest rates associated with more
recent sedimentary outcrops. Rates of denudation for
forested environments, including the present study, appear
towards the low end of the range (Likens et al., 1977,
Verstraten, 1977). A similar pattern of geological control
emerges, however, with lower rates on quartzites, than
sandstones. Thus the present result is high in contrast
with other investigations but may be explained by the
appreciable contribution of Ca^{2+} and HCO$_3^-$ derived from
the calcareous facies.

CONCLUSIONS

Catchment experiments can provide valuable information regarding both rates and mechanisms of geomorphological
processes, and the catchment ecosystem approach permits
identification of the significant biogeochemical inter-

Table 22.6. Some rates of chemical denudation
on non-carbonate rocks

Region	Land use/Geology	Rate of ground lowering (B*)	Source
S.W. England	Various	10 - 60	Webb and Welling, 1980
S.W. England	Cereal crops, sandstone & conglomerates	23	Foster, 1980
S.W. England	Bracken/heather/ sandstone	1.6	Waylen, 1976
N.E. England	Various	9.7 - 42.8	Arnett, 1978
Poland	Non carbonate rocks	4.3 - 36.3	Pulina, 1972
California white Mtns.	Various	1.4 - 2.1	Marchand, 1971
Queensland Australia	Various	3.8 - 80.0	Douglas, 1973
New South Wales Australia	Various	1.6 - 10.6	"
Scandinavia	Metamorphic	5 - 10	Rapp, 1960
Maryland, U.S.A.	Micaceous schist	2.7	Cleaves *et al.*, 1970
New Mexico U.S.A.	Mixed forest/dwarf trees Quartzite	0.8 - 1.5	Miller, J., 1961
	Mixed forest/dwarf trees Granite	0.5 - 5.5	
	Mixed forest/dwarf trees Sandstone	5.8 - 21.5	
Luxembourg	Mixed oak-beech shales/ quartzites	4.9	Verstraten, 1977
New Hampshire U.S.A.	Northern hardwood Till	2.9	Likens *et al.*, 1977
Midland England	Mixed oak-alder Sandstones/shales	11.0	Present study

*B = Bubnoff units = m^3 km^2 $year^{-1}$ or ground lowering in mm $year^{-1}$ $x\,10^3$

actions and controls on such processes. Although these frameworks have been adopted here, and in a number of other recent investigations the limitations and potential of such an approach has yet to be fully appreciated.

Detailed studies of solutional processes are important to the geomorphologist since spatial variations in stream solute concentrations are frequently linked to a broad geological control and may be used to emphasise the unequal distribution of the chemical denudation process (eg. Webb and Walling, 1980). However, calculation of input and output budgets at the meso- or micro-scale, provides little information on sediment and solute sources. Analyses of soil, slope and canopy leaching processes, highlighted in this paper, are essential if such sources are to be identified and quantified, although the techniques themselves still require considerable refinement. Furthermore, consideration must be given to the accuracy of budget calculations if environmental or climatic comparisons of denudation rates are to be realistic.

Investigation of both pedological and biological interactions with solute movement is also essential, but has been largely omitted from geomorphological research. The fate of many ions available from a bulk precipitation input, or derived from weathering, may be affected by adsorption onto the soil ion exchange complex or uptake by the biomass in nutrient cycling. These two processes may modify both the rate and magnitude of denudation of specific ions, and the return of solutes to the slope subsystem, by foliar leaching and/or litter fall, may further enhance solutional denudation at certain times of the year. Detailed chemical analyses are required to provide an insight into the complexity of many processes involved, especially for ions which occur in many different forms, such as nitrogen. Furthermore, organic fraction losses, also closely linked to the biosphere, may have an important bearing not only on total loadings but on the complexing which may occur with other inorganic molecules (eg. Arnett, 1978; Foster and Grieve, 1982).

ACKNOWLEDGEMENTS

The authors are indebted to a number of organisations and individuals for assistance. Sir William Dugdale generously permitted access to the monitoring sites, and Mr. H.B. Bladon (Estate Manager) provided much valuable information and on-site guidance. Coventry (Lanchester) Polytechnic have financed a research assistant for the project and we are also indebted to Mrs. S. Addleton for diagram preparation.

REFERENCES

Anderson, D.H. and Hawkes, H.E., 1958, Relative mobility of the common elements in weathering of some schist and granite areas, *Geochim. et Cosmochim. Acta*, 14, 204-210.

Arnett, R.R., 1978, Regional disparities in the denudation rate of organic sediments, *Zeitschrift für Geomorphologie Supplement Band*, 29, 169-179.

Cleaves, E.T., Godfrey, A.E. and Bricker, O.P., 1970, The geochemical balance of a small watershed and its geomorphic implications, *Bulletin of the Geological Society of America*, 81, 3015-3032.

Douglas, I., 1973, Rates of denudation in selected small catchments in Eastern Australia, *University of Hull, Occasional Papers in Geography*, no. 21.

Foster, I.D.L., 1980, Chemical yields in runoff and denudation in a small arable catchment, East Devon, England, *Journal of Hydrology*, 47, 349-368.

Foster, I.D.L., Grieve, I.C. and Christmas, A.D., 1981, The use of specific electrical conductance in studies of natural waters and soil solutions, *Hydrological Sciences Bulletin*, 26, 257-269.

Foster, I.D.L. and Grieve, I.C., 1982, Short term fluctuations in dissolved organic matter concentrations in streamflow draining a forested watershed and their relation to the catchment budget, *Earth Surface Processes and Landforms*, 7, 417-425.

Garrels, R.M. and Christ, C.L., 1965, *Solutions, minerals and equilibria*, (Harper and Row).

Garrels, R.M. and Mackenzie, F.T., 1971, *Evolution of sedimentary rocks*, (Norton).

Gregory, S., 1955, Some aspects of the variability of annual rainfall over the British Isles for the standard period 1901-1930, *Quarterly Journal of the Royal Meteorological Society*, 81, 257-262.

Gregory, S., 1976, Regional climates, in: *The climate of the British Isles*, ed. Chandler T.J. and Gregory, S., (Longman, London), 330-342.

Hesse, P.R., 1971, *A textbook of soil chemical analysis*, (Murray).

Likens, G.E., Bormann, F.H., Pierce, R.S., Eaton, J.S. and Johnson, N.M., 1977, *Biogeochemistry of a forested ecosystem*, (Springer Verlag).

Loughnan, F.C., 1969, *Chemical weathering of the silicate minerals*, (Elsevier).

Maciolek, J.A., 1962, Limnological organic analyses by quantitative wet oxidation, *United States Fisheries and Wildlife Service Report*, No. 60.

MAFF, 1967, *Potential Transpiration*, Ministry of Agriculture, Fisheries and Food. Technical Bulletin 16. (H.M.S.O., London).

Marchand, D.E., 1971, Rates and modes of denudation, White Mountains, Eastern California, *American Journal of Science*, 270, 109-135.

Miller, J.P., 1961, Solutes in small streams draining single rock types, Sangre de Cristo Range, New Mexico, *United States Geological Survey Water Supply Paper*, 1535-F.

Parizek, R.R. and Lane, B.E., 1970, Soil water sampling using pan and deep pressure-vacuum lysimeters, *Journal of Hydrology*, 11, 1-21.

Piper, A.M., 1944, A graphic procedure in the geochemical interpretation of water analysis, *Transactions American Geophysical Union*, 25, 914-923.

Pulina, M., 1972, A comment on present day chemical denudation in Poland, *Geographia Polonica*, 23, 45-62.

Rapp, A., 1960, Recent development of mountain slopes in Karkevagge and surroundings, Northern Scandinavia, *Geografiska Annaler*, 42, 65-200.

Reid, J.M., Macleod, D.A. and Cresser, M.S., 1981, Factors affecting the chemistry of precipitation and river water in an upland catchment, *Journal of Hydrology*, 50, 129-145.

Russell, E.W., 1973, *Soil conditions and plant growth*, (Longmans), 10th ed.

Verstraten, J.M., 1977, Chemical erosion in a forested watershed in the Oesling, Luxembourg, *Earth Surface Processes*, 2, 175-184.

Walling, D.E., 1978, Reliability considerations in the evaluation of river loadings, *Zeitschrift für Geomorphologie Supplement Band*, 29, 29-42.

Water Data Unit, 1978, *Surface Water United Kingdom, 1971-73*, (H.M.S.O., London).

Waylen, M.J., 1976, Aspects of the hydrochemistry of a small drainage basin, Unpublished Ph.D. Thesis, University of Bristol.

Waylen, M.J., 1979, Chemical weathering in a drainage basin underlain by old red sandstone, *Earth Surface Processes*, 4, 167-178.

Webb, B.W. and Walling, D.E., 1980, Stream solute studies and geomorphological research: some examples from the Exe Basin, Devon, U.K., *Zeitschrift für Geomorphologie*, 36, 245-263.

Young, A., 1974, The rate of slope retreat, in: *Progress in Geomorphology*, ed. Brown, E.H. and Waters, R.S., Institute of British Geographers Special Publication no.7, 65-78.

23 Hydrochemical characteristics of

a Dartmoor hillslope

A.G. Williams, J.L. Ternan and M. Kent

INTRODUCTION

Many field investigations of subsurface runoff have been
concerned only with hydrological aspects, in relation to,
for example, storm runoff generation and baseflow sources
(eg. Hewlett and Hibbert, 1963; Weyman, 1974; Anderson and
Burt, 1978). More recently however, several studies have
stressed the importance of soil water in understanding
stream water chemistry (eg. Burt, 1979; Verstraten, 1980).
Both lines of research have been conducted under a variety
of field conditions, reviewed by Dunne (1978), ranging from
thin soils less than 1 m deep developed on bedrock, to
areas with soil up to 2.5 m deep. Additionally, studies
have been carried out under a variety of vegetation types,
from grassland and bracken (Arnett, 1976) to forest
(Hewlett, 1961; Ragan, 1968).

This paper proposes a hydrological model for a thick
regolith characteristically found on hillslopes of the
Dartmoor granite, where slope regolith thicknesses in
excess of 20 m have been recorded in reservoir site investigations. The model is based on research in the
Narrator Brook Catchment, S.W. Dartmoor and considers the
effects of several alternative interflow pathways on the
stream water chemistry.

THE NARRATOR CATCHMENT

The Narrator Catchment is located 17 km north east of
Plymouth on the south west margin of Dartmoor. Geologically
the drainage basin is underlain by the Dartmoor granite,
which is essentially a porphyritic type. Like many basins
on Dartmoor it is overlain by a variable thickness of decomposed granite or growan and within the Narrator valley
the depth of unconsolidated material is known to be greater
than 35 m.

The soils in the Narrator Catchment are representative
of many Dartmoor valleys. On the higher flatter areas the
principal type is the iron pan stagnopodzol with a peaty

surface horizon and a thin iron pan (figure 23.1). The
steeper hillslopes are dominated by brown podzolic soils
which are characterised by a humose topsoil and ochreous
subsoil (figure 23.2). Within the valley floor there are
extensive areas of oligo-amorphous peat which are associated
with the sites of springs and seepages.

Vegetation in the catchment comprises a semi-natural
mosaic of communities typical of much of Dartmoor. Kent
and Wathern (1980) identified eleven community types of
which the four major ones are:-

1. Sitka spruce forest

2. Acid grassland

3. Acid grassland extensively invaded by bracken

4. Molinia-dominated blanket bog.

This paper is principally concerned with flow in the acid
grassland.

HYDROLOGICAL MODEL FOR A DARTMOOR HILLSLOPE

Four principal interflow paths through the soil may be
identified and these are associated with different soil
conditions. They are:-

Pathway 1. Flow above an iron pan

Pathway 2. Flow above a fragipan

Pathway 3. Saturation upwards from the fragipan

Pathway 4. Saturation within the fragipan

In addition to these pathways within the soil, an overland
flow pathway (Pathway 5) operates.

1. Flow above an iron pan (figures 23.1a and 23.1b)

When water has infiltrated through the peaty surface horizon
of the stagnopodzol soil, it is forced to flow laterally by
the iron pan, which is totally impermeable (figure 23.1a).
As the storm continues, the peat in the upper horizon becomes completely saturated to the surface. Little further
rain can then be absorbed and water begins to flow across
the surface as saturation overland flow (figure 23.1b).
The importance of these two flow types has been noted by
Weyman (1974) and Waylen (1979) for the upper basin of the
East Twin catchment, Mendip Hills, Somerset.

2. Flow above a fragipan (figure 23.2)

The fragipan is an indurated horizon with a platy structure,
found in the brown podzolic soils at about 60-90 cm depth.
It is of localised occurrence on Dartmoor and associated
with areas of former periglacial activity. Fitzpatrick
(1956), in a comprehensive discussion of their origin,
stressed the importance of these horizons to water movement. When the fragipan is highly indurated it can completely prevent the vertical percolation of water. As a

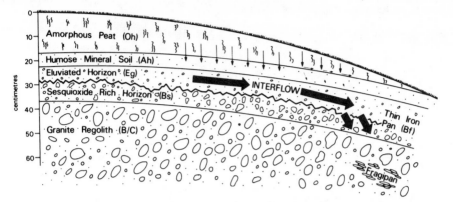

Figure 23.1a.   Hydrological model for a Dartmoor hillslope,
Pathway 1: unsaturated peat, flow above an iron pan.

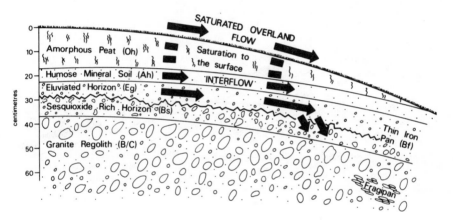

Figure 23.1b.   Pathway 1: saturated peat, flow above an
iron pan.

result, soil water moves laterally downslope above this
horizon (figure 23.2).   The flow of water takes place
through textural pores and where such a major horizon of
interflow occurs, the pores in the regolith contain no
fines.   These pores are not structural ones, the sandy
loam being only very weakly structured, but may be considered as micropipes, able to conduct water laterally down
the hillslope.

3. Saturation upwards (figure 23.3)

The depth of flow above the fragipan will increase through
the storm if the downslope throughput is less than the
amount of water coming from upslope together with that
percolating through the soil (figure 23.3).   The depth of
this saturated layer will depend on many factors including

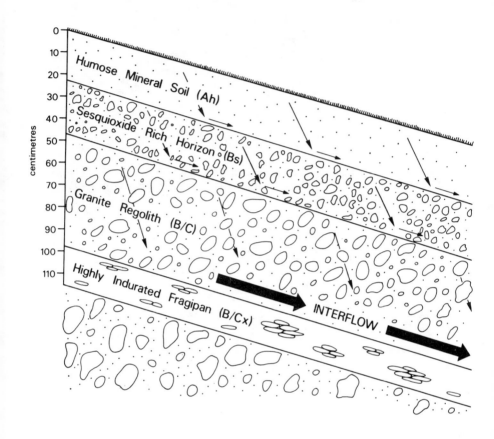

Figure 23.2. Hydrological model for a Dartmoor hillslope, Pathway 2: flow above a fragipan.

total amount of rain, storm intensity, antecedent soil moisture content and the nature of the indurated horizon, in addition to topographic features such as slope angle and distance down the slope. This process of saturation upwards was first noted by Hewlett and Hibbert (1967) and later described by Weyman (1974).

4. Flow within an impeding layer (figure 23.4)

In this instance, a wetting front moves downwards through the impeding layer (figure 23.4). Initially this occurs very rapidly to satisfy the capillary pores within the

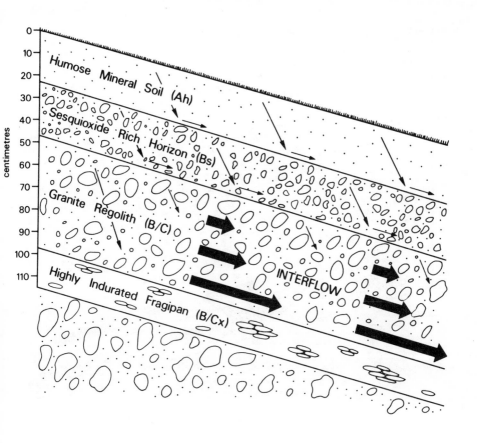

Figure 23.3.   Hydrological model for a Dartmoor hillslope,
    Pathway 3: saturation upwards.

soil.   Most of the water is retained by the soil or may
flow in an unsaturated state downslope.   Thus discharge
amounts are very small compared with pathways 2 and 3.
This model was suggested by Whipkey and Kirkby (1978) but
has not previously been described in the field.   This pathway was only recorded at a site within the forested area
of the catchment, although water collected in a pan below
the fragipan on a grassland pit indicates that this type
is not restricted to the forest.

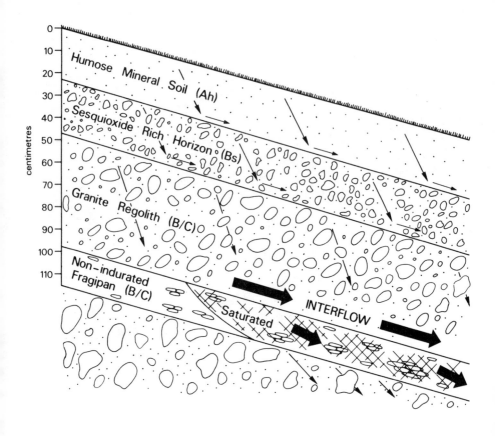

Figure 23.4. Hydrological model for a Dartmoor hillslope, Pathway 4: saturation within an impeding layer.

5. Overland flow

There are two types of overland flow; saturation overland flow which results both from saturation upwards from the iron pan as described above (figure 23.1a) and from direct precipitation on wet areas and flushes in the valley bottom. In addition infiltration excess overland flow (figure 23.1b) occurs when rainfall intensity is greater than the infiltration capacity of the soil so that the un-absorbed excess flows across the surface.

The infiltration capacity of the hillslope was between 4.8 mm hr^{-1} and 6.4 mm hr^{-1} (H. Williams, personal

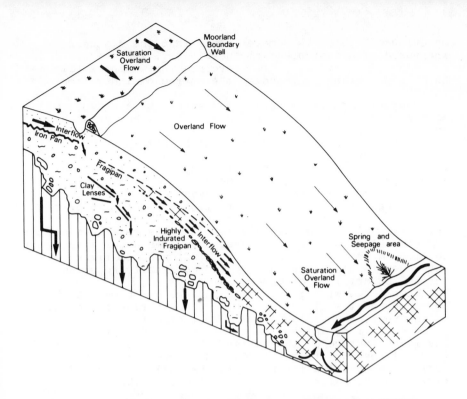

Figure 23.5. Hydrological model for a Dartmoor hillslope.

communication) and so not greatly in excess of typical
rainfall intensities. These infiltration values for the
soil may however be higher than those of the grass turf mat
which seems to control overland water movement.

Overall model

The five components can be integrated in one overall model
(figure 23.5) that exists at a hillslope scale, although
not at a point scale. On the moorland, the iron pan
stagnopodzol soil is saturated for much of the year.
Saturation upwards to the surface leads to overland flow
which feeds tributaries and accounts for the rapid response
of Dartmoor streams.
     On the steeper sloping hillside where the brown pod-
zolics occur, the major zone of flow within the soil is
above the fragipan (Pathway 2). At the upper part of the
slope, again on the brown podzolics, the thickness of the
saturated water layer is small because the only source of
water is that percolating through the soil. However further
downslope where additional water is being supplied from
upslope, the thickness of the saturated layer increases
(Pathway 3) as shown in figure 23.5. Pathway 4 is controlled by the size of pores and head of water above it and
so is also dependent on the position on the slope. Pathway
5, infiltration excess overland flow, was a widespread

occurrence. There is some evidence that the discharge increased with distance from the divide. The overall flow pattern on the hillslope is therefore very complex (figure 23.5) and further modified by the degree of induration of the fragipan.

Subsurface water moves to the stream by a network of springs and seepages (figure 23.5). These appear to be of several different types. Firstly, many springs and seepages occur at slope base concavities which correspond to the classic interflow sites of Kirkby and Chorley (1967). Secondly, there are springs which may be related to joints in the granite, these springs being associated with gullies arising from past mineral vein excavation. Finally, there are springs due to vertical variations in regolith permeability. These springs issue at one level and begin to lose water almost immediately into their stream beds and so are probably supplied by perched saturated zones within the decomposed granite. Clay lenses, indicated in figure 23.5 have been found within the decomposed granite regolith during the drilling of boreholes.

Associated with many of the springs are areas of oligoamorphous peat soil. These soils are waterlogged for much of the year, responding quickly to incident rain and hence provide an important source of storm runoff to the stream. The flashy nature of the Narrator Brook has been described in detail by Murgatroyd (1980) and is due partly to flow from these contributing areas in the valley floor and partly to the saturated overland flow from the moorland.

INTERCEPTION OF SOIL WATER AND OVERLAND FLOW

A series of soil water interception pits were established along transects through the acid grassland, acid grassland extensively invaded by bracken and the forest. These pits were arranged downslope at approximately equal distances from one another. In this paper the pathways in the acid grassland community which include all except Pathway 4 are considered. Pathway 4 was found only in the forest transect.

Interflow was collected using a Whipkey type gutter system (Whipkey, 1965) with plastic guttering installed just below the litter layer and each of the major horizons. Polythene sheeting 1 m long was inserted about 2 cm into the soil face to lead water to the gutters and then into 1.15 l collecting bottles. In heavy storms, discharge exceeded the capacity of the bottles thus imposing certain limitation on the analyses. Weekly bulk samples were taken at all sites for the period 18th January 1977 to 27th March 1978. In addition to the interflow sites, the sampling network included 23 springs and seepages together with 8 hr samples from the main gauging station at the catchment exit.

Pathway discharges

Table 23.1 gives the total volumes of flow recorded for each of the pathways over a period of 1 year. These represent the minimum quantities of discharge, because of the

Table 23.1. Event frequency and volume of
flow for each pathway from 14/2/77 to 6/2/78

Description	Event Frequency	Minimum for year (1)	Mean Weekly discharge (ml)*
Pathway 1	56% (29 weeks)	21.0	724
Pathway 2	82% (43 weeks)	40.0	930
Pathway 3	44% (23 weeks)	19.5	848
Pathway 4	27% (14 weeks)	7.0	501
Pathway 5** (Overland flow)	88% (46 weeks)	31.4	682

* Discharge per 1 m of contour length of hillslope

** Data for infiltration excess overland flow (brown podzolics).

limitations of the collection system. Hence the quantities
shown in table 23.1 are appreciably lower than the actual
amount of interflow.

Discharge from Pathway 1 was low (table 23.1) because
the flow was monitored at a site just downslope from a
ditch at the moorland boundary. However, flow rates in a
moorland pit upslope of the ditch were up to 180 l hr^{-1}.
Regular monitoring of this upper pit was not possible because it was usually filled with water moving off the iron
pan. Pathway 2 discharges were about 5 l hr^{-1} during wet
conditions.

The volumes recorded for Pathways 3 and 4 in table 23.1,
are probably close to the actual amounts of interflow discharged. When flow took place the quantities of discharge
were quite high, being 850 and 500 ml respectively. However as Atkinson (1978) notes, the construction of a pit
increases the flow of water because of the formation of a
saturated wedge upslope. Thus the figures in table 23.1
are merely a guide to the importance of each interflow
pathway.

Pathway 5 discharge was 31.4 l for the year. The frequency of flow in table 1 is greatest for Pathway 5.

Consideration of several small storms (table 23.2)
shows that the discharges are in the same order as table
23.1.

Temporal distribution of flow

The temporal distribution of flow along the five pathways
is shown in figure 23.6 along with rainfall occurrence.
The diagram shows that different pathways came into

Table 23.2. Discharge* for selected storms (ml)

Date	31/1/77	14/3/77	5/9/77	26/9/77
Total Weekly Rainfall (mm)	54.7	39.5	37.0	34.5
Maximum 2 hr Rainfall Intensity mm hr^{-1}	10.5	5.5	7.5	5.8
Pathway 1	376	98	655	364
Pathway 2	1150**	1150**	754	1150**
Pathway 3	289	604	423	247
Pathway 4	55	-	-	123
Pathway 5*** (Overland flow)	1150**	520	1150**	1150**

* Per 1 m contour length of hillslope

** Capacity of collecting bottle exceeded

*** Data for infiltration excess overland flow

operation or ceased according to the season. Rainfall occurred on 49 weeks of the year. Overland flow, although not the greatest in terms of discharge, was the most frequently occurring at 46 weeks. Within the soil, the flow above the fragipan (Pathway 2) was the most important, being almost continuous with the exception of August (figure 23.6). Flow along Pathway 1, although also frequent, was more intermittent.

Weekly discharge volumes for Pathways 1 and 2 were very closely correlated with maximum daily total rain in the preceding week ($r>0.75$, 99.9% significance level) demonstrating the importance of storm rainfall amounts. Flow along Pathways 3 and 4 was less regular than along the other flow paths and ceased to operate in early summer, commencing again in the autumn (figure 23.6). Weekly discharge volumes were significantly correlated with total weekly rainfall ($r>0.53$, 95% significance level) only, demonstrating the importance of total amount of water in producing saturation upwards (Pathway 3) and saturation within the impeding layer (Pathway 4).

INTERFLOW CHEMISTRY

Interflow chemistry is fundamental to an understanding of the dissolved load of streams. Additionally, solutes derived from soil sources provide a major component of solutional losses from drainage basins and can exceed that

Figure 23.6. Occurrence of rain, overland flow and discharge along the four interflow pathways from 14/2/77 to 6/2/78.

of all the mechanical processes combined (Carson and Kirkby, 1972). However, with the exception of Burt (1979) few direct studies have been made of interflow chemistry. Reliance has generally been placed on the use of porous cup samplers which obtain water held at a tension of up to 0.8 bars, rather than interflow gutters which collect only free flowing water (Verstraten, 1980). Field tile drains have also been used by Foster (1979) as a means of

389

Table 23.3. Mean weighted solute concentrations for the atmosphere and the five pathways, from 14/2/77 to 6/2/78

Hydro-logical Pathway	pH	Na mg l^{-1}	K mg l^{-1}	Ca mg l^{-1}	Mg mg l^{-1}	SiO_2 mg l^{-1}	Cl mg l^{-1}	Mean SEC μScm^{-1}
Precipi-tation	4.22	3.94	0.41	0.34	0.45	0.01	5.67	45
1	4.49	5.33	3.72	1.35	0.95	1.33	8.22	75
2	5.32	4.86	0.95	2.00	0.54	1.64	6.80	54
3	4.94	4.91	1.83	1.94	0.63	1.44	7.68	68
4	4.36	6.11	1.64	1.66	0.67	5.88	9.57	116
5	4.75	4.56	8.41	2.72	1.28	3.62	9.34	114

collecting free flowing soil water. The analysis of such free flowing water is much more central to an understanding of stream water chemistry than the analysis of soil water held at higher tensions.

This study is part of a systematic investigation of the solutes in interflow. All samples were analysed for specific electrical conductance, pH, sodium, potassium, calcium, magnesium, silica and chloride. The mean weighted solute concentrations for precipitation and the five flow types are presented in table 23.3.

Precipitation

The chemical composition of 'bulk precipitation' (Whitehead and Feth, 1964) in table 23.3 is dominated by sodium and chloride as might be expected for a catchment only 20 km from the sea. The pH was generally low, the weighted mean level being 4.22 with a range from 3.62 to 6.18. Silica concentrations were also usually below detectable limits with a weighted mean of 0.01 mg l^{-1} for the year.

Flow above an iron pan (figure 23.1)

Except for hydrogen, all ions measured showed an increase in concentration over precipitation (table 23.3). Silica, for example increased from 0.01 mg^{-1} to 1.33 mg^{-1}. Potassium increased (x 9), calcium (x 4), and magnesium (x 2.1). Sodium and chloride increased by only 1.4 and this increase is explicable in terms of evapotrans-piration alone. Specific electrical conductance, which may be used as a descriptive measure of overall chemical quality, was generally between 50-85 μScm^{-1} at 25 °C but occasionally marked fluctuations were recorded. These fluctuations probably reflect the flushing of accumu-lated material from the soil (Foster, 1979), an effect seen

for example in July when conductivity reached its maximum
of 163 μScm^{-1} after four dry weeks.

Flow above a fragipan (figure 23.2)

Except for hydrogen, chemical concentrations above the
fragipan were again always greater than in bulk precipitation. Concentrations of silica (x160), calcium (x6) and
potassium (x2) all increased markedly, whereas only small
increases were recorded for sodium, chloride and magnesium
(x1.2). One feature of this pathway was the very stable
pattern of silica concentrations, probably limited by
equilibrium constraints (Bricker and Garrels, 1967).
There was a slight seasonal silica trend, reaching a
minimum of 0.8 mg l^{-1} in early winter and increasing to
about 1.4 mg l^{-1} in summer. Conductivity showed a maximum
in the summer of 65 μScm^{-1} and an autumn minimum of about
35 μScm^{-1}.

Saturation upwards (figure 23.3)

Again hydrogen ion concentration decreased and concentrations of all the other ions increased (table 23.3). It
is not possible to identify seasonal trends because of the
intermittent nature of flow (figure 23.6) although a summer
maximum in conductivity of 352 μScm^{-1} was recorded and a
winter minimum of 38 μScm^{-1}.

Flow within an impeding layer (figure 23.4)

The chemical concentrations of all the ions measured within
the forest were high (table 23.3). In particular, the low
pH is due to organic acids derived from the spruce litter
during decomposition. These may enhance the weathering of
soil minerals. As with Pathway 1, the conductivity levels
showed wide fluctuations throughout the year due to
irregular flushing of accumulated salts.

Overland flow (figure 23.5)

The chemical concentration in overland flow is influenced
by solutes supplied from the atmosphere and from biological
sources. Sodium and chloride are derived from the
atmosphere; the increase in sodium concentration (x1.2)
may be attributed to evaporation alone, whereas that of
chloride (x1.7) may also reflect the impaction of aerosols
and dust on the grass (Eriksson, 1955). The great increase
in silica (up to 3.6 mg l-) potassium and calcium (x8) concentrations (table 23.3) is attributed to leaching and
biological decomposition of the vegetation. Conductivity
shows a seasonal trend with a minimum in winter and maximum
in spring and summer.

Summary of major patterns observed

The chemical composition of overland flow depends mainly on
the atmosphere and biological patterns of solute supply,
whereas that of the four interflow pathways is determined

by the hydrological regime. Thus Pathway 2 generally has the lowest chemical concentrations because of the greater frequency and volumes of flow (table 23.3, figure 23.6). In contrast, Pathway 4 has the highest solute concentrations because the flow is infrequent and the residence time of the water and hence contact time with the minerals is therefore high. Part of the higher concentration may also be accounted for by the greater evapotranspiration taking place in the forest between each flush event. Because the discharge amounts are very low, the total quantities of salts removed are considerably lower than the more frequently flushed horizons.

Although mean concentration for Pathways 2 and 3 are similar (table 23.3), data for individual events show significant differences related to the frequency of flushing of the upper horizon (Pathway 3) and the volume of water moving through that horizon. During small storm events or after prolonged dry periods, the solute concentrations of Pathway 3 were generally greater than those recorded for Pathway 2, 15 cm lower down the profile (table 23.4). During major storm events, however, the solute concentrations of Pathway 3 are lower than those of Pathway 2 (table 23.4). This may be attributed to the very low residence time of the storm water and the dilution effects of high volumes of water overriding any initial flushout of accumulated salts.

Table 23.4. Differences in chemical concentrations for Pathways 2 and 3 between high and low discharges

Flow Classification	Path-way type	Cond. μScm^{-1}	pH	Na $mg\,l^{-1}$	K $mg\,l^{-1}$	Ca $mg\,l^{-1}$	Mg $mg\,l^{-1}$	SiO_2 $mg\,l^{-1}$	Cl $mg\,l^{-1}$
High Flow 22/1/78	2	41.2	5.67	4.00	0.27	2.34	0.65	1.46	7.8
	3	38.4	5.28	4.00	0.43	1.60	0.60	0.70	7.6
Low Flow 26/9/77	2	37.8	5.73	5.43	0.27	1.54	0.47	1.60	3.6
	3	92.0	4.64	5.70	7.43	5.12	0.92	1.20	7.6

Although hydrological factors are of major importance to an understanding of the concentration of solutes, the chemical sources and behaviour of the ions can often be crucial. Potassium concentrations, for example, were highest in overland flow, (table 23.3), being derived from the decomposition of the grass. These ions were then supplied to interflow and the highest potassium concentration in the soil was for Pathway 1 (table 23.3) which was nearest the ground surface (32 cm). The lower concentrations in the deeper types of flow are due to the uptake by plant roots and adsorption by the clay colloids.

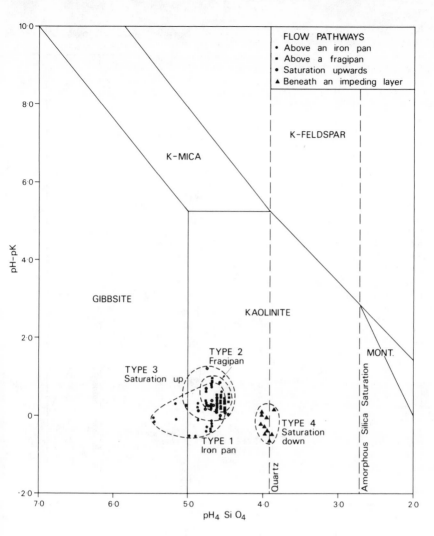

Figure 23.7.   A comparison of stability fields among some
    silicate minerals and gibbsite at 25°C and 1 atm.
    pressure with water quality analyses from the four
    interflow pathways.

A COMPOSITE CHEMICAL MODEL

Stability diagrams provide a useful way of classifying and
presenting data when considering interflow pathways through
the soil.   Bricker and Garrels (1967) have also shown the
usefulness of these diagrams in examination of the chemical
genesis of water.   Figures 23.7 and 23.8 show the relationship between gibbsite, kaolinite and three silicate
minerals, assuming that aluminium is conserved in the
chemical reactions.   The ratio of K^+, H^+ and SiO_2 for the
four pathways (figure 23.5) have been plotted onto the

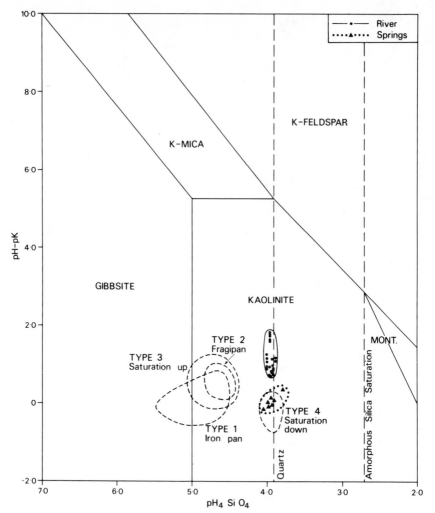

Figure 23.8. A comparison of stability fields among some
    silicate minerals and gibbsite at 25°C and 1 atm.
    pressure with water quality analyses from the springs
    and river.

diagram (figure 23.7) which clearly demonstrates the contrasts between their chemical characteristics.
    The interflow water above the iron path (Pathway 1,
figure 23.1) is moving very rapidly in winter with the
result that the decomposition is tending towards gibbsite
(figure 23.7). In summer, water movement is less rapid
and decomposition tends towards kaolinite. Pathways 2 and
3 (figures 23.2 and 23.3) are similar, the water being in
equilibrium with kaolinite and with none of the other
phases depicted on figure 23.7. Pathway 4 is also in equilibrium with kaolinite but there is a great difference
between this and the other pathways for reasons outlined

above. Thus various hydrological pathways are of major importance in the breakdown of silicate minerals and in determining the secondary weathering products. This agrees with the observation of Loughnan (1962) who observed that the quantity of water leaching through the profile is the most important single factor controlling the breakdown of parent materials and the genesis of secondary products.

The presence of gibbsite in the clay fraction has been confirmed at the interflow sites (Pathways 1 and 3), where it may result from weak weathering and strong leaching. However, the dominant mineral found in the soil profile was kaolinite as would be expected from the phase diagram (figure 23.7).

Consideration of spring waters poses a problem as to whether they are fed by shallow interflow or deeper sources. Work by Ternan and Williams (1979) has shown that the chemical composition of Dartmoor springs is influenced primarily by hydrological characteristics. The work suggests that springs and seepages emerging from the base of steep slopes are supplied from deeper sources. When analyses of spring waters are plotted on the stability diagram (figure 23.8) the stability field is similar to that of Pathway 4. The effect of the various hydrological pathways on the chemical composition of springs is clear and must therefore depend on the chemical reactions within the soil. This result accords with Smith and Dunne (1977) who noted that the solute concentrations of a stream in siliceous material are determined by rapidly occurring soil-water reactions.

River samples lie well within the kaolinite stability field (figure 23.8). The river data are higher on the pH-pK axis than the springs because the river has a higher pH, potassium concentrations being similar. The stream water has equilibriated with the CO_2 partial pressure in the atmosphere whereas the spring water contains more dissolved CO_2 and thus has a lower pH. The elongated shape of the river data is due to the seasonal variation in air temperature which control the partial pressure of CO_2 through the year.

CONCLUSION

The hydrological model presented for the Dartmoor hillslope demonstrates the main pathways over the soil surface and through the soil body to the stream. Overland flow was widespread and of frequent occurrence on both the moorland and the acid grassland hillslope. Saturation overland flow from moorland areas and footslope locations contributed directly to the stream. Infiltration excess overland flow appeared to increase with distance downslope, providing an additional source of water to footslope areas of saturation overland flow. Interflow occurrence was highly variable in both space and time. Not all the interflow pathways operated at one point in space because of the variable nature of the soil and slope conditions. Similarly, not all these pathways functioned throughout the year.

The model provides a useful framework for considering the effects of the various hydrochemical characteristics of different water pathways on the stream. Saturation overland flow entered the stream directly and changed its chemistry during flood events. Infiltration excess overland flow did not affect the stream chemistry directly but was more important in its influence of the chemistry of shallow interflow. Flows above the iron pan, Pathway 1, and above the fragipan, Pathway 2, were of prime importance and each one controlled the local weathering conditions, weathering being most active in the freely drained locations. There is thus one weathering system on the moorland plateau, where water moved along the iron pan, and another on the hillslope, where most water movement was above the fragipan. A similarity between deep interflow, Pathway 4, and the springs has been demonstrated. Thus the springs may be supplied from deep interflow as well as groundwater sources.

Further investigation of the water pathways, their role in supplying water to the stream, and their solute composition should enable the construction of more realistic weathering models for Dartmoor and more detailed assessment of contemporary weathering processes.

ACKNOWLEDGEMENTS

The authors wish to acknowledge the co-operation of the South West Water Authority for permitting us to instrument the Narrator Valley for experimental work and to thank Plymouth Polytechnic for its continued financial support for the project. We would also like to thank Sarah Webber for cartographic assistance, David Bosworth for technical assistance, Seana Doyle for typing the manuscript and Gerry Taylor the local South West Water Authority manager for his co-operation on many occasions.

REFERENCES

Arnett, R.R., 1976, Some pedological features affecting the permeability of hillside slopes in Caydale, Yorkshire, *Earth Surface Processes*, 1, 3-16.

Atkinson, T.C., 1978, Techniques for measuring subsurface flow on hillslopes, in: *Hillslope hydrology*, ed. Kirkby, M.J., (Wiley, Chichester), 73-120.

Anderson, M.G. and Burt, T.P., 1978, The role of topography in controlling throughflow generation, *Earth Surface Processes*, 3, 331-344.

Bricker, O.P. and Garrels, R.M., 1967, Mineralogic factors in natural water equilibria, in: *Principles and applications of water chemistry*, ed. Faust, S.D. & Hunter, J.V., (Wiley, London), 449-469.

Burt, T.P., 1979, The relationship between throughflow generation and the solute concentration of soil and streamwater, *Earth Surface Processes*, 4, 257-266.

Carson, M.A. and Kirkby, M.J., 1972, *Hillslope form and process*, (Cambridge University Press).

Dunne, T., 1978, Field studies of hillslope flow and processes, in: *Hillslope hydrology*, ed. Kirkby, M.J., (Wiley, Chichester), 227-294.

Eriksson, E., 1955, Air-borne salts and the chemical composition of rivers, *Tellus*, 4, 215-232.

Fitzpatrick, E.A., 1956, An indurated soil horizon formed by permafrost, *Journal of Soil Science*, 7, 248-254.

Foster, I.D.L., 1979, Intra-catchment variability in solute response: an East Devon example, *Earth Surface Processes*, 4, 381-394.

Hewlett, J.D., 1961, Soil moisture as a source of base flow from steep mountain watersheds, *U.S. Dept. Agriculture Forest Service, Eastern Forest Experimental Station, Asheville, North Carolina, Station Paper* no. 132.

Hewlett, J.D. and Hibbert, A.R., 1963, Moisture and energy conditions within a sloping soil mass during drainage, *Journal of Geophysical Research*, 68, 1081-1087.

Hewlett, J.D. and Hibbert, A.R., 1967, Factors affecting the response of small watersheds to precipitation in humid area, in: *Forest Hydrology*, ed. Sopper, W.E. and Lull, H.W. (Pergamon Press), 275-290.

Kent, M. and Wathern, P., 1980, The vegetation of a Dartmoor catchment, *Vegetatio*, 43, 163-172.

Kirkby, M.J. and Chorley, R.J., 1967, Throughflow, overland flow and erosion, *Bulletin International Association of Scientific Hydrology*, 12, 5-21.

Loughnan, F.C., 1962, Some considerations in the weathering of silicate minerals, *Journal of Sedimentary Petrology*, 32, 284-290.

Murgatroyd, A.L., 1980, Fluvial transport in the Narrator Brook, Devon; a summary of sources, dynamics and controls. Unpublished Ph.D. thesis, Plymouth Polytechnic.

Ragan, R.M., 1968, An experimental investigation of partial area contributions, *Bulletin International Association of Scientific Hydrology*, 13, 241-249.

Smith, T.R. and Dunne, T., 1977, Watershed geochemistry: the control of aqueous solutions by soil materials in a small watershed, *Earth Surface Processes*, 2, 421-425.

Ternan, J.L. and Williams, A.G., 1979, Hydrological pathways and granite weathering on Dartmoor, in: *Geographical approaches to fluvial processes*, ed. Pitty, A.F., (GeoBooks, Norwich), 5-30.

397

Verstraten, J.M., 1980, *Water-Rock interactions: a case study in a very low grade metamorphic shale catchment in the Ardennes, N.W. Luxembourg*, (Geo Abstracts, Norwich).

Waylen, M.K., 1979, Chemical weathering in a drainage basin underlain by Old Red Sandstone, *Earth Surface Processes*, 4, 167-178.

Weyman, D.R., 1974, Runoff process, contributing area and streamflow in a small upland catchment, in: *Fluvial processes in instrumented catchments, Institute of British Geographers, Special Publication No. 6*, ed. Gregory, K.J. and Walling, D.E., 33-43.

Whipkey, R.Z., 1965, Subsurface stormflow on forested slopes, *Bulletin International Association of Scientific Hydrology*, 10, 74-85.

Whipkey, R.Z. and Kirkby, M.J., 1978, Flow within the soil, in: *Hillslope hydrology*, ed. Kirkby, M.J., (Wiley, Chichester) 121-144.

Whitehead, H.C. and Feth, J.H., 1964, Chemical composition of rain, dry fallout and bulk precipitation, California, *Journal Geophysical Research*, 69, 3319-3333.

24 Magnitude and frequency characteristics of suspended sediment transport in Devon rivers

B.W. Webb and D.E. Walling

THE BACKGROUND

There have been relatively few studies of the magnitude and frequency of sediment transport by rivers during the two decades which have elapsed since the pioneering work of Wolman and Miller (1960). Nevertheless, a number of investigations have sought to establish for different environments the characteristics of fluvial transport in headwater streams (eg. Kennedy, 1964; Finlayson, 1977) and major rivers (eg. Grove, 1972; Schmidt, 1981), to isolate contrasts in the timing of bed load, suspended sediment and dissolved load yields (eg. Douglas, 1964; Walling, 1971; Smith and Newson, 1974) and to assess the geomorphological implications of the magnitude and frequency properties of fluvial systems (eg. Andrews, 1980; Webb and Walling, 1982). However, several theoretical and practical considerations have limited the development of traditional magnitude and frequency studies. In the former context, many recent geo-morphological investigations of drainage basins have emphasised the effectiveness of fluvial events in modifying landform and landscape (Baker, 1977; Wolman and Gerson, 1978; Anderson and Calver, 1980; Newson, 1980; Beven, 1981) rather than their role in removing material from the river system. Considerable attention has, therefore, been given to the formulation of new concepts based on lability and sensitivity of landscape to change (Trudgill, 1976; Brunsden and Thornes, 1979), transgression of external and internal thresholds (Harvey, 1977; Schumm, 1979), the occurrence of complex responses (Schumm, 1977) and the operation of recovery and repair processes (Anderson and Calver, 1977; Wolman and Gerson, 1978). Practical diffi-culties of collecting good quality information on the sedi-ment yields of rivers have also restricted the number of studies seeking and able to elucidate the magnitude and frequency characteristics of fluvial transport. Fisk (1977) has bemoaned the lack of appropriate data, especially in areas outside the USA, and the limitations of existing sediment records reflect both their quality and length. Recent studies have emphasised the inaccurate and imprecise

nature of sediment load estimates based on limited river
sampling (Walling and Webb, 1981, 1982), and further
problems arise from the short-term nature of many previous
research investigations of sediment dynamics which clearly
provide very little scope for magnitude and frequency
analysis.

Unlike many catchment experiments in geomorphology
reported in this volume, where fluvial processes have been
successfully investigated in a relatively short period
through the application of careful experimental design,
studies of the magnitude and frequency characteristics of
fluvial transport require a different approach in order to
be valuable. Once the study basin has been identified, the
design of a magnitude and frequency investigation is relatively straightforward, since the aim is to collect good
quality information on sediment transported past the catchment outlet, rather than to establish a network of instrumentation within the drainage basin which will be capable of
elucidating a particular fluvial process. However, magnitude and frequency studies may ultimately require more
effort to execute, because it is only through a long-term
commitment to detailed river monitoring that suitable
records of fluvial transport can be obtained. Lack of good
quality sediment information collected over a period of
appropriate length not only precludes the accurate definition of magnitude and frequency properties, but also
jeopardises comparison of results from different drainage
basins and, in turn, assessment of the factors influencing
this aspect of fluvial behaviour. With only a few exceptions (eg. Walling, 1971; Fisk, 1977; Dickinson and Wall,
1978), inter- and intra-basin contrasts in magnitude and
frequency characteristics and associated controls have been
largely neglected in drainage basin studies.

In the present study, attention is directed to the
transport of suspended sediment in Devon rivers, and continuous monitoring at three sites has provided information
for a total of 19 years. These data, to the knowledge of
the authors, represent the longest detailed records of
suspended sediment transport available in Britain, and provide a unique opportunity not only to define precisely the
magnitude and frequency properties of suspended solids
removal in a humid temperate environment, but also to
evaluate contrasts in suspended sediment transport within a
Devon river system.

THE STUDY CATCHMENTS

The study catchments comprise three west-bank tributaries
of the River Exe system in Devon, which vary in topographic,
geological and hydrometeorological character (figure 24.1).
The River Barle at Brushford drains 128 km^2 of upland moorland and hill farming country in the northwest of the Exe
Basin; it is underlain by massive sandstones, siltstones
and slates of Devonian age, and exhibits the highest mean
annual precipitation and runoff totals which exceed 1800
and 1500 mm respectively in the headwaters of the catchment.
The River Dart at Bickleigh (46 km^2) is the smallest of the

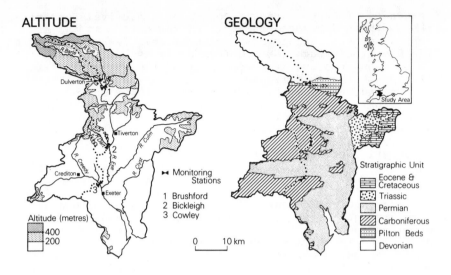

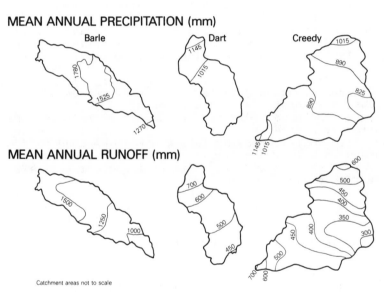

Figure 24.1.   Characteristics of the study catchments.

study catchments and drains Upper Carboniferous sandstone
and shale sequences in the central part of the Exe Basin.
The catchment is relatively steep, it is dominated by agricultural land use of permanent pasture and some limited
arable, and exhibits a range of mean annual precipitation
and runoff from in excess of 1100 and 700 mm respectively
on highest ground to less than 900 and 450 mm respectively
at the basin outlet.   The River Creedy at Cowley drains the
largest catchment area (262 km^2) and has lower relief than
the other study basins.   Arenaceous and argillaceous

sequences of Upper Carboniferous age respectively underlie
the northern and southern parts of the catchment and are
separated by a variety of Permian lithologies preserved in
the Crediton trough. Mixed farming is associated with the
Permian outcrop, but the remainder of the basin is, like
that of the Dart, dominated by permanent pasture. Precipitation and runoff totals are generally lower in the River
Creedy basin and mean annual levels fall to less than 800
and 300 mm respectively near the outlet. None of the three
study catchments are affected by major domestic or industrial pollution.

Continuous records of discharge for the Rivers Creedy
and Barle are available from the South West Water Authority,
and flow data for the River Dart have been collected by the
authors. Detailed information on suspended sediment loads
for the three catchments has been obtained by combining
discharge data with continuous records of suspended sediment concentration. The latter have been produced using
photoelectric turbidity sensors mounted directly in the
river channels (Walling, 1978). This equipment has proved
well suited to the predominantly fine-grained particulate
load of Devon rivers, and generates continuous records of
turbidity which are readily converted to values of average
suspended sediment concentration (total filterable solids)
in the measuring cross-sections using calibration relationships derived from field sampling (Walling, 1977). Abstrac-
tion and combination of hourly instantaneous values of
suspended sediment concentration and discharge from the
chart records produces detailed records of suspended sediment transport for the study catchments.

Hourly suspended sediment load data for the River Creedy
are currently available for a 9 year period from October
1st, 1972 to September 30th, 1981 and will be used to define unequivocally the medium-term magnitude and frequency
characteristics of suspended sediment transport in a sizeable British drainage basin. Hourly load values have also
been assembled for the Rivers Barle and Dart for a 5 year
period from October 1st, 1976 to September 30th, 1981.
These will be compared with equivalent results from the
River Creedy in order to evaluate the contrasts in suspended
sediment transport which occur within a single Devon river
system.

MAGNITUDE AND FREQUENCY CHARACTERISTICS OF
SUSPENDED SEDIMENT TRANSPORT

Total, annual and monthly yields

The total runoff from the River Creedy during the 9 year
study period exceeded 1012×10^6 m^3 and was associated with
a total suspended sediment load of 91 639 t and a discharge-
weighted mean suspended sediment concentration of 90.6 mgl^{-1}.
Annual water and suspended sediment yields are listed in
table 24.1 and it is clear that suspended sediment loads
exhibit greater variation from year to year than the runoff
totals. The coefficient of variation of annual suspended
sediment yield (55.3%) is considerably greater than that for
annual runoff totals (35.1%) and indicates that suspended

Table 24.1.   Annual water and suspended sediment
yields from the River Creedy at Cowley

Water* Year	Runoff ($m^3.10^6$)	Suspended Sediment Load ($t.10^3$)
72/73	80.362	7.482
73/74	142.321	20.619
74/75	119.104	10.546
75/76	35.573	1.938
76/77	176.934	16.234
77/78	129.043	10.222
78/79	100.190	4.717
79/80	116.951	11.109
80/81	111.582	8.771
Mean	112.451	10.182
Standard Deviation	39.499	5.631

* Water year runs from 1st October to 30th September

sediment transport is not solely controlled by water yield
on an annual basis.

A plot of annual load against annual runoff for individual water years (figure 24.2) reveals a strong positive
relationship between suspended sediment yield and discharge,
but the slope of 1.4 fitted to the relationship by least-
squares regression (figure 24.2) demonstrates that the
increase in transport during wetter years is more than proportional to the increase in runoff. Furthermore, considerable scatter is apparent in the relationship between
annual load and discharge and this reflects the dependence
of suspended sediment transport in a given year on not only
the total or peak discharge, but also the number, origin
(whether rain or snow-generated) and sequence of major
storm events (Webb and Walling, 1982).

Nine years of data also allow a meaningful evaluation
of the annual water discharge and suspended sediment transport regimes, and the average monthly yields together with
their associated coefficients of variation are presented in
figure 24.3. The general forms of the discharge and load
regimes are similar, but average monthly runoff yields peak
in February whilst sediment yields peak in January.
Furthermore, sediment transport is also greater in December
than February. This may point to an exhaustion effect,
whereby the large flow totals of December and January re-
move readily available sediment accumulated during the
summer and autumn period, so that loads during February are
relatively depleted despite peak runoff yields. The annual

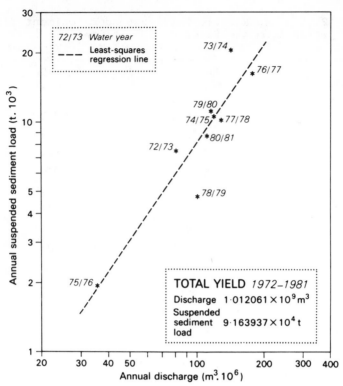

Figure 24.2. Relationship between annual runoff and annual suspended sediment yield in the River Creedy at Cowley.

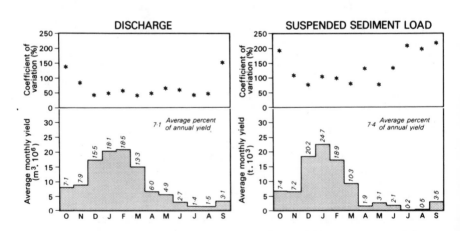

Figure 24.3. Average monthly yields of runoff and suspended sediment and associated coefficient of variation values for the River Creedy at Cowley.

regime of suspended sediment yield is also more extreme
than that of discharge. The period from December to
February accounts for nearly 64% of the annual suspended
sediment transport compared with 52% of the annual water
yield. During the summer months, suspended sediment transport is very restricted and less than 12% of the annual
load is removed in the 6 months from April to September,
compared with nearly 20% of the annual runoff during the
same period. The extremely low proportion of annual suspended sediment yield associated with the predominantly
baseflow period of July and August emphasises that removal
of suspended sediment is essentially a storm-based process.
The distribution of average monthly yields also shows a
less regular pattern for suspended sediment load than for
discharge. The march of monthly flow values evidences a
progressive increase from a July minimum to a February
maximum and an uninterrupted decline through to June,
whereas the distribution of mean monthly suspended sediment
loads is complicated by a small secondary maximum in May
and minimum in November.

The greater variability of monthly suspended sediment
loads may contribute to the more irregular nature of the
average annual regime. Coefficient of variation values
(figure 24.3) derived from monthly yields are appreciably
lower for discharge than suspended sediment load. Only for
September, October and November does the coefficient of
variation of monthly runoff exceed 75% and this reflects
the variable timing of the onset of major storms in the
autumn and early winter period. In contrast, suspended
sediment loads in all months are associated with a coefficient of variation greater than 75% and for the period
between June and October values approach or exceed 200%.
Since suspended sediment transport is largely accomplished
during relatively short storm events, which in turn have an
essentially random occurrence, monthly load values can be
expected to vary considerably from year to year. Large
variations occur especially during the low flow and low
sediment transport period of late summer and early autumn
when the chance occurrence of an intense storm may greatly
influence the monthly load and cause considerable deviation
from the average monthly yield.

Load duration data

Construction of simple and cumulative load duration curves,
based on hourly load values (figure 24.4), provides a
further perspective on suspended sediment transport in the
River Creedy during the 9 year study period. The considerable variability of suspended sediment loads noted in
previous studies of British rivers (eg. Douglas, 1964;
Smith and Newson, 1974) is also characteristic of the
present study basin and duration data reveal that hourly
sediment yields range over more than six orders of magnitude in comparison with between three and four orders for
hourly discharge values. Relatively high suspended sediment yields of 1 thr^{-1} were equalled or exceeded during
only 10% of the study period and for 50% of the time loads
were less than 5 kghr^{-1}. However, the importance of high

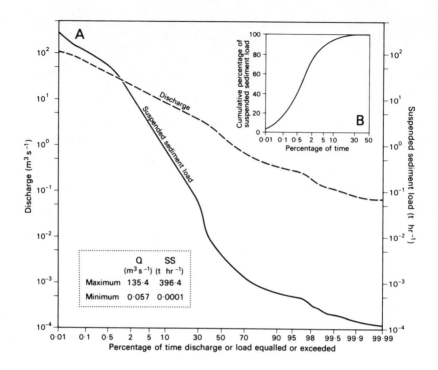

Figure 24.4. Runoff and suspended sediment load duration
   curves (A), and cumulative sediment load duration curve
   (B) for the River Creedy at Cowley.

yields, associated with short-lived storm events, to total
suspended sediment transport is more evident from load
duration information expressed in cumulative form (figure
24.4B). These data indicate that 50% of the total load was
removed in only 26.3 days, which in turn represents a mere
0.8% of the total study period or an average of 3 days per
year. A large part (90%) of the total suspended sediment
transport during the study period was accomplished in only
7% of the time, and the ineffectiveness of suspended sediment removal during lower flows is clearly apparent from
figure 24.4B which indicates that loads recurring 20% of the
time or more frequently contribute less than 1.3% of the
total suspended sediment yield.

Extremes of suspended sediment transport

The importance of major storms to the removal of suspended
sediment in the River Creedy is apparent from table 24.2
which lists for each water year the maximum daily yield of
runoff and sediment. These values have been calculated
from the hourly instantaneous data and are also expressed
as a percentage of the total annual yield. The most
extreme daily event of each year has a much greater impact
on suspended sediment transport than water yield and
although proportions vary considerably from year to year,

Water year	Maximum daily yield			
	Runoff ($m^3.10^6$)	%*	Suspended sediment load ($t.10^3$)	%*
72/73	4.598	5.7	1.363	18.2
73/74	4.777	3.4	2.547	12.4
74/75	2.683	2.3	1.666	15.8
75/76	1.156	3.3	0.233	12.0
76/77	3.901	2.2	1.523	9.4
77/78	6.688	5.2	1.134	11.1
78/79	2.381	2.4	0.696	14.8
79/80	4.080	3.5	2.530	22.8
80/81	2.365	2.1	0.610	7.0
Mean		3.3		13.7

* Percentage of total annual yield

the maximum daily yield accounts on average for 13.7% of the annual suspended sediment load in comparison with only 3.3% of the annual runoff.

The annual series of maximum daily yields also shows more variability for suspended sediment transport than discharge. The greatest annual maximum daily yield of suspended sediment is nearly 11 times larger than the smallest, whereas for peak daily runoff totals the comparable range is less than six times. This result indicates that events of increasing recurrence interval become more extreme in terms of suspended sediment transport than for flow volume.

CATCHMENT VARIATIONS IN MAGNITUDE AND FREQUENCY PROPERTIES

The data collected for the River Creedy since 1972 have allowed the magnitude and frequency characteristics of suspended sediment transport to be precisely defined for a major tributary of the Exe Basin. Further detailed records of suspended sediment load, available for the Rivers Barle and Dart from the period October 1st, 1976 to September 30th, 1981, enable the extent to which magnitude and frequency properties of suspended sediment transport vary across a single British river system to be investigated.

Contrasts are evident in several aspects of suspended sediment removal including average annual yield per unit drainage area. The River Barle, although situated in the wettest part of the Exe Basin and experiencing high runoff totals (figure 24.1), exhibits the lowest specific suspended sediment yield (16.4 tkm^{-2} year^{-1}) for the study period.

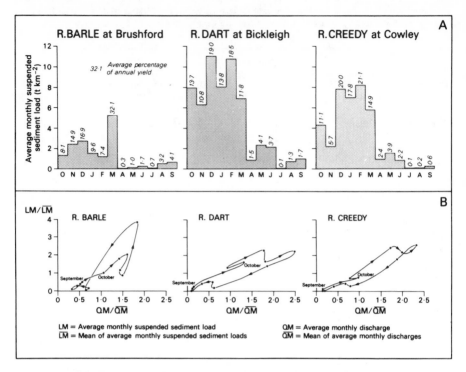

Figure 24.5. Annual regime of suspended sediment transport
    in the study catchments evidenced by the march of
    average monthly specific sediment yields (A), and the
    relationship between monthly load and runoff (B).

Potential for erosion and sediment generation in the Barle
catchment is limited by the resistant nature of the underlying Devonian strata and the lack of intensively cultivated
land, and in consequence suspended sediment concentrations
rarely exceed 100 mgl^{-1}. Higher values of peak suspended
sediment concentration (> 3000 mgl^{-1}) and specific suspended
sediment yield (39.0 tkm^{-2} year^{-1}) were encountered in the
River Creedy and reflect more readily eroded Carboniferous
and Permian rock types and a larger proportion of agricultural land. In addition to geological and land-use
factors, steep topography and limited catchment area
enhance erosion and sediment delivery in the River Dart
and give rise to the highest mean annual sediment yield
(58.3 tkm^{-2} year^{-1}) for the study period.
    The annual regime of suspended sediment yield also
evidences considerable variation between the three study
catchments, and average monthly yields per unit area (figure
24.5A) exhibit clear differences in the number and timing
of maxima in the annual distribution. Monthly loads, for
example, peak in December for the River Dart, in February
for the River Creedy and in March for the River Barle and
also show important secondary maxima at Bickleigh and
Cowley, but not at Brushford. These differences in regime

in part reflect spatial variation in the incidence of storm activity between the northern upland part of the Exe Basin and the remainder of the river system. A further aspect of the annual march of suspended sediment yield which varies between the study catchments is the trend of monthly load values in relation to monthly runoff. Fournier and Henin (1962) have devised a method of plotting this relationship which standardises values of average monthly load and discharge in terms of the mean of the average monthly yields for the whole year, and this technique has been applied to the data collected for each study catchment (figure 24.5B). The resulting plots have a looped form, which reveal hysteresis in the annual march of monthly suspended sediment yields, but the graphs, according to Fournier and Henin's classification, also show a complex rather than a simple relationship between monthly suspended sediment loads and discharge. Furthermore, a predominantly clockwise trend is apparent in the graphs for the Rivers Dart and Creedy, which suggests some exhaustion of sediment source areas during storms of the autumn and early winter period. In contrast, the loop for the River Barle has a major anticlockwise trend during the winter months and indicates that the large flows of March have access to considerable supplies of suspended sediment.

Strong contrasts are also evident between the study basins in the detailed timing of suspended sediment removal. Load duration curves constructed in cumulative form (figure 24.6) reveal that significant sediment transport is accomplished in a shorter period by the River Barle than by the other two catchments. 50% of the total suspended sediment load transported during the study period was removed in only 0.2% (88 hours) of the time at Brushford in comparison with 0.35% (153 hours) and 0.75% (329 hours) respectively for Bickleigh and Cowley, and equivalent duration figures for 90% of the load are 2.5% (46 days) of the period in the River Barle, 2.75% (50 days) in the River Dart and 7% (128 days) in the River Creedy. The extremely episodic nature of suspended sediment transport at Brushford is highlighted by the removal of more than 10% of the total load in only 0.01% of the time, and it appears that the physiographic characteristics of the Barle catchment severely limit sediment availability so that significant source areas are only accessed or eroded during the most extreme events. In contrast, significant supplies of suspended sediment are available to a wider range of flows in the River Creedy and sediment removal is a more protracted process. The cumulative load duration curve for Bickleigh plots in between that for Brushford and Cowley (figure 24.6), although the timing of suspended sediment transport in the River Dart is closer to that of the River Barle than the River Creedy. Unlike the River Barle, the form of the cumulative load duration curve for the River Dart is not related to conditions of limited sediment availability, but reflects the flow distribution of a small and steep catchment which is considerably skewed towards higher discharges and enhances the role of peak flows in suspended sediment transport (Wolman and Miller, 1960).

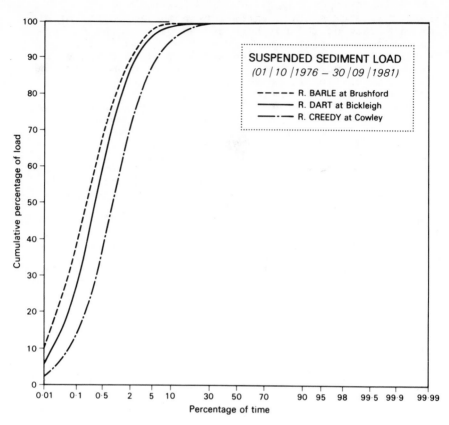

Figure 24.6. Cumulative suspended sediment load duration
   curves for the study catchments.

   The reaction of suspended sediment yields to extreme
events provides further evidence of differences in sediment
generation and transport between the three study basins.
On an annual basis, the relationship between suspended
sediment and runoff yields (figure 24.7A) evidences a considerable increase in slope from the Rivers Dart and Creedy
to the River Barle, so that the contrast in suspended sediment transport between wettest and driest years is more
dramatic at Brushford than at Cowley or Bickleigh. It is
also apparent from table 24.3, which lists the annual
maximum daily suspended sediment load at each monitoring
station, and from figure 24.7B, which illustrates the
percentages of the total annual loads contributed by the
maximum daily events, that the three study basins also
respond differently to the impact of individual major
events.   There is no correspondence between catchments in
the incidence of peak transport rates, since the highest
value of maximum daily sediment yield occurred in the
1980/81 water year at Brushford, in 1976/77 at Bickleigh
and in 1979/80 at Cowley. This variation between tributaries arises in a catchment the size of the Exe Basin

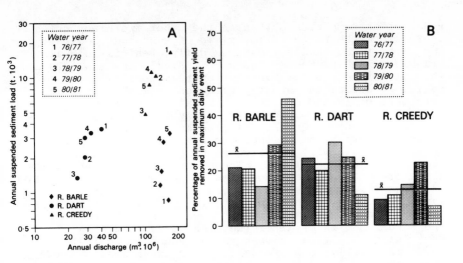

Figure 24.7. Relationships between annual runoff and annual suspended sediment yield (A), and maximum daily suspended sediment loads of individual water years (B) in the study catchments. x̄ refers to average percentage from five years of record.

Table 24.3. The annual series of maximum daily suspended sediment yields in the study catchments

Water Year	Maximum daily yield (tkm^{-2})		
	R. Barle	R. Dart	R. Creedy
76/77	1.414	19.267	5.812
77/78	1.864	8.983	4.327
78/79	1.675	8.949	2.658
79/80	6.236	18.048	9.655
80/81	15.252	7.350	2.329

(1500 km^2) because major storm events may be geographically restricted to one part of the river system or display considerable lateral variation in intensity if the whole basin is affected. Furthermore, figure 24.7B indicates that major storms have the greatest influence on suspended sediment transport at Brushford, where maximum daily yields account on average for 26.3% of the annual load and in excess of 45% for the 1980/81 water year. In the Rivers Dart and Creedy the maximum daily events transport respectively 22.2% and 13.0% on average of the annual yield of suspended sediment. The River Barle also exhibits a greater

411

range of extreme events than in the other two study catchments since the ratio of the yield of suspended sediment in
the largest and smallest maximum daily event is 10.8 at
Brushford, 4.2 at Cowley and 2.6 at Bickleigh (table 24.3).
This difference in the range of extreme conditions experienced during the study period can be linked to both
catchment scale and sediment availability. In the small
basin of the River Dart, it is likely that events of relatively frequent occurrence will be able to readily tap
sediment sources, since there is a high probability that
the whole basin will be affected by even small storms
(Wolman and Gerson, 1978). The occurrence of rarer events,
therefore, may not be associated with a substantial increase
in accessibility to sediment source areas. However, in
larger catchments, such as the River Creedy, there may be
appreciable increases in the sediment contributing area as
the recurrence interval of the event becomes greater. The
very large variation in peak yields encountered in the River
Barle may partly reflect the large drainage area ($> 100\,km^2$),
but also is related to the limited sources of suspended
sediment in comparison to the Rivers Dart and Creedy. In
these circumstances, it seems that sediment sources are
only accessed significantly during the most extreme events
and the contrast in sediment loads associated with rare and
more frequent storms becomes considerable.

CONCLUSIONS

This contribution has focussed on the longer-term operation
of fluvial processes and has sought to define the magnitude
and frequency properties of suspended sediment transport in
three Devon rivers. Nine years of continuous flow and
suspended sediment concentration record for the River Creedy
at Cowley has allowed accurate quantification of total
yields, the relationship of annual load to annual runoff,
and the regime of flow and suspended sediment transport.
The timing of suspended sediment removal has been investigated by the construction of duration curves which show
that 90% of the total load was removed in only 7% of the
study period. The importance of short-lived and infrequent
storm events to sediment transport is emphasised by an
analysis of maximum daily yield values, which on average
account for more than 13% of the annual load in the River
Creedy.
    Results for 5 years from two other Devon rivers confirm
the general characteristics of suspended sediment transport
recorded in the Creedy but also reveal significant contrasts
in magnitude and frequency properties within a single river
system. Average annual specific suspended sediment yields
both less and greater than that of the River Creedy
($39.0\ tkm^{-2}\ year^{-1}$) were encountered respectively in the
River Barle ($16.4\ tkm^{-2}\ year^{-1}$) and River Dart ($58.3\ tkm^{-2}$
$year^{-1}$), and the three tributaries exhibited different
annual suspended sediment transport regimes. Furthermore,
the timing of suspended sediment transport and the role of
extreme events in sediment removal varied appreciably
between the three catchments and were most extreme in the

River Barle at Brushford. It is thought that variations in catchment scale and in sediment availability are responsible for contrasts in the magnitude and frequency properties, and that further attention needs to be given to the precise interaction of runoff contributing zones and sediment source areas before contrasts in suspended sediment transport between catchments can be fully explained.

ACKNOWLEDGEMENTS

The authors gratefully acknowledge the financial support for work on suspended sediment dynamics provided by a Natural Environment Research Council research grant, and the assistance of the South West Water Authority in supplying river flow data.

REFERENCES

Anderson, M.G. and Calver, A., 1977, On the persistence of landscape features formed by a large flood, *Transactions of the Institute of British Geographers*, New Series 2, 243-254.

Anderson, M.G. and Calver, A., 1980, Channel plan changes following large floods, in: *Timescales in geomorphology*, ed. Cullingford, R.A. *et al.*, (Wiley-Interscience, Chichester), 43-52.

Andrews, E.D., 1980, Effective and bankfull discharges of streams in the Yampa River Basin, Colorado and Wyoming, *Journal of Hydrology*, 46, 311-330.

Baker, V.R., 1977, Stream-channel response to floods, with examples from central Texas, *Geological Society of America Bulletin*, 88, 1057-1071.

Beven, K., 1981, The effects of ordering on the geomorphic effectiveness of hydrologic events, in: *Erosion and sediment transport in Pacific Rim Steeplands*, IAHS Publication no.132, 510-526.

Brunsden, D. and Thornes, J.B., 1979, Landscape sensitivity and change, *Transactions of the Institute of British Geographers*, New Series 4, 463-484.

Dickinson, W.T. and Wall, G.T., 1978, Temporal and spatial patterns in erosion and fluvial processes, in: *Research in fluvial geomorphology*, ed. Davidson-Arnott, R. and Nickling, W., (Geo-Abstracts, Norwich), 133-148.

Douglas, I., 1964, Intensity and periodicity in denudation processes with special reference to the removal of material in solution by rivers, *Zeitschrift für Geomorphologie*, 8, 453-473.

Finlayson, B., 1977, Runoff contributing areas and erosion, *Research Papers of the School of Geography, University of Oxford*, no.18.

Fisk, H.N., 1977, Magnitude and frequency of transport of solids by streams in the Mississippi Basin, *American Journal of Science*, 277, 862-875.

Fournier, F. and Henin, S., 1962, Etude de la forme de la relation existant entre l'écoulement mensuel et le débit solide mensuel, in: *Symposium of Bari*, IAHS Publication no.59, 353-358.

Grove, A.T., 1972, The dissolved and solid load carried by some west African rivers: Senegal, Niger, Benue and Shari, *Journal of Hydrology*, 16, 277-300.

Harvey, A.M., 1977, Event frequency in sediment production and channel change, in: *River channel changes*, ed. Gregory, K.J., (Wiley-Interscience, Chichester), 301-315.

Kennedy, V.G., 1964, Sediment transported by Georgia streams, *US Geological Survey Water Supply Paper*, 1668, 101 pp.

Newson, M.D., 1980, The geomorphological effectiveness of floods - a contribution stimulated by two recent events in mid-Wales, *Earth Surface Processes*, 5, 1-16.

Schmidt, K.H., 1981, Der Sedimenthaushalt der Ruhr, *Zeitschrift für Geomorphologie, N.F.*, Supp. 39, 59-70.

Schumm, S.A., 1977, *The fluvial system*, (Wiley-Interscience, New York).

Schumm, S.A., 1979, Geomorphic thresholds: the concept and its applications, *Transactions of the Institute of British Geographers*, New Series 4, 485-515.

Smith, D.I. and Newson, M.D., 1974, The dynamics of solutional and mechanical erosion in limestone catchments on the Mendip Hills, Somerset, in: *Fluvial processes in instrumented watersheds*, ed. Gregory, K.G. and Walling, D.E., Institute of British Geographers Special Publication no.6, 155-167.

Trudgill, S.T., 1976, Rock weathering and climate: quantitative and experimental aspects, in: *Geomorphology and climate*, ed. Derbyshire, E., (Wiley-Interscience, London), 59-99.

Walling, D.E., 1971, Sediment dynamics of small instrumented catchments in south east Devon, *Reports and Transactions of the Devonshire Association for the Advancement of Science Literature and Art*, 103, 147-165.

Walling, D.E., 1977, Limitations for the rating curve technique for estimating suspended sediment loads, with particular reference to British rivers, in: *Erosion and solid matter transport in inland waters*, IAHS Publication no.122, 34-48.

Walling, D.E., 1978, Suspended sediment and solute response characteristics of the River Exe, Devon, England, in: *Research in fluvial geomorphology*, ed. Davidson-Arnott, R. and Nickling, W., (Geo-Abstracts, Norwich), 169-197.

Walling, D.E. and Webb, B.W., 1981, The reliability of suspended sediment load data, in: *Erosion and sediment transport measurement*, IAHS Publication no.133, 177-194.

Walling, D.E. and Webb, B.W., 1982, The design of sampling programmes for studying catchment nutrient dynamics, in: *Proceedings of the International Symposium on hydrological research basins*, (Mitteilungen, Landeshydrologie Sonderheft, Bern), 747-758.

Webb, B.W. and Walling, D.E., 1982, The magnitude and frequency characteristics of fluvial transport in a Devon drainage basin and some geomorphological implications, *Catena*, 9, 9-23.

Wolman, M.G. and Gerson, R., 1978, Relative scales of time and effectiveness of climate in watershed geo-morphology, *Earth Surface Processes*, 3, 189-208.

Wolman, M.G. and Miller, J.C., 1960, Magnitude and frequency of forces in geomorphic processes, *Journal of Geology*, 68, 54-74.

PART IV

HILLSLOPE AND CHANNEL PROCESSES

25 The relationship between soil creep rate and certain controlling variables in a catchment in upper Weardale, northern England

Ewan W. Anderson and Nicholas J. Cox

INTRODUCTION

Soil creep is agreed by geomorphologists to be a relatively slow process. Indeed, some workers have gone further and described it as almost imperceptible. Over the last 25 years, however, accurate and reliable measurement procedures have been developed, and it has been shown that measurement errors can be reduced to acceptable magnitudes (Anderson and Finlayson, 1975; Anderson and Cox, 1978). There is a continuing need for rate measurements from a variety of environments, preferably over extended time spans (eg. Young, 1978). A further aim of research is to relate creep rates to associated controlling variables. This may be at a largely empirical level, in which analysis of observational data leads to the identification of the strongest relationships and the development of predictive statistical models. It may also include attempts to test theoretical speculations about the mechanisms underlying creep; or conversely it may lead to suggestions of variables which should be included in models of creep mechanisms.

Over the last 100 years, as a result of both theoretical arguments and empirical studies, many variables have been suggested as possible controls of the rate of soil creep. Meteorological cycles (wetting and drying, heating and cooling, freezing and thawing) have been regarded as producing net downslope movement when combined with the vertical action of gravity. This also implies that slope angle may be an important control. The possible effects of plant roots, surface vegetation cover and animals have been discussed, together with distance from the divide (or from the slope base) and slope aspect. Many properties of the soil itself have been invoked as controls, including soil moisture, cohesion, shear strength, plasticity, shrinkage, texture, organic content, density and thickness.

In this paper we report on an investigation of the relationship between soil creep rate and controlling variables, which forms part of a larger study of a small instrumented catchment near Rookhope in upper Weardale in northern England (Anderson, 1977). The emphasis is on relationships inferred from plot-to-plot variation within the catchment in a particular period, and the methods used for data analysis have been applied to data for continuous (ie. quantitative) variables. Some controlling variables do not lend themselves to this approach. It is not practicable to monitor meteorological cycles separately at each plot in the catchment, while some properties are not readily quantified by single measures appropriate to all plots (eg. texture, because not all soils possess mineral horizons; and vegetation, because the taxa present vary from plot to plot). It may seem from this that the requirements of statistical techniques have been allowed to dictate the form of the analysis: however, the influence of other variables has been discussed elsewhere in detail (Anderson, 1977), and it may be hoped that the effects of variables such as vegetation and texture will be captured through their relationship with other variables.

CATCHMENT DESCRIPTION AND DATA COLLECTION

The study area is a small catchment near Rookhope in upper Weardale within the northern Pennines. Measurements were taken in an area of 15 ha, which is bounded by two small streams, a quarry railway embankment and a divide, and is thus strictly speaking an interfluve. Altitudes vary between 367 m and 487 m. Three hillslope profiles (967 angle readings) were recorded using a pantometer (Pitty, 1968): the median angle was 10°, and more than half the readings fell between 7° and 13°. The bedrock consists of Carboniferous sedimentaries of varying resistance (mudstones, shales and sandstones). Solifluction deposits are present. On wetter sites peaty gleys and surface-water gleys can be distinguished, while iron podzols and peaty podzols occur on drier sites.

Three main vegetation types can be identified in a complex mosaic showing sharp changes over short distances. Each type is dominated by a single genus, *Nardus*, *Pteridium* or *Juncus*. All three occur on the gentlest and the steepest slopes, but the *Nardus* grassland shows the greatest tolerance of drainage conditions. The *Pteridium* heath occurs everywhere on dry slopes with thin iron podzols, while the *Juncus* bog is limited to the wetter parts of the catchment in which the peaty gleys are also developed. Hence in broad terms a classification of vegetation types also amounts to a classification of soil and moisture conditions.

The observational design adopted after pilot studies can be described as a three-stage procedure. First, a cross-classification of three slope classes (1-8°, 9-16°, 17-24°) and three vegetation types (as above) yielded nine site categories. Secondly, 20 plots were chosen so that each category was represented by at least one plot: varying areal extent of categories provided constraints on the choice of plots, but efforts were made to ensure as close an approach to random selection as possible. Thirdly, measurements of creep and associated controlling variables were taken at each plot. A grid of nine points with rows

and columns spaced 0.6 m apart was used for creep measuring
devices, while other properties were generally measured
immediately downslope in order to minimise the possibility
of disturbing these devices. The spacing of measurements
at each plot represented a compromise between the need for
a relatively homogeneous plot and the need to take readings
from each instrument without disturbing neighbouring
apparatus.

Methods for soil creep measurement have been discussed
in detail elsewhere (Anderson and Finlayson, 1975; Anderson
and Cox, 1978). Six methods were used at each plot:
Anderson's inclinometer pegs, Anderson's tubes, aluminium
pillars, Young's pits, dowelling pillars and Cassidy's
tubes. A detailed analysis of linear creep rates (units:
mm year^{-1}) measured in 1973-74 was given by Anderson and
Cox (1978). The general pattern was one of consistency
among different instruments. When taken in pairs, the 15
Pearson correlations between instruments ranged from 0.76
to 0.98 with a median of 0.92. The first principal component in a principal components analysis of the standardised data matrix for six instruments and 20 plots accounted
for 91.9% of the total variance. In more detail, Cassidy's
tube stands out as discordant: it contributes the five
lowest correlations (r between 0.76 and 0.87) and it loads
most highly on the second principal component. There are
substantial doubts about this method since the tubing may
be too rigid to allow both downslope and upslope movement.
It seems unlikely that the strong agreement between the
remaining methods is fortuitous, and it is regarded as an
indication that measurement errors are relatively small.
This interpretation is supported by results from an independent analysis using median polish (Tukey, 1977), a
robust alternative to two-way analysis of variance. On
various grounds, technical and statistical, the best methods
appear to be Anderson's inclinometer pegs and Anderson's
tubes. These are very highly correlated with each other
(r = 0.98). Results for inclinometer pegs are analysed
below in relation to controlling variables. We also report
on results for volumetric rates (volume per unit contour
width per unit time, units: cm^2 year^{-1}), which have been
estimated using data on linear rates and on maximum depth
of movement as recorded by Cassidy's tubes and dowelling
pillars.

Some twenty controlling variables were considered in
relation to soil creep. Outline notes on measurement procedure follow: two general biases deserve mention. It was
felt to be important to adopt, as far as possible, methods
which could be used easily and rapidly in the field,
especially since some standard laboratory procedures require
large soil samples. Given the number of plots studied,
the distortions involved in collection and transfer to the
laboratory, the requirements of landowners, and the known
heterogeneity of soil, such methods may be impracticable,
impolite and spuriously precise. Similarly, it was thought
important to measure properties for conditions which
actually occur in the field, or are very likely to occur,
rather than for simulated conditions which may not obtain
in the field. This explains the use of levels such as

'field capacity' or 'winter moisture', which are admitted to be fairly crude and arbitrary, rather than some more precisely defined condition reproduced in laboratory measurement.

Controlling soil variables measured.

Sine of slope angle: Based on average of three angle measurements from Abney level placed on a 1 m long blade on the surface.

Soil depth (cm): Average of results from augering and an excavated pit.

Organic depth (cm): Average of results from excavated pit.

% organic: % loss of mass on heating for 16 hr at 375°C: results from different horizons combined in weighted mean.

Field capacity: Mass of water as % of total. Soil samples collected in 50 ml tins, soaked in water for 2 days, allowed to drain for 30 min before oven drying for 24 hr at 105°C: results from different horizons combined in weighted mean.

Winter moisture and summer moisture: Mass of water as % of total (or of dry soil) for wettest and driest conditions observed in the field. Measured with Speedy flask (Akroyd, 1964): results from different horizons combined in weighted mean.

Saturation level (cm above arbitrary datum): Highest level reached by water as indicated by debris mark inside Anderson's tube.

Plasticity index = liquid limit - plastic limit (all in % of mass of dry soil): Using tests as in British Standard 1377. Separately for whole soil and for organic layer.

Linear shrinkage: % reduction of length on air drying (48 hr) and oven drying (24 hr at 65°C, 24 hr at 105°C) of soil cores 5 cm in length and 2.5 cm in diameter extracted at winter moisture: results from different horizons combined in weighted mean.

Density (g cm^{-3}): Measured using method of Chepil (1950) for dry soil and from soil core immersed in water for 72 hr for soil at field capacity: results from different horizons combined in weighted mean.

Apparent cohesion (kg cm^{-2}), angle of internal friction (degrees), unconfined compression (stress, kg cm^{-2}; strain, cm): Maxima observed with combined unconfined compression and shear test apparatus (King and Cresswell, 1954): results from different horizons combined in weighted mean.

Penetration (kg cm^{-2}): Resistance to penetration by hand penetrometer. Modal reading for horizon with maximum resistance at winter moisture.

DATA ANALYSIS

The two measures of creep rate are strongly related to each other as would be expected. Volumetric (v) is given as a function of linear (ℓ) by

$$v = -0.11 + 0.66\ell$$

with a correlation of 0.94. Logarithmic transformation of both variables gives a power function fit, which is

$$v = 0.52\ell^{1.23}$$

Correlation between logged variables is 0.96. This function goes through the origin, which makes physical sense, and it appears from examination of residuals to be marginally preferable. This very high correlation between linear and volumetric rates implies that analyses of the two as response variables are essentially parallel.

Linear rate ranges from 0.3 to 2.4 mm year^{-1} with a mean of 1.3 mm year^{-1}. Volumetric rate ranges from 0.1 to 1.7 cm^2 year^{-1} with a mean of 0.75 cm^2 year^{-1}.

The correlations between linear and volumetric creep rates and the set of controlling variables are given in table 25.1. Almost all controls which have been measured correlate strongly with both measures of creep. The strongest correlations (above 0.85 in magnitude) are with winter and summer moisture, field capacity, plasticity and dry bulk density (negative). The correlations with sine of slope angle stand out as the weakest observed (-0.44, -0.35).

From these correlations it follows that the bivariate regressions for several controlling variables have r^2 of 0.72 to 0.77. There would seem to be little to choose between them. Moisture as a proportion of dry mass and as a proportion of wet mass are one-to-one functions of each other: we prefer the less common wet mass measures since the dry mass measures have skewed frequency distributions in the catchment (plasticity measures also have skewed distributions). Winter moisture (% wet) emerges as one of the best predictors when it is recalled that field capacity is a level that might not be attained under field conditions at all plots.

Extension from bivariate to multiple regression might seem natural in view of the multivariate character of the problem: we expect several different variables to be at work influencing soil creep. The statistical model entertained now has the form

$$y = b_0 + \sum_{j=1}^{p} b_j x_j + e$$

where y denotes data for creep rate, the response variable; $x_1, \ldots, x_p$ are predictors or controlling variables; $b_0, \ldots, b_p$ are parameters estimated from the data; and e the residual for each data point. The advantages and limitations of this method are well known (eg. Kendall,

423

Table 25.1. Correlations between creep rates and controlling variables

	Linear rate	Volumetric rate
Sine of slope angle	-0.44	-0.35
Soil depth	0.75	0.76
Organic depth	0.70	0.74
% organic	0.81	0.82
Field capacity	0.85	0.87
Winter moisture (% wet)	0.86	0.88
Winter moisture (% dry)	0.88	0.86
Summer moisture (% wet)	0.86	0.87
Summer moisture (% dry)	0.85	0.79
Saturation level	0.66	0.63
Plasticity index	0.85	0.85
Plasticity index (organic)	0.77	0.76
Shrinkage	0.84	0.81
Density (dry bulk)	-0.87	-0.85
Density (field capacity)	-0.82	-0.80
Apparent cohesion	0.68	0.75
Angle of internal friction	-0.81	-0.78
Unconfined compression (stress)	-0.63	-0.66
" " (strain)	-0.70	-0.70
Penetrometer resistance	-0.83	-0.82

1975; Mosteller and Tukey, 1977) and only a few points need be emphasised briefly. As in bivariate regression, the linear structure of the model is not derived theoretically, and need not make physical sense: at best we have merely an empirical description of relationships. In multiple regression, no allowance is made for interactions among the controlling variables, although in practice if they are highly correlated with each other, then considerable technical problems arise in parameter estimation, and it may be very difficult to interpret results. Certainly the multiple regression equation must then be taken as a whole:

it makes no sense to try to split variation into different components attributable to different controls. With p controlling variables, there are $2P - 1$ possible regressions ($2^{20} - 1$ in our case) which cannot all be examined: hence much thought must be given to the selection of predictors. As the number of parameters approaches the number of data points then the coefficient of determination (denoted R^2 in multiple regression) inevitably approaches 1: we are bound to get a good fit if we use enough parameters. We might avoid this by modifying R^2 for degrees of freedom, by invoking thresholds based on significance levels, or merely by leaning towards simpler models. In any case measures of fit and numerical criteria should be supplemented with graphical examination of data and residuals (Cox and Jones, 1981).

Before considering models with several predictors it is important to examine the correlations among predictors. The general pattern is one of strong correlations. Of 190 correlations among 20 predictors, the quartiles are 0.78 and 0.90 and the octiles 0.63 and 0.93 (signs ignored). A few of these correlations are inbuilt: the exact nonlinear relations between the two measures of winter moisture and summer moisture produce linear correlations of 0.93 and 0.94. (The rank correlations would be 1 exactly.) The inequalities - soil depth $\geqslant$ organic depth, winter moisture $\geqslant$ summer moisture, saturated density $\geqslant$ dry bulk density - also lead to increased correlations. But on the whole the strength of correlations is acceptable as a reflection of field relationships.

If we look at the 50 strongest correlations ($r^2 > 0.8$) we find a cluster of variables closely related to each other (table 25.2). Field capacity, winter and summer moisture, density (negative), plasticity, shrinkage, and % organic together define one moisture-related dimension of variation, from wet, less dense soils with high organic matter, plasticity and shrinkage to dry, denser soils with low organic matter, plasticity and shrinkage. Wetter soils also tend to be deeper, with thicker organic horizons, and lower angles of internal friction.

Looking at weaker correlations, apparent cohesion (positive), unconfined compression and penetration resistance measures (negative) emerge as correlated with the basic moisture dimension, but less strongly among themselves. Saturation level is expectably correlated with other moisture variables, but not very strongly: since we have satisfactory moisture variables in abundance, we need not consider it further. Sine of slope angle is not closely correlated with any other control: a broad general tendency for plots on gentler slopes to be wetter is not very strong. No correlation involving sine is above 0.47 in magnitude ($r^2 \leqslant 0.22$).

Principal components analysis of subsets of variables supplements this picture without modifying it, and the results need not therefore be reported here in detail.

Bringing together correlations between creep and controlling variables, and among controlling variables, it may be seen that those variables which are strongly correlated with creep are in general strongly correlated with

Table 25.2. Correlations among controlling variables

1. 20 variables give 190 correlations

2. 50 strongest: $r^2 > 0.80$

3. 36 of these: all possible correlations among field capacity, winter moisture (% wet), winter moisture (%dry), summer moisture (% wet), summer moisture (% dry), plasticity index, shrinkage, % organic, dry bulk density (negative correlations)

4. 14 remaining:
   - soil depth with field capacity, winter moisture (% wet), summer moisture (% wet), shrinkage
   - angle of internal friction (negative correlations) with winter moisture (% wet), summer moisture (% wet), winter moisture (% dry), summer moisture (% dry), plasticity, shrinkage
   - plasticity index (organic) with plasticity index, shrinkage
   - organic depth with % organic
   - density at field capacity with dry bulk density

5. Variables not involved in any correlation stronger than $r^2 = 0.8$:

   sine of slope angle, saturation level, apparent cohesion, unconfined compression (stress & strain), penetrometer resistance

each other. This implies that any attempt to go beyond single-predictor regressions is unlikely to be very successful, since there does not appear to be any component of variation in creep that is uncorrelated with moisture-related variables. Various multiple regression trials carried out support this conjecture, bearing in mind that unadjusted R^2 inevitably drifts towards 1 as the number of parameters increases. For example, adding penetrometer resistance and saturated density to winter moisture as predictors of volumetric rate increases R^2 from 0.77 to 0.81, but this increase seems to be offset by the loss of simplicity and the arbitrariness in choice of variables (the attained significance levels for adding each variable are 0.14 for penetrometer resistance and 0.41 for saturated density: many workers would judge these to be too high for inclusion in the model, which would again leave us with winter moisture alone).

The single-predictor models which are favoured therefore include winter moisture (% wet) (figures 25.1 and 25.2):

linear rate = 0.055 + 0.022 (winter moisture) $R^2 = 0.74$

volumetric rate = -0.14 + 0.015 (winter moisture) $R^2 = 0.77$

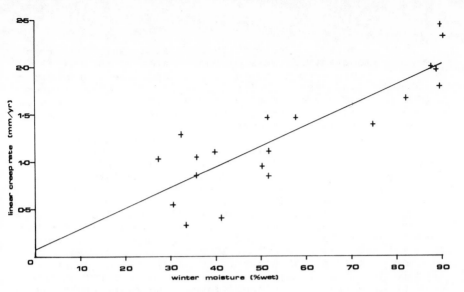

Figure 25.1.   Scatter plot: linear creep rate versus winter
    moisture (% wet) with regression line.

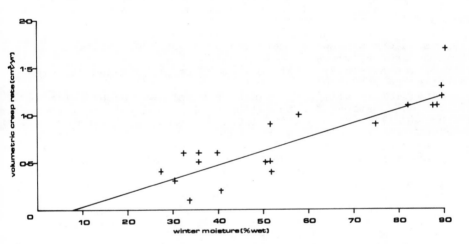

Figure 25.2.   Scatter plot: volumetric creep rate versus
    winter moisture (% wet) with regression line.

Graphical examination suggests that little would be gained
by attempting transformation of either creep measures or
winter moisture.   Conversely, the intercepts are fairly
small, and little would be lost by fitting regression lines
through the origin.

DISCUSSION

The major finding is that moisture-related variables are the most important controls of creep rates in the catchment. It would seem to be largely an open question whether moisture measures such as field capacity or winter moisture are emphasised rather than related material properties such as plasticity or shrinkage, because in practice such variables are all highly correlated with each other.

The most notable other finding is the unimportance of sine of slope angle, despite theoretical arguments for its role in influencing soil creep rate (Kirkby, 1967). Empirical measurements in many other environments have failed to demonstrate a relationship between creep and slope (Young, 1974). The description of the observational design showed that plots at several different slopes were chosen. Nevertheless moisture seems to swamp any influence that slope may have. However, no slopes above 23.5° were studied, and only four out of 20 plots are on slopes greater than 12.5°. Hence this is hardly the best data set to determine the strength and character of any relationship between creep rate and slope, since slope is fairly constant within the catchment. On the other hand, Finlayson (1981) also reports a lack of relationship between total soil movement (volumetric measure) and slope angle for the East Twin Brook catchment in the Mendip Hills of Somerset. Five out of 14 sites listed in his Table II have slopes between 25.5° and 29°. Hence the Weardale results cannot be dismissed out of hand as merely a by-product of relative constancy of slope.

The hypothesis that soil creep is proportional to sine of slope angle (or to tangent: the two are approximately equal for relatively gentle slopes) allows the use of diffusion equations for modelling hillslope development (see esp. Culling, 1960, 1963, 1965). A broad general prediction from these diffusion models is that there is a tendency over geological time for upper convexities to form. The short-term process evidence contradicts this hypothesis, and might lead some to reject diffusion models, but the frequent occurrence of convex divides still requires explanation. Quaternary glacial and cryonival modification (or even Tertiary planation) can readily be invoked in an area such as upper Weardale: whether this is the whole story remains an open question.

The limitations of an exercise of this kind are clear. It produces indications about relationships, not proofs or empirical laws. First, relationships are being inferred from spatial variation from plot to plot within the catchment: the analysis is not couched in terms of variations over time. Nevertheless detailed examination of the relations between meteorological cycles and soil movement (Anderson, 1977) tallies with the picture emphasising moisture which is presented here. Second, only one time scale is studied, and the relative importance of controls may vary for different time scales. Third, over substantial lengths of time soil movement will influence the other variables. This feedback aspect is implicitly ignored by any approach based on regression. Fourth, despite attempts

428

to gather data for different site conditions in a definite observational design, variation is, in large degree, uncontrolled by the observers. In particular, the varying extent of different combinations of conditions makes it difficult to determine some relationships. We would like to have sites which are very steep and very wet, and sites which are very steep and very dry, but neither kind exists in our catchment. A larger study area would include not only a greater variety of sites but also uncontrolled variations in other properties of less interest. Fifth, for rigorous testing data from new, previously unobserved plots are needed to determine the magnitude of prediction error. Since the creep rates are known, the exercise is strictly one of 'postdiction': the parameter estimates determined by least squares lean towards the data as far as they can, and the regression models seem better than they really are. Sixth, the number of plots (20) is fairly small: the authors are planning a more extensive long-term experiment. Seventh, the controls measured are really proxy or surrogate variables related to mechanisms rather than direct measures of these mechanisms. It may be that physical explanations should ideally be sought in terms of properties at the scale of individual pores and particles, or of soil aggregates, whereas many measures used here are defined at much coarser scales. Eighth, interaction between controlling variables remains a serious problem: the mechanisms contributing to soil creep seem difficult to distinguish in principle, let alone in practice.

CONCLUSION

A detailed analysis of plot-to-plot variation in soil creep rate and associated controlling variables has shown that most of the variation in creep is predictable in terms of moisture-related variables. Whether soil moisture itself, as linked to wetting and drying cycles in particular, or material properties such as plasticity and shrinkage should be emphasised in physically-based explanations is largely an open question and should be one focus of research in the immediate future. Similarly the character and strength of any relationship between creep and slope gradient requires future research, given the lack of fit between theoretical models and field observations. In such work, observational designs involving stratification and replication, controlled laboratory experimentation and physically-based mathematical modelling should all prove complementary.

REFERENCES

Akroyd, T.N.W., 1964, *Laboratory testing in soil engineering*, (Soil Mechanics Ltd., London).

Anderson, E.W., 1977, Soil creep: an assessment of certain controlling factors with special reference to upper Weardale, England. Unpublished Ph.D. thesis, University of Durham.

Anderson, E.W. and Cox, N.J., 1978, A comparison of different instruments for measuring soil creep, *Catena*, 5, 81-93.

Anderson, E.W. and Finlayson, B.L., 1975, Instruments for measuring soil creep, *British Geomorphological Research Group Technical Bulletin* no.16, 51 pp.

Chepil, W.S., 1950, Methods of estimating apparent density of discrete soil grains and aggregates, *Soil Science*, 70, 351-61.

Cox, N.J. and Jones, K., 1981, Exploratory data analysis, in: *Quantitative geography*, ed. Wrigley, N. and Bennett, R.J., (Routledge and Kegan Paul, London), 135-43.

Culling, W.E.H., 1960, Analytical theory of erosion, *Journal of Geology*, 68, 336-44.

Culling, W.E.H., 1963, Soil creep and the development of hillside slopes, *Journal of Geology*, 71, 127-61.

Culling, W.E.H., 1965, Theory of erosion on soil-covered slopes, *Journal of Geology*, 73, 230-54.

Finlayson, B.L., 1981, Field measurements of soil creep, *Earth Surface Processes and Landforms*, 6, 35-48.

Kendall, M.G., 1975, *Multivariate analysis*, (Griffin, London).

King, J.H.G. and Cresswell, D.A., 1954, *Soil mechanics related to building*, (Pitman, London).

Kirkby, M.J., 1967, Measurement and theory of soil creep, *Journal of Geology*, 75, 359-78.

Mosteller, F. and Tukey, J.W., 1977, *Data analysis and regression*, (Addison-Wesley, Reading, Mass.).

Pitty, A.F., 1968, A simple device for the field measurement of hillslopes, *Journal of Geology*, 76, 717-20.

Tukey, J.W., 1977, *Exploratory data analysis*, (Addison-Wesley, Reading, Mass.).

Young, A., 1974, The rate of slope retreat, *Institute of British Geographers Special Publication* no.7, 65-78.

Young, A., 1978, A twelve-year record of soil movement on a slope, *Zeitschrift für Geomorphologie* Supplement-band 29, 104-10.

26 Patterns of hillslope solutional denudation in relation to the spatial distribution of soil moisture and soil chemistry over a hillslope hollow and spur

T.P. Burt, R.W. Crabtree and N.A. Fielder

INTRODUCTION

The study of hillslope solutional denudation has received
much less attention by geomorphologists than the study of
hillslope hydrological processes. This is, in part, understandable, since it is clearly important to know the hydrological processes operating on a hillslope in order to be
able to interpret the patterns of solutional activity.
However, the process of throughflow and the variables which
control the size and extent of the saturated zone on the
hillslope have now been described in some detail (Weyman,
1973; Anderson and Burt, 1978). It therefore seems appropriate to begin to relate the hydrological processes to the
patterns of solutional activity operating on the hillslope.
    Most ideas of solute removal from hillslopes have been
inferred from the study of solute concentrations in stream
water, particularly during storm events (eg. Walling and
Foster, 1975). In some cases, the solute concentrations
of throughflow have also been examined, in an attempt to
interpret the solutional processes operating on the hillslope (Burt, 1979; Pilgrim, Huff and Steele, 1979). Carson
and Kirkby (1972, p 238) note that the literature contains
few measurements specifically related to the problem of
subsurface erosion and slope development. Two lines of
evidence are presented here in an attempt to provide direct
observations of solutional denudation; first, the processes
of soil formation and soil profile development; and,
secondly, evidence of the hydrochemical processes operating
on the slope as indicated by the use of micro-weight loss
techniques. The results provided by these two direct
methods, in combination with a knowledge of soil moisture
patterns on the hillslope, allow the relative patterns of
solutional denudation and hillslope development to be
tentatively estimated. Such conclusions will not provide
absolute estimates of erosion rates, but are hopefully an
improvement on the indirect evidence provided by observations of water quality made at the slope base.

431

THE STUDY SITE

Observations of soil moisture patterns, soil profile development, and hydrochemical soil processes were made at Bicknoller Combe, Somerset, England, (NGR 312140). Bicknoller Combe is a small catchment (0.6 km^2) on the western escarpment of the Quantock Hills. A quarry at the mouth of the Combe shows the bedrock to be massively bedded reddish purple metamorphosed quartzites dipping at 30° to the north-west. The rocks belong to the Hangman's Grits succession of the Middle Devonian (Webby, 1965). The succession visible in the quarry was assumed to be lithologically similar to the rock underlying the study area, about 200 m along the strike to the north-east. Upon this impermeable bedrock has developed a freely draining stony sandy loam soil, up to 2 m deep. The valley sides have a slope angle of about 25° throughout the catchment, although this reduces to 20° at the base of hollows, and approaches 30° where spur bases have been oversteepened by channel incision. Heathland vegetation, including *Erica tetralix*, *Vaccinium myrtillus* and *Ulex*, dominates the flat interfluve areas, merging into *Pteridium aquilinum* with grasses and *Galium saxatile* on the steep slopes.

A single hillslope hollow and its adjacent spurs were the site for all observations described in this paper (figure 26.1). Here the steep slope section is about 250 m long; above this it quickly merges into a flat interfluve zone 100 m wide. The base of the hillslope system was the site for a detailed study of soil moisture conditions and throughflow generation (Burt, 1978a; Anderson and Burt, 1978). Subsequently a broader study of soil moisture patterns over the whole slope has been made. The observations of soil profile characteristics (Fielder, 1981) and rock tablet weight loss (Trudgill, 1975; Crabtree, 1981) were made after the main soil moisture study had been completed. The three components of this report were conducted independently of one another, but the results are complementary and allow some discussion of hillslope solutional denudation patterns to be attempted, in relation to spatial variations of soil moisture and soil chemistry.

SOIL MOISTURE PATTERNS IN THE HILLSLOPE HOLLOW SYSTEM

Soil moisture conditions were monitored in detail at the base of the hollow and spurs using a grid of tensiometers, extending 70 m upslope with a maximum grid width of 50 m. Two automatic Scanivalve fluid scanning switch tensiometer systems were employed (Burt, 1978b). A later study of soil moisture conditions over the entire slope section used manual tensiometers (Burt, 1978c). The automatic tensiometer systems provided information on soil water potential over the lower slope (at a depth of 60 cm) every hour and allowed detailed soil water changes during storms and subsequent delayed throughflow events to be defined (Burt, 1978a; Anderson and Burt, 1978). The typical response of the stream to a significant input of rainfall involves two separate discharge peaks: the first occurs very soon after

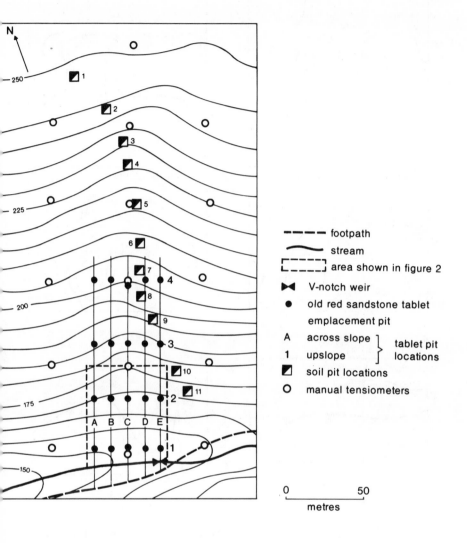

Figure 26.1. The instrumented hollow and spurs showing the location of the experimental observations.

the rainfall and is due to a combination of processes - Hortonian overland flow off the valley footpath, saturation overland flow off the marshy headwater area, and an increased throughflow discharge caused by growth of the saturated wedge in the hollows in response to rapid transmission of rainwater through the soil to the already saturated soil. Following the first peak, stream discharge rapidly subsides but rises again to form a second 'rounded'

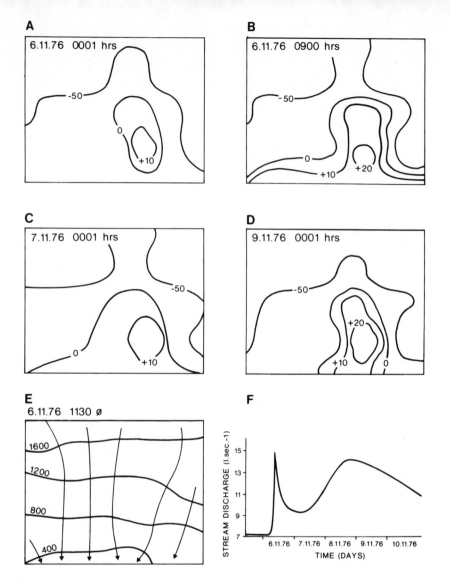

Figure 26.2.   Soil water conditions in the lower hollow
    during a throughflow event.

peak 1-3 days after the rainfall input. This delayed peak
is caused by throughflow: the delay occurs because of the
time taken for rainwater to infiltrate into the soil and
then flow downslope to the stream.
     Figure 26.2 shows five maps of soil water potential for
the instrumented hillslope section during a throughflow
event.   The small saturated wedge existing prior to the
storm (figure 26.2a) expands rapidly during the rainfall to
provide an extensive, though not deep, saturated wedge
(figure 26.2b).   Soil water potentials on the spurs become

less negative at this time indicating an increase in soil
water content. In contrast to the extensive and long-
lasting saturation of the hollow, only the spur bases indicate any signs of saturation, this being confined to the
period soon after the rainfall input. After the period of
maximum areal extension, the saturated wedge contracts in
size and the spurs also begin to drain (figure 26.2c). The
delayed peak in stream discharge is accompanied by deepening
of the saturated wedge in the lower section of the hollow
due to the convergent movement of throughflow into the
lower hollow from the adjacent spurs and from the upper
hollow (figure 26.2d). Following the peak in throughflow
generation, the hollow drains rapidly and the soil water
potential distribution soon approaches the pre-storm state.
The throughflow convergence is indicated by the map of
total hydraulic potential (figure 26.2e); orthogonals to
total potential isolines indicate soil moisture flow direction. On a slope of 25°, elevation potential will dominate
the total potential distribution; thus the flow net is
strongly correlated with the contour pattern.

Because of the low velocity of throughflow, only convergence in the lower hollow is likely to contribute to the
generation of the delayed peak in stream discharge. Further
up the hollow, similar processes of convergence will occur
with the result that generally wetter conditions are maintained for long periods in the hollow at the expense of the
spur zones. An extensive saturated wedge will be maintained
along the hollow in all but the driest summers (Burt and
Anderson, 1980). The maintenance of wetter soil in the
hollow may allow more rapid vertical and lateral transmission of soil water than can occur on the spurs.

The preservation of wetter conditions along the entire
hollow length was confirmed by an extensive survey using
manual tensiometers. Figure 26.3 shows a map of soil water
potential for the entire hollow system, for a depth of
40 cm. This confirms that saturated conditions are maintained quite close to the surface up much of the hollow;
170 cm upslope from the stream the soil water potential at
40 cm depth is only -5.4 cm. By contrast, the spurs indicate rather drier conditions, with no potentials higher
than -20 cm, and in two locations potentials are below
-100 cm. The situation illustrated on figure 26.3 does
represent relatively wet soil conditions (stream discharge
= 15 l s^{-1}), relating to a rainfall input of 8 mm, four
days before. Under 'normal' conditions it might be expected that the saturated zone would be rather less extensive (at a depth of 40 cm) than that mapped on 5.3.81.
Even so, it can be assumed that the hydrological contrast
between spurs and hollow, which has been demonstrated for
the lower hollow (Anderson and Burt, 1978), will be preserved over the entire hillslope area.

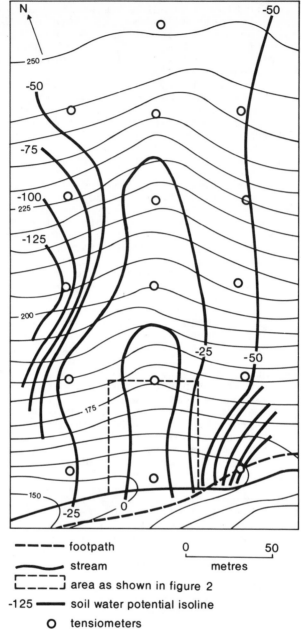

Figure 26.3. Soil water conditions over the entire hollow
and spur system.

PATTERNS OF HILLSLOPE SOLUTIONAL DENUDATION:
THE RESULTS OF THE MICRO-WEIGHT LOSS TECHNIQUE

The relative pattern of solutional denudation over the
slope was investigated by emplacing micro-weight loss
tablets, made from Old Red Sandstone, over the slope
(Crabtree, 1981). Rock material removed from the quarry
was cored parallel to the bedding and the cores sliced into
standard sized tablets (0.031 m diameter; 0.007 m thickness
- see Trudgill, 1975). Analysis of the rock under thin-
section optical microscopy revealed a composition of 75%
very fine grained quartz crystals. Haematite staining and
a clay matrix (probably kaolinite and chlorite) comprised
15%, very small feldspars and micas 5%, and about 5% was
very fine grained high birefringence calcite. X-ray diffraction confirmed the presence of these constituents. The
pattern of tablet emplacement involved five sets of pits
across the slope, centred on the hollow, 10 m apart. Ten
tablets were emplaced in each pit at a depth of 0.4 m.
Four identical rows were employed up the slope 30 m apart
(figure 26.1). It was not possible to place the tablets
at the soil-bedrock interface because of the deep, stony
nature of the soil which was consequently difficult to
excavate. The results therefore show only the relative
solutional denudation rate of the bedrock material incorporated within the soil matrix at a fixed depth. However,
the soil moisture conditions at this depth should be
analogous with moisture conditions at the bedrock interface (Burt, 1978a), although the frequency of saturation
will obviously differ. Thus it may still be possible to
apply the tablet results to the soil-bedrock interface,
although only to show relative variations in the rate of
solution. The tablets remained in the soil for 15 months
(May 1979 - August 1980). Soil samples were taken from the
base of each pit to show changes in soil chemistry over the
slope. Following retrieval of the tablets from the pits,
they were cleaned and reweighed.

Finlayson (1977) found that the weight loss of Old Red
Sandstone tablets represented the loss of material due
both to solution and also to granular disintegration resulting from the reduction of cohesion of the rock matrix
by solution. A comparison of fresh and excavated tablet
surfaces showed no signs of granular disintegration. The
measured weight loss of the tablets was considered to be
due entirely to solution processes, involving the loss of
silica by hydrolysis. The pattern of relative solutional
denudation over the slope will reflect changes in soil
acidity. Neutralisation of acid infiltrating soil water
will take place by solute uptake from the soil under
alkaline conditions, or by bedrock dissolution under acid
soils.

The upslope changes in weight-loss and soil chemistry
showed a complex pattern over the slope (figure 26.4).
Significant upslope increases in relative solutional denudation rates were found on the downstream spur and spur
flank and in the hollow, but not on the upstream spur or
spur flank. The general increase in weight loss upslope is
shown by the result of a linear regression analysis between

437

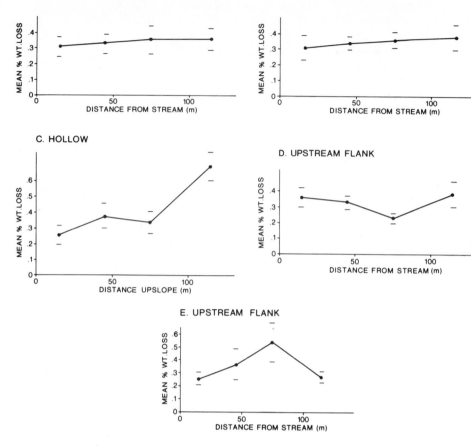

Figure 26.4. Upslope changes in tablet weight loss and
    soil chemistry for the five profiles.

tablet percentage weight loss (P) and distance upslope (D);
the data from the five upslope profiles were combined in
this case:

$$P = 0.31 + 0.0015D$$

$$r = 0.973; \ n = 186 \quad \text{correlation significant at 99\%.}$$

The results show a relative upslope increase in solutional
denudation for the first 115 m of the slope above the
stream.   This relates to the increase in soil acidity upslope and to the general decrease in soil carbonate content.
The downslope movement of throughflow may act to reduce
solutional denudation rates downslope: increased amounts of
carbonate in the soil will help to reduce soil acidity
downslope; in addition, throughflow reaching the base of
the slope will have a long residence time and will therefore have a lower potential for solutional erosion.

The effects of the varying hillslope topography can be seen by comparing the mean weight loss for each upslope profile:

	mean % weight loss (4 pits)
Downstream spur	0.34 ± 0.07
Downstream spur flank	0.35 ± 0.07
Hollow	0.42 ± 0.09
Upstream spur flank	0.33 ± 0.07
Upstream spur	0.31 ± 0.15

The greatest weight loss is in the hollow. The downstream spur and spur flank show slightly higher weight loss than the upstream spur and spur flank. The wetter downstream spur zone (figure 26.3) is slightly more acid which may indicate that shorter residence time water is present, compared to the upstream spur zone. Examination of the results of the cross-slope rows of five pits also illustrates the change in solutional denudation rates for the spurs and hollow. Figure 26.5 shows the cross-slope weight loss results at 43 m and 115 m upslope. In both cases the hollow shows the maximum weight loss, though the hollow-spur contrast clearly increases further upslope. This is understandable, since the hydrological contrast between the hollow and spurs also increases upslope (figure 26.3). The high upslope increase in relative solutional denudation in the hollow may therefore relate to the arrival of large quantities of short residence time acid throughflow, which is channelled rapidly down the hollow from the upper spur zones and particularly from the interfluve areas. Such rapid and large volumes of throughflow will be absent from the spur zones.

The weight loss tablet experiment described here is only for the lower half of the slope. However, the results

Figure 26.5.   Cross-slope weight loss profiles at 43 m and 115 m upslope.

do suggest an increase in solutional denudation upslope which is greatest in the hollow. Burt (1979) argued that continued development of the contrasting slope forms of the hollow and spur were connected in some way to the saturated conditions which are always present in the hollow but seldom present on the spur; removal of solutes from the lower part of the hollow due to expansion of the saturated wedge during delayed throughflow events was the erosional mechanism postulated. The rock tablet weight loss experiments presented here show that the main focus of solutional activity in the hollow is further upslope than Burt (1979) suggested. However, the generally wetter conditions present in the hollow are still shown to be dominant, since this allows rapid transmission of aggressive water into the central hollow zone.

THE DISTRIBUTION OF EXTRACTABLE IRON
IN SOIL PROFILES UP THE HOLLOW

Eleven soil profiles were examined between the summit and base of the hollow (figure 26.1). Surface litter was removed from each profile and soil was sampled from within each 5 cm depth interval, or from each 10 cm interval near the base of the profile. Fresh soil samples were tested for pH (0.01 M $CaCl_2$ at sticky point) and the remaining soil was air-dried and sieved (2 mm) to retain the fine earth. Pyrophosphate extractable iron (Fe p) and residual dithionite extractable (Fe res) in the fine earth were determined (Bascomb, 1968; Avery and Bascomb, 1974), Pyrophosphate extractable iron was separated into iron precipitated by ammonia (Fe ppt) and iron soluble in ammonia (Fe sol).

Substances washed from the leaves of plants (especially polyphenols) enter the soil and under acidic conditions can form stable water-soluble complexes with iron. Held in such complexes iron may be mobilised down through the soil until the complex breaks down and amorphous hydroxides are released. Amorphous hydroxides may gradually dehydrate and develop a crystalline structure. Extraction of soil with pyrophosphate at pH 10 (Fe p) provides a good estimate of 'active' soil iron. Fe ppt (precipitated by ammonia) represents amorphous 'gel' hydroxides, and iron precipitated from pH-dependent complexes, while Fe sol, (soluble in ammonia) represents iron held in particularly stable organic complexes. Following pyrophosphate extraction residual 'aged' iron oxides with developing crystalline structure may be extracted with sodium dithionite (pH 3.8). Iron bound within the lattices of clay minerals is not normally removed.

The iron ratio Fe p/Fe p + Fe res (Avery, 1980; Fielder, 1981) expresses complexed and amorphous iron as a proportion of total non-silicate iron, and is a useful criterion for soil classification. In podzolic soils the iron ratio increases down the soil profile, and in an illuvial B_s horizon the value of this ratio should be at least 0.3 and normally greater than 0.5 (Avery, 1980). Changes in the iron ratio with depth can also be used to identify wet or

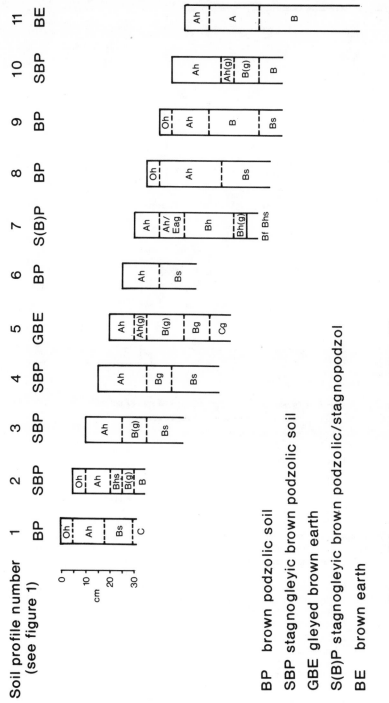

Soil profile number
(see figure 1)

| 1 | 2 | 3 | 4 | 5 | 6 | 7 | 8 | 9 | 10 | 11 |
| BP | SBP | SBP | SBP | GBE | BP | S(B)P | BP | BP | SBP | BE |

BP   brown podzolic soil

SBP   stagnogleyic brown podzolic soil

GBE   gleyed brown earth

S(B)P stagnogleyic brown podzolic/stagnopodzol

BE   brown earth

Figure 26.6.   The sequence of soil profiles down the hollow.

441

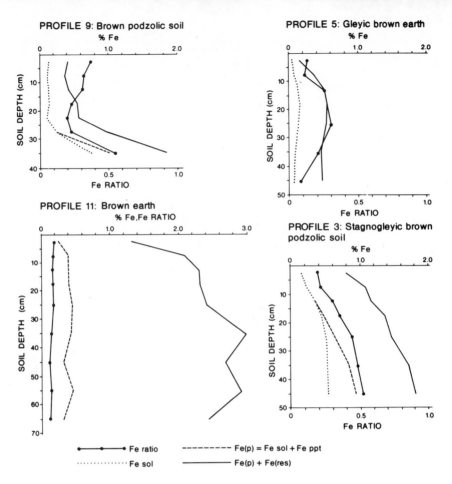

Figure 26.7.   Distribution of iron in selected soil
        profiles.

gleyed horizons (Fielder, 1981).   In periodically wet
stagnogleyic soils, reducing conditions facilitate the
formation of Fe sol.   Fe ppt is often completely depleted
and Fe res may become the principal form of soil iron.   The
distribution of the iron ratio becomes 'C' shaped, declining
away from the surface but increasing again at depth.
        The sequence of soils between the summit and the base
of slope at the study site included brown podzolic soils,
stagnogleyic brown podzolic soils, brown earths and gleyic
brown earths (figure 26.6).   Profile 3 (figure 26.7a) was
a stagnogleyic brown podzolic soil with some mottling in
the B horizon.   Fe ppt was detected only at the base of
this profile and the iron ratio showed a typical 'C' shaped
distribution.   Profile 5 (figure 26.7b) was the wettest
profile encountered and was gleyed particularly towards the
base of the profile.   No Fe ppt was detected.   Profile 9
(figure 26.7c) was a brown podzolic soil.   This profile was
drier than profile 3 (saturated either less frequently or

or for shorter duration) and Fe ppt was absent only in the
surface organic horizon. Profile 11 (figure 26.7d)
illustrates a typical brown earth. Iron was predominantly
in the form of Fe res (as in gleyed soils) giving a low
iron ratio, but, (unlike gleyed soils) Fe ppt was present.
There was little or no Fe sol in this soil and the iron
ratio remained almost constant with depth, implying no redistribution of iron. This was in contrast to profile 9 in
which a regular process of redistribution was indicated and
the iron ratio reached 0.52 in the B_S horizon.

The distribution of iron in the soil profiles sampled
down the hollow confirms the hydrological and solutional
importance of the hollow feature (figure 26.6). In the
stagnogleyic soils, features associated with gleying were
most pronounced 10-20 cm above the base of the profile;
this suggests that this was the zone through which soil
water moved most frequently. Soils that remain saturated
for the greatest length of time would show most evidence of
gleying (eg. profiles 3, 5 and 7). It is not possible to
distinguish between the effects of either standing water or
a large volume of water moving slowly in the lateral plane,
though it is thought that the latter case is most probable
on the steep slopes being discussed. In the hollow soils,
low iron ratio values in B horizons confirm the continued
presence of saturated conditions, a factor which must
favour higher rates of solutional denudation in that
location relative to the spurs. Better drained soils would
be expected on the summit plane and spurs similar to
profile 11, with relatively poorly-drained soils within
the hollow; the hollow profiles examined reflect this
distribution, although further samples are required from
the spur zone to confirm the topographic contrast. Profiles
near the base of the slope evidenced the greatest extractable iron values. This suggests that iron removed in
solution from gleyed horizons is accumulating further downslope.

SPATIAL VARIATIONS IN SOLUTIONAL DENUDATION ON HILLSLOPES

It is common for soil types and soil chemistry on slopes
underlain by single rock types to change upslope; in general
these upslope soil changes involve an increase in acidity
and a decrease in carbonate content, as found, for example,
by Furley (1971). In this situation it is likely that
solutional denudation rates will increase upslope. The use
of weight loss tablets on the hillslope at Bicknoller
Combe has shown a relative increase in solutional denudation upslope. The immediate cause of this variation can
be related to the pattern of soil conditions on the slope,
in particular the spatial variation of soil acidity.

The hydrological conditions on the hillslope provide
the framework within which the solutional denudation takes
place. The general increase in the carbonate content of
the soil downslope, and the associated decrease in acidity,
relate largely to the increased contact time of throughflow
at the base of the slope. The variable effects of evaporation over the slope may also be important in providing

larger amounts of available solutes at the base of the hillslope and in the hollow, due to increased evaporation rates from the wetter soil (Carson and Kirkby, 1972). The hydrological contrast between the hollow and spurs may also result in real differences in the rate of solution. There is no difference in acidity between the upper hollow and its adjacent spurs, and indeed the hollow soil has a higher carbonate content. Even so, the hollow exhibits much higher solutional denudation rates which must be related to its continued saturated or near-saturated state. The increased volumes of throughflow, which will almost certainly be of shorter residence time than the spur soil water, produces this increased solution. The importance of moisture conditions in the hollow has also been shown with respect to the mobilisation of iron, relocating iron down the soil profile and downslope. The results suggest that the topographic contrast between spur and hollow will continue to develop within the general context of hill-slope decline caused by increasing solutional denudation upslope.

The results provide no answer to the problem of interpreting the solute variations exhibited in throughflow leaving the base of the slope. Rock tablet experiments and soil profile analysis provide only a relatively lengthy time-averaged view of processes operating on the hillslope. Whilst such experiments are of great importance to the study of hillslope development under solutional denudation, shorter period measurements (for example, soil water solutes) will be required if the hillslope soil system is to be fully understood. Geomorphologists therefore need to adopt the same short-time scale approach to hillslope solutional erosion as has been employed in studies of run-off processes. The results of such studies may then allow more successful prediction of long-term hillslope evolution - in combination with the types of longer-term experimental data described here - and at the same time will allow short-term processes, such as solute leaching, to be more fully evaluated.

REFERENCES

Anderson, M.G. and Burt, T.P., 1978, The role of topography in controlling throughflow generation, *Earth Surface Processes*, 3, 331-344.

Avery, B.W., 1980, Soil classification for England and Wales (Higher categories), *Soil Survey Technical Monograph*, 14.

Avery, B.W. and Bascomb, C.L., 1974, Soil survey laboratory methods, *Soil Survey Technical Monograph*, 6.

Burt, T.P., 1978a, Runoff processes in a small upland catchment with special reference to the role of hillslope hollows, Unpublished PhD Thesis, University of Bristol.

Burt, T.P., 1978b, An automatic fluid scanning switch tensiometer system, *British Geomorphological Research Group Technical Bulletin*, no.21.

Burt, T.P., 1978c, Three simple and low-cost instruments for the measurement of soil moisture properties, *Huddersfield Polytechnic, Department of Geography Occasional Publication*, no.6.

Burt, T.P. and Anderson, M.G., 1980, Soil moisture conditions on an instrumented slope, Somerset, March to October 1976, in: *Atlas of Drought in Britain 1975-76*, (Institute of British Geographers, London), 44.

Carson, M.A. and Kirkby, M.J., 1972, *Hillslope Form and Process*, (Cambridge University Press).

Crabtree, R.W., 1981, Hillslope solute sources and solutional denudation on Magnesian Limestone. Unpublished PhD Thesis, University of Sheffield.

Fielder, N.A., 1981, Distribution of iron and aluminium in some acidic soils. Unpublished PhD Thesis, University of Bristol.

Finlayson, B., 1977, Runoff contributing areas and erosion, *School of Geography, University of Oxford, Research Paper*, no.18.

Furley, P.A., 1971, Relationships between slope form and soil properties over chalk parent materials, in: *Institute of British Geographers Special Publication*, no.3.

Pilgrim, D.H., Huff, D.D. and Steele, T.D., 1979, The use of specific conductance and contact time relations for separating flow components in storm runoff, *Water Resources Research*, 15, 329-339.

Trudgill, S.T., 1975, Shorter Technical Methods (1), *British Geomorphological Research Group Technical Bulletin*, no.17.

Walling, D.E. and Foster, I.D.L., 1975, Variations in the natural chemical concentration of river water during flood flows, and the lag effect: some further comments, *Journal of Hydrology*, 26, 237-244.

Webby, B.D., 1965, The stratigraphy and structure of the Devonian rocks in the Quantock Hills, West Somerset, *Proceedings of the Geological Association*, 65, 321-344.

Weyman, D.R., 1973, Measurement of the downslope flow of water in a soil, *Journal of Hydrology*, 20, 267-288.

27 Some relationships between debris flow motion and micro-topography for the Kamikamihori Fan, North Japan Alps

Setsuo Okuda and Hiroshi Suwa

INTRODUCTION

Debris flows transport large amounts of debris from a valley to a fan, producing rapid topographic changes on the fan, and sometimes cause serious damage to the people living there. Thus, it is very important to be able to predict danger zones at the mountain front which may be attacked by debris flows, and this is an important problem in the field of disaster prevention science. From the geomorphological viewpoint, it is also interesting to develop a quantitative understanding of the processes of fan development.

In Japan, various methods have been adopted to study this problem, and these include hydraulic model experiments (Takahashi, 1980) for investigating debris flow deposition on a fan, and stochastic models of debris deposition based on Random Walk theory (Imamura *et al.*, 1980). However, field experiments undertaken on the fan play a very important role in studying the problem, because the fundamental assumptions of a realistic model should be founded on careful observations of actual phenomena. The effectiveness of an assumed model can only be judged by comparing results from the model with those from field experiments.

Field observations on debris flow motion and on changes in valley and fan morphology have been carried out over the last decade at Kamikamihori Valley and Fan on the eastern slope of Mt. Yakedake, in the Northern Japan Alps, Central Japan. For the past few years, the investigation has focused on the depositional processes of debris flows, and in particular the relationships between debris flow motion and the micro-topography of the fan. In this paper, the field observation methods are briefly described and the main results obtained from the observations are reported.

MEASUREMENT TECHNIQUES

Some new measurement techniques for debris flow phenomena incorporating automatic and telemetering systems have been developed and these have been reported at a previous

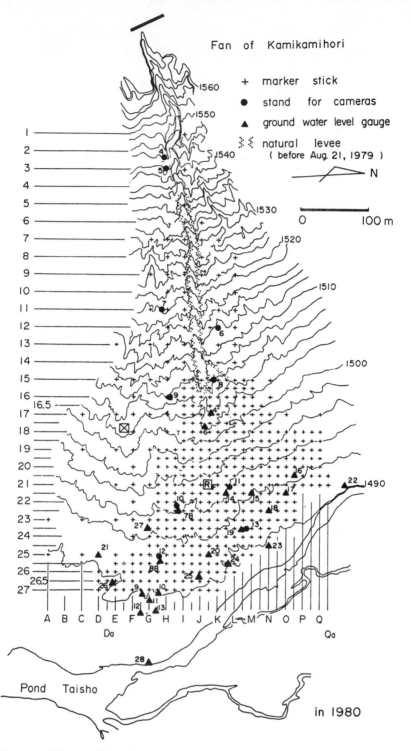

Figure 27.1.   The observation network on the Kamikamihori fan.

Symposium on Field Experiments in Geomorphology (Okuda
et al., 1980; Suwa and Okuda, 1980). In this paper, the
description of the measurement techniques will therefore be
reduced to the minimum necessary for understanding our
present work. Figure 27.1 illustrates the recent location
of the observation system on the fan. The motion of debris
flows has been recorded with a network of various types of
camera. Thus, the frontal speed of any flow containing
large rock fragments can be calculated from an exact record
of the arrival times of the flow front at each sensor,
which were positioned about every 100 m along the flow
direction in the valley and every 50 m on the fan. Changes
in ground water level in the fan were observed automatically
or manually at a network of bore holes using water level
gauges.

Changes in ground surface levels on the fan were
surveyed repeatedly at many fixed points using marker rods,
and sudden and large changes in surface topography were
measured by detailed level surveys or topographic surveys
before and after every debris flow or flood. The size
distribution of debris fragments in debris flow deposits
has been investigated using photographs taken with portable
cameras or large scale airphotos, but the fabric of debris
flow deposits in the fan will be investigated in the near
future.

RESULTS OF THE FIELD OBSERVATIONS

A considerable volume of data on various aspects of debris
flow behaviour have accumulated during the past 11 years,
but it has not been possible to derive precise conclusions
from these data, because the various processes contributing
to debris flow motion and topographic change on the fan are
so complex and mutually dependent that precise general rules
or quantitative relationships among the controlling factors
are difficult to establish.

However, after careful study of the large amount of
data collected, some relationships between the physical
factors involving strong correlations have been detected
and partially described by quantitative expressions. It
is hoped that further progress with such efforts will produce the basis for a general theory or a worthwhile model.
Accordingly, some of the relationships between the physical
factors associated with debris flow depositional processes
and micro-topographic changes within the fan are described
below.

Debris flow motion and deposition on the fan

Several records of moving debris flows on the fan have been
obtained with the automatic recording system employing
cameras. The nature of debris flow movement at a specified
section (check dam No.6 upper) has been recorded using
various instruments including a 35 mm camera, a cine camera,
a TV camera and special speedometers. These document movement at the entrance to the fan, an important initial
condition which controls the flow downstream on the fan.
Typical examples of such records are shown in figure 27.2.

A, at
16^h 07^m 35^s

B, at
16^h 07^m 37^s

C, at
16^h 07^m 39^s

Figure 27.2.   A sequence of photographs of the debris flow
    August 23, 1980, passing over dam No. 6 upper; from a
    series of photographs taken every one second.

D, at
16h 07m 41s

E, at
16h 07m 43s

Figure 27.3.   Surface view of the debris flow, July 19,
    1980, above dam No. 6 upper.   Block indicated by
    arrow is 40 cm long.

451

The debris flow front travels down with a growing top containing many large rock fragments (in this case the maximum diameter was 3.3 m) as described in previous reports. In a recently recorded event (figure 27.3) a block of mortar (10 cm x 20 cm x 40 cm, arrow mark in the photo) was found floating on the surface of the front, and this indicates that the front had scoured the channel floor where several pieces of block had been buried in a column to determine the scouring depth of the debris flow. It can therefore be seen that a strong vertical mixing within the front brought the block to the surface of the flow.

The most recent records (video tape recording) show that the front begins to reduce speed on arrival at the fan and finally stops, while the muddy flow which follows the front changes its direction, moves around the front deposit, and continues to flow downstream, leaving behind the front deposits as a pile of large debris fragments. This behaviour is also reflected by the sudden change of stream direction on the lower fan caused by the blocking of the former stream course by the deposits of debris from the front and the appearance of a new stream course produced by the muddy flow diverted by the debris deposits.

Physical factors controlling the distance travelled by the debris flow

From 1970 to 1980, the behaviour of 14 debris flows with different scales was observed on the fan and data concerning the various physical factors influencing the occurrence or motion of the flow were collected. Several important characteristics which seem to have a direct influence on the distance of travel of the debris flow, i.e. the distance between the fan head (check dam No.6 upper) and the farthest downstream edge of the debris flow deposit, are listed in table 27.1. The estimation of distance of travel of debris flows under various conditions is a very important problem in predicting the danger zone at a mountain front and in investigating the topographic changes on the fan.

Obviously, the initial flow conditions measured at the upstream section have a great effect on the ultimate distance of travel. The hydrological variables in table 27.1 have been specially introduced in order to express the effective total discharge or rainfall and are defined below that table. The correlation matrix calculated for the variables in table 27.1 is presented in table 27.2. The results are limited by the restricted number of observations (14) but certain trends are evidenced by this table.

Dynamic analysis of distance of travel

As a relatively high correlation is evident in table 27.2 between travel distance L and the front velocity at the fan head V_f, it is necessary to consider the theoretical background to the relationship between these two physical quantities. A number of quantitative studies on the travel distance of debris flows have been carried out in Japan using hydraulic experiments or stochastic models, but these have included so many parameters or assumptions that it is

Table 27.1. Debris flow characteristics

	'76				'78		'79					'80		
date	July 19	Aug. 3	Aug. 14	Aug. 17	Sep. 4	Sep. 29	Aug. 21	Aug. 21	Aug. 22	Sep. 21	Oct. 1	July 19	Aug. 23	Sep. 7
time	7^h51^m	17^h30^m 22^h53^m	14^h00^m	14^h00^m	7^h58^m	19^h13^m	17^h11^m	4^h43^m	7^h49^m	16^h41^m	1^h27^m	9^h30^m	16^h02^m	19^h20^m
scale	L	L	L	L	S	M	S	L	Ex L	L	M	S	M	S
above the fan head (at the dam No. 6 upper)														
Q_I (m³)	7700	4800	—	6500	330	—	—	—	—	12600	—	—	3700	—
Q_{II} (m³)	8300	5700	9000	9000	490	—	—	—	—	13500	—	—	4600	—
Q_{III} (m³)	9200	6400	11500	11500	710	—	—	—	—	14800	—	—	5100	—
q_{max} (m³/sec)	124	103	98	98	24	(22)	1.3	(5.5)	(166)	98	2.2	(5.1)	65	1.0
v_f (m/sec)	3.4	3.8	4.0	5.1	1.9	2.2	—	—	(6.4)	3.3	—	0.88	4.3	—
A_{max} (m²)	17.0	25	—	12.9	11.0	10.1	—	—	27	24	—	5.8	21	—
h_{max} (m)	3.4	3.2	—	3.8	1.5	1.9	—	—	4.0	3.4	—	1.0	3.7	—
d_{max} (m)	2.3	3.1	—	6.0	1.9	0.5	—	—	(6)	5.6	—	1.2	3.3	—
rainfall condition														
R_{10} (mm)	11.5	6	12.5	12.5	3.5	4	8	11.5	8	11	6.5	8	5	8.5
R_{60} (mm)	14.5	14	21.5	15	13.5	20.5	14	15	51	21	28.5	20.5	20.5	19.5
R_{6h} (mm)	16	46.5	50	19	48.5	32	34.5	47	102	21	42.5	57	57	26.5
R_{10a} (mm)	8.5	2.5	3.5	7.5	1	3.5	1.5	2.5	11	8	1.5	5	2.5	1.5
R_{ib} (mm)	12.1	8.6	15.2	13.1	6.6	7.6	9.8	13.2	18.3	13	11.3	11.7	9.3	10.9
R_{iab} (mm)	16.4	9.9	17	16.8	7.1	9.4	10.6	14.5	23.8	17	12.1	14.2	10.5	11.6
about the terminal position														
L (m)	607	718	670	627	599	565	184	680	789	755	415	116	321	9
T_{20}	0.064	0.029	0.065	0.065	0.065	0.059	0.115	0.066	0.023	0.029	0.111	0.093	0.132	0.261
T_{50}	0.068	0.037	0.063	0.065	0.067	0.068	0.105	0.060	0.023	0.042	0.129	0.105	0.110	0.223
T_{100}	0.078	0.054	0.059	0.067	0.072	0.067	0.112	0.073	0.035	0.049	0.117	0.107	0.095	0.172
T_{200}	0.081	0.062	0.073	0.072	0.077	0.078	0.104	0.068	0.054	0.062	0.111	0.115	0.088	0.162

total discharge Q_I (q > 20 m³/sec), Q_{II} (q > 10 m³/sec), Q_{III} (q > 5 m³/sec); q_{max} : maximum discharge,
v_f : front velocity, A_{max} : maximum cross sectional area of flow head, h_{max} : maximum height of flow head,
d_{max} : diameter of largest stone, R_{10} : 10 minutes rainfall before debris flow occurrence, R_{60} : hourly rainfall b.,
R_{6h} : 6 hours rainfall b., R_{10a} : 10 minutes rainfall after debris flow occurrence,
10 minutes rainfall index before debris flow occurrence : $R_{ib} \equiv R_{10} + \frac{1}{5}(R_{60} - R_{10}) + \frac{1}{30}(R_{6h} - R_{60})$,
index before and after d. f. occurrence : $R_{iab} \equiv R_{ib} + \frac{1}{2}R_{10a}$, L : runout distance from the fan head (from the dam No. 4),
T : tangent of slope angle before debris deposition (T_{20} : averaged value over 20 meters section upstream of the snout,
T_{50} : over 50 meters section, T_{100} : over 100 meters section, T_{200} : over 200 meters section)

Table 27.2. Correlation matrix calculated from table 27.1.

	$\ln Q_I$	$\ln Q_{II}$	$\ln Q_{III}$	q_{max}	v_f	h_{max}	d_{max}	R_{10}	R_{60}	R_{6h}	R_{ib}	R_{iab}
q_{max}	0.89	0.85	0.88									
v_f	0.65	0.69	0.70									
A_{max}	0.60	0.57	0.53									
h_{max}	0.90	0.91	0.91		0.95							
d_{max}				0.35	0.76	0.65						
R_{10}	0.77	0.78	0.81	0.78	0.39							
R_{60}	0.45	0.44	0.42	0.03	-0.12							
R_{6h}	-0.60	-0.61	-0.64	-0.69	-0.06							
R_{ib}	0.85	0.87	0.90	0.75	0.61							
R_{iab}	0.83	0.84	0.86	0.77	0.63							
L	0.25	0.24	0.26	0.38	0.69	0.55	0.47	0.23	0.21	0.15	0.33	0.41

Kamikamihori Valley,   about the debris flows   '76 - '80

very difficult to apply their results to observations of
natural phenomena at the present time. As a simple model,
Okuda has suggested that the deceleration of the debris
flow front (excluding the muddy tail flow) can be approximated as the motion of a mass point sliding down a slope,
taking into consideration the apparent friction and drag
force obtained from observations on the fan, (Okuda *et al.*,
1973).

      According to mass point dynamics, the travel distance
L is given by the following expression

$$L = \frac{1}{2k} \ln \frac{k\, V_f^2 + A}{A} \tag{1}$$

where V_f is an initial velocity at the starting point and
k is a drag coefficient in the case of drag proportional to
the square of velocity. The resultant acceleration A is
defined by:

$$A \equiv g\,(\mu \cos\theta - \sin\theta)$$

where g is the acceleration of gravity, θ is the mean slope
angle and μ is a dynamic friction coefficient.

If the term $k\, V_f^2$ is very small in comparison with A,
equation(1) may be modified into a simpler approximate form

$$L = V_f^2 / \, 2A \tag{2}$$

which can be derived directly from elementary dynamics.

      The relationship between V_f and L observed in the field
is shown in figure 27.4 by the small circles. The lines
1 and 2 are given by equation(1) and line 3 by equation(2),
with values of k = 0.0005 / m, μ = 0.0952 (line 1),

454

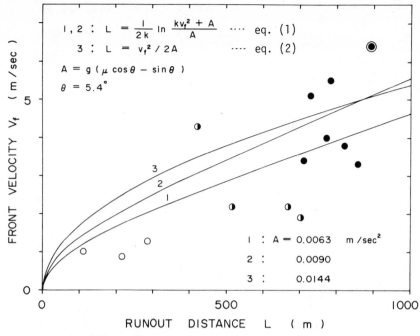

Figure 27.4.   The relationship between front velocity and
        travel distance for debris flows on the Kamikamihori
        fan.

0.0946 (line 2) and 0.0960 (line 3) and a mean slope angle
$\theta = 5.4°$.   The result indicates that dynamic analysis,
assuming a simple mass point system, cannot provide a good
explanation of the actual relationship between V_f and L.
This suggests that the travel distance depends not only on
the initial front velocity, but also on other elements,
possibly including the scale of the debris flow and the
rheological properties of the flowing material, and that
further quantitative study is needed to estimate the travel
distance more exactly for various initial conditions.

Estimation of the equivalent coefficient of friction
from the course profile

As another empirical method for predicting the travel
distance, knowledge of the equivalent coefficient of friction of a massive flow can be used.   Heim (1932) and
Scheidegger (1973) showed that an equivalent coefficient
of friction of a massive flow of debris could be estimated
from a course profile on the basis of mass point dynamics,
in which the dynamic friction coefficient corresponds to
the equivalent coefficient of friction derived from the
observation of the course, i.e. the course along which the
debris mass has travelled from a higher starting point to

a lower settling point. The equivalent coefficient of friction μ_e is calculated from the observed value of the angle of elevation θ looking up from the final resting point to the starting or rupture point, using the relationship

$$\mu_e = \tan \theta.$$

In the field, the final resting point can be determined easily from the pattern of deposition of the debris mass, but the starting point cannot be determined exactly, because a massive failure or slide does not always start from one point, but often extends over an extensive region. A detailed survey of the scoured area in an upstream valley or slope involves very difficult field work.

In the case of debris flows which reach a large size only after travelling some distance, the starting point of the debris flow front cannot be determined exactly. In this case a specified section where the front velocity of the moving debris flow can be measured exactly with photographic recording is allocated as the starting point. Because the front is already moving with the speed V in this section, the elevation of the starting point can be calculated by adding the kinetic energy term $V^2/2g$ to the ground level of the section.

When many observations on the speed of various fronts passing through the specified section and the final resting point of the debris flow have been assembled, the equivalent coefficient of friction can be calculated for every debris flow front.

Some examples of the course profiles of debris flows are shown in figure 27.5 along with the calculated values of μ_e. In these figures, the vertical distance between the energy line and the ground surface at any place should represent the term $V^2/2g$ for each point. However, the observed speeds in the middle reach of the course are much smaller than those expected from the figure, and this discrepancy cannot be fully explained at present. It may be associated with a fan head incision phenomenon or vertical drops at check dams producing large energy losses.

The value of μ_e calculated from the course profile provides only an average value for the total reach between the starting and stopping points, and local variations in μ_e along the course are not considered. A frequency distribution of μ_e values obtained from the course profile on the Kamikamihori fan is shown in figure 27.6 in which the values of μ calculated from the dynamic equations outlined previously are depicted by the symbol (D). The result shows that in most debris flows in our experimental area, the coefficient of friction ranges between 0.09 and 0.12. These values are much smaller than those for dry grain flows cited by Scheidegger (1973) or Hsü (1975), being about 20% of the cited values, and this means that the wet grain flow, i.e. debris flows, can travel much longer distances across the fan than dry mass motion. The close approach of the energy line to the ground surface near the final resting point in the above figure indicates that a slight change in the surface topography or friction coefficient on a fan may cause a large change in the travel distance of the debris flow. Although the physical

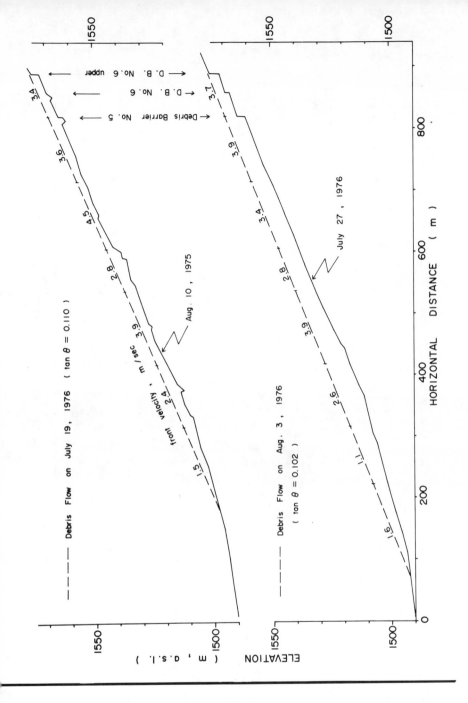

Figure 27.5. Energy lines of debris flow motion and longitudinal profiles of the Kamikamihori fan along its central line. (tanθ of the energy line corresponds to the eqivalent co-efficient of friction)

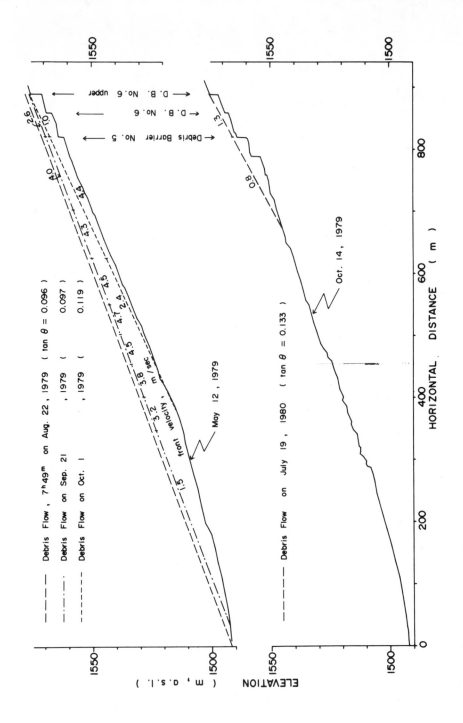

Figure 27.5.   continued.

458

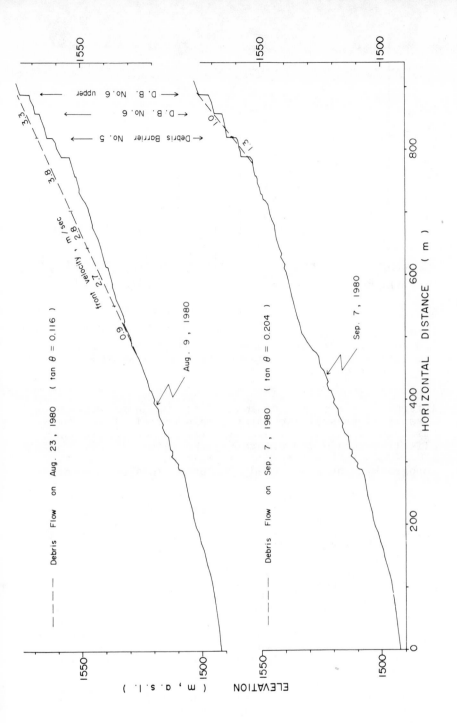

Figure 27.5.   continued.

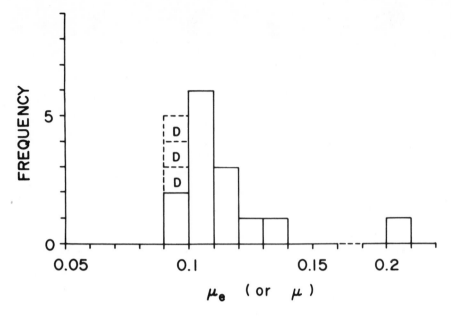

Figure 27.6. Frequency distribution of the equivalent
    coefficient of friction μ_e (D was calculated using the
    dynamic equation).

interpretation of the dynamic motion of debris masses on
fans is at present incomplete, this empirical approach
based on field experiments may offer a convenient and
effective method for predicting the travel distance.

Topographic changes caused by debris flows on the fan

As explained in previous reports, the frequent occurrence
of debris flows causes a considerable change in topography
of the fan. Especially at the fan head, a rapid change
involving serious scouring or deposition may be caused by
even a single debris flow.
    Figure 27.7 shows typical changes in the transverse
profile along several survey lines near the fan head (line
numbers correspond to the figures written on the left side
of figure 27.1) from 1979 to 1980. During this period,
most of the topographic change was produced by the debris
flow on August 23, 1980. Scouring of the central bed and
lateral deposition are found along line 1, and flattening
has taken place along lines 2 and 3. From line 4 to 6, a
considerable increase in bed elevation was produced by
rapid deposition of debris, and along line 7, an increase
in elevation occurred in the central part of the original
water course. Such topographic changes in the transverse
profiles, especially the rapid infilling or sometimes
scouring of the water course, often cause a considerable
movement of the initial point of lateral flooding of the
debris flow and affect the flow direction of succeeding

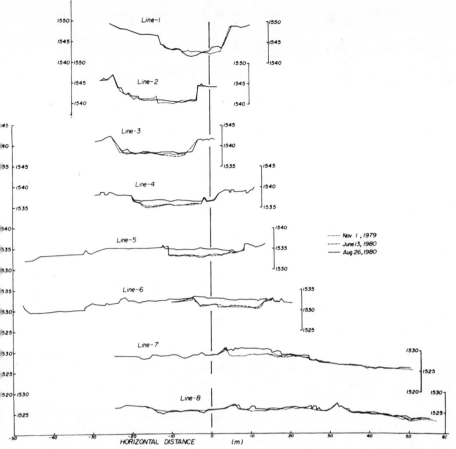

Figure 27.7. Changes in cross sections of the fan head caused by debris flows.

debris flows near the fan head. This in turn can cause un-expected large changes in the flow path of the debris flow in the lower region at the fan. Careful and repeated observations of micro-topography are therefore necessary for the exact prediction of areas of debris deposition on fans where debris flows occur frequently.

Figure 27.8 shows the sequential change in ground sur-face level for various points along the central line of the fan from 1962 to 1980 (right), and a comparison of two longitudinal profiles in 1976 and 1980 along the same line (left). On the right hand profiles, the dates of occurrence of the debris flows are indicated by the arrows along the top and their scale by L (large), M (middle) and S (small). Upstream or near the fan head (survey points 1 to 8), a remarkable lowering of ground level accompanied the very large debris flow of August 22, 1979. This lowering corresponds to so-called 'fan head incision'. Further down the fan, a progressive increase in ground surface level

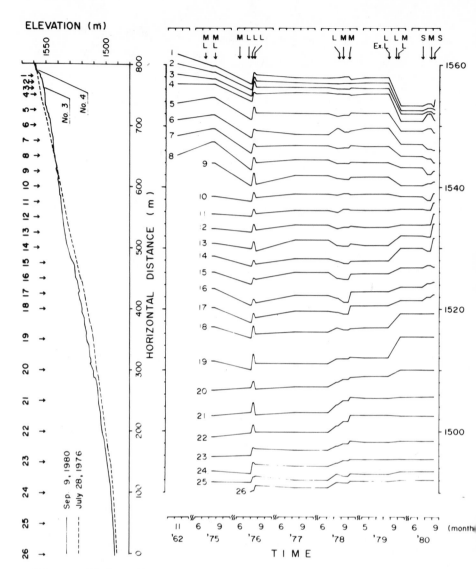

Figure 27.8.   Changes in ground level at selected measurement points along a longitudinal profile of the fan.

occurs at nearly all points. This feature is also clearly
shown by the change in the longitudinal profile, which
exhibits a general tendency for scouring to proceed in the
upper region, whilst deposition occurs in the lower region.
Careful inspection of the change in the profile, however,
indicates the existence of a somewhat sudden change in the
thickness of deposits and in surface ruggedness at a
boundary near points 20 to 21. In the region above this
point, the deposits that accumulated during these 4 years
are about 5 m thick and the surface is very rugged, while

462

in the lower region, the deposition increment is only about
2 or 3 m thick and the surface is rather smooth. These
differences may be ascribed to contrasts in the depositional
processes and in the total amount of additional deposits
between the two regions. In the upper region, the debris
lobes consist of larger boulder-type rock deposits with a
sand and pebble matrix, and the lobes tend to assume a
thick and swollen shape. In the lower region, the debris
flow lobes consist primarily of sand and pebbles with only
a small proportion of cobbles and boulders and tend to
assume a flatter shape. In addition to the debris flow
lobes, thinner deposits of sand and silt are deposited by
the frequent floods which traverse the lower region as far
as the margin of the fan.

Further interesting features are shown by the right hand
side of figure 27.8. For example, the deposition by debris
flows during 1976 occurred predominantly within the reach
from point 23 to 26. In 1978 this shifted to the reach
from point 18 to 24, in 1979 to the reach between points
13 and 20, and in 1980 to the reach between points 9 and 17.
It seems that the dominant depositional area has been
gradually moving upstream along this central survey line
during these four years.

CONCLUSIONS

Based on observations of debris flow motion and micro-
topography undertaken on the Kamikamihori fan at the eastern
foot of Mt. Yakedake, quantitative analyses concerned with
the estimation of travel distances of debris flows and the
characteristic topographic changes caused by debris flows
have been undertaken to clarify the relationships between
debris flow motion and fan topography. The main conclusions are as follows:
1) A correlation matrix of the physical factors which
   seemed to influence the travel distance L was established and it was shown that the frontal velocity V_f
   at the entrance to the fan was the most important
   control of the travel distance.
2) Using a theoretical relationship between V_f and L based
   on a simple assumption of mass point dynamics, the
   apparent friction coefficient was calculated.
3) From the profiles of the debris flow course on the fan,
   the apparent friction coefficient was estimated taking
   the front velocity V_f into consideration.
4) The estimated values of the friction coefficient
   obtained from both methods (2) and (3) mostly range
   from $0.095\sim0.11$, indicating that the travel distances
   of the debris flows are much longer than those of dry
   mass movement with the same mass volume.
5) Direct observations of the debris flow motion and
   sequential surveys of the longitudinal profiles of the
   fan show that fan head incision occurs during the
   passage of a debris flow.
6) Repeated surveys of the depositional form of the debris
   flows, demonstrated the existence of two different
   patterns of debris flow deposition on the fan. Thick

463

swollen lobes containing large boulders are deposited
in the upper region, while thin flat lobes containing
pebbles and cobbles are primarily deposited in the
lower region.
Further studies are needed to clarify the quantitative
relationships between debris flow motion and fan topography.

REFERENCES

Heim, A., 1932, *Bergsturz und Menschenleben*, (Fretz und
Wasmuth, Zürich).

Hsü, K.J., 1975, Catastrophic debris stream generated by
rockfalls, *Geological Society of America Bulletin*,
86, 129-140.

Imamura, R. and Sugita, M., 1980, On the simulation of
depositional processes of debris flow by Random
Walk model, *Shinsabo*, No.114, 17-26, (in Japanese).

Okuda, S. *et al.*, 1973, Field surveys on debris flows,
Disaster Prevention Research Institute Annuals,
No.16A, 53-69, (in Japanese).

Okuda, S. *et al.*, 1980, Observations on the motion of a
debris flow and its geomorphological effects,
Zeitschrift für Geomorphologie, Suppl.Bd.35,
142-163.

Scheidegger, A.E., 1973, On the prediction of the reach and
velocity of catastrophic landslides, *Rock Mechanics*,
5, 231-236.

Suwa, H. and Okuda, S., 1980, Dissection of valleys by
debris flows, *Zeitschrift für Geomorphologie*,
Suppl.Bd.35, 164-182.

Takahashi, T., 1980, Debris flow in prismatic open channels,
*Journal of the Hydraulics Division, American Society
of Civil Engineering* 106, 381-396.

28 Precise measurement of microforms and fabric of alluvial cones for prediction of landform evolution

Masashige Hirano and Takayuki Ishii

INTRODUCTION

The execution of field experiments represents a recent and important trend in geomorphology and measurements of stress, sediment yield, and other variables have frequently been discussed. However, detailed topographic measurements can also provide important information on landform evolution and for quantifying the rate of denudation or transportation, as, for example, illustrated by Schumm (1956) at Perth Amboy.

Detailed mapping of landforms, especially of microforms, often requires very considerable effort, and it is usually difficult to produce an appropriate map quickly. Therefore, it has been necessary to develop specialised measuring devices. The slope profiler devised by Ishii (1980) has been used to measure the detailed form of alluvial fans in order to assess its usefulness for rapid and accurate mapping of microforms. The surface fabric of cones and fans, reflected in the orientation and grain-size distribution of surface stones, is also an important feature which characterizes particular forms and processes. Measurement of fabric is therefore also referred to here. The fieldwork described was carried out at two sites on Honshu Island, Japan (figure 28.1).

MEASUREMENT OF RELIEF AND FABRIC

Measurement of relief was carried out using Ishii's slope profiler, which is shown in figure 28.2. The details of the instrument have been discussed elsewhere (Ishii, 1980; 1981), and only a brief outline of the instrument and of the measuring procedure is provided here. The profiler when placed on a slope gives the values of a=BE and b=BF, as shown in figure 28.2. The instrument has two legs of equal length connected at the hinge A, and the lengths L=AC=AD and l=AB=AE are always constant. If the plumb line AG is kept vertical, the horizontal and vertical distances between the points C and D are given by:

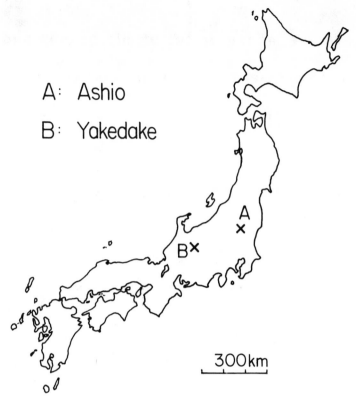

A: Ashio

B: Yakedake

300km

Figure 28.1. Location of the study sites.

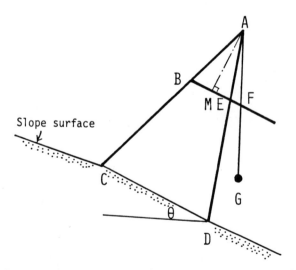

Figure 28.2. Schematic representation of the slope
    profiling instrument. Point A indicates a hinge con-
    necting two legs, AC and AD; the bar passing the points
    B, E and F is a scale fixed at B; G is a weight giving
    the plumb line AG; Point M gives the middle point of BE.

466

$$x = \frac{aL}{l} \cos \theta, \qquad\qquad\qquad (1)$$

$$y = \frac{aL}{l} \sin \theta, \qquad\qquad\qquad (2)$$

respectively, where the angle θ is calculated from

$$\theta = 90° - \tan^{-1}(\frac{\sqrt{4\,l^2 - a^2}}{2b - a}). \qquad\qquad (3)$$

In order to measure microforms, a tape is laid between two bench-marks and the lengths a and b are read successively along the tape, where the leg C is moved relative to fixed D. If it is no longer possible to widen the leg C, the position of the leg D is shifted to the last point of the leg C. Then profiles along the line are obtained by calculation of x and y based on equations (1), (2) and (3), successively. Densely-spaced profiles allow a contour map to be drawn.

Field observations, including a, b, microforms, and general remarks on slope materials, were recorded on a magnetic tape in a cassette recorder with an attachable tie-pin-type microphone, so that it was possible to survey alone and even during rain.

Measurement lines were laid out by triangulation with a hand compass at the Yakedake site. Vertical and horizontal correction of the survey is an important step to produce a precise map. On this account, survey stakes sited by the Disaster Prevention Institute of Kyoto University were helpful and effective as reliable bench marks.

Measurement of fabric is usually difficult, but for small areas in the Ashio Mountains, it was possible to use a hexagonal wire net with a mesh size of about 1.2 cm diameter, in order to obtain an accurate record of the fabric pattern. The wire net was put on the land surface and photographed. The photographs were then analysed in the laboratory. Debris (rock fragments) larger than the wire net diameter were noted and marked. Their distribution was then overlaid on the contour map produced from the measured profiles. The method mentioned above would have been difficult to apply to a large fan such as that developed at the foot of Mt. Yakedake. Therefore, in that case, a fabric sketch was carried out along the measured lines. The distribution of stones larger than 20 cm in diameter was mapped exactly at a scale of 1:100 or 1:200.

Contour lines were drawn based on the data converted to numerical values giving the position and height of the fan surface along the profiles. In the case of Mt. Yakedake the fabric sketch played an important role in drawing the contour lines exactly to fit to micro-topography, including cliffs, ravines, and big stones.

RESULTS FOR A TALUS SLOPE AT ASHIO

The locality of the talus slopes studied at Ashio is shown
in figure 28.1 (Ishii, 1978). A number of units (lobes)
of debris flow deposits are found here, and they are
distinguished clearly on the detailed map obtained by the
method described here. Figure 28.3 is an example of such
a map which shows some of these lobes.

A typical lobe is usually 2.5 to 3 m in width and 3-5 m
in length. The steep front of the lobe has an inclination
of 23-45°, and the gentle portion behind it has an angle of
about 10°. Detailed maps of the sub-units (individual
lobes) obtained here are shown in figures 28.4 and 28.5 as
examples. It was ascertained by analysis of meteorological
data that the lobes were mobilized by heavy rain on October
19th, 1979, when 110 mm was recorded in 1 day.

As shown by figures 28.4 and 28.5, where the fabric
pattern has been overlaid on the contour lines, there is
a tendency for large stones to be located at the front of
a lobe formed as a flow unit (for example, lobe B-2 on
figure 28.4) though this is less noticeable in the case
of lobe A-1 (figure 28.5). The average size of debris com-
posing the lobe is 5-10 cm with the maximum stone reaching
65 cm on the longest axis.

The subsurface structure of flow units was observed in
a trench, where it was seen that individual units frequently
attained thicknesses of 20-50 cm. The coarse portion near
the surface lacks a matrix and is distinguished clearly
from matrix-filled deeper part. The coarse portion at the
surface usually occupies about 20% of the total thickness.

As shown by these examples, the lobe units documented
by precise measurement can be related to such meteorological
data as daily precipitation. From this point of view,
detailed correlation between individual topography and the
specific daily climatic record becomes possible if the
topography is recognised precisely as shown here.

RESULTS FOR AN ALLUVIAL FAN AT MT. YAKEDAKE

Alluvial fans formed by debris flows have developed at the
foot of Mt. Yakedake, an active volcano near Kamikochi,
Central Japan. The area has been regularly surveyed for
more than 10 years and a number of reports have been
published (eg. Suwa and Okuda, 1980, and references contained therein). As reported in those reports, prediction
of the direction and travel distance of debris flows is
necessary to counter the hazard associated with debris
flows moving from the upstream gully across the fan. It
is therefore essential to study the characteristics of
micro-topography near the outlet of the gully. Measurements
at Kamikamihori (Toge) valley were carried out on September
29 to 30, 1980 and June 1 to 3, 1981. The relief map was
obtained using the slope profiler as described above.
Fabric was recorded by sketching along the measurement
lines, because the fan was so large that the method used
at Ashio was difficult to apply. Altitudinal data were
overlaid on the fabric pattern and contour lines were

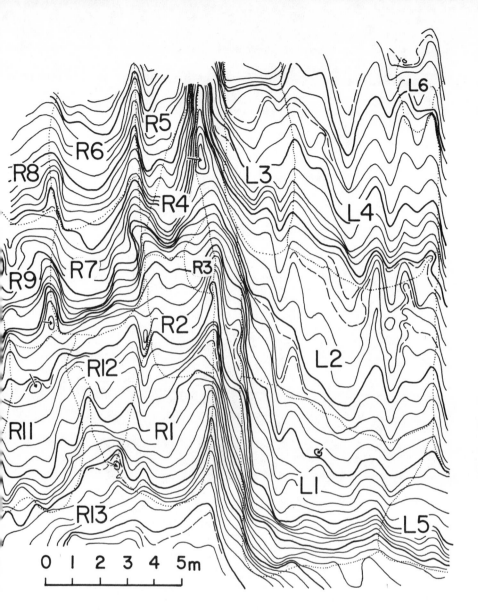

Figure 28.3. Detailed contour map of rock debris deposits
on the Ashio Mountains. A number of lobe units can be
distinguished. The contour interval is 10 cm.

constructed. The result is shown in figure 28.6. The
sketches of fabric pattern provided important guides in
the construction of the contour lines.
    A number of lobes were detected in the mapped area.
The lobes have various ages, as shown by different degrees
of dissection. The most recent one was formed in August,
1980. Some of the lobes are cut by the present river

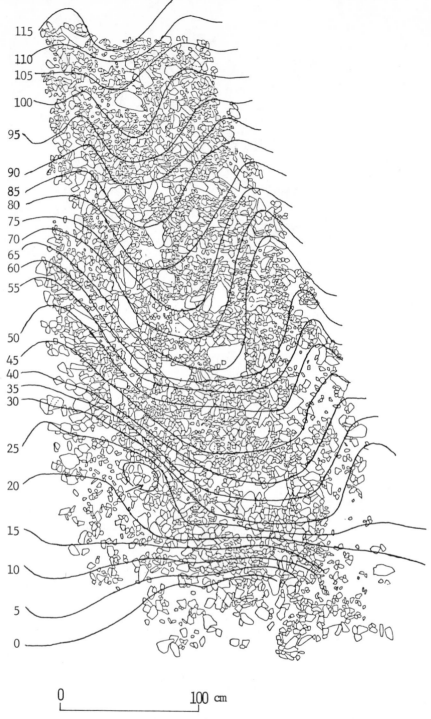

115
110
105
100
95
90
85
80
75
70
65
60
55
50
45
40
35
30
25
20
15
10
5
0

0   100 cm

Figure 28.4.   An example of micro-topography of an indi-
     vidual lobe overlain on the fabric pattern.   Lobe B-2
     with contour interval 5 cm.

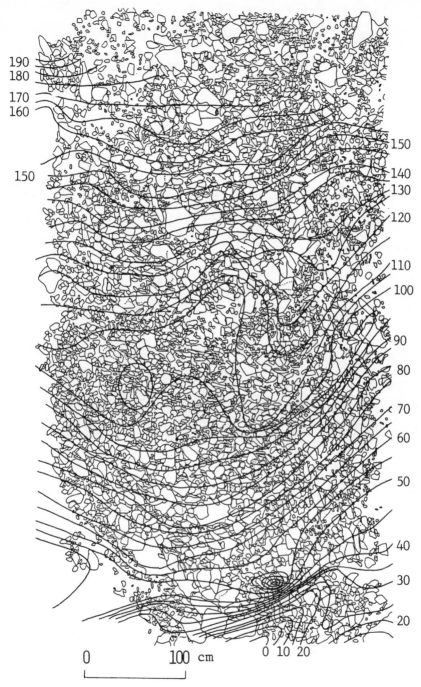

Figure 28.5. An example of the micro-topography of an individual lobe overlain on the fabric pattern. Lobe A-1 is shown with contour interval 5 cm.

471

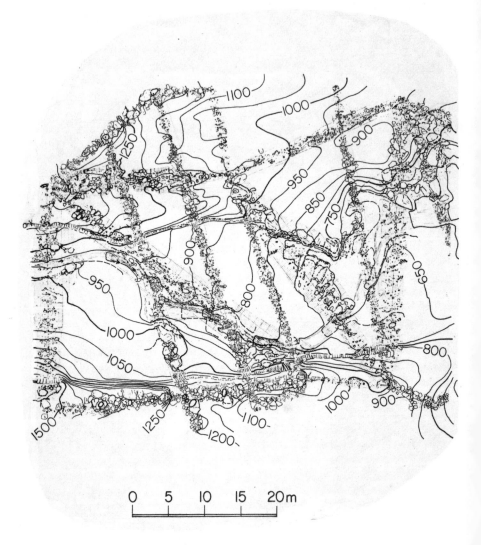

Figure 28.6. Detailed contour map of the area near the
top of the fan at the foot of Mt. Yakedale; overlain
on the fabric pattern obtained along survey lines.
Contour interval 50cm.

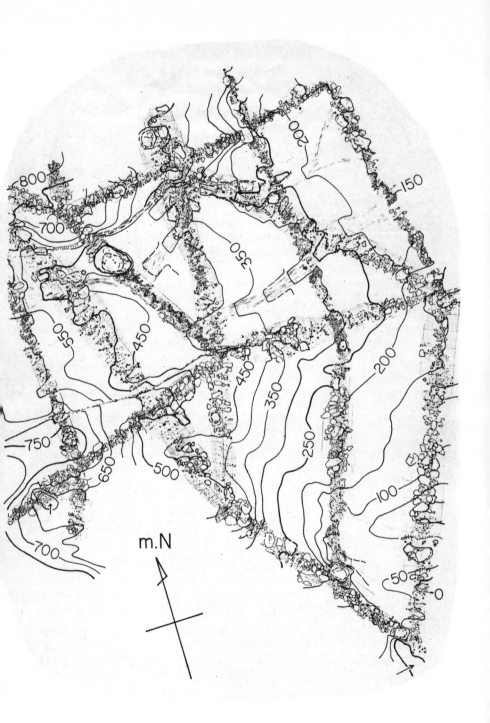

m.N

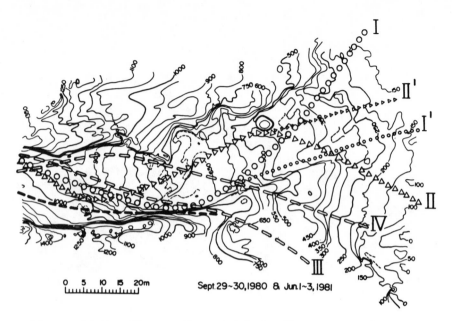

Figure 28.7. Flow paths of debris flows assumed on the
   basis of micro-topography and fabric near the top of
   the fan.

course forming clear-cut cliffs at the boundaries. As
shown by the relationship between the contour map of micro-
topography and the very large stones detected on the fabric
map, river and valley meanders have been affected by these
stones. Eyebrow-shaped cliffs terminate at big stones and
these cliffs have been produced chiefly by erosion associated with the debris flow. This fact suggests that
the orientation or course of the debris flow must be refracted or at least controlled by these stones forming the
cliffs.
   If the volume and/or rate of future debris flows are
tentatively assumed, prediction of orientation and travel
distance is to some degree possible. Figure 28.7 shows
several flow paths obtained in this way, where path I
represents the case where a mudflow avoids the large stones,
and case II represents a path which is strongly controlled
and refracted by the stones. The case I flow path seems
to be most probable for a small debris flow, though it is
expected that such a flow might stop in the middle part of
the flow path. On the contrary, paths III and IV might
correspond to larger and more rapid flows, and, thus,
straighter paths than the others are assumed, although
cases III and IV differ in the phase of their meandering
in the illustrated area.

SUMMARY

Precise measurement of topography can be useful for
analysing the characteristic processes and forms producing
the present landform. The measurements undertaken at Ashio
demonstrated that a number of lobes (flow units) can be
distinguished and that a group of individual lobes may be
correlated with meteorological events.
    The same measurements of topography and fabric under-
taken at Mt. Yakedake demonstrated that characteristic
erosional topography produced by debris flows exhibits a
definite pattern of meandering which has been controlled
by the distribution of larger stones. In addition, it is
possible to distinguish individual lobes produced on different occasions on the basis of their state of dissection.
The surface fabric of the cones and fans provides important
information for understanding the processes that shaped the
present landform, in addition to the information for com-
posing contour lines. From both these points of view,
precise measurement of three-dimensional topography and
fabric using appropriate and rigorous techniques, such as
those described here, are shown to provide a sound basis
for field experiments in geomorphology.

ACKNOWLEDGEMENTS

The authors wish to express their thanks to Professor Setsuo
Okuda of the Disaster Prevention Institute of Kyoto University for generous assistances during survey at Mt.
Yakedake. Thanks are also due to Professor Thomas Dunne
and Professor Bernard Hallet of the Department of Geological Sciences and the Quaternary Research Center of
University of Washington for critical reading of the
manuscript.

REFERENCES

Ishii, T., 1978, Influences of the grain size of rock
        fragment and the slope length on the development
        of talus slope, *Geographical Report, Osaka Kyoiku
        University*, 17, 35-46.

Ishii, T., 1980, Simple measurement in field work, *Chiri*,
        25-5, 52-62, (in Japanese).

Ishii, T., 1981, Microforms and slope processes of the
        Ashio Mountains in Central Japan, *Transactions of
        the Japanese Geomorphological Union*, 2-2, 279-290.

Schumm, S.A., 1956, Evolution of drainage systems and
        slopes in badlands at Perth Amboy, New Jersey,
        Bulletin of the Geological Society of America, 67,
        597-646.

Suwa, H. and Okuda, S., 1980, Topographical changes in
        Kamikamihori Valley by debris flows at Mt. Yakedake,
        Transactions of the Japanese Geomorphological Union,
        1-1, 43-55.

29 An analysis of sediment transport by debris

flows in the Jiangjia Gully, Yunnan

Kang Zhicheng and Zhang Shucheng

INTRODUCTION

The debris flow is a natural disaster phenomenon which is
common in the mountainous regions of China. This two phase
flow of water and earth, sand and stones is characterised
by sudden occurrence, rapid movement and short duration,
and represents conditions of turbulent or laminar flow
intermediate between mass sliding and hydraulic motion.
Most of the depositional forms appear as ridges, islands
(tongues), or sheets (figures 29.1, 29.2, 29.3 and 29.4),
and the deposits exhibit an unsorted texture, including all
sizes of material (figure 29.5). The geomorphological end
product of the debris flow process is the production of a
debris flow fan at the mountain front (figure 29.6).
Debris flows have caused considerable damage to the industrial and agricultural activities in our mountainous
regions and in order to reduce their impact and establish
control measures there is a need for further investigations,
observations and studies of this phenomenon.

The Jiangjia Gully, located on the right bank of the
Xiaojiang River valley in northeast Yunnan, is one of the
large-scale debris flow gullies that occur in China (figure
29.7). It has a drainage area of 47.1 km^2, and the trunk
gully is 12 km long. The maximum and the minimum altitudes
are 3269 m and 1088 m respectively. There are 178 large
and small branch gullies (counting the ones over 550 m
long), and the main branch gullies include the Menqian
Gully, the Duozhao Gully, the Daoazi Gully, and the Chajing
Gully (figure 29.8). Many years of observation indicate
that a considerable number of debris flows occur in the
Jiangjia Gully each year (table 29.1) and that it provides
a very convenient location to observe and study them. In
the trunk gully, debris flows appear as surges with an
interval of several seconds or more between every two
surges (figure 29.9). Each debris flow lasts for 4-5
hours, during which period more than 100 surges may occur.
The height of their fronts range from 1 or 2 m to 4.7 m
(figure 29.10). The maximum velocity is 15 m s^{-1} and the
maximum momentary discharge is 2420 m^3 s^{-1}. The moving

Figure 29.1. Ridge-like debris flow deposits.

Figure 29.2. Island-like debris flow deposits.

Figure 29.3.   Sheet-like debris flow deposits.

Figure 29.4.   Leaf-like debris flow deposits.

debris flow is similar to a concrete slurry and has a
density between 1.8 and 2.37 t m^{-3}. Original soil masses
4-6 m in diameter have been found drifting down the debris
flow, and boulders 4-5 m in diameter have been carried
downstream (figure 29.11).
     The debris flow is a rapid geologic and geomorphic
process and is the most rapid exogenic planation process.
The upstream erosion and downstream deposition associated
with the debris flow basin greatly elongate the equilibrium
longitudinal slope and shorten the eroded longitudinal slope,
leading the debris flow gully to equilibrium and stability.

Figure 29.5.   The texture of debris flow deposits.

Figure 29.6.   The depositional fan of the debris flow.

Figure 29.7.   A view of the Jiangjia Gully basin.

Table 29.1.   Frequency of occurrence and sediment
discharge of debris flows in the Jiangjia Gully

Year	1965	1966	1967	1973	1974	1975
Times of occurrence	28	17	15	14	22	12
Discharges of solid material (m^3 x 10^4)	353	180	194	246	387	211

Using observations on the Jiangjia Gully, the annual
supply of sediment and the total potential supply of solid
material can be estimated by considering the equilibrium
longitudinal slope of the debris flow basin.   These data
are of value in analysing the development history of the
debris flow and the potential for establishing control
measures.

THE ANNUAL SUPPLY OF SOLID MATERIAL

Although the controls of debris flow occurrence are complex
in detail, the supply of water and solid material exert a
basic influence.   In view of the location of debris flows
in this region and their potential evolution, the supply
of water may be viewed as relatively constant, but the
supply of solid material is changeable, and variations in
the longitudinal slope of the gully bed influence the movement of the debris flow.   Therefore, the development of the
debris flow is closely related to the supply of solid
material and the longitudinal slope of the gully bed.
    In the drainage basin of the Jiangjia Gully solid
material is derived largely from the Menqian Gully and the
Duozhao Gully, ie. the eroding portion of the basin.   The

481

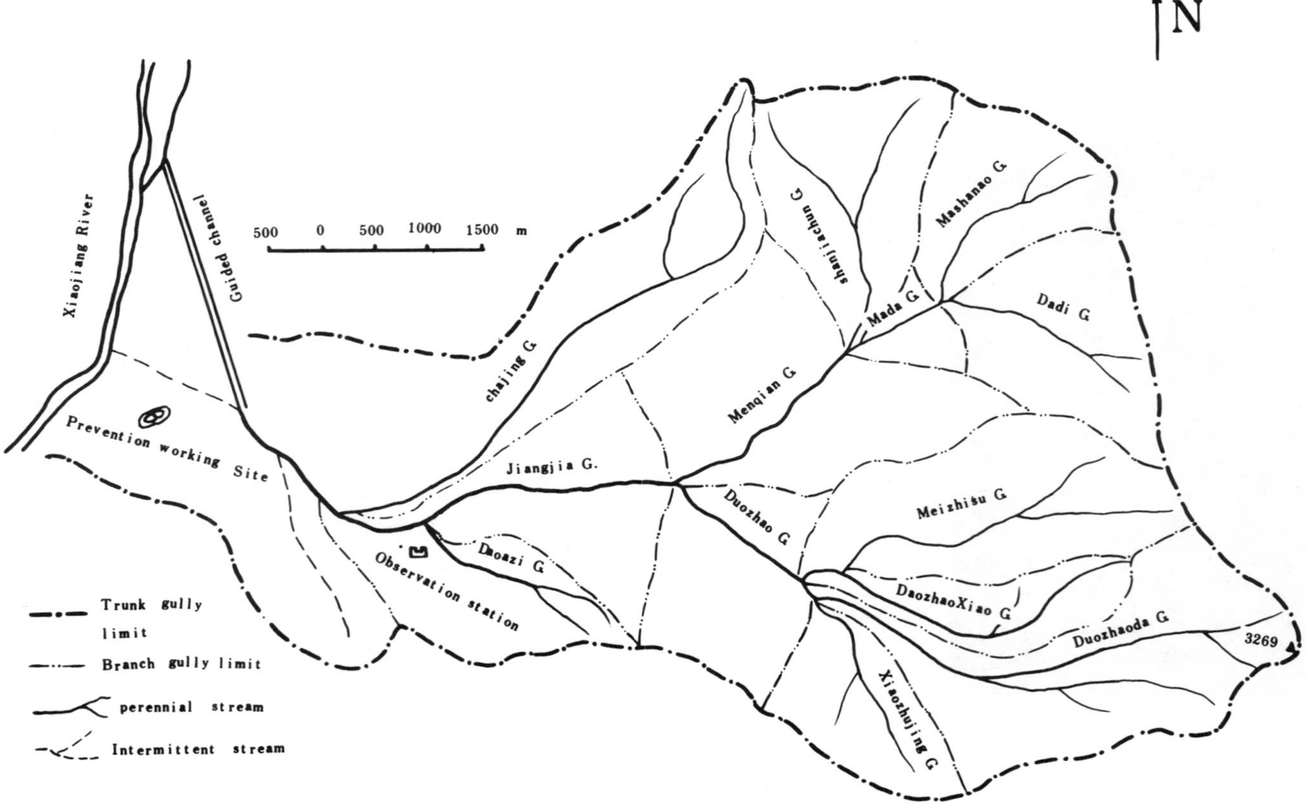
N
Xiaojiang River
Guided channel
500 0 500 1000 1500 m
chajing G
shanjiachun G
Mashanao G
Mada G
Dadi G
Menqian G
Jiangjia G.
Meizhisu G
Duozhao G
DaozhaoXiao G
Duozhaoda G
3269
Xiaozhujing G
Daoazi G
Observation station
Prevention working Site
Trunk gully limit
Branch gully limit
perennial stream
Intermittent stream

Figure 29.8. The drainage basin of the Jiangjia Gully.

Figure 29.9.   Surges in a debris flow in the Jiangjia Gully.

Figure 29.10.   The leading edge of a debris flow surge.

catchment area of the two gullies is 29.2 km^2 (13.7 km^2 and
15.5 km^2 respectively), and accounts for two thirds of the
total area of the basin.   The collapsing zone that directly
feeds the debris flow is 7 km^2 (3.921 km^2 and 3.073 km^2
respectively), and represents 25% of the catchment area of
the eroding portion.   The total length of the large branch
gullies in the eroding portion is 28.96 km^2, providing a
density of approximately 1 km km^{-2}.
    The longitudinal slope of these branch gullies is
considerable.   In general, all the gullies are cutting
down and data for the period 1957 to 1973 indicate that the

Figure 29.11. A boulder transported by the debris flow.

Table 19.2. Annual supply of solid material

Gully name	Direct supply area $A(m^2)$	Yearly downcut depth $d(m)$	Yearly supply amount $V(m^3)$	Percentage of total supply by eroding portion
Sanjiachun	813 000	0.5	406 500	17%
Mashanao	876 000	0.4	350 400	14%
Dadi	680 000	0.4	272 000	12%
Mada	223 000	0.4	89 200	4%
Menqian	132 900	0.3	398 700	17%
Totals			1 516 800	64%
Meizisu	952 000	0.3	285 600	12%
Daozhaoxiao	307 000	0.3	92 100	4%
Duozhaoda	589 000	0.3	176 700	8%
Xiaozhujing	360 000	0.3	108 000	5%
Duozhao	865 000	0.2	173 000	7%
Totals			835 400	36%
Sum total			2 352 200	100%

annual depth of downcutting averages 0.3 m in the Menqian Gully, 0.5 m in the Sanjiachun Gully and 0.4 m in the Mashanao Gully and the Dadi Gully. The annual downcutting in the Duozhao Gully, where downcutting is not readily apparent, is approximately 0.2 m and its branch gullies exhibit annual downcutting of less than 0.3 m.

Because collapse is the major direct source of supply of solid material, the collapsed area of each branch gully can be estimated from field mapping evidence (table 29.2). As the gullies cut down the slopes on both sides that were formerly in a critical condition become unstable. The annual supply of solid material V (table 29.2) may therefore be estimated from measurements of the direct supply area A and the annual depth of downcutting d. The formula is:

$$V = dA \qquad (1)$$

Table 29.2 indicates that the annual supply of solid material to the debris flow from the eroding portion of the basin is about 2.35 million m^3. The data obtained from direct measurements of debris flows between 1965 and 1967 and between 1973 and 1975 (table 29.1) show that the annual transport of solid material is 2.62 million m^3 (assuming that there is 83% solid material in the debris flow). That the former estimate is 90% of the value obtained by direct measurement implies that the solid material associated with the generation of the debris flows is largely derived from the collapsed masses on both banks of the gullies, and that only a small proportion (10%) originates from slope wash.

Table 29.2 also indicates that 64% of the material transported by the debris flows in the Jiangjia Gully originates from the Menqian Gully and that 36% originates from the Duozhao Gully. This means that debris carried in the Jiangjia Gully is generated mainly in the Menqian Gully.

TOTAL POTENTIAL SUPPLY OF SOLID MATERIAL

The longitudinal slope of the gully bed in the source zone of the debris flow is relatively steep. In the course of debris flow development, however, the longitudinal slope of the gully bed will decrease until a stable longitudinal slope is produced as the inevitable end result of debris flow processes. It is necessary to consider how much solid material will be transported between the present stage and the final extinction of the debris flow. This question may be approached as follows. Disregarding other factors, when the longitudinal slope of the gully bed and the slopes on both sides of the gully finally attain stability, the amount of solid material supplied by the whole basin during the intervening period will represent the total potential supply of solid material.

First of all, it is necessary to estimate the amount of solid material supplied as the longitudinal slope tends to stability. According to measurements obtained in recent years, the average longitudinal slopes of the Menqian Gully and the Duozhao Gully are 22% and 20% respectively. Field investigations indicate that their average stable longitudinal slope is approximately 17%*. Accordingly, the difference between the current average longitudinal slope and the future longitudinal slope is 5% in the Menqian Gully and 3% in the Duozhao Gully.

Using the difference between the present average longitudinal slope s_1 and the future stable longitudinal slope s_2, and the gully length L, it is possible to calculate the probable maximum depth of downcutting as:

$$d_{max} = L(\tan s_1 - \tan s_2) \qquad (2)$$

In view of the unequal depth of downcutting along the gully bed, the formula to calculate the total supply of solid material V, must employ the average depth of downcutting viz:

$$V = \tfrac{1}{2}(d_1 + d_2)A \qquad (3)$$

where d_1 is the maximum depth of downcutting at the gully mouth, and d_2 is the maximum depth of downcutting at the gully head.

Using this approach, the total volume of solid material that will be supplied by all the branch gullies in the eroding part of the basin was estimated at 748.3 million m^3, of which 523.8 million m^3 will be supplied by Menqian Gully and 224.5 million m^3 by the Duozhao Gully (table 29.3).

THE ACTIVE LIFE OF A DEBRIS FLOW AND THE ESTABLISHMENT OF CONTROL MEASURES

The active life of a debris flow represents the time taken for the longitudinal slope of the bed and the slopes on both sides of the debris flow gully to develop from the unstable stage to the stable stage. In the light of the total storage (about 750 million m^3) and the annual rate of export (about 2.35 million m^3) of solid material in the debris flow, debris flows in the Jiangjia Gully are likely to persist for more than 350 years. After that the gully will tend to stability. Without any source of solid material, only runoff supplied by rainfall and ground water will pass through the basin, and no debris flows will occur.

As seen above, the debris flow is a natural disaster phenomenon. In the last 100 years, debris flows in the Jiangjia Gully have frequently caused damage to the local industrial and agricultural production. In order to reduce the scale of debris flows and to speed up their extinction, debris flow control schemes must be introduced and these must be based on a knowledge of the causes of debris flow formation. Since 90% of the solid material in the debris flows in the Jiangjia Gully originates from side-slope collapse caused by gully erosion, and only 10% is produced by slope wash, the debris flow control schemes introduced mainly involve the construction of a ladder dam system to check the sediment in the debris flow and to reduce the

* The percentage cited here refers not to the actual stable longitudinal slope, but to the average value of the longitudinal slope. The actual stable longitudinal slope should be < 17% in the lower reaches, > 17% in the upper reaches and 17% in the middle reaches.

Table 29.3. Total potential supply of solid material

Gully name	Length of gully $L(m)$	Difference in longitudinal slope (%)	Maximum depth of downcutting $d_{max}(m)$	$\frac{1}{2}(d_1 + d_2)$ (m)	Direct supply area $A(m^2)$	Potential supply of solid material $V=\frac{1}{2}(d_1+d_2)A(m^3)$
Menqian	2280	3	68	34	1 329 000	45 200 000
Sanjiachun	4780	5	239	153.5	813 000	124 800 000
Mada	3260	5	163	115.5	223 000	25 800 000
Mashanao	2260	5	263	213	876 000	186 600 000
Dadi	5060	5	253	208	680 000	141 400 000
Duozhao	1400	3	42	21	865 000	18 200 000
Meizisu	4700	3	141	99	952 000	94 200 000
Dzozhaoxiao	4500	3	135	96	307 000	295 500 000
Duozhaoda	7500	3	113	78	589 000	55 600 000
Xizozhujing	3600	3	108	75	360 000	27 000 000

487

longitudinal slope of the gully bed and stabilize the gully. In order to prevent soil and water losses from forming incipient debris flows in the upper reaches, afforestation is to be carried out, reclamation is to be prohibited on steep slopes, and check dams (generally 2-3 m high) are to be built across the tributary gullies. If these control schemes can be introduced, they will reduce and eliminate the source of solid materials, and speed up the extinction of debris flows.

Because debris flow control schemes are costly to implement and the practical effects of the schemes were uncertain, a pilot project was conducted between 1973 and 1974, and relatively satisfactory results have been achieved. For example, a debris basin with an area of 0.36 km^2 built on the deposition fan in the lower reaches of the Jiangjia Gully retained 2.25 million m^3 of solid material from the debris flows between 1973 and 1975. A 5 m high silt arrester constructed across the trunk gully has retained 2.5 million m^3 of solid material so that the length of the stable longitudinal slope now reaches 5 km and the original longitudinal slope has been reduced by 20%. The afforestation and check dam project carried out in parts of the upper reaches has also achieved some success. The practical effectiveness of the pilot debris flow control project has proved that control schemes designed on the basis of the source of solid material supplied to debris flows are correct.

In order to provide a sound scientific basis for the control projects in the Jiangjia Gully and other debris flow gullies, we shall continue our observations and research on debris flows in the Jiangjia Gully.

REFERENCES

Kang Zhicheng and Zhang Shucheng, 1980, A preliminary analysis of fluid properties of mud-rock flows, (in Chinese), *Reports on the Interantional Congress on River Silt*, (Guanghua Publishing House), 1, 213.

Tang Banxing, Du Ronghuan, Kang Zhicheng and Zhang Shuchang, 1980, Research on mud-rock flows in China, (in Chinese), *Dili Xuebao (Acta Geographica Sinica)*, 35(3), 259-264.

30 Flow processes and river channel

morphology

R.D. Hey and C.R. Thorne

CHANNEL FORM AND PROCESSES

Empirical models of the type developed by Leopold and
Maddock (1953) and Wolman (1955) have traditionally been
used to define the hydraulic geometry of alluvial channels.
Although there are general similarities between the equations that have been derived, no universal model has yet
been established. This principally arises from the failure
to include all the controlling variables in the regression
models (Hey, 1982a). Clearly regime-type equations, even
if they include all the controlling factors as independent
variables and had general application, do not explain how
and why rivers adjust their overall shape and dimensions.
These equations simply describe channel morphology,
numerically.

In contrast process-response models offer both prediction and explanation. To construct this type of model it
is first necessary to identify the variables which define
the hydraulic geometry of alluvial channels by considering
how a river responds to erosional and depositional activity.
This indicates that alluvial channels have nine degrees of
freedom because they are able to adjust their velocity (V),
hydraulic radius (R), wetted perimeter (P), maximum flow
depth (d_m), slope (S), bed form height and wavelength
(Δ, λ), sinuosity (p) and meander arc length (Z) in response
to changes in discharge (Q), sediment load (Q_s), calibre
and characteristics of the bed and bank material (D_b, D_r, D_l)
and the valley slope (S_v). For stable channels the former
can be regarded as dependent variables, and the latter as
independent variables. Each dependent variable (degree of
freedom) must have an associated governing equation. Nine
equations are required and these define the processes of
continuity, flow resistance, sediment transport, bed form
development (2), bank erosion, bar deposition and meander
mechanisms (2)(table 30.1) (Vanoni, 1971; Hey, 1974; 1978).

Given general equations defining these processes, the
hydraulic geometry of a channel can be determined, for any
prescribed number of degrees of freedom, through the
simultaneous solution of the requisite number of process

Table 30.1. Channel morphology and process equations (after Hey, 1982a)

Degrees of freedom	Dependent variables	Fixed variables	Independent variables	Type of flow	Governing equations
1	V	$R,S,P,d_m,\Delta,\lambda,p,Z$	Q	FIXED BED	1 CONTINUITY
2	V,R	$S,P,d_m,\Delta,\lambda,p,Z$	Q,D,D_r,D_l	FIXED BED	1 CONTINUITY
3	V,R,S	P,d_m,Δ,λ,p,Z	Q,Q_s,D,D_r,D_l	MOBILE BED	1 CONTINUITY 2 FLOW RESISTANCE 3 SEDIMENT TRANSPORT
5	V,R,S,Δ,λ	P,d_m,p,Z	Q,Q_s,D,D_r,D_l	MOBILE BED	1 CONTINUITY 2 FLOW RESISTANCE 3 SEDIMENT TRANSPORT 4 BED FORM HEIGHT 5 BED FORM WAVELENGTH
7	$V,R,S,\Delta,\lambda,P,d_m$	p,Z	Q,Q_s,D,D_r,D_l	MOBILE BED	1 CONTINUITY 2 FLOW RESISTANCE 3 SEDIMENT TRANSPORT 4 BED FORM HEIGHT 5 BED FORM WAVELENGTH 6 BANK EROSION 7 BAR DEPOSITION
9	$V,R,S,\Delta,\lambda,P,d_m,p,Z$	–	Q,Q_s,D,D_r,D_l,S_v	MOBILE BED	1 CONTINUITY 2 FLOW RESISTANCE 3 SEDIMENT TRANSPORT 4 BED FORM HEIGHT 5 BED FORM WAVELENGTH 6 BANK EROSION 7 BAR DEPOSITION 8 SINUOSITY 9 RIFFLE SPACING

equations. Theoretically, therefore, it should be possible
to predict the hydraulic geometry of stable channels and
explain the adjustment process. In practice the technique
has still to be fully developed due to deficiencies in our
understanding of the basic flow processes operating in
alluvial channels.

FLOW PROCESSES IN GRAVEL-BED RIVERS

Much of the research carried out in the United Kingdom is
concerned with flow in gravel-bed rivers. As bed forms, on
the scale of ripples, dunes and antidunes, are not observed
in gravel-bed rivers under natural flow conditions plane
bed conditions can be assumed to apply.

Secondary flows

Flow in alluvial channels is usually three dimensional and
may be resolved into primary and secondary components.
Secondary currents are defined to be in the plane normal to
the local axis of primary flow. They distort the distributions of primary velocity and boundary shear stress from
those in simple flows and have important effects on sedimentary processes.

To investigate secondary flows, data were collected
from five sites on the River Severn in mid-Wales. Long and
cross stream velocities were measured simultaneously using
a two component electro-magnetic current meter. Primary
and secondary components were derived from these data on
the assumption of zero net secondary discharge (Bathurst,
Thorne and Hey, 1977, 1979; Thorne and Hey, 1979).

Secondary circulation at a bend is dominated by a large
cell driving fast, surface water towards the outer bank and
slow, near bed water towards the inner bank (figure 30.1).
This cell, caused by skewing of cross stream vorticity into
a streamwise direction has been widely observed (Hawthorne,
1951; Rozovskii, 1957; Perkins, 1970). The data also show
that there can be a small cell of reverse rotation next to
the outer bank. This outer bank cell has also been reported
in some earlier studies (Einstein and Harder, 1954;
Rozovskii, 1954; Hey and Thorne, 1975). It is caused by
interaction of the main cell with the outer bank and
appears where the bank is steep but not where it shelves.
Although small compared with the main cell, the outer bank
cell is important because it strongly affects processes of
bank erosion and planform change.

Between bends, the secondary circulation of the upstream bend decays, to be replaced by a new circulation
for the downstream bend. In channels with high width to
depth ratios the new skew-induced cell originates near the
bed and develops upwards, thereby displacing the old cell
(Chacinski and Francis, 1952; Chacinski, 1954). Flow at
the inflexion point between bends therefore consists of two
cells, one above the other (figure 30.2). This flow
pattern is probably the result of bed topography. In bends
the cross section is highly assymetrical with the thalweg
close to the outer bank (figure 30.1). At the inflexion

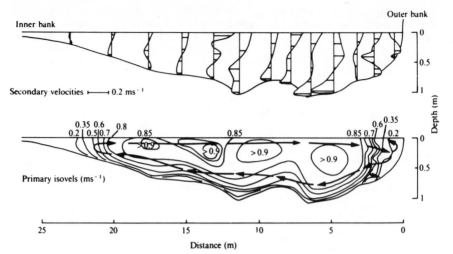

Figure 30.1. Primary and secondary velocities at a bend.
Penstrowed, River Severn, Wales. Discharge 13 m^3 s^{-1}.
(from Thorne and Hey, 1979).

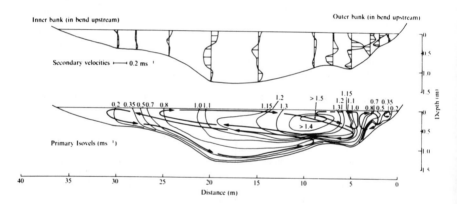

Figure 30.2. Primary and secondary velocities at an
inflexion point. Penstrowed, River Severn, Wales.
Discharge 22 m^3 s^{-1}. (from Thorne and Hey, 1979).

point the cross section is more symmetrical, coinciding
with the switch of thalweg from one bank to the other
(figure 30.2). This downstream change in bed topography
causes curving of the streamlines, especially near the bed,
and skewing of the shear field, thereby generating the new
secondary circulation.

Boundary Shear Stress

To investigate the effect of secondary flows on the boundary
shear stress distribution at each site, longstream velocities

were also measured close to the bed using an Ott C1 current
meter. Point values of boundary shear stress were then
calculated from the velocity gradient using the Prandtl-von
Karman flow law (Bathurst, Thorne and Hey, 1979; Bathurst,
1979). Peak values of boundary shear stress occur where
primary velocity is relatively high and where isovels are
compressed near the bed by downwelling. Low values occur
where primary velocity is relatively low and where there is
upwelling.

Straight reaches show contrasting stress patterns
between pools and riffles. At a pool the distribution shows
peaks and troughs with no clear maximum (figure 30.3a).
The ratios of peak to mean values are between 1.5 and 2.
This pattern might result from a multicell stress-induced
secondary circulation as described by Perkins (1970). This
circulation would produce alternate regions of upwelling
and downwelling between cells. At a riffle the distribution
shows a central peak below the core of maximum velocity and
smaller shoulder peaks (figure 30.3b). Ratios of peak to
mean stress for this and other sections lie in the range
1.5 to 2.5. This pattern may result from acceleration of
primary flow over the riffle and secondary circulation
associated with flow divergence.

Channel bends show two peak values of boundary shear
stress associated with the core of maximum velocity and
with the region of downwelling between the main and outer
bank cells (figure 30.4). Peak values below the core of
maximum velocity were in the range 1.5 to 2.5 times the
section mean, which agrees with observations by Apmann
(1972) and by Hooke (1975). Downwelling produces peaks up
to four times the section mean depending on the strength of
the outer bank cell. Where the outer bank cell is absent,
downwelling is too diffuse to produce any peak. The relative size of the two peaks varies with discharge. At low
discharge both primary and secondary velocities are small
and either peak may be the larger. At medium discharges
the effects of secondary discharge seem to be at their
strongest and often the outer bank peak is the greater.
However, at high discharges primary flow effects overshadow
secondary flow effects and the stress peak below the core
of maximum velocity is the larger.

The location of the core of maximum velocity and the
shear stress peak below it change not only with bend geometry but also with discharge. At the bend entrance the
core of maximum velocity is located between centre channel
and the inner bank due to a tendency towards free vortex
flow. In the bend the skew-induced secondary circulation
develops and eventually breaks down the free vortex carrying
the core of maximum velocity and its associated stress peak
towards the outer bank (Ippen and Drinker, 1962; Dietrich,
Smith and Dunne, 1979). Consequently, the positions of the
core of maximum velocity and stress peak depend on the
strength of secondary circulation. At medium discharges
the secondary circulation is relatively strong and so the
point at which the core of maximum velocity crosses over
towards the outer bank is located early in the bend. However, as discharge increases to high values this point
drifts downstream toward the bend exit (Bhowmik and Stall,
1978).

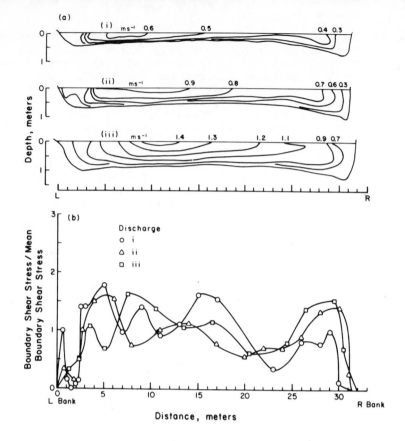

Figure 30.3a. Isovel pattern and boundary shear stress
distributions (ratio of point to mean values) at a
pool. Railway Straight, River Severn, Wales.
Discharges (i) 3.8, (ii) 11.5, (iii) 25.8 m³ s⁻¹.
Patterns affected by weeds at left bank and by trees
trailing in water at right bank (from Bathurst, 1979).

Flow resistance

Several equations have been proposed in the past to account
for the effect of the resistance to flow on the average
velocity of flow. The Manning, Chezy and D'arcy-Weisbach
equations are probably the most familiar and widely used
examples of such equations. The D'arcy-Weisbach equation

$$f = \frac{8gRS}{V^2} \qquad (1)$$

in which f = the D'arcy-Weisbach friction factor and g = the
acceleration due to gravity, is now generally recommended,
principally because it is dimensionally correct and
theoretically sound (Silberman, 1963). Any flow resistance
equation relies on the establishment of a friction factor
and it is the establishment of this factor which is of
prime concern.

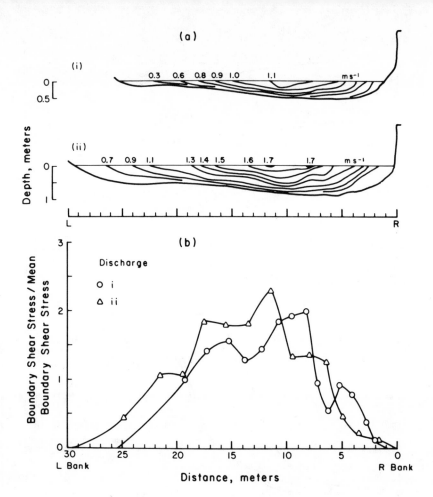

Figure 30.3b. Isovel pattern and boundary shear stress distribution at a riffle. Maes Mawr, River Severn, Wales. Discharges (i) 5.7, (ii) 17.6 m³ s⁻¹ (from Bathurst, 1979).

A theoretical equation defining the D'arcy-Weisbach friction factor was developed by Keulegan (1938) on the assumption that the log-velocity law applied and that flow was two dimensional and uniform. His analysis indicated that the resistance to flow in fixed bed channels was basically dependent on the relative roughness of the flow, namely the ratio of the hydraulic radius (R) to the grain diameter (D) of the uniform bed material. Experimental results confirmed this theoretical equation even though flow in polygonal channels is three dimensional and shear stresses are non-uniformly distributed. Subsequent research has shown that for graded gravel-bed material the effective grain diameter is 3.5 D_{84} (84% finer than given size, grid by number sampling of the surface material) and that cross sectional shape of the channel and differences

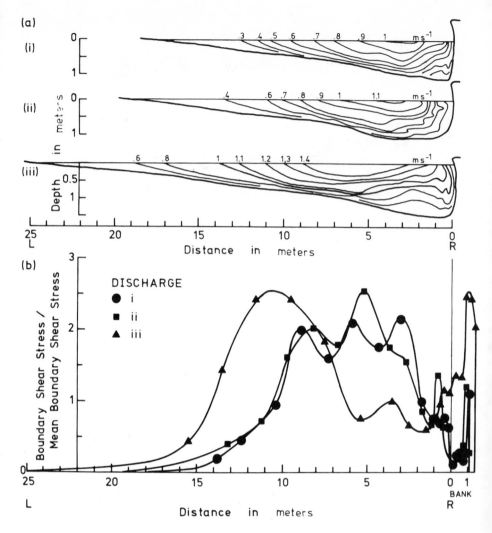

Figure 30.4. Isovel pattern and boundary shear stress
distribution at a bend. Maes Mawr Bend, River Severn,
Wales. Discharges (i) 5.2, (ii) 6.8, (iii) 15.5 m^3 s^{-1}.
(from Bathurst, Thorne and Hey, 1979).

in bed and bank roughness can also have a significant
effect on flow resistance (Hey, 1979a). Until more is
known about the roughness characteristics of vegetated
surfaces, the best practical equation for defining the
friction factor in channels where the relative roughness
($R/_{D_{84}}$) exceeds 3 is given by

$$\frac{1}{\sqrt{f}} = 2.03 \log \left[\frac{aR}{3.5D_{84}} \right] \tag{2}$$

496

where $a = f\left[R/d_m\right]$ and can be defined graphically (Hey, 1979a). An approximate solution for 'a' is given by $a = 11.1 \left(R/d_m\right)^{-0.314}$ (Bathurst, 1982). This equation has been successfully applied to riffle sections in gravel-bed rivers where flow is approximately uniform. The standard error of estimate of flow resistance was ±12.7%, and ±4.7% for estimating discharge. As expected, it is less accurate for pool sections, standard error of estimate of flow resistance ±153.7% and ±30% for discharge, due to the development of backwater ponding effects (Hey, 1979a).

Although the equation was developed for fixed bed channels, it appears to work satisfactorily under mobile bed conditions. Even if the full width of the channel is active the equation should still apply because the larger material, which has a dominant effect on flow resistance, has a similar frequency of occurrence in the subsurface and surface bed material.

When the relative roughness is less than 3, the equation no longer applies because large scale roughness becomes effective and the log velocity equation no longer accurately defines the flow phenomenon. For relative roughness less than 1.2, Bathurst, (1977, 1978) has developed a theoretically based equation which was calibrated with field data from the UK. This is defined by

$$\left(\frac{8}{f}\right)^{0.5} = \left(\frac{R}{0.365D_{84}}\right)^{2.34} \left(\frac{W}{d}\right)^{7(\mu - 0.08)} \tag{3}$$

$$\mu = 0.139 \log \left[\frac{1.91\ D_{84}}{R}\right]$$

where W = channel width and d = average flow depth. Data from the USA confirms the general form of this equation. At transitional relative roughnesses, between 1.2 and 3, a hybrid equation based on equations (2) and (3) appears to be appropriate.

Bed load transport

Experiments have been carried out at a number of riffle sites on the Rivers Wye, Severn and Dulas into the initiation of gravel bed material movement. Surface bed material was sampled at each site using a grid sampling procedure to obtain a representative sample of 100 stones. This material was painted and re-inserted in a line perpendicular to the channel banks. Observations were made of the critical conditions for movement of the painted stones and the associated water surface slopes and flow depths were monitored. After disregarding stone movements of less than 1 m as being due to re-inserted material finding a new niche in the bed, the results indicate that the critical Shields entrainment function for platey type bed material was approximately 0.03, significantly lower than the generally accepted level of 0.056 (Hey, 1980). Similar low values have been obtained by Neill (1968) and Church (1972). Church suggested that the critical entrainment function is dependent on the degree of imbrication of the bed material, although the size distribution of the bed material and the

497

shape of the individual particles also appears to influence
its value. At every site, the intermediate sized material
was the first to be transported. The finer material was
sheltered by the coarse fraction while the larger was too
heavy to be entrained.
    Observations have also been made on the operation of
transport processes through pool-riffle sequences. These
confirm the hypothesis that the pools scour at high flow
stages and fill at lower stages, and vice versa for riffles.

Bank erosion

Bank erosion is one of the mechanisms responsible for
changes in channel width. Laursen (1958) was the first to
recognise the need to develop a bank competence equation
to define the adjustment process. This will vary depending
on the structure of the banks. The banks of gravel-bed
rivers are often composite in form, with noncohesive sandy
gravel overlain by cohesive sandy silt/clay.
    Observations were made at 14 sites on the River Severn
in Wales to investigate the mechanisms controlling the
erosion of composite river banks. Erosion pins (Wolman,
1959) were installed in the upper and lower portions of
the banks to monitor rates of retreat by fluvial erosion
and these showed that the lower bank sandy gravel was easily
eroded by the flow, but that the cohesive sandy silt/clay
was quite resistant. Rates of erosion for the two materials
were in the range 350-600 mm year^{-1} for the gravel and less
than 28 mm year^{-1} for the silt/clay.
    This disparity in the rates of fluvial erosion led to
undercutting that generated cantilevers in the upper bank
(figure 30.5). Cantilevers were possible because of the

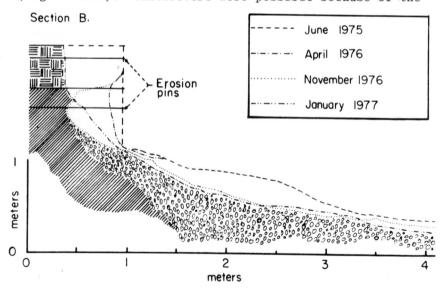

Figure 30.5. Generation of a cantilever by fluvial under-
    cutting of a composite bank. Morfodion, River Severn,
    Wales. (from Thorne and Lewin, 1979).

tensile strength of the cohesive soil of the upper bank and reinforcing by plant roots. Undercutting continued and cantilever width increased until a state of limiting stability was reached. Mechanical collapse then occurred by either shear, beam or tensile failure depending on cantilever geometry. The rate of upper bank retreat was necessarily matched to the rate of undercutting and was observed to be up to 600 mm year^{-1}.

Cantilever stability was analysed using static equilibrium theory taking into account the potential for tension and desiccation cracking. Dimensionless charts developed from the cantilever stability analysis can be used to predict the mode of failure and the limiting cantilever width to within about 10% (Thorne, 1978; Thorne and Tovey, 1981).

After failure, soil blocks came to rest on the lower bank or on the bed at the foot of the bank. Unless removed by the flow they tended to produce a cohesive piedmont that protected the lower bank gravel from fluvial erosion and reduced the rate of undercutting. Bank retreat rate was therefore controlled by fluvial scouring of bank erosion debris and failed material from the toe, even though the dominant failure mechanism of the upper bank (cantilever failure) was not fluvial in nature.

Bar deposition

Point bar and medial bar development influence the hydraulic characteristics of a river by adjusting its width, depth and plan shape. While considerable research has been carried out into the operation of depositional processes in sand-bed channels, as for example by Bridge and Jarvis (1976), relatively little is known about transport processes and bar development in gravel-bed rivers.

Scour and deposition were investigated at several sites on the River Severn by tracer experiments using bed material sampled from the channel and painted yellow (Thorne, 1978; Thorne and Lewin, 1979). At Maes Mawr Bend, five sets of 100 tracers were placed along sections around the bed (figure 30.6a). Their subsequent movement was recorded after a period of medium flows up to about half bankfull capacity and again after a period that included flows up to bankfull (figures 30.6b-f).

The results show that, under medium flows, scour was concentrated on the outer bank around the bend apex. Material entrained there moved laterally up the point bar as it was transported along the channel, so that tracers deposited at the inflexion point tended to accumulate at the channel centre. This deposition was usually only temporary, however, and material eroded from the outer bank was finally deposited on the point bar in the bend downstream, on the same side of the channel. This pattern of lateral movement was caused by the skew-induced secondary circulation described in the section on flow hydraulics. During higher flows scouring occurred all along the outer bank and around the channel centre line in the bed. Little scouring occurred at the inner bank. Entrained material was deposited at the inflexion point and at the point bar in the downstream bend.

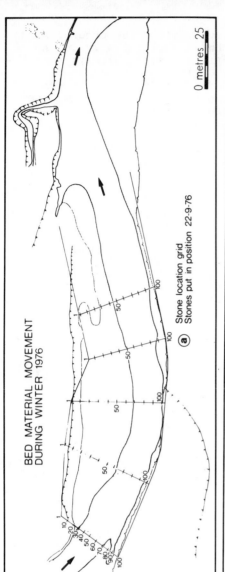

Figure 30.6. Movement of bed material tracers at Maes Mawr Bend, River Severn, Wales. (from Thorne and Lewin, 1979).

Figure 30.6a. Initial location of tracers.

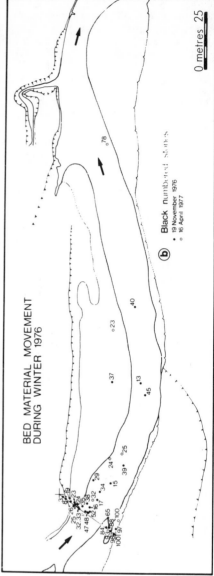

Figure 30.6b. Section 1 (Upstream Inflexion Point).

500

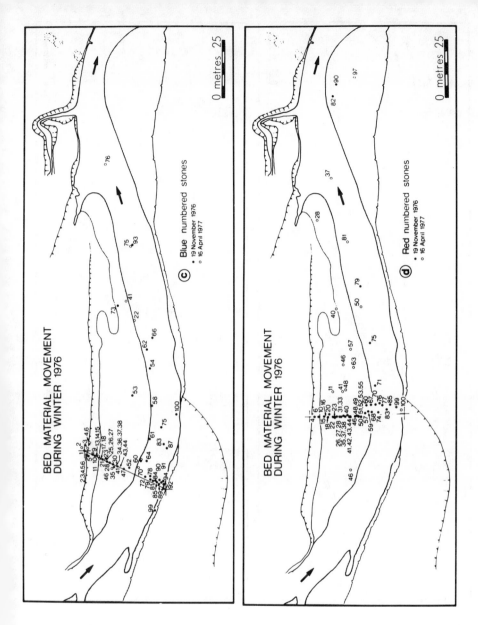

Figure 30.6c.
Section 2 (Bend
entrance).

Figure 30.6d.
Section 3 (Bend
apex).

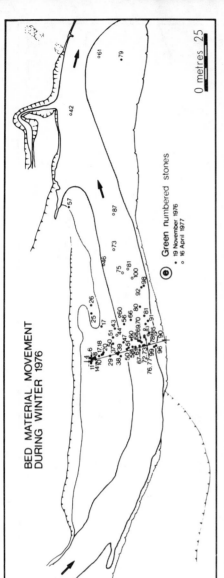

Figure 30.6e.
Section 4 (Bend exit).

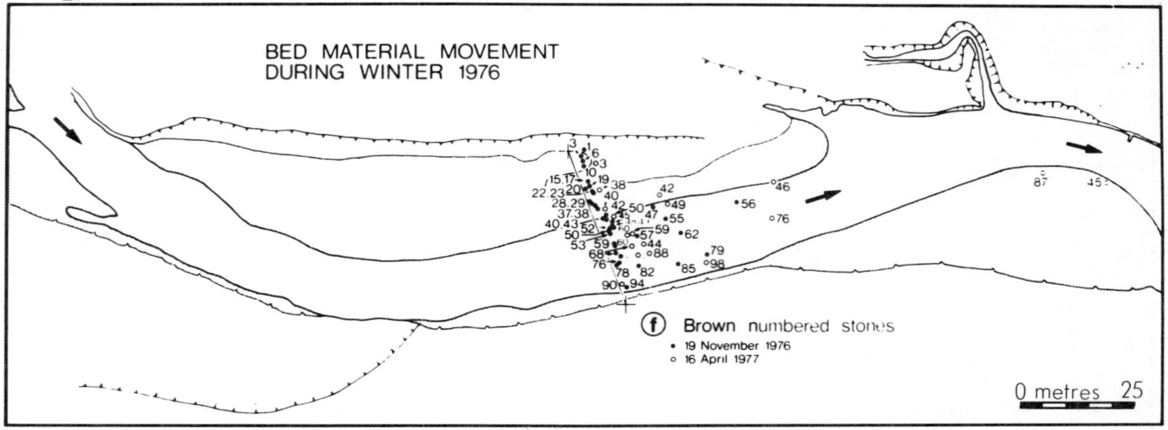

Figure 30.6f.
Section 5 (Down-stream Inflexion Point).

Rates of bank retreat around the bend reflected the differing frequencies of basal scour and undercutting. At the bend apex, where flows of only half bankfull caused basal scour, the retreat rate was 500 mm year^{-1}. At the inflexion points, where much larger flows of lower frequency were required to scour the bank base, the retreat rate was only 200 mm year^{-1}.

The distribution of scour and deposition at the bend may be explained by the flow hydraulics. For scour to occur at a point, the flow away from that point must be able to transport more sediment than the flow approaching the point. Consequently, scour results from a longstream increase in boundary shear stress. Similarly, deposition results from a longstream decrease in stress. At a bend there is a longstream increase in stress along the outer bank as the stress peak below the zone of downwelling develops and where the stress peak associated with the core of maximum velocity crosses the channel. This cross-over causes a longstream decrease in stress on the point bar at the inner bank. Deposition therefore occurs on the point bar since, although stresses and sediment transport rates are high, there is decreasing transport capacity. Scour occurs at the outer bank depending on the location of the cross-over and the strength of downwelling. At medium flows, when the effects of secondary flows are at their greatest, cross-over occurs at about the bend apex and so scour is concentrated there. At high flow, larger stress levels produce basal scour along the outer bank, but down-stream movement of the cross-over region results in the locus of basal scour moving downstream towards the bend exit. Deposition is still favoured on the point bar by the longstream decrease in transport capacity, since the locus of deposition also moving downstream with the cross-over region.

Meander mechanisms

A variety of explanations have been propounded to explain the initiation of meanders. To date no entirely adequate explanation is available, although there appears to be a consensus of opinion that pool-riffle and meander wave-lengths are related to a basic instability which results in local flow convergency and divergency (Yalin, 1971; Bagnold, 1980; Leopold, 1982). Until more is known about secondary flows and transport processes in natural channels, much of this must be speculative.

Nevertheless it is possible to define two relationships which enable the meander geometry to be uniquely defined. These relationships give some insight into possible meander mechanisms. The first equation is provided by the definition of channel sinuosity (p), the ratio of channel length to valley axis length, as this is also given by

$$p = S_v/S \qquad (4)$$

where S_v is the valley axis slope and S the channel slope. Provided valley slope is considered to be an independent variable, which is generally the case for engineering

Table 30.2. Relationships between meander wavelength (Λ) radius of curvature (r) and bankfull channel width (W) for various arc angles (θ)* (from Hey, 1976).

Arc Angle $\theta°$	Value of K in r = kw	Value of m in Λ = mr	Value of n in Λ = n W
270	1.33	2.83	3.77
225	1.60	3.70	5.91
180	2.00	4.00	8.00
150	2.40	3.86	9.27
135	2.67	3.70	9.85
120	3.00	3.46	10.39
90	4.00	2.83	11.31
60	6.00	2.00	12.00
45	8.00	1.53	12.25
30	12.00	1.04	12.42
15	24.00	0.52	12.53
10	36.00	0.35	12.55
5	72.00	0.17	12.56
1	360.00	0.04	12.57

*k = 360/θ;   m = 4 Sin $(\theta/2)$;   n = (1,440/θ) Sin $(\theta/2)$

design purposes, then any change in channel slope through erosion or deposition to a value defined by the bed load transport equation will result in a new plan form. Unfortunately there are an infinite number of plan shapes for a given sinuosity and to obtain a unique solution it is also necessary to define one other measure of meander plan form (Hey, 1976). Possible equations include those relating meander wavelength, amplitude or radius of curvature to channel width. However, it has been shown that these empirical equations are specific to a particular meander arc angle and are not generally applicable (table 30.2) (Hey, 1976).

The second general equation enables the length of the meander arc (Z), the thalweg distance measured between successive inflexion points or adjacent riffles in a straight channel, to be determined. This is defined by

$$Z = 2\pi W \tag{5}$$

where W is the bankfull channel width. Field data indicate that this applies irrespective of the size of the channel or its degree of sinuosity (Leopold, Wolman and Miller, 1964; Hey, 1976; Richards, 1976). Any change in channel width resulting from bank erosion or deposition will influence the riffle spacing and the meander pattern.

Secondary circulation probably accounts for this general
relationship as it has been shown, using spatial correlation
techniques to predict channel response to a localised increase in bank roughness, that the meander arc length is
defined by $(2\pi W)$ provided that two secondary flow cells are
present and that their average size is defined by half the
width of the channel (Hey, 1976). Measurement of secondary
flows in natural channels suggests that this is probably a
valid assumption.

MORPHOLOGY OF STABLE GRAVEL-BED RIVERS

Mathematical models

Ideally the simultaneous solution of the seven process
equations would determine the hydraulic geometry of stable
gravel-bed rivers (table 30.1). Current deficiencies with
bed load transport equations and incomplete understanding
of the factors controlling channel width hamper the development of this type of process-response model for gravel-bed
rivers.

Empirical models

Regime type analyses, using multiple regression techniques
produce general prediction equations for gravel-bed rivers.
To ensure that the equations have general application it
is essential that all the independent variables controlling
channel morphology (table 30.1) are used to derive the
equations and that data are obtained which maximises the
variability within and between the independent variables.
Ideally the independent variables should be uncorrelated.
    Many existing hydraulic geometry equations exclude
important independent variables. This immediately restricts
their application to a limited and often unspecified range
of channel types. For example, many equations are of the
form

$$V = 0.61 \ Q^{0.33} \qquad m \ s^{-1} \qquad (6)$$

$$W = 2.99 \ Q^{0.5} \qquad m \qquad (7)$$

$$d = 0.55 \ Q^{0.17} \qquad m \qquad (8)$$

where V = average bankfull velocity, W = bankfull channel
width, d = average bankfull depth and Q the bankfull discharge (Nixon, 1959). If no information is available on
the associated bed load transport rates, bed sediment size,
bank material type and valley slope it is impossible to
interpret the exponents and coefficients in these equations.
Unless the unmeasured independent variables are constant or
uncorrelated with discharge, then the exponent of discharge
will reflect the values of the unmeasured variables and not
simply the effect of discharge in isolation, on velocity,
width and depth. Even if the other unmeasured independent
variables are constant, their values will still influence
the coefficients in these equations.
    In an attempt to avoid these problems and to develop
general regime type equations for gravel-bed rivers data

505

obtained from 66 riffle sites on gravel-bed rivers in the
UK were subject to multiple regression analysis (Hey,
1982b). The dependent variables were wetted perimeter (P),
hydraulic radius (R), maximum flow depth (d_m), average
velocity (V), channel slope (S), sinuosity (p) and meander
arc length (Z) for bankfull stage and the independent variables bankfull discharge (Q), associated bed load transport
(Q_S), median grain size surface bed material (D_{50}), variability of surface bed material ($\sigma_D = D_{84}/D_{16}$) and the
percent of silt clay in the banks (C). After removal of
independent variables which made an insignificant statistical
contribution to the explanation of the dependent variables,
the following equations (Hey, 1982b) were obtained.

$$P = 2.20 \ Q^{0.54} \ Q_S^{-0.05} \qquad m \qquad\qquad r^2 = 95.20\% \qquad (9)$$

$$R = 0.161 \ Q^{0.41} \ D_{50}^{-0.15} \ m \qquad\qquad r^2 = 94.21\% \qquad (10)$$

$$d_m = 0.252 \ Q^{0.38} \ D_{50}^{-0.16} \ m \qquad\qquad r^2 = 87.45\% \qquad (11)$$

$$S = 0.68 \ Q^{-0.53} \ Q_S^{0.13} \ D_{50}^{0.97} \qquad r^2 = 95.67\% \qquad (12)$$

$$V = QPR \qquad\qquad\qquad m \ s^{-1} \qquad\qquad\qquad\qquad (13)$$

$$p = S_V/S \qquad\qquad\qquad\qquad\qquad\qquad\qquad\qquad\qquad (4)$$

$$Z = 2\pi W \qquad\qquad\qquad m \qquad (W = f \left[P, R, d_m\right]) \quad (5)$$

for a data range given by

Q   2.12 - 820 $m^3 \ s^{-1}$

Q_S 7.5 x 10^{-5} - 3.74 x 10^{-2} $m^3 \ s^{-1}$

D   0.021 - 0.190 m

σ_D 2.03 - 8.52

C   7.92 - 83.76%

S_V 0.000334 - 0.0215

As the independent variables were uncorrelated with each
other, the exponents reflect the physical control of the
independent variable on the dependent one. Bank material
type appears to have no significant effect on the hydraulic
geometry but this is mainly due to the restricted range of
bank types in the data base. All the banks were composed
of fine cohesive alluvium overlying coarse gravel. The
coefficients, however, will reflect the nature of the bank
type. Equally, bed load transport rates do not appear to
have a major influence on the hydraulic geometry, but this
probably reflects the limited range of bed load transport
at the sampled locations. Sites with a greater range of
transport rates would probably produce a greater variability
of channel dimensions.

Dominant discharge

Although rivers experience a range of flows, it has been
suggested that they adjust to a single dominant discharge
(Inglis, 1946; Nixon, 1959), this being defined as the
constant flow that develops the same gross shapes and
dimensions as the natural sequence of discharges. Flume

506

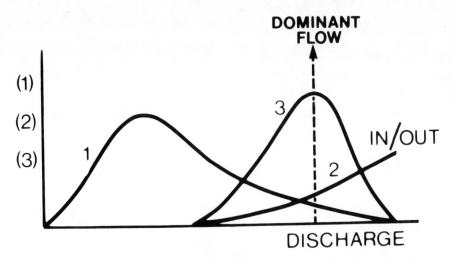

Figure 30.7. Flow transporting most sediment: hydraulic
    conditions limiting. Key: 1) frequency 2) instan-
    taneous sediment discharge 3) collective sediment
    discharge (from Hey, 1975).

and field experiments suggest that this is the flow at or
about the bankfull stage (Ackers and Charlton, 1970; Hey,
1972).

    This association between dominant discharge and bank-
full flow has been explained by studies of the magnitude
and frequency of sediment transport processes (Wolman and
Miller, 1955; Andrews, 1980) because the frequency of flow
which transports most sediment in the long term equates
with the frequency of bankfull flow (Wolman and Leopold,
1959). Therefore bankfull flow can be regarded as the
dominant regime discharge, because it is responsible for
the transport of the greatest volume of sediment through
the section. The magnitude and frequency of the flow
transporting most sediment is dependent on the flow regime
and the nature of the sediment transport processes (figure
30.7). For stable gravel-bed rivers in the UK, this is the
1.5 year flood (annual series). Consequently this flow
level can be used as the design discharge in the regime
equations and its associated bed load transport rate as the
design sediment discharge (Hey, 1975).

    For unstable sections, the frequency of bankfull flow
is modified, occurring less frequently in eroding sections
and more often in depositing ones. Bankfull discharge is
then determined by the flow doing most erosion or deposition (Hey, 1975). Stabilization of unstable sections is
achieved either by use of bankfull discharge and bed load
transport rates associated with similar neighbouring stable
sections in the regime equations or by the use of the flow
and input sediment load associated with the appropriate
flood return period for a stable channel at that site.

507

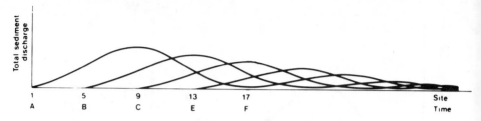

a) theoretical model of spatial and temporal changes
   (from Hey, 1979b).

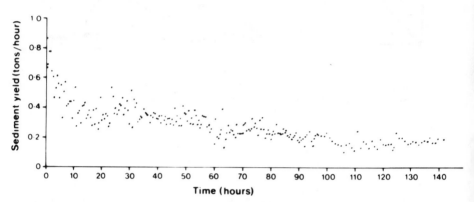

b) sediment yield from experimental model given
   rejuvenation at time 0 (from Schumm, 1976).

Figure 30.8. Variations in sediment discharge due to
   erosion and deposition.

DYNAMIC MODELLING OF CHANNEL MORPHOLOGY

Conceptual models.

Natural channels are rarely stable over long time periods
(> 100 years) due to the interaction of erosional and
depositional activity in time and space. A simple qualitative model, based on the operation of sediment transport
processes and upstream and downstream feedback mechanisms,
has been developed from consideration of spatial and
temporal changes in bankfull values of channel slope, flow
depth, bed material size, bed shear stress and sediment
transport rates during periods of erosion and deposition
(Hey, 1979b). The model predicts a damped oscillation
between erosional and depositional activity in time and
space as the river responds to an initial instability resulting from changes in either climate, sediment yield,
runoff, land or sea levels (figure 30.8a). The initial
cut and fill phases are responsible for valley incision and
flood plain development, while secondary and subsequent
activity can produce river terraces. Morphological changes

508

and variations in sediment yield predicted by the model
compare favourably with laboratory (Schumm, 1976)(figure
30.8b) and field observations (Born and Ritter, 1970;
Small, 1973).

Mathematical models.

Potentially, mathematical modelling techniques offer the
best solution to the problem of predicting channel response
to changes in discharge and sediment load. These models
use the governing flow equations (table 30.1) and feedback
mechanisms to route water and sediment through a series of
channel reaches. It is the feedback mechanisms which effectively control the dynamic adjustment of the river system
and these operate upstream, through drawdown and backwater
effects and downstream through sediment supply. Operation
of the model is illustrated in figure 30.9.
    Simple one dimensional models have been developed to
predict changes in flow depth, slope and velocity using the
continuity, flow resistance and sediment transport equations
(US Army Corps of Engineers, 1977). However, it is not
currently possible to model width and plan shape changes
due to lack of appropriate process equations. Even the one
dimensional models can be inaccurate when applied to gravel-
bed rivers due to deficiencies in the flow resistance and
bed load transport equations used in the models.

Empirical models

Empirical models of channel development require information
on morphological and sedimentological changes in space and
time. This information can be obtained by making detailed
measurements of flow and channel processes over a short
period and calibrating the measured rates of change against
long term observational data from historical maps and
aerial photographs. In this way it is possible to identify
the processes and formative discharge responsible for
particular types and rates of channel change. For example
Thorne and Lewin (1979) investigated short and long term
channel development at Maes Mawr on the River Severn in
Wales. Short term process measurements (reported previously
in the sections on flow hydraulics, bank erosion and bar
sedimentation) were successful in explaining the mechanics
of the sediment transfer system at a bend under flows up
to bankfull. Rates of bank recession measured in the
field (0.5 m year^{-1}) and observed historically (0.7 m year^{-1})
for that bend showed good agreement. However, the
historical record showed that, in addition to the ordered
and progressive meander development associated with flows
up to bankfull discharge, there were other types of rapid
channel change, associated with overbank events of high
magnitude but low frequency. The effects of these high
magnitude events were in some cases long lasting. Not only
did they influence subsequent progressive channel development under flows up to bankfull, but also they produced
particular fluvial landforms that would not have been
formed by in-bank flows. However, it is extremely difficult
to record the mechanics involved with any degree of accuracy.

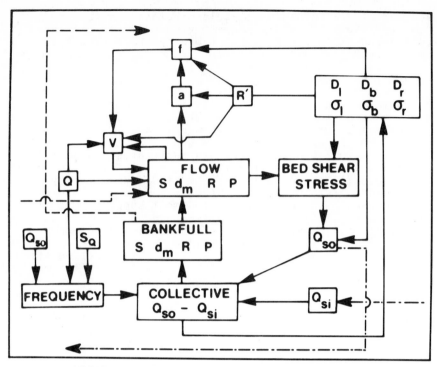

– – – UPSTREAM FEEDBACK   –·–·– DOWNSTREAM FEEDBACK

Figure 30.9.   Dynamic process-response model based on
    operation of continuity, flow resistance and sediment
    transport equations and upstream and downstream feed-
    back mechanisms. (after Hey, 1979b). (Q discharge,
    V average velocity, R hydraulic radius, P wetted
    perimeter, d_m maximum flow depth, S channel slope,
    f flow resistance, a hydraulic shape factor,
    R' effective hydraulic radius, D_b, D_r, D_l mean grain
    size bed, right and left bank sediment, σ_b, σ_r, σ_l
    standard deviation bed, right and left bank sediment,
    Q_{50} median flow, S_Q flow variability, Q_{so} sediment
    output, Q_{si} sediment input).

It is necessary, therefore, in developing empirical
models of channel development to use both field process
measurements and documentary historical analyses to develop
a good understanding of the way in which channel form
relates to channel processes.

ACKNOWLEDGEMENTS

Most of the research outlined in this paper was carried out
with the financial support of the Natural Environment
Research Council, UK. The authors also gratefully acknowledge the following publishers for permission to reproduce
figures:

Nature (figures 30.1 and 30.2, table 30.2), American
Society of Civil Engineers (figure 30.4), Kendall/Hunt
(figures 30.3, 30.5 and 30.6), Earth Surface Processes
(figures 30.8a and 30.9), Allen and Unwin (figure 30.8b),
Wiley (table 30.1).

REFERENCES

Ackers, P. & Charlton, F.G., 1970, Dimensional analysis of
    alluvial channels with special reference to meander
    length, *Journal of Hydraulic Research*, 8, 287-315.

Andrews, E.D., 1980, Effective and bankfull discharge of
    streams in the Yampa basin, Colorado & Wyoming,
    Journal of Hydrology, 46, 311-330.

Apmann, R.P., 1972, Flow processes in open channel bends,
    Journal of the Hydraulics Division, ASCE, 98,
    795-810.

Bagnold, R.A., 1980, An empirical correlation of bedload
    transport rates in flumes and natural rivers,
    Proceedings of the Royal Society of London, 372A,
    453-473.

Bathurst, J.C., 1977, Resistance to flow in rivers with
    stony beds, Unpublished PhD thesis, University of
    East Anglia, Norwich, UK.

Bathurst, J.C., 1978, Flow resistance of large scale rough-
    ness, *Journal of the Hydraulics Division, ASCE*,
    104, 1587-1603.

Bathurst, J.C., 1979, Distribution of boundary shear stress
    in rivers, in: *Adjustments of the fluvial system*,
    ed. Rhodes, D.D. and Williams, G.P., (Kendall/Hunt
    Publishing Co., Dubuque, Iowa, USA), 95-116.

Bathurst, J.C., 1982, Theoretical aspects of flow
    resistance, in: *Gravel-Bed Rivers*, ed. Hey, R.D.,
    Bathurst, J.C. and Thorne, C.R., (Wiley, London),
    chapter 5.

Bathurst, J.C., Thorne, C.R. and Hey, R.D., 1977, Direct
    Measurements of Secondary Currents in River Bends,
    Nature, 269, 504-506.

Bathurst, J.C., Thorne, C.R. and Hey, R.D., 1979, Secondary
    flow and shear stress at river bends, *Journal of
    the Hydraulics Division, ASCE*, 105, 1277-1295.

Bhowmik, N.G. and Stall, J.D., 1978, Hydraulics of flow in
    the Kaskaskia River, in: *Proceedings of the
    Hydraulics Division Speciality Conference on Veri-
    fication of Mathematical and Physical Models in
    Hydraulic Engineering, ASCE*, 79-86.

Born, S.M. and Ritter, D.F., 1970, Modern terrace develop-
    ment near Pyramid Lake, Nevada, and its geologic
    implications, *Bulletin of the Geological Society
    of America*, 81, 1233-1242.

Bridge, J.S. and Jarvis, J., 1976, Flow and sedimentary processes in the meandering River South Esk, Glen Cova, Scotland, *Earth Surface Processes*, 1, 303-336.

Chacinski, T.M., 1954, Patterns of motion in open-channel bends, *International Association of Scientific Hydrology Publication* no. 38, 311-318.

Chacinski, T.M. and Francis, J.R.D., 1952, Discussion of 'On the origin of river meanders' by P.W. Werner, *Transactions of the American Geophysical Union*, 33, 771-773.

Church, M., 1972, Baffin Island Sandurs: a study of Arctic fluvial processes, *Bulletin of the Geological Survey of Canada*, no.216.

Dietrich, W.E., Smith, J.D. and Dunne, T., 1979, Flow and sediment transport in a sand bedded meander, *Journal of Geology*, 87, 305-315.

Einstein, H.A. and Harder, J.A., 1954, Velocity distribution and the boundary layer at channel beds, *Transactions of the American Geophysical Union*, 35, 114-120.

Hawthorne, W.R., 1951, Secondary circulation in fluid flow, *Proceedings of the Royal Society of London*, no.A1086, 374-387.

Hey, R.D., 1972, Analysis of some of the factors influencing the hydraulic geometry of alluvial channels, Unpublished PhD, University of Cambridge, UK.

Hey, R.D., 1974, Prediction and effect of flooding in alluvial systems, in: *Prediction of Geological Hazards*, ed. Funnell, B.M., (Geological Society of London, Misc. Paper no.3), 42-56.

Hey, R.D., 1975, Design discharge for natural channels, in: *Science, Technology and Environmental Management*, ed. Hey, R.D. and Davies, T.D., (Saxon House), 73-88.

Hey, R.D., 1976, Geometry of river meanders, *Nature*, 262, 482-484.

Hey, R.D., 1978, Determinate hydraulic geometry of river channels, *Journal of the Hydraulics Division, ASCE*, 104, 869-885.

Hey, R.D., 1979a, Flow resistance in gravel-bed rivers, *Journal of the Hydraulics Division, ASCE*, 105, 365-379.

Hey, R.D., 1979b, Dynamic process-response model of river channel development, *Earth Surface Processes*, 4, 59-72.

Hey, R.D., 1980, Final Report on channel stability, Craig Goch Joint Committee.

Hey, R.D., 1982a, Gravel-bed rivers: form and processes, in: *Gravel-Bed Rivers*, ed. Hey, R.D., Bathurst, J.C. and Thorne, C.R., (Wiley, London), Chapter 1.

512

Hey, R.D., 1982b, Design equations for mobile gravel-bed rivers, in: *Gravel-Bed Rivers*, ed. Hey, R.D., Bathurst, J.C. and Thorne, C.R., (Wiley, London), Chapter 20.

Hey, R.D. and Thorne, C.R., 1975, Secondary flows in river channels, *Area*, 7, 191-196.

Hooke, R. Le B., 1975, Shear stress and sediment distribution in a meander bend, *UNGI Rapport 30*, University of Uppsala, Department of Physical Geography, Uppsala, Sweden.

Inglis, Sir C., 1946, Meanders and their bearing on river training, *Maritime Paper No. 7*, Institution of Civil Engineers, London.

Ippen, A.T. & Drinker, P.A., 1962, Boundary shear stress in curved trapezoidal channels, *Journal of the Hydraulics Division, ASCE*, 88, 143-179.

Keulegan, G.H., 1938, Laws of turbulent flow in open channels, *Journal of Research of the National Bureau of Standards*, 21, Research Paper 1151, 707-741.

Laursen, E.M., 1958, Sediment transport mechanics in stable channel design, *Transactions of the American Society of Civil Engineers*, 123, 157-174.

Leopold, L.B., 1982, Water surface topography in river channels and some implications for meander development, in: *Gravel-Bed Rivers*, ed. Hey, R.D., Bathurst, J.C. and Thorne, C.R. (Wiley, London), Chapter 13.

Leopold, L.B. and Maddock, T., Jr., 1953, The hydraulic geometry of stream channels and some physiographic implications, *United States Geological Survey, Professional Paper 252*.

Leopold, L.B., Wolman, M.G. and Miller, J.P., 1964, *Fluvial processes in geomorphology*, (Freeman, San Francisco).

Neill, C.R., 1968, A re-examination of the beginning of movement for coarse granular bed materials, *Report No. INT 168, Hydraulics Research Station*, Wallingford, UK.

Nixon, M., 1959, A study of the bankfull discharge of rivers in England and Wales, *Proceedings Institution of Civil Engineers*, 12, 151-174.

Perkins, H.J., 1970, The formation of streamwise vorticity in turbulent flow, *Journal of Fluid Mechanics*, 44, 721-740.

Richards, K., 1976, The morphology of riffle-pool sequences, *Earth Surface Processes*, 1, 71-88.

Rozovskii, I.L., 1957, Flow of water in bends of open channels, *Academy of Sciences of Ukrainian S.S.R.* Kiev, USSR. (Translated by Y. Prushansky, Israel programme for scientific translations, S. Monson, Jerusalem, Israel, 1961, P.S.T. Cat. no.363).

Schumm, S.A., 1976, Episodic erosion: a modification of the geomorphic cycle, in: *Theories of Landform Development*, ed. Melhorn, W.N. and Flerual, R.C., (SUNY, Binghampton, New York), 69-85.

Silberman, E., 1963, Friction factors in open channels, *Journal of the Hydraulics Division, ASCE*, 89, 97-143.

Small, R.J., 1973, Braiding terraces in the Val D'Herens, Switzerland, *Geography*, 58, 129-135.

Thorne, C.R., 1978, Processes of bank erosion in river channels, Unpublished PhD Thesis, University of East Anglia, Norwich, UK.

Thorne, C.R. and Hey, R.D., 1979, Direct measurements of secondary currents at a river inflexion point, *Nature*, 280, 226-228.

Thorne, C.R. and Lewin, J., 1979, Bank erosion, bed material movement and planform development in a meandering river, in: *Adjustments of the fluvial system*, ed. Rhodes, D.D. and Williams, G.P., (Kendall Hunt Publishing Co., Dubuque, Iowa, USA), 117-137.

Thorne, C.R. and Tovey, N.K., 1981, Stability of composite river banks, *Earth Surface Processes and Landforms*, 6, 469-484.

United States Army Corps of Engineers, 1977, *HEC-6, Scour and deposition in rivers and reservoirs*, (Hydrologic Engineering Center, Davis, California).

Vanoni, V.B., (Chairman of Task Committee on Preparation of Sedimentation Manual), 1971, Sediment Transportation Mechanics; Fundamentals of Sediment Transportation, *Journal of the Hydraulics Division, ASCE*, 97, 1979-2022.

Wolman, M.G., 1955, The natural channel of Brandywine Creek, Pennsylvania, *United States Geological Survey, Professional Paper 271*.

Wolman, M.G., 1959, Factors influencing erosion on a cohesive river bank, *American Journal of Science*, 257, 204-216.

Wolman, M.G. and Leopold, L.B., 1959, River flood plains: some observations on their formation, *United States Geological Survey, Professional Paper 282-C*.

Wolman, M.G. and Miller, J.P., 1955, Magnitude and frequency of forces in geomorphic processes, *Journal of Geology*, 68, 54-74.

31 The influence of vegetation on

stream channel processes

A.M. Gurnell and K.J. Gregory

INTRODUCTION

Although a considerable amount of research has been devoted
to the influence of vegetation on hydrology and, more
recently, on solute production, there has been less attention accorded to ways in which vegetation can directly
affect, and also can act as an indicator of, fluvial processes. Vegetation pattern and composition are controlled
by a broad spectrum of factors including macro- and micro-
climate, supply of nutrients, interspecific competition,
management practices, amount and variability of soil
moisture and the erosive and depositional effects of
fluvial processes. These factors all have different
effects in relation to the germination and growth of different species and they provide a complex system of controls
upon vegetation at a specific site. This paper explores
the significance of vegetation influence in relation to
stream channel processes in an experimental area which
includes both heathland and woodland, and then proceeds to
focus particularly upon heathland vegetation and vegetation
associated with river channels.

The effects of factors influencing the composition of
heathland vegetation have been reviewed by Gimingham (1972)
who showed the great importance of the water regime in
heathland areas. The water relations of heathland vegetation have been the subject of much research. Rutter
(1955), for example, demonstrated the effects of water
table depth and fluctuation on the percentage of the total
biomass of foliage occupied by different heathland species
and Loach (1966) investigated the combined effects on
vegetation of water table depth and fluctuation and of the
nutrient content of the soil. In addition many researchers
have studied the response of particular heathland species
to variations in water regime, sometimes including the
impact of the solute content of soil moisture (eg. Bannister
1964a, b, c; Jones and Etherington, 1970; Jones, 1971;
Daniels, 1975; Rahman, 1976; and Rahman and Rutter, 1980).

In addition to the effect of soil moisture on heathland
vegetation there is also evidence to suggest that heathland

vegetation can exert a variable control on the supply of
moisture to the soil. Different species can have varying
impacts on interception, (Aranda and Coutts, 1965; Leyton
et al., 1967) throughfall and soil moisture content. Near
the study area four stands of Calluna vulgaris of different
ages (6, 10, 14, 18 years) have each shown significant
differences in rates of interception, throughfall, transpiration and associated soil moisture content (Hughes, personal
communication).

Much previous research on the water relations of heathland vegetation has been carried out either in controlled
laboratory conditions or on relatively small field plots,
but this paper proposes that such relationships are not
only useful when studying fluvial processes over larger
areas but can actually be used to estimate catchment
behaviour.

Studies of river channels have not given the attention
to vegetation that has been accorded to the other factors
controlling channel morphology. In their study of forested
areas in the Appalachians, Hack and Goodlett (1960) showed
that vegetation distribution can be largely explained in
terms of different moisture regimes and they also identified
the significance of vegetation in valley floor development
during a major flood event. In a classic paper on the
Sleepers River Basin in Vermont, Zimmerman, Goodlett and
Comer (1967) analysed the influences of vegetation upon
channel morphology and showed how channel width was substantially controlled by the incidence of vegetation at
drainage areas less than 2 km^2, that width increased downstream for drainage areas between 2 and 12 km^2 with variations due to the incidence of vegetation, and that the
vegetation influence was less significant for larger
drainage areas. In a study of gravel rivers in Britain it
was shown (Charlton, Brown and Benson, 1978) that channels
with grassy banks are 30% wider and tree-lined channels are
30% narrower than average. In south west USA, riparian
vegetation has been shown (Graf, 1980) to have a significant
influence upon channel morphology and an understanding of
this influence is an essential for good management.
Vegetation influence on channel processes has been scruti-
nised in at least two ways. First, there have been some
studies showing the general influence upon channel processes (eg. Rachocki, 1978) and most recently there have
been valuable studies investigating the manner in which
large organic debris influences channel morphology. Thus,
with reference to contrasted low and high gradient streams,
Keller and Swanson (1979) showed how debris accumulations
have effects upon channel morphology including bank stability, and development of the channel pattern. In the
coastal Redwood environment of California, Keller and Tally
(1979) showed that 60% of the fall along the long profile
of a stream in a 4.9 km^2 basin was the result of organic
debris stepping the long profile and that, in a 19.8 km
area, at least 50% of the pools were influenced by the
incidence of large organic debris. In this area the residence time of the vegetation debris could often be more
than 100 years and as much as 220 years and there was
comparatively little evidence of channel change.

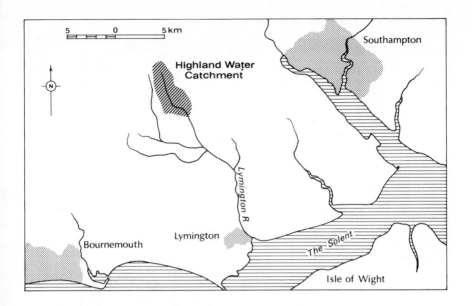

Figure 31.1.   Location of the study area.

    Secondly, there have been attempts to quantify the
influence of biomass upon channel morphology and process.
Thus in the semi arid environment of north western Colorado,
Begin and Schumm (1979) used relations between drainage
area, discharge and flow depth to develop a shear stress
indicator which could be used to identify the channel
reaches most likely to fail by gullying. A similar approach
had been employed by Graf (1979a, 1979b) who developed an
expression for biomass on the valley floors which could be
related to tractive force in a critical function which gave
a basis for considering the thresholds beyond which arroyos
and gullies may develop. Because of the way in which
vegetation management is often used as the key to river
channel control, it is desirable to know the exact ways in
which channel morphology is influenced by vegetation. Such
influence is not spatially uniform and Mosley (1981) noted
that the effects of different vegetation and root network
characteristics along stream banks probably account for a
large amount of the unexplained variability in his data
set of river channels from South Island, New Zealand. In
the New Forest channels described in this paper, in addition to the influence of vegetation upon channel morphology,
there are indicators that channel adjustment is taking
place and this may be inextricably related to the controlling influence of the riparian vegetation.
    The significance of vegetation with respect to fluvial
processes has been investigated in the basin of the Highland
Water (11.4 km^2) in the New Forest, Hampshire (figure 31.1).
The catchment is underlain by Barton sand and Barton clay,
both of Eocene age and almost horizontally bedded, with a
Pleistocene gravel capping on the interfluves (figure 31.2B).

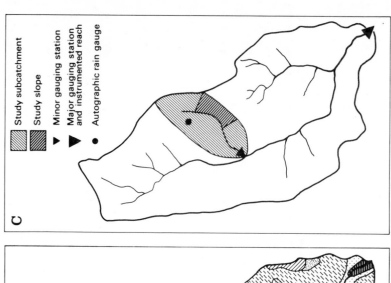

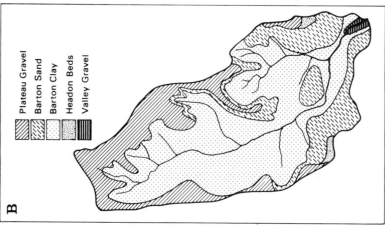

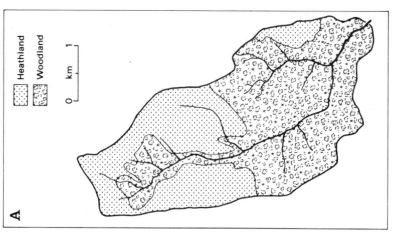

Figure 31.2.   The Highland Water catchment.   A. Vegetation.   B. Geology.   C. Research design.

The vegetation (figure 31.2A) can be broadly subdivided
into heathland, occupying most of the northern half of the
catchment, and mixed woodland which is located in the
southern half of the catchment. The significance of vegetation has been investigated at three scales (figure 31.2C);
firstly, in relation to drainage network dynamics and discharge within a subcatchment (1.28 km^2); secondly, in
relation to soil moisture variations and soil permeability
on a single hillslope; and, finally, in relation to the
components of the main drainage network, with particular
emphasis on channel form and change within the Highland
Water itself.

THE SUBCATCHMENT SCALE

A subcatchment of the Highland water has been the subject
of detailed fieldwork since 1974. Gurnell (1978) investigated the dynamics of the drainage network and their relationship with rock type and vegetation composition.
Variations in drainage network extent and areas of standing
water were monitored weekly from March 1975 to February
1976. A large scale base map of the catchment was constructed using aerial photographs and this was used to map
the extent of the active drainage network, where water
could be seen to be moving, and areas where water was
standing on the surface. At the same time discharge was
continuously monitored at a gauging station at the catchment outlet and an autographic raingauge was used to record
precipitation.

Temporal variations in drainage density have now been
reported in several studies, as reviewed by Gardiner and
Gregory (1981). In the study catchment such temporal
fluctuations ranged from densities of 0.4 to 17.1 km km^{-2}
during the period of observation (figure 31.3A). The
relationship between drainage density and discharge (figure
31.3A) shows the importance of the extent of the drainage
network in streamflow generation, and figure 31.3B demonstrates that the area of standing water, indicating soil
saturation, is also significant. By using the weekly maps
of the drainage network and the graphs relating drainage
density and area of standing water to discharge, it was
possible to delimit areas which were likely to contribute
significantly to discharge at particular discharge levels
(figure 31.4B). The pattern which emerged is quite complex
and does not conform to any simple concept of an expanding
contributing area although the areas nearest to the stream
network are obviously important at most discharge levels
since this was the major factor used in producing the map.
However, there does seem to be correspondence between areas
of wet heathland vegetation, underlain by Barton clay
(figure 31.4A), and those parts of the catchment which
form a part of the estimated surface 'contributing area'
for observed discharges under a wide range of flow conditions. At the scale of a small catchment it was evident
that the drainage network and contributing area were very
variable and that the area of the catchment covered by wet
heathland and underlain by Barton clay seemed to be the

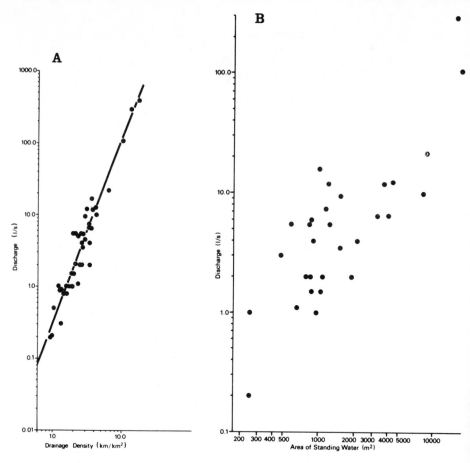

Figure 31.3. A. The relationship between drainage density and discharge. B. The relationship between area of standing water and discharge.

most significant control upon runoff from the catchment. In fact the boundary between the wet and dry heathland very closely follows the boundary between the Barton clay and the overlying Barton sand and plateau gravels. At the junction between the two types of vegetation, water which has seeped through the more permeable sands and gravels appears on the surface in many places and this seepage has morphological expression in the form of a seepage step (Tuckfield, 1973) produced by water sapping. Many of the drainage lines terminated abruptly at this seepage step during wet conditions and so the relationship between vegetation and the drainage network at this scale closely reflects rock type. However further more detailed analysis showed that within the wet heathland area of the catchment there were marked differences in drainage network dynamics, and that on an individual hillslope these variations could be associated closely with vegetation composition.

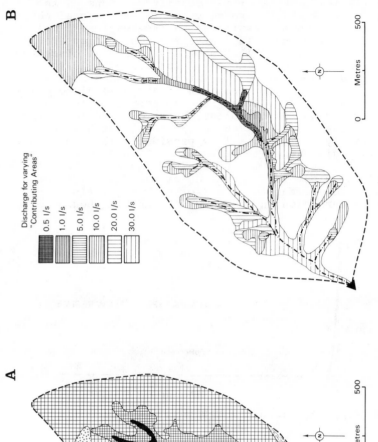

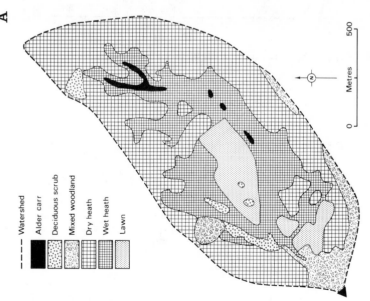

Figure 31.4. A. Vegetation pattern.
          B. Estimated surface 'contributing area' in relation to discharge level.

521

A section of the north-west facing part of the catchment
was studied in greater detail with emphasis upon vegetation
characteristics. In order to identify sections of the
drainage network which were very sensitive to local characteristics, a network classification was developed employing
the catchment flow duration curve for hourly flows for 1975
to 1976 (figure 31.5). The changes of slope at discharges
of approximately 0.9 to 1.0 l s^{-1} and 10 to 15 l s^{-1} were
used to subdivide flows and related mapped drainage networks into perennial, intermittent and ephemeral networks
(Gregory and Walling, 1973). The ephemeral regime was
further subdivided into two parts according to the point
at which the slope of the curve apparently stabilised
(approximately 30 l s^{-1}). This provided a four fold subdivision of discharge, termed regimes 1 to 4 (see figure
31.5). The previously mapped drainage networks that existed
on the study slope for discharges of over 10 l s^{-1} were
then analysed, because such networks, with little or no
permanent expression in the form of channels should be very

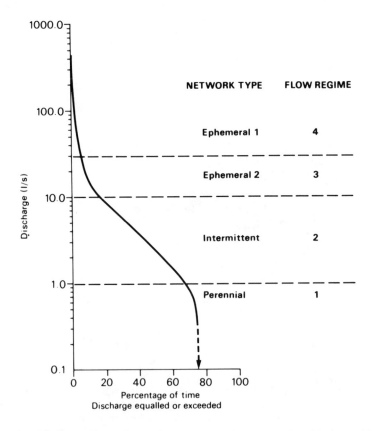

Figure 31.5. Flow duration curve for hourly flows, 1975 to
1976, and flow classification.

Table 31.1. Percentage cover of commonly occurring species on plots A to D

Plot	A	B	C	D
Species:				
Calluna vulgaris	1	16	50	34
Erica tetralix	33	38	23	34
Myrica gale	5	11	3	3
Pteridium aquilinum		1	6	
Molinia caerulea	66	66	69	69
Agrostis spp.		3	8	
Trichophorum cespitosum	4	2	1	1
Narthecium ossifragum	13	12	1	
Juncus spp.	18	11	3	5
Sphagnum spp.	22	10		14

sensitive to local, rather than to catchment wide, characteristics.

The study slope had an almost continuous cover of dry heathland, dominated by *Calluna vulgaris* and *Molinia caerulea*, on the gravel capping above the seepage step. Below the seepage step the slope had a constant angle of 6 degrees, was underlain by Barton clay and had a cover of wet·heathland vegetation. However, this vegetation varied greatly and four areas of markedly different vegetation composition were identified and their characteristics quantified using a pin frame, to estimate percentage cover of the different species. Mean percentage cover of each species was calculated for each of the four plots (table 31.1) by sampling at 15 m intervals along four equally spaced downslope transects. The plots ranged from humid to wet heaths (Gimingham, 1972) and it was possible to qualitatively differentiate between soil moisture conditions on the different plots. Although many species are tolerant over a wide range of soil moisture conditions, *Calluna vulgaris* tends to occur on less moist sites than *Erica tetralix* and *Sphagnum* and *Juncus* species are indicative of even wetter conditions. Thus, the vegetation surveys suggested that plot A had the highest soil moisture content, plot C the lowest soil moisture content and plots B and D were intermediate. Using the weekly drainage network maps, drainage densities were calculated for each plot for all sets of observations with an active perennial network (Discharge > 10 l s^{-1}; regime 3 or 4). The results are very different for the four plots in spite of their topographic similarity (figure 31.6). The lowest densities occurred on the wettest plot, as indicated by vegetation composition; the highest densities occurred on the driest

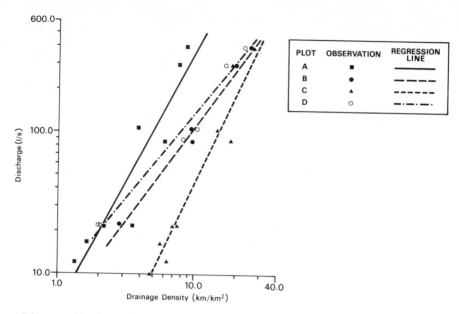

Figure 31.6. The relationship between catchment discharge
and drainage density within plots covered by vegetation
of different composition.

plot and the intermediate plots had densities which covered
the complete range between plots A and C.

Further detail on the significance of the wet heathland
vegetation composition was subsequently gathered using 61
piezometers arranged in downslope transects within plots
A, B and C. The piezometers were installed to monitor
pressure head in a band 55 to 65 cm below the ground surface
and readings of pressure head were taken manually over the
entire piezometer network on 74 occasions during 1978. The
discharge at the gauging station was used to assign the
data for each set of pressure head readings to one of the
four flow regimes already identified, so that average
pressure head could be calculated for each of the four
regimes at every piezometer site.

The detailed sampling programme, data analysis and
discussion of the relationships between these mean pressure
head values and the overlying vegetation have been reported
elsewhere (Gurnell 1981), but in general a complex pattern
emerged with no simple downslope trend in pressure head.
Simple regression relationships calculated between mean
pressure head for each regime and both the vertical
distance above the base of the slope and the vertical
distance below the seepage step showed little upslope trend.
The highest coefficient of determination was 0.262. However, once vegetation related variables were introduced
into the analysis, the coefficient of determination increased dramatically, confirming the correspondence between
vegetation pattern and variation in pressure head.

524

It is interesting to note the lack of correspondence
between pressure head and position on the slope, since we
have already noted the same characteristic with respect to
variations in drainage density along the slope in the wet
heathland zone. With an impermeable underlying rock type,
water must move over the surface or through the soil to
roughly balance the difference between input at the seepage
step and output at the base of the slope. However, observed
variations in pressure head were so abrupt that it is
difficult to explain them in terms of a simple transfer of
water between soil and surface and so it seemed very likely
that significant local variations in soil permeability
existed on the slope. The saturated hydraulic conductivity
of the material surrounding each piezometer was estimated
using a technique proposed by Luthin and Kirkham (1949).
This involved pumping water out of the piezometers and
observing the rate of recovery of the water.

Table 31.2. The relationship between saturated
hydraulic conductivity and pressure head

Number of piezometers		12	18	11	9
Saturated hydraulic conductivity (k) (m day^{-1})		k<0.2	0.2≤k<0.5	0.5≤k<1.2	k≥1.2
Mean pressure head (cm from ground surface)	regime 1	-14.4	-26.8	-36.6	-41.4
	regime 2	-7.1	-18.3	-21.8	-29.4
	regime 3	-2.7	-7.7	-5.5	-10.0
	regime 4	-2.7	-4.1	-4.1	-4.8

The results showed a correspondence between pressure
head and permeability (table 31.2) and thus indicated that
the composition of wet heathland on the study slope could
be used as a surrogate for permeability and as a very
logical group of independent variables for inclusion in
regression analyses of soil moisture variables.
Having established the importance of vegetation composition as an indicator of fluvial processes on hillslopes
it is now necessary to consider how we might most simply
collect the vegetation data to extrapolate the results of
the regression analyses over a larger area. Detailed
measurements using a pin frame provide the most unbiased
estimates of vegetation cover but the method is very time-
consuming and a quicker way of gaining this information is
to use a visual quadrat technique. In order to minimise
field work such a quadrat sampling technique could just be
employed as ground control for aerial photograph analysis
and this was the method used to produce the map shown in
figure 31.7A using a ground control of 95 quadrat sites.
One verification of the map's accuracy is to compare it
with the relevant section of the surface contributing area
map which is reproduced in figure 31.7B. The correspondence

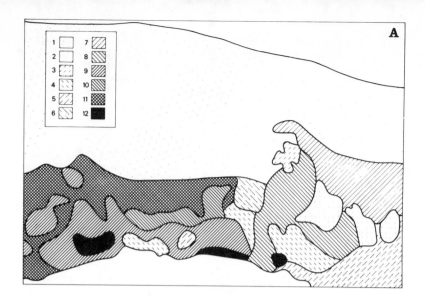

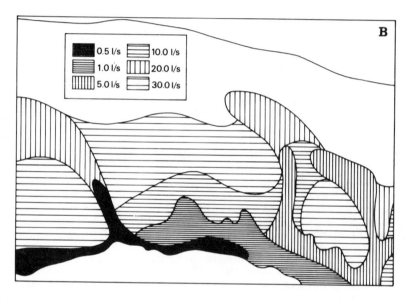

Figure 31.7. A. Vegetation map of the study slope.
Categories 1 to 12 reflect the increasingly wet soil
conditions. For vegetation composition see table 31.3.
B. Estimated surface 'contributing area' on the study
slope in relation to catchment discharge level.

between estimated surface contributing area and vegetation
composition is striking and encouraging, even if rather
complex, and it points towards the further potential of
this type of study in different areas and at different
scales.

Table 31.3. Percentage cover of selected species in relation to the categories of vegetation composition mapped in figure 31.7A.

Species	1 DRY HEATH	2	3	4	5	6	7	8	9	10	11	12 ALDER CARR
					W E T		H E A T H					
Calluna vulgaris	78	79	33	21	53	2	18	14	5	4	5	
Erica tetralix	2	9	6	9	21	43	28	16	7	8	18	
Myrica gale	-	-	-	7	8	4	-	13	10	22	7	
Pteridium aquilinum	5	-	1	23	1	-	-	-	9	-	-	
Juncus spp	-	-	-	-	-	-	-	6	3	6	4	
Sphagnum spp	-	-	-	3	4	3	6	5	4	2	5	
Narthecium ossifragum	-	-	-	1	1	3	5	3	-	-	20	

Based upon a vegetation survey by P.A. Hughes

THE CHANNEL SCALE

One of the greatest impediments to the investigation of channel morphology has been the difficulty encountered in relating data from representative cross sections to the spatial pattern of channel morphology. It has, therefore, been necessary to achieve a better understanding of the relationship between network character, channel morphology and channel cross section. Some progress has been made with respect to morphology by relating channel cross sectional area to upstream channel length (Gregory, 1979a) and then deriving network volumes which relate well to precipitation characteristics especially intensity, and to peak discharge as represented by an index of mean annual flood (Gregory and Ovenden, 1979). However, in addition to the morphological capacity of the network, it is also necessary to include channel typology and this has been conceived (Gregory, 1979b) using Manning roughness and further developed (Gardiner and Gregory, 1981) employing the Darcy-Weisbach friction factor. Absolute values of roughness and relative variations throughout a channel system occur in response to factors including vegetation influence and this has been investigated in the Highland Water by survey of the drainage network as a whole, by scrutiny of selected reaches, and by detailed investigation of specific cross sections.

The spatial pattern of channels within a basin is usually depicted assuming that all channels are essentially similar in character, but in the study basin channel typology is particularly related to vegetation. Field mapping of the channel network (figure 31.8) allowed the

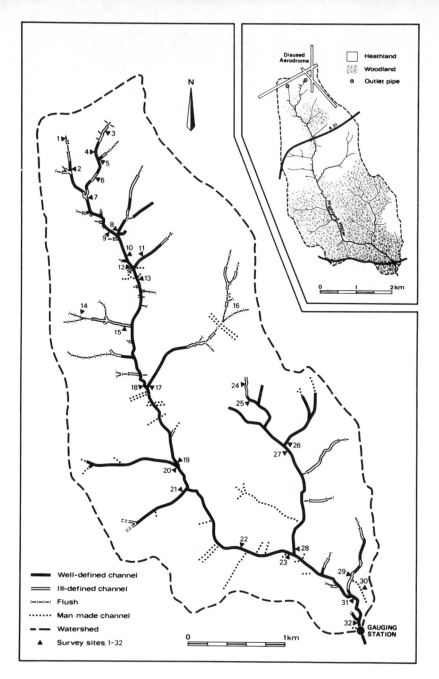

Figure 31.8. A stream channel typology for the network of
    the Highland Water. Definitions of the channel types
    are given in the text and roughness was estimated at
    low flows at sites 1-32 using the Manning equation.

identification of channels according to categories of bank
material and vegetation influence and three categories were
finally mapped as, firstly, well-defined channels (mean
width-depth ratio for alluvial sections = 16.6); secondly,
ill-defined channels (mean W/D = 20.8); and thirdly, flushes
(mean W/D = 40.2). Differentiation of the three types was
based upon existence of clear channel margins in relation
to the vegetation. A method of quantitatively describing
channel typology was sought and roughness was adopted by
making measurements at 32 sites throughout the network
(figure 31.8). Measurements were made at low flows by
dilution gauging at each site and, together with field
survey of channel dimensions and slope, it was possible to
estimate a Manning roughness value and Darcy-Weisbach friction factor for each channel cross section. These values
were necessarily computed during low flows and are therefore
lower than the values which would obtain at higher discharges (eg. Richards and Hollis, 1980). The average
Manning roughness value for well-defined alluvial channels
was 0.1216 compared with 0.2379 for ill-defined channels
and 0.5580 for flushes. The roughness value for flushes
corresponds to ranges of values obtained for different types
of grass cover (eg. Ree, Wimberley and Crow, 1977), but the
channel values would inevitably decrease at higher discharges.

Channel typology thus varies with vegetation character;
roughness affects the routing of flows along the network,
but mean roughness values are accompanied by considerable
variations from site to site. An empirical method is
therefore required to investigate vegetation influence upon
channel form. In the Highland Water basin this has been
achieved by employing a monitored 50 m reach at each end of
which is a gauging station with records since 1978. The
channel is being studied to investigate the way in which
flow is routed through the reach and this has shown that
vegetation exercises an influence not only directly upon
the banks and channel morphology but indirectly as a result
of the presence of vegetation debris dams. Such debris
dams although studied in North American experiments (Keller
and Swanson, 1979; Keller and Tally, 1979) have not been
elaborated in detail in Britain. In June 1981 in the
0.75 km immediately upstream from the gauging station,
there were five active and nine denuded debris accumulations (figure 31.9). The active ones are responsible
for a step in the channel profile, and they are composed of
fallen trees and branches with sediment and organic debris
infill. In most cases they appear to build up to a complete obstruction, and to break partially after a high
discharge. Material then accumulates again to restore
the obstruction.

A method was required to show how some 19 dams per km
of channel related to stream channel process. Deliberate
release of one dam in November 1980 indicated that a
distinctive trace was produced on the hydrograph recorded
at the upper and lower gauging stations and the discharge
record for 2.5 years was examined for such occurrences
(figure 31.10). Against the background of an average of
78 hydrograph rises each year, there are many instances

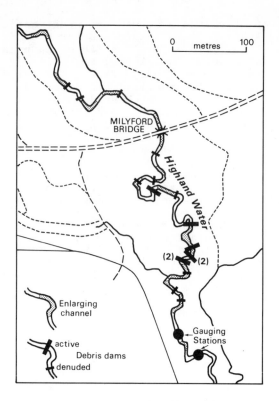

Figure 31.9. Debris dams in July 1981 in the reach
 immediately upstream from the instrumented reach.
 A denuded dam is still recognisable but no longer
 provides a continuous obstruction to flow.

where the bursting of vegetation dams has sent pulses of
flow along the channel. Considering only pulses which were
very clearly recorded at both gauging stations, there
appear to have been at least 16 events per year when vegetation dams have affected the discharge record. Although
such events have tended to occur in the winter half of the
year; they have not been exclusive to the winter six months.
    The influence of such vegetation dams is now being
monitored in relation to detailed surveys of the morphology
of the reach. Channel cross sections are being resurveyed
to identify amount of change and the way in which it relates to vegetation control. Preliminary surveys have
shown that the channel is enlarging over considerable
lengths probably due to improved drainage and to road runoff. The distribution of actively eroding reaches is
clearly related to the distribution of active and denuded
debris dams (figure 31.9) and enlarged channel sections
tend to occur immediately upstream of the vegetation
accumulations. It appears likely that the vegetation
accumulations act as the triggers for channel adjustment
and hence in studies of channel morphology comparing

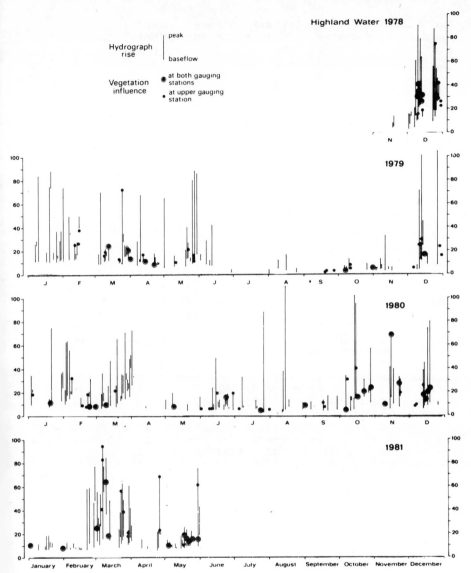

Figure 31.10. Hydrographs in the instrumented reach from 1978 to 1981. The vegetation influence exerted by the failure of debris dams is most clearly shown when re-corded on the trace at both gauging stations and the influence is possible but not conclusive when recorded only at the upper gauging station.

adjusting and stable channels, the influence of vegetation debris is particularly significant. A further influence of vegetation dams relates to overbank deposition. Along this channel there is no simple distinction between channel and overbank sedimentation because vegetation dams tend to instigate local avulsion over small areas of flood plain.

The accumulation of overbank sediment therefore occurs
sequentially over small areas and not as more extensive
sediments arising from specific large events.

CONCLUSION

The influence of vegetation upon fluvial processes can
profitably be investigated within an experimental catchment
if attention is focused upon analysis at several scales
and upon the inter-relationship of the results. In the
Highland Water basin, subcatchment, hillslope, instrumented
reach and total basin studies are in progress to document
the significance of vegetation. Results indicate that at
the subcatchment scale a complex pattern of runoff generation can be illuminated using the vegetation pattern and
particularly the distribution of the wet heathland.
Detailed measurements of pressure head at the hillslope
scale show how composition of wet heathland on the study
slope can be used as a surrogate for permeability. At the
catchment scale, it is necessary to investigate the relation
between channel morphology and drainage network, and a
channel typology has been developed based upon estimates
of roughness. Although vegetation contributes to the
roughness values through the detail of bank vegetation, a
substantial additional effect upon stream channel processes
is exercised by accumulations of vegetation debris as
debris dams. Such dams may provide the trigger for channel
enlargement which is occurring throughout much of the basin
and they may also exercise an important influence upon
sedimentation inducing localised overbank inundation.
Whereas results from small catchment experiments have often
led to inferences about the significance of vegetation, it
is only by use of an hierarchical investigation encompassing
several complimentary scales that the significance of
vegetation may be identified. It is clear that just as
vegetation can be an indicator of runoff production at the
subcatchment and hillslope scale, it can provide an influence upon the flow and sediment routing and storage at
the catchment scale and an influence upon channel morphology, particularly where stream channel adjustment is
taking place.

REFERENCES

Aranda, J.M. and Coutts, J.R.H., 1963, Micrometeorological
    observations in an afforested area in Aberdeenshire:
    rainfall characteristics, *Journal of Soil Science*,
    14, 124-133.

Bannister, P., 1964a, The water relations of certain heath
    plants with reference to their ecological amplitude.
    I. Introduction: Germination and establishment,
    Journal of Ecology, 52, 423-432.

Bannister, P., 1964b, The water relations of certain heath
    plants with reference to their ecological amplitude.
    II. Field studies, *Journal of Ecology*, 52, 481-497.

Bannister, P., 1964c, The water relations of certain heath plants with reference to their ecological amplitude. III. Experimental studies: General conclusions, *Journal of Ecology*, 52, 499-509.

Begin, Z.B. and Schumm, S.A., 1979, Instability of alluvial valley floors: A method for its assessment, *Transactions of the American Society of Agricultural Engineers*, 22, 347-350.

Charlton, F.G., Brown, P.M. and Benson, R.W., 1978, *The hydraulic geometry of some gravel rivers in Britain*, Report INT-180, (Hydraulics Research Station, Wallingford).

Daniels, R.E., 1975, Observations on the performance of *Narthecium ossifragum* (L.) Huds. and *Phragmites communis* Trin., *Journal of Ecology*, 63, 965-977.

Gardiner, V. and Gregory, K.J., 1981, Drainage density in rainfall-runoff modelling, in: *International Symposium on Rainfall-Runoff Modelling*, ed. Singh, V.P., (Mississippi State University), 449-476.

Gimingham, C.H., 1972, *Ecology of heathlands*, (Chapman and Hall, London).

Graf, W.L., 1979a, Mining and channel response, *Annals of the Association of American Geographers*, 69, 262-275.

Graf, W.L., 1979b, The development of montane arroyos and gullies, *Earth Surface Processes*, 4, 1-14.

Graf, W.L., 1980, Riparian management: A flood control perspective, *Journal of Soil and Water Conservation*, 35, 158-161.

Gregory, K.J., 1979a, Drainage network power, *Water Resources Research*, 15, 775-777.

Gregory, K.J., 1979b, Changes in drainage network composition, *Acta Universitatis Ouluensis*, Series A, 82, 19-28.

Gregory, K.J., 1980, Reply to discussion, *Water Resources Research*, 16, 1130.

Gregory, K.J. and Ovenden, J.C., 1979, Drainage network volumes and precipitation in Britain, *Transactions of the Institute of British Geographers*, New Series 4, 1-11.

Gregory, K.J. and Walling, D.E., 1973, *Drainage Basin Form and Process*, (Arnold, London).

Gurnell, A.M., 1978, The dynamics of a drainage network, *Nordic Hydrology*, 9, 293-306.

Gurnell, A.M., 1981, Heathland vegetation, soil moisture and dynamic contributing area, *Earth Surface Processes and Landforms*, 6, 553-570.

Hack, J.C. and Goodlett, J.C., 1960, Geomorphology and forest ecology of a mountain region in the central Appalachians, *United States Geological Survey Professional Paper*, 347.

Jones, H.E. and Etherington, J.R., 1970, Comparative studies of plant growth and distribution in relation to waterlogging. I. The survival of *Erica cinerea* L. and *Erica tetralix* L. and its apparent relationship to iron and manganese uptake in waterlogged soils, *Journal of Ecology*, 58, 487-496.

Jones, H.E., 1971, Comparative studies of plant growth and distribution in relation to waterlogging. III. The response of *Erica cinerea* L. to waterlogging in peat soils of differing iron content, *Journal of Ecology*, 59, 583-591.

Keller, E.A. and Swanson, F.J., 1979, Effects of large organic material on channel form and fluvial processes, *Earth Surface Processes*, 4, 361-380.

Keller, E.A. and Tally, T., 1979, Effects of large organic debris on channel form and fluvial processes in the coastal redwood environment, in: *Adjustments of the Fluvial System*, ed. Rhodes, D.D. and Williams, G.P., 169-197.

Leyton, L., Reynolds, E.R.C., and Thompson, F.B., 1967, Rainfall interception in forest and moorland, in: *International Symposium on Forest Hydrology*, ed. Sopper, W.E. and Lull, H.W., 163-177.

Loach, K., 1966, Relations between soil nutrients and vegetation in wet-heaths. I. Soil nutrient content and moisture conditions, *Journal of Ecology*, 54, 597-608.

Luthin, J.N. and Kirkham, D., 1949, A piezometer method for measuring permeability of soil *in situ* below a water table, *Soil Science*, 68, 349-358.

Moseley, M.P., 1981, Semi-determinate hydraulic geometry of river channels South Island, New Zealand, *Earth Surface Processes and Landforms*, 6, 127-138.

Rachocki, A., 1978, Wptyw roslinnosci na ksztattowanie koryt i brzegow rzek, *Przeglad Geograficzny* 3, 469-481.

Rahman, M.S., 1976, A comparison of the ecology of *Deschampsia cespitosa* (L.) Beauv. and *Dactylis glomerata* L. in relation to the water factor. I. Studies in field conditions, *Journal of Ecology*, 64, 449-462.

Rahman, M.S. and Rutter, A.J., 1980, A comparison of the ecology of *Deschampsia cespitosa* and *Dactylis glomerata* in relation to the water factor. II. Controlled experiments in glasshouse conditions, *Journal of Ecology*, 68, 479-491.

Ree, W.O., Wimberley, F.L. and Crow, F.R., 1977, Manning n and the overland flow equation, *Transactions of the American Society of Agricultural Engineers*, 20, 89-95.

Richards, G. and Hollis, G.E., 1980, Relationship of the roughness coefficient and Manning's n and discharge in an urban river, *Journal of the Institution of Water Engineers and Scientists*, 34, 357-360.

Rutter, A.J., 1955, The composition of wet-heath vegetation in relation to the water table, *Journal of Ecology*, 43, 507-543.

Tuckfield, C.G., 1973, Seepage steps in the New Forest, Hampshire, England, *Water Resources Research*, 9, 367-377.

Zimmerman, R.C., Goodlett, J.C. and Comer, G.H., 1967, The influence of vegetation on channel form of small streams, *International Association of Scientific Hydrology Publication*, 75, 255-275.

32 Stream response to flash floods in upland Scotland

A. Werritty

INTRODUCTION

The maximum rates of fluvial erosion and sedimentation
within Britain are associated with the powerful and active
gravel-bed rivers of the uplands. The scientific study of
such rivers is of major importance in terms of engineering
problems such as river training and bank stabilisation, the
genesis of coarse-grained fluvial sediments and the maintenance of regionally significant fisheries. However, our
understanding as to how such rivers adjust to the great
variety of natural and regulated runoff regimes is at
present extremely imperfect. The initial focus for such
work in Britain has concentrated on Welsh gravel-bed rivers
which have attracted a number of research projects over the
last five years. As a result notable advances have been
made in the study of flow resistance (Hey, 1979; Bathurst,
1982), secondary circulation (Bathurst et al., 1977), bank
stability (Thorne and Lewin, 1979), sediment transport
(Newson, 1980) and channel migration and bar development
(Lewin, 1976). Comparable work in Scotland has thus far
only embraced two topics: sedimentation in coarse-grained
alluvium (Bluck, 1976; 1979) and channel processes in
divided channels of low sinuosity (Werritty and Ferguson,
1980).
  Until relatively recently, research into the morphology
and associated processes in gravel-bed rivers has largely
centred on powerful and very active braided channels in
proglacial environments (Church, 1971; Smith, 1974).
Braiding on this scale is not characteristic of upland
Britain; instead the dominant channel type is a transitional one with elements common to both braiding and
meandering (Ferguson and Werritty, 1982). The resulting
channel pattern is best described by the separate class of
'wandering gravel rivers' (Church et al., 1981). Such
rivers are typified by high width-depth ratios, and divided
flow around significant mid-channel bars. Whilst these
rivers continue to display the familiar pool-riffle sequence
of fine-grained alluvial channels, there is a meander-like

alternation of riffle orientation controlled by the location
of the major bar units. These rivers are transitional in
terms of channel pattern since they have lower sinuosities
than in fine-grained meandering rivers and lack the degree
of channel sub-division characteristics of braided channels
in proglacial environments. Wandering gravel rivers of
this type are common throughout much of upland Scotland.

In addition to their distinctive morphology, in Scotland
these rivers appear to be associated with a pattern of
formative discharges dominated by brief but intense convective summer storms or long periods of cyclonic rainfall
during the autumn or early winter. The convective storms
are especially significant in that they generate a runoff
response which is so rapid that it can only be described
as a 'flash flood'. In a further contrast to the runoff
regime of braided rivers in proglacial environments, the
role of the annual snowmelt is very modest. As a result of
this rather irregular and episodic pattern of formative
discharges, the floodplains of such rivers are typically
well-vegetated leaving only small areas of exposed gravel
adjacent to the river which can be readily reworked.
Typically, the bed material of these channels is armoured
thereby further accentuating the role of extreme events
which are able to transform the active area. Thus, in
magnitude-frequency terms, a relatively high threshold for
sediment entrainment, coupled with a right-skewed distribution of formative discharges tend to emphasise the role
of major floods (cf. the analysis of Baker (1977) for
streams in Texas).

No long term time-lapse analyses of gravel-bed rivers
have yet been attempted in upland Britain despite an increasing realisation that the timescales appropriate to
recording significant fluvial adjustment are more appro-
priately measured in decades rather than years (Church,
1981; Newson, 1979). In the absence of such long term
synoptic studies, plus the lack of information concerning
the role of extreme events in such rivers, I wish to examine
the following questions:
1. What is the scale and pattern of channel adjustment in
    active Scottish gravel-bed rivers? What are the
    associated rates of erosion?
2. What is the geomorphic significance of flash floods in
    these rivers? Are such floods the major formative
    agent shaping the floodplain?
3. What is the pattern of recovery following a flash flood?

METHOD

In selecting a suitable research strategy two major
problems were immediately apparent. Firstly, intense
summer convective storms in upland Scotland are highly
localised, of short duration and relatively rare. Secondly,
the small size of the drainage basin selected for detailed
analysis, and the extreme instability of the channel pattern
make continuous gauging of runoff extremely difficult.
Since analysis of short-term channel changes and bar
development is already in progress on the nearby but much

larger River Feshie (Ferguson and Werritty, 1982), it was
decided to attempt a long term time-lapse study at the
chosen site on Dorback Burn. This involves regular surveys
over a period of at least 10 years based upon the Vigil
Network methodology developed by L.B. Leopold and his
associates in the U.S. Geological Survey (Leopold, 1962;
Emmett and Hadley, 1968). The results reported in this
paper are from the first three years of the project and
must therefore be regarded as provisional.

Data are collected within the following framework. A
network of fixed points along the edge of the active area
was established by ground survey in November 1978, and
initially five channel cross-sections (increased to nine
in June 1980) have been documented. Resurvey by levelling
is undertaken every six months, with additional surveys
immediately following a major event. Planimetric control
is provided by repeated plane-table mapping using a self-
reducing alidade (accurate to within ± 0.2 m) and a detailed
photographic record which is maintained via low-angle
oblique photo-mosaics from a suitable vantage point on the
summit of a kame-like ridge (figure 32.5). The pattern,
but not the absolute rate, of bedload transport is monitored
by painted pebbles which have been injected both in the
channels and on bar surfaces within the active area.
Finally, this essentially morphological approach is complemented by some process data. Discrete events can be
identified from the continuous discharge record maintained
by the Northeast River Purification Board on the River
Nethy at Nethybridge (figure 32.1), from the autographic
raingauge located on the summit of Cairngorm, and an intermittent record of daily rainfall at Nethybridge.

CATCHMENT PROPERTIES AND RUNOFF REGIME

The River Nethy is a right-bank tributary of the River Spey
which flows within a trough aligned southwest-northeast
between the Cairngorm and Monadhliath Mountains in the
Central Highlands of Scotland (57°16' N, 3°40' W; National
Grid Reference NH 9921). The drainage basin of the River
Nethy upstream of the gauging station is 101 km^2, whereas
that for the study reach on Dorback Burn, a major tributary
of the River Nethy, is only 18.6 km^2 (figure 32.1). The
highest ground within the overall drainage basin delineates
the southern and eastern part of the watershed (1245 m at
the summit of Cairngorm) with a general reduction in
elevation to the north and west reaching 200 m at the confluence with the River Spey. Annual rainfall ranges from
> 2000 mm on the highest ground to < 800 mm at Nethybridge.
The bedrock geology is predominantly Moinian schist with
localised granite outcrops of the Cairngorm batholith
exposed on the high plateau in the southern part of the
catchment. More significant in terms of recent fluvial
activity is the extensive drift cover of well-stratified
fluvio-glacial sands and gravels which give rise to large
kame-like ridges (Young, 1975). These locally constrain
the lateral migration of the present-day channel and provide a major source for the gravel and cobble bed material.

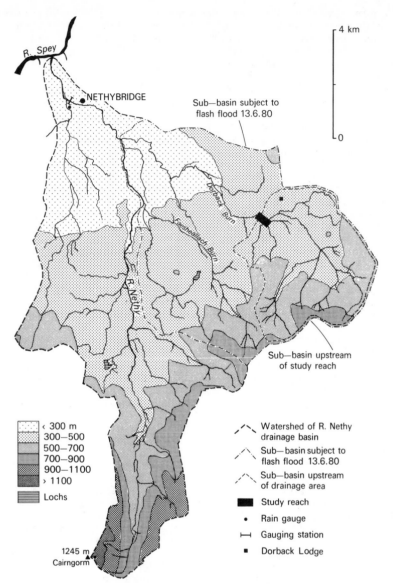

Figure 32.1. Topographic map of the River Nethy basin, hydrometric stations and location of study reach on Dorback Burn.

Since the river system is still dominated by this sediment supply and the reworking of these outwash deposits, the landscape is essentially a relict one in that present-day fluvial activity is still strongly constrained by the legacy of the lateglacial. There is thus little coupling of slope and channel processes. The vegetation is an open moorland dominated by wet and dry heath communities

540

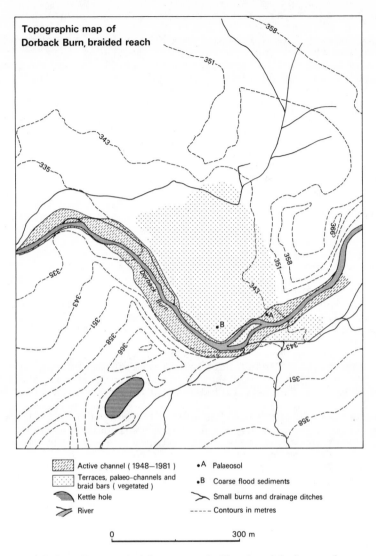

Figure 32.2. Topographic map of the braided reach on
    Dorback Burn.

(*Calluna vulgaris* and *Erica tetralix*) with occasional
remnants of the native Scots Pine Forest (*Pinus sylvestris*)
which was extensively cleared in this area in the early
nineteenth century (O'Sullivan, 1970).
    The study reach is a 250 m length of channel located at
a site where Dorback Burn dissects a kame-like ridge
(figure 32.2). The pattern of recent channel migration is
indicated by the absence of a 200 year old palaeosol
beneath the most recent terraces (A in figure 32.2), together with the occurrence of 40 year old lichens (*Rhizo-
carpon geographicum*) on flood gravels just beyond the

currently active area (B in figure 32.2). A continuous
record of discharge on the River Nethy has been obtained at
Nethybridge by the Northeast River Purification Board since
June 1975; this being preceded by a record of mean daily
flows back to 1964. Flow duration analysis based on these
data reveal a median flow of c. 2 m^3 s^{-1} with flows
> 12 m^3 s^{-1} only occurring for 1% of the record. However,
between June 1975 and June 1981 there have been two out-
standing flash floods. On July 3rd, 1978 a 24 hour rainfall of 89.5 mm was reported for the summit of Cairngorm
including 7 hours during which rainfall intensities
fluctuated between 5 and 10 mm hr^{-1}. The same storm recorded a 24 hour rainfall of 27.5 mm at Nethybridge with
76.5 mm at Coire Cas and 79 mm at Glenmore Lodge, both
stations located in the contiguous basin of the Allt Mor
draining the northern slopes of the Cairngorms. The
resulting flash flood recorded a peak instantaneous discharge of 105 m^3 s^{-1} and flow above 20 m^3 s^{-1} for 26 hours.
The second event of June 6th, 1980 substantially exceeded
the first in terms of a peak instantaneous discharge of
202 m^3 s^{-1}, but recorded only 9 hours of flow above
20 m^3 s^{-1}. This storm was very much more localised than
the first one and no rainfall records are available. The
shape of the two storm hydrographs were markedly different
(figure 32.3). Whereas the first storm covered a large
portion of the catchment and generated very high flood
flows on all the major tributaries of the River Nethy; the
second storm was much more localised within the Dorback
Burn sub-basin. Field checking of the whole catchment
revealed extensive flood damage only within the area
drained by the Faesheallach Burn and Dorback Burn (figure
32.1). It seems likely that the much more flashy runoff
response associated with the June 1980 event resulted from
a highly localised convective storm, which despite its
brief duration nevertheless registered a very high total
rainfall. A similar storm has been reported for the Allt
Mor when 52.5 mm fell on August 4th, 1978 with a peak
recorded intensity of 33.5 mm $hour^{-1}$ (McEwen 1981). However, in terms of the geomorphic impact of the two storms
over Dorback Burn, the much greater duration of the 1978
event above a threshold of 20 m^3 s^{-1} is probably of greater
significance than the fact that its peak value was only
half that of the 1980 event.
    Since the gauging station is considerably downstream
of the study reach, an estimate of the peak discharge at
the site was calculated for both storms using the slope-
area method (table 32.1). Well-defined trashlines on both
sides of the channel enabled the surface water slope to be
reconstructed at several cross sections. Pebble counts on
the bed material at the selected sections were used to
determine D_{84} values in order to calculate suitable values
for Darcy Weisbach's f (Hey, 1979). The resulting peak
discharges for each flood were 26.4 m^3 s^{-1} (or in terms of
unit discharge 1.42 m^3 s^{-1} km^{-2}) for 1978, and 78.9 -
80.4 m^3 s^{-1} (or 4.24 - 4.32 m^3 s^{-1} km^{-2}) for the 1980
storm. The much higher unit discharge for the second
storm is a relatively high value for upland Scotland and
reflects the very localised nature of that event.

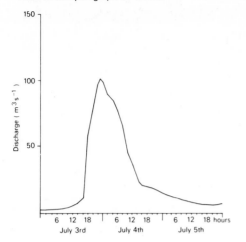

Flash flood hydrograph 3—4.7.78

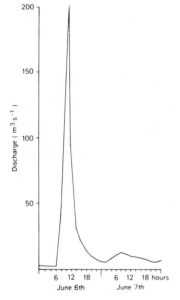

Flash flood hydrograph 6.6.80

Peak discharges above 20 m^{3}s^{-1}
on R. Nethy at Nethybridge
May 1975—June 1981

Date of event	Peak discharge (m^{3}s^{-1})
17.9.75	22.0
15.10.76	41.5
2.3.77	24.0
6.5.77	26.0
3.7.78	43.0
4.7.78	105.0
4.8.78	59.0
First survey	
15.11.78	30.0
8.12.78	23.5
26.2.79	36.3
27.2.79	23.1
2.3.79	22.4
11.4.79	39.7
12.4.79	28.2
13.4.79	24.2
Second survey	
17.5.79	27.3
17.8.79	38.0
Third survey	
17.12.79	21.6
9.2.80	29.2
12.2.80	22.4
6.6.80	202.0
Fourth survey	
19.11.80	21.6
Fifth survey	

Figure 32.3. Hydrographs at Nethybridge gauging station for the flash floods on July 3, 1978 and June 6, 1980.

THE CHANGING MORPHOLOGY OF THE ACTIVE AREA: 1948-76, 1978-81

The changing channel morphology within the active area will be analysed in detail over the period October 1978 to July 1981 using the methods described earlier (section A). But in order to place this recent survey within its wider context the first part of this section will be concerned with channel change as recorded in seven sets of aerial photographs.

Table 32.1. Estimated discharge values for the 1978 and 1980 flash floods

	River Nethy catchment (a)			Dorback Burn study reach (b)		
	Catchment area (km²)	Peak discharge (m³s⁻¹)	Unit discharge (m³s⁻¹km⁻²)	Catchment area (km²)	Peak discharge (m³s⁻¹)	Unit discharge (m³s⁻¹km⁻²)
July 4 1978	101	105	1.04	18.6	26.4 (c)	1.42
June 6 1980	101	202	2.00	18.6	80.4 (d)	4.32
					78.9 (e)	4.24

(a) Northeast River Purification Board flow record at Nethybridge

(b) Estimated by a slope-area method

(c) Cross-section 7

(d) Cross-section 1

(e) Cross-section 5

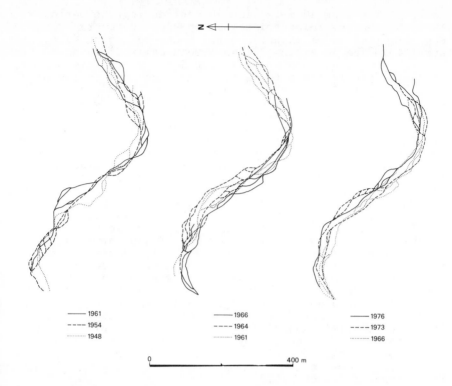

Figure 32.4. Channel pattern change on Dorback Burn
1948-76, based upon air photo interpretation.

Channel changes over the period 1948-76

Detailed analysis of channel change from aerial photographs
is often difficult (Werritty and Ferguson, 1980). At this
specific site the problems include the wide range of scales
(1:11 000 to 1:50 000) and the very varied tonal contrast
on individual prints. Nevertheless, it has proved possible
to map the major channel systems and thus define the zone
of active channel migration over the selected period
(figure 32.2). Several interesting features emerge when
successive channel patterns are superimposed (figure 32.4).
Whereas the period between 1948 and 1954 was characterised
by a downstream shift in the most intensely braided zone,
the ensuing 7 years up to 1961 were associated with a
marked increase in the local sinuosity. During the three
years following 1961 a further shift of the main channel to
the northeast occurred causing erosion of the right-bank
low terrace. Thereafter, between 1964 and 1966 the main
channel switched its alignment 30-40 m in a major avulsion
which on this occasion resulted in erosion of the left-bank
terrace. It is likely that this adjustment occurred during
September 1965 when mean daily flows greater than 30 m^3 s^{-1}
were recorded twice on the River Nethy. Channel migration

545

locally exceeding 40 m is also evident during the period
1966-73. During this period the imperfect discharge record
suggests that major channel change occurred during March/
April 1970 and November 1971 when mean daily flows on
several occasions were greater than 30 m^3 s^{-1}. This period
of relatively frequent formative discharges culminated on
April 15th, 1970 with a mean daily flow of 45 m^3 s^{-1}, a
value comparable to that of the flood in 1978. Over the
next 8 years no single event approached the size of this
flood and the channel became less sinuous and relatively
stable particularly over the period 1973-78. Thus there was
no discernible change detected by the author in three low
angle oblique photo-mosaics of the study reach dated May
1976, May 1977 and April 1978. This implies a relative
absence of competent discharges, which is confirmed by the
record of only five instantaneous peaks above 20 m^3 s^{-1}
over the period June 1975 - June 1978 (table 32.2).

This record of channel change based upon aerial photographs is largely qualitative because of the nature of the
source materials. Nevertheless it clearly identifies the
width and extent of the corridor periodically occupied by
this wandering gravel river, and demonstrates that the
river's recent history is punctuated by phases in which
both extensive braiding and significant meandering tendencies can be identified. There is also the suggestion
that preferential channel locations are re-occupied and
that channel change occurs by avulsion rather than lateral
migration as in fine-grained alluvial channels.

First survey: October 1978

Following the July 1978 flood an initial base-line survey
was completed the following October. This would appear to
postdate a second substantial flood of 59 m^3 s^{-1} recorded
at Nethybridge on August 4th (table 32.2). However, the
rainfall records indicate that this storm was localised
over Cairngorm and the adjacent basin of the Allt Mor, and
thus the resulting flood flow was along the north-flowing
trunk stream of the River Nethy rather than Dorback Burn.
Thus the two low angle oblique photographs (figures 32.5a
and 32.5b) taken 'before' and 'after' provide a visual
summary of the geomorphic impact of the July storm in
isolation. The planimetric map from this survey reveals a
sinuous but divided main channel with two major bars as
depositional forms (A and B in figure 32.6). The channel
further divides at C and spills over into a chaotic area of
micro-braiding along the eastern side of the active area.
At the proximal end of this divided channel diagonal bars
have evolved with chutes into the main channel (figure
32.7a); whereas at the distal end the flow appears to be
controlled by the chaotic pattern of falling stage sedimentation. On the opposite side of the study reach, aggradation at D in the main channel resulted in a local ponding-
back of flow which in turn caused overbank flow at E
(figure 32.7b). Extensive bank erosion on the left-hand
side of the main channel up to cross-section 5 provided the
source for numerous turf blocks which are littered over the
bar surfaces and acted as sediment traps for the wash load

Table 32.2. Peak discharges above 20 m³ s⁻¹ on the
River Nethy at Nethybridge May 1975 - June 1981

Date of event	Peak discharge $(m^3\ s^{-1})$
17.9.75	22.0
15.10.76	41.5
2.3.77	24.0
6.5.77	26.0
3.7.78	43.0
4.7.78	105.0
4.8.78	59.0
First survey	
15.11.78	30.0
8.12.78	23.5
26.2.79	36.3
27.2.79	23.1
2.3.79	22.4
11.4.79	39.7
12.4.79	28.2
13.4.79	24.2
Second survey	
17.5.79	27.3
17.8.79	38.0
Third survey	
17.12.79	21.6
9.2.80	29.2
12.2.80	22.4
6.6.80	202.0
Fourth survey	
19.11.80	21.6
Fifth survey	

during the falling stage of the flood (figure 32.7c).
There was also extensive overbank sedimentation of this
washload resulting in alluviation of former channels within
the active area (eg. F in figure 32.6). A synopsis of the
hydraulic data defining peak flow conditions are summarised
in table 32.3. They are consistent with the independent
observation that boulders with intermediate axes of 33-49 cm
were entrained during the flood.

Second survey: May 1979

Following a period of relatively high runoff values (seven
discrete events above 20 m³ s⁻¹), major adjustments to the
immediate post-flood channel pattern occurred during the
six months up to May 1979 (figure 32.6). Incision of the
eastern channel on cross-section 5 resulted in the isolation
of the complex zone of micro-braiding downstream of C.
Substantial enlargement of the same channel at G (0.5 m of

547

(a) 25 May 1977, prior to first flood;

(b) 29 October 1978, after first flood;

(c) 1 May 1979, before second flood.

Figure 32.5.   Views across part of the active area towards
   northern end of cross sections 6, 7 & 8 & low terrace.

Dorback Burn braided reach

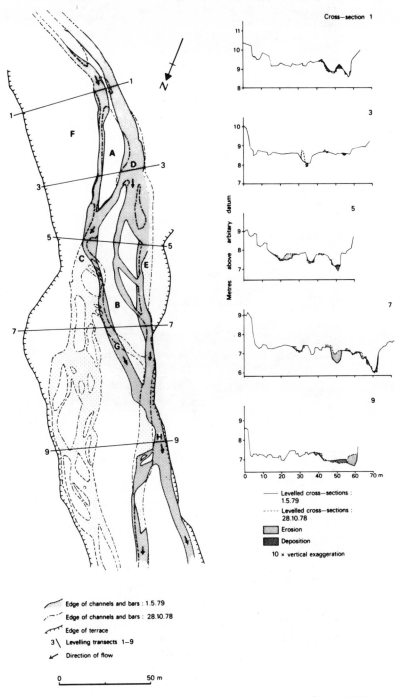

Figure 32.6.   Channel change: October 1978 – May 1979.
Labels A, B etc. denote locations referred to in the
text.

549

Figure 32.7. (a) Diagonal bar with chutes dissecting the bar front. Turf blocks anf fine grained sediment in the foreground.

Figure 32.7. (b) Overbank flow at E (see figure 6) on former flood plain returning to main channel: persisted for nearly 2yr after the first flood. Undercut face of well stratified kame-like ridge in background.

Figure 32.7. (c) Turf blocks and falling-stage sedimentation embroidering bar surface opposite diagonal bar in figure 32.7 (a).

Table 32.3. Summary of hydraulic data for the two flash floods

	July 4, 1978 (Cross-section 7)	June 6, 1980 (Cross-section 1)	June 6, 1980 (Cross-section 5)
Cross-sectional area (m²)	14.8	27.9	24.2
Hydraulic radius (m)	0.425	0.603	0.524
Width (m)	34.8	46.3	46.2
(1) D_{84} for intermediate axis (mm)	71	50.6	71.8
(2) Channel slope	0.014	0.0171	0.0333
Darcy Weisbach's f	0.147	0.0973	0.1294
Average velocity (m s⁻¹)	1.78	2.88	3.253
(3) Average shear stress (Nm⁻²)	5.95	10.31	17.44

(1) Calculated from pebble count of 50 clasts
(2) Derived from gradient of trash lines
(3) Based upon Duboys equation

incision) has given rise to a more acutely-angled confluence
downstream which in turn has caused 5 m of bank erosion at
F and the abandonment of the October channel at cross-
section 9. The overall appearance of the channel system is
a simpler and much more integrated one than in the previous
survey, (cf. figures 32.5b and 32.5c), ie. there has been
a rapid phase of post-flood recovery in which some of the
flood-induced changes have themselves been modified.

Third survey: October 1979

The six months preceding this survey were remarkably un-
eventful with only two events above 20 m^3 s^{-1} being recorded
at the gauging station. The two longitudinal bars immediately upstream and downstream of cross-section 1 have
united into a single form (figure 32.8), and there are minor
adjustments to the shape of individual bar units especially
downstream of cross-section 9. But generally there is very
little change in either the planimetric or cross-sectional
channel forms. This configuration probably represents a
new locally 'stable' situation which would have persisted
for some time had it not been completely destroyed by the
second flash flood in June 1980. This 'stable' phase from
May 1979 to June 1980 is similar to that already noted in
the aerial photographs between 1973 and July 1978.

Fourth survey: June 1980

Following an estimated discharge of 80 m^3 s^{-1} in the study
reach on June 6th, 1980 an initial resurvey was completed
within 7 days. The active area had been completely transformed (figure 32.9), and a new channel excavated at A
partly along the former alignment of a very small inactive
channel. A marked westward migration of the main channel
between cross-sections 1 and 5 resulted in 0.7 m of incision and 3-5 m of bank erosion. In contrast the former
channels and bar systems at B, C, D and E were subject to
0.5-0.7 m of aggradation completely masking the pre-flood
micro-relief. As in the first survey a zone of shallow
highly divided flow was established between F and G, with
a new distributary H constituting the route back to main
channel at I. The mid-channel bar at J with divergent flow
off the distal end (figure 32.10a) is typical of many of
the depositional forms within the channel. In contrast the
'island' at K is primarily an erosional feature caused by
enlargement of the two adjacent channels. As in the
previous flood there was also extensive overbank sedimentation with 12 cm deposited on the floodplain at L (figure
32.10b).
    The hydraulic data used to estimate the peak discharge
are summarised in table 32.3. In terms of average velocity,
hydraulic radius and cross-sectional area, this second
flood was substantially larger than that in 1978. The entrainment of boulders with intermediate axes of 41-58 cm
again provides independent confirmation of the values
listed in table 32.3. Despite much higher average shear
stresses the second flood appears to have had a more modest
impact in terms of remodelling the active area in comparison

Dorback Burn braided reach

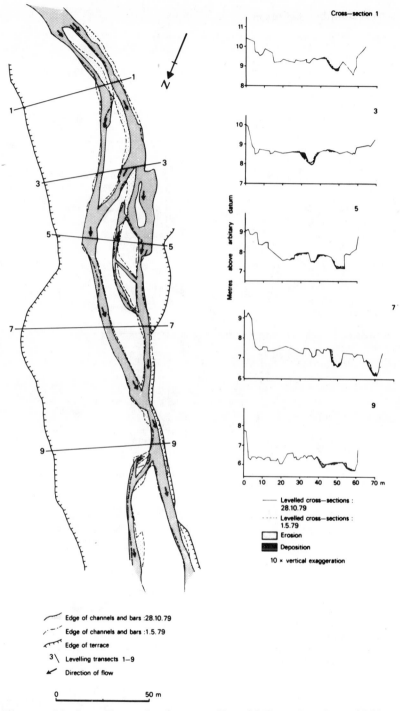

Figure 32.8.   Channel change: May 1979 – October 1979.

Dorback Burn braided reach

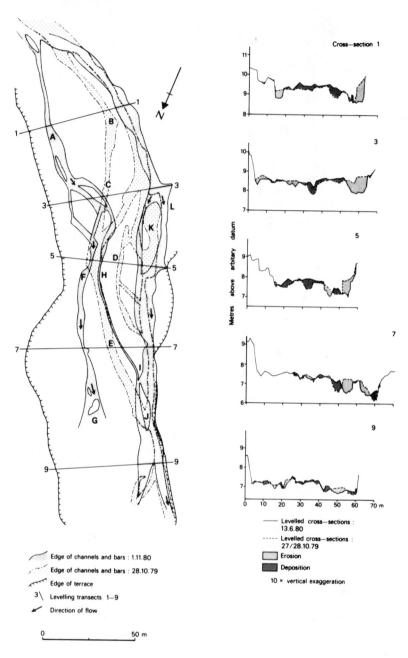

Figure 32.9. Channel change: October 1979 - June 1980.
Labels A, B etc. denote locations referred to in text.

Figure 32.10. (a) Mid-channel bar J (figure 9) with
divergent flow off distal margin (flow away from view).
Formed on June 6, 1980, destroyed by April 1981.

Figure 32. 10. (b) Fine overbank sediments at L (figure 9),
site for grain size data in figure 12.

Figure 32.10. (c) New bar complex at D (figure 11),
arising from enlargemnet of distributary channel
upstream. In right foreground a series of coalescing
lobate gravel deltas prograding into former main channel.

with the 1978 flood. One possible explanation is the much shorter period when flow was above the threshold of sediment transport, implying that the important control determining the geomorphic significance of an event is the duration of flow above this critical value rather than the maximum instantaneous shear stress.

Fifth survey: June/July 1981

Despite a period of relatively low flow at Nethybridge over the 12 months up to June 1981, considerable post-flood adjustment is again apparent. Photographic evidence demonstrates that the channel pattern on November 1st, 1980 was indistinguishable from that of June 1980, but by June 1981 a single and almost undivided channel of low sinuosity had been established (figure 32.11). The former main channel A, B is now inactive because of channel adjustments immediately upstream which have caused a diversion of the main flow into the once narrow distributary channel C. Since this channel now accommodates the total flow it has been enlarged by up to 0.6 m in depth and 2-3 m in width between cross-sections 1 to 5. The resulting sediment transport appears to have been very localised since extensive aggradation has occurred downstream at D where a completely new bar complex has evolved between cross-sections 7 and 8. The inactivity of the former main channel is further demonstrated by 0.4 m of aggradation and the upstream prograding gravel deltas at E (figure 32.10c). As in the survey of May 1979 the phase of micro-braiding at F was short-lived as a result of incision of the main channel at cross-section 5.

DISCUSSION

Three years' observations on a short reach of Dorback Burn have demonstrated great complexity in the pattern of channel adjustment to the flow regime. Nevertheless, some generalisations can be derived from this case-study.

Only two extreme events, each less than 24 hours in duration, have occurred over the past 6 years, and both completely remodelled the active area. It is difficult to make reliable estimates of the recurrence intervals for each storm because of the short runoff record; but the addition of the imperfect record of daily mean flow suggests that these two floods were the largest over the period 1964-1981. However, in magnitude-frequency terms whilst these two events were of major geomorphic significance, they were by no means the only competent flows during this period. The air photo evidence and detailed ground survey clearly indicate that channel adjustment has occurred during flows of 20-30 m^3 s^{-1} at Nethybridge. However, it seems likely that there is no simple threshold for formative discharges; much depends upon the state of the channel immediately prior to the event. Thus both the 1978 and 1980 floods established a potentially unstable channel pattern arising from the chaotic nature of falling stage sedimentation and the formation of under-sized distributaries.

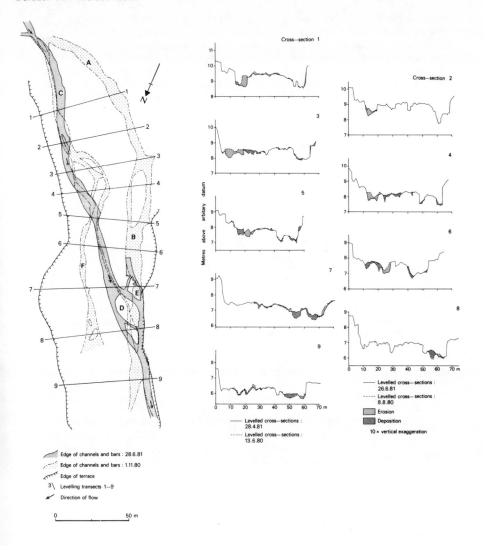

Figure 32.11.   Channel change: June 1980 - June/July 1981.
Labels A, B, etc. denote locations referred to in the
text.

Thereafter relatively modest post-flood flows resulted in
a rapid simplification of the channel system restoring it
to a potentially more stable long-term configuration. This
pattern of relatively rapid post-flood recovery confirms
the recent revision of magnitude-frequency concepts by
Wolman and Gerson (1978). But an important qualification
is necessary on account of the relatively high threshold
for sediment entrainment in wandering gravel rivers. This
arises because the erosional and depositional imprint of
major floods is relatively long-lived over much of the
active area; and it is only in the vicinity of the current

557

channel that major adjustments take place in terms of localised incision and aggradation. However, this is sufficient to cause the periodic abandonment and re-occupation of preferential channel locations, often within less than 6 months, thus giving rise to a channel pattern which is remarkably unstable. The rates of erosion associated with these less than half-yearly channel changes are very much lower than the 5 m of bank erosion often associated with the rare floods.

In terms of post-flood recovery Dorback Burn differs from several of Wolman and Gerson's examples from fine-grained channels in the eastern United States. Whilst it is possible to detect a relatively rapid overall simplification of the channel form, this is not at the expense of obliterating the more widespread imprint of the flood across the active area. There is also the strong suggestion over the period 1948-81 that whereas braiding is the characteristic pattern after a major flood, meandering typically develops during long periods of non-competent flows. Such a cyclic tendency has already been noted by Ferguson and Werritty (1982) in the River Feshie, a much larger wandering gravel river flowing from the Cairngorms. Since such rivers occupy a transitional class between braided and meandering patterns it is not surprising that these rivers should display both tendencies over a sufficiently long period of time. In this situation the distinction between meandering and braiding is far from clear suggesting that a more flexible river classification scheme is required along the lines being developed by Church *et al.* (1981).

The final conclusion relates to the timescales appropriate for the analysis of this type of channel. This study has demonstrated that synoptic analysis of channel change is potentially misleading unless it is coupled with the antecedent history of competent flows and related to a timescale which will include a wide range of channel conditions. In this context the Vigil Network approach adopted here seems well-suited to the questions posed. However, the wider question of the recent geomorphic history of the floodplain and the adjacent low terraces is still problematic; the scale of aggradation and erosion reported above being insufficient to explain the scale of the palaeoforms and former braid bars which embroider the terraces. This is either because the record analysed thus far has been atypical or discharges in the immediately recent past were higher and subject to greater fluctuation. The solution to such a problem requires the extension and replication of this type of study on other wandering gravel rivers in a variety of non-tropical piedmont zones, together with a greater emphasis on the operation of fluvial processes over recent decades and even centuries. Only then will we be able to place the operation of present-day processes within a context and timescale which are geomorphically significant.

ACKNOWLEDGEMENTS

I wish to acknowledge the generous assistance of Keith
Gribben of the Northeast River Purification Board in making
flow records available, the landowners for permission to
work at this site, and the Carnegie Trust and University
of St. Andrews for financial assistance. The great bulk of
the survey work was ably undertaken by members of the 1979,
1980 and 1981 Junior Honours classes in Geography at the
University of St. Andrews. Unwary research students, Peter
Thorne, Mike Acreman and Lindsey McEwen have also generously
acted as field assistants. I have also profited from
discussions in the field with Rob Ferguson, John Lewin and
Asher Schick.

REFERENCES

Baker, V.R., 1977, Stream channel response to floods, with
    examples from Central Texas, *Bulletin of the
    Geological Society of America*, 88, 1057-71.

Bathurst, J.C., 1982, Theoretical aspects of flow resis-
    tance, in: *Gravel-bed rivers: fluvial processes,
    engineering and management*, ed. Hey, R.D.,
    Bathurst, J.C. and Thorne, C.R., (Wiley, London),
    Chapter 5.

Bathurst, J.C., Thorne, C.R. and Hey, R.D., 1977, Direct
    measurement of secondary currents in river bends,
    Nature, 269, 504-6.

Bluck, B.J., 1976, Sedimentation in some Scottish rivers of
    low sinuosity, *Transactions of the Royal Society of
    Edinburgh*, 69, 425-56.

Bluck, B.J., 1979, Structure of coarse-grained braided
    stream alluvium, *Transactions of the Royal Society
    of Edinburgh*, 70, 181-221.

Church, M., 1972, Baffin Island sandurs: a study of Arctic
    fluvial processes, *Canadian Geological Survey
    Bulletin*, 216.

Church, M., 1981, On experimental method in geomorphology.
    Paper presented at '*Catchment experiments in
    fluvial geomorphology*', IGU Commission on Field
    Experiments in Geomorphology, UK meeting at
    Huddersfield, August, 1981.

Church, M., Moore, D. and Rood, K., 1981, *Catalogue of
    alluvial river regime data*, Dept. of Geography,
    University of British Columbia, Canada.

Emmett, W.W. and Hadley, R.F., 1968, The Vigil Network:
    preservation and access of data, *U.S. Geological
    Survey Circular* 460-C.

Ferguson, R.I. and Werritty, A., 1982, Bar development and
    channel changes in the gravelly River Feshie,
    Scotland, in: *Modern and ancient fluvial systems*,
    ed. Collinson, J., (Special Publication of the
    International Association of Sedimentologists), (in
    press).

559

Hey, R.D., 1979, Flow resistance in gravel-bed rivers, *Journal of the Hydraulics Division, ASCE*, 105, 365-79.

Leopold, L.B., 1962, The Vigil Network, *Bulletin of the International Association of Scientific Hydrology*, 7, 5-9.

Lewin, J., 1976, Initiation of bedforms and meanders in coarse-grained sediment, *Bulletin of the Geological Society of America*, 87, 281-5.

McEwen, L.J., 1981, An assessment of the geomorphic impact and significance of the flash flood occurring on 4 August 1978 on the Allt Mor, Glenmore, Inverness-shire. Unpublished M.A. dissertation, Dept. of Geography, University of St. Andrews.

Newson, M.D., 1979, A framework for field experiments in mountain areas of Great Britain, *Studia Geomorphologica Carpatho-Balcanica*, XIII, 163-74.

Newson, M.D., 1980, The erosion of drainage ditches and its effect on bedload yields in mid-Wales, *Earth Surface Processes*, 5, 275-90.

O'Sullivan, P.E., 1970, The ecological history of the Forest of Abernethy, Inverness-shire. Unpublished D.Phil. thesis, New University of Ulster.

Smith, N.D., 1974, Sedimentology and bar formation in the upper Kicking Horse River, a braided outwash stream, *Journal of Geology*, 82, 203-23.

Thorne, C.R. and Lewin, J., 1979, Bank processes, bed material movement and planform development in a meandering river, in: *Adjustments of the fluvial system*, ed. Rhodes, D.D. and Williams, G. (Kendall-Hunt, Dubuque, Iowa), 117-37.

Werritty, A. and Ferguson, R.I., 1980, Pattern changes in a Scottish braided river over 1, 30 and 200 years, in: *Timescales in geomorphology*, ed. Cullingford, R. et. al., (Wiley, London), 53-68.

Wolman, M.G. and Gerson, R., 1978, Relative scales of time and effectiveness of climate in watershed geomorphology, *Earth Surface Processes*, 3, 189-208.

Young, J.A.T., 1975, A re-interpretation of the deglaciation of Abernethy Forest, Inverness-shire, *Scottish Journal of Geology*, 11, 193-205.

PART V

PERSPECTIVE

33 On experimental method in geomorphology

Michael Church

INTRODUCTION

Almost any deliberately planned programme of field observations in geomorphology - certainly any programme that
yields comparative measurements of landforms or of land
surface conditions - has come to be designated a 'geomorphological experiment'. A scientific experiment is an
operation designed to discover some principle or natural
effect, or to establish or controvert it once discovered.
It differs from casual observation in that the phenomena
observed are, to a greater or lesser extent, controlled by
human agency, and from systematically structured observations in that the results must bear on the existence or
verity of some conceptual generalization about the
phenomena.

An experiment therefore exhibits several distinctive
characteristics:

 i) there must be a conceptual model of the processes or
     relations of interest which will ultimately be
     supported or refuted by the experiment. The model
     is usually rather general and is not directly tested.
     It gives rise to

 ii) specific hypotheses about landforms or land forming
     processes that will be established or falsified by
     the experiment. If the conceptual model represents
     a well-developed theory, the hypotheses will consti-
     tute exact predictions. To test the hypotheses, we
     require three further conditions:

iii) definition of explicit geomorphological properties of
     interest and operational statements of the measure-
     ments that will be made of them. (These definitions
     must be sufficiently complete that replicate measure-
     ments may be made elsewhere, perhaps in a different
     environment.)

 iv) a formalized schedule of measurements to be made in
     conditions that are controlled insofar as possible to
     ensure that the remaining variability be predictable
     under the research hypotheses;

v)   a specified scheme for analysis of the measurements
     that will discriminate the possibilities in (ii):
     a successful experiment must lend unequivocal support
     to or clearly reject the research hypotheses. To
     achieve this, there must be
vi) a data collection and management system designed for
     the purpose.

By this rubric, very little geomorphological work
qualifies as properly experimental. It is most easy to see
how the requirements for experimental work may be met in a
classical laboratory context. Nonetheless, the possibility
to make deliberate progress in the elucidation of geomorphological principles probably depends in major degree
upon how successfully we can adapt this rubric to field
experiments. This paper will examine some aspects of the
conduct of field experiments in geomorphology.

                        CONCEPTUAL BASIS

An experiment is directed at critical assessment of a conceptual model or generalization about nature. The model
may derive from previously developed theory, from more
general scientific principles, or more informally from
prior systematic field study or from casual observations.
More than one model may be of interest: experiments are
often designed to discriminate amongst competing models.
     Such models are not directly tested. They are usually
rather general and not susceptible to explicit field
examination. Rather, the model is used to derive specific
hypotheses or predictions that can be tested directly. The
particular hypotheses for examination are chosen by the
investigator, and this is a critical step in the design of
an experiment. One seeks 'sharp' or 'critical' predictions: that is, ones that will permit unequivocal acceptance
or rejection of the conceptual model, or clear choice between competing models. Such outcomes have been characterized as 'strong inferences' by Platt (1964). Experiments
that yield strong inferences lead to rapid progress in the
discipline. Such experiments most often lead to contro-
version of a concept. That is because no amount of confirming evidence may establish any proposal as ultimately
true, but a single counter-example will indicate the in-
correctness of a theory.
     Sometimes, more than one conceptual model is consistent
with a given research hypothesis. In that case, experimental confirmation of the hypothesis does not discriminate
the concepts. A strong or critical experiment will not
lead to this situation. However, if no such experiment is
feasible, a particular conceptual model will appear to be
the most consistent description of nature as the number of
confirmed research hypotheses that it generates outgrows
that of any competing model.
     Two further features of conceptual design contribute to
efficient work. Chamberlin's (1897) method of multiple-
working hypotheses not only guards against the distortion
of results to fit a ruling hypothesis, it leads to rapid

                              564

discrimination amongst alternative models. Second, the
development of non-contradictory hypotheses about several
aspects of a problem, or about several closely related
problems, for examination in one experiment, is more economical than the conduct of a sequence of single purpose
experiments. This is particularly true of experiments on
complex landforms, such as catchments.

It is critical - and sometimes difficult - to select
hypotheses which are susceptible to testing because the
indicated observations are feasible to make. Hence, steps
(ii), (iii) and (iv) above are closely linked. The key to
the experimental solution of some problems lies in an
appropriate representation of them (step (ii)). Here,
analogy with other problems that are better known is often
a powerful tool. In other cases, finding a method to get
the answer (steps (iii) and (iv)) constitutes the critical
step. It is here that geomorphological experiments are
most likely to be original and innovative because of the
spatial and temporal scales of the phenomena in question.
It is rather difficult to invent critical experimental
tests of either the Davisian or equilibrium 'grand theories'
of fluvial landscape development for this reason.

ESTABLISHING EXPERIMENTAL CONTROL

This is possibly the most difficult issue of all for the
geomorphologist, again because of the scale of most
phenomena of interest. We will consider the problem under
three headings: experimental manipulation of the landscape,
manipulation of models, and the use of statistical criteria.

Experimental manipulation of the landscape

This entails intentional, controlled interference with the
natural conditions of the environment in order to obtain
unequivocal results about a limited subset of the processes
that change the landscape. This is the type I experiment
of Slaymaker et al. (1980) and the only type of geomorphological experiment recognized by Ahnert (1980). It
certainly yields the most elegant results because the
principal effects are observed directly. It is also an
extremely difficult exercise because of the time and energy
costs that are apt to be involved.

Relatively few full-scale type I experiments, therefore, have ever been attempted. By far the most common
type is the catchment experiment. A closely related
strategy is the use of experimental 'plots'. Appropriate
selection or preparation of plots achieves one of the main
requirements for experimental control: the phenomena are
simplified sufficiently that extraneous environmental
variability is eliminated. (cf. Bryan et al., 1978; Dunne
and Dietrich, 1980, for examples of results presented to
this Commission). On the other hand, this strategy may
introduce, for the geomorphologist, a problem of interpretation: if runoff and sediment yield observations are
carried out at plot scale, the results may be controlled
sufficiently well to discriminate amongst research

565

hypotheses. However, extrapolation of results to larger
units of landscape is made difficult by just that variability it was desired to avoid for experimental purposes.

Scale models

One way to overcome the time and energy costs of experimentation is to reduce the scale of the phenomena concerned.
Notable exponents of this approach in recent years have
been Stanley Schumm and his coworkers, who have examined
many aspects of river channel patterns and alluvial landforms from small models (cf. summary of work in Schumm,
1977). No attempt has been made to present scaling
criteria; hence the results remain qualitative insofar as
their transfer to full-scale field examples is concerned.
This raises an interesting question about the nature of
'qualitative experimentation' to which we shall return.
Consider, now, the application of similarity criteria in
order to obtain a quantitatively scaled model. The problem
is germane to the subject of field experimentation since
workers often choose a small example in the field to investigate phenomena that may occur over a wide range in
scale. Again, the conduct of catchment studies exemplifies
the situation.
    Similarity under scale change may follow geometrical
criteria (lengths are proportional: angles identical),
kinematic criteria (velocities are proportional) or
dynamical criteria (forces are proportional). The subject
of scale modelling is highly developed and we will consider
here only some examples that illustrate the issues that
arise in the context of field experimentation. These issues
are crucial for comparison of experiments conducted at
appreciably different scales and for transfer of results
to obtain predictions in a new geographical setting.
    In a river channel, the major criterion for dynamical
similarity is that the ratio of inertial to gravity forces
be preserved: that is, that the Froude number remain
constant. Hence, $v_1^2/gL_1 = v_2^2/gL_2$, where v is velocity,
g is the acceleration of gravity, and L is the length scale.
We may deduce from this that

$$Q_1/Q_2 = (L_1/L_2)^{5/2} \qquad (1a)$$

$$\text{and} \quad t_1/t_2 = (L_1/L_2)^{1/2} \qquad (1b)$$

where Q is discharge and t is time. From (1a) we see that
$L \propto Q^{0.4}$ for simple Froude equivalence in rivers. But we
know from downstream hydraulic geometry (which is the
proper comparison to make, since we are then comparing
channels of varying size under equivalent flows) that

$$d_* \propto Q^f; \qquad 0.3 < f < 0.4 \text{ usually} \qquad (2a)$$

$$w_S \propto Q^b; \qquad b \sim 0.5 \text{ always} \qquad (2b)$$

where $d_* = A/w_S$ is mean depth of flow and w_S is surface
width of the flow for a stream of cross-sectional area A.
We conclude from this that alluvial rivers are not simple

models of each other; that there is more than one length
scale for the channel; and that hydraulic geometry expresses
the distortion that occurs under scale change. The well-
known distortion of aspect ratio, w_s/d_*, is the consequence
of this.

From (1b) we note that the time scale in a Froude model
varies as $\sqrt{L}$. This often does not provide a sufficiently
drastic reduction in scale to provide workable models of
geomorphological processes that occupy decades. It is
expedient, then, to increase the tempo of the forcing function in the model (ie., increase the frequency of high
flows). The existence of significant thresholds for sediment mobilization allows this with no untoward consequences,
but processes observed in the model still conform with
equation 1b.

Froude-scaled models of alluvial river channels are
possible and have been achieved recently by Ashmore (1982)
to study channel braiding, and by Southard et $al.$ (1980) to
study sand transport and ripple formation (neither of these
was a field model). Both of these experiments introduce a
new feature of scale change, that new physics may appear.
The change of sediment grain size in Ashmore's experiment
is limited by some differences in behaviour between gravel
and sand. In Southard's experiment, it was desirable also
to scale viscous effects: the experiment was operated at
85°C! The achievable scale ratios remain rather modest,
then. In the very restrictive case explored by Southard
et $al.$, the ratio $L_1/L_2 \leqslant 2.5$ in practice. Still, the
results indicate that homology can be established in a well
designed field experiment at reduced scale.

Different problems are introduced by attempting to
scale landscape phenomena. Arthur Strahler (1958) introduced a set of criteria which have been reexamined by David
Mark and I (Church and Mark, 1980). Considering only the
geometrical expression of landscape properties, we concluded
that, in general, extensive properties (dimensioned
quantities, such as lengths, areas, volumes) are not simply
proportional under scale change, but that certain intensive
properties (ratios, including angles) are. I refer to one
example in catchment studies: it is of interest to know
whether drainage density changes with catchment size.
Under geometrical similarity, it must, since area, A_d, with
dimension L^2, will grow more quickly than channel length,
L_d, a linear property. Common sense tells us that the
functional requirements of drainage demand that drainage
density remain constant for uniform climate and surface
properties. The empirical resolution of this issue is unexpectedly complicated by a rather subtle sampling problem.
In the only unbiased sample known to us (Wood and Snell,
1957), common sense wins: drainage density is constant and
drainage networks are not self-similar under scale change.

This has implications for catchment experiments on runoff and sediment yield. Intensive processes, such as runoff generation and sediment mobilization, may be expected
to behave similarly at all scales in comparable terrain,
but flow routing must take cognizance of network distortion
(which has analogous consequences, presumably, right back
onto hillside slopes). The situation applies a $fortiori$

for sediment transfer studies, since reservoirs (intermediate storage points) become increasingly common in large
basins. Our usual methods for measurement - outlet gauges -
are, of course, subject to the effects introduced by travel
distance.

Statistical methods

Another means to control environmental variability is to
examine a number of samples in a manner that permits evaluation of or removal of variability that is not of central
concern to the experiment. This represents the type III
experiments of Slaymaker *et al.* (1980). This strategy
avoids the problem of trying to manipulate large forces and
masses over long periods, but it introduces possibly large
sampling costs and difficult judgements about classification, comparability and resolution. Four techniques, of
increasing sophistication, will be discussed.

Replication simply involves observation of an experimental
treatment applied to many different experimental units
(plots; hillsides; catchments,...). It is hoped that
extraneous natural variability will be controlled by becoming proportionally well distributed over all the outcomes of interest in the large sample. The measure of
additional variability is reduced as $\sqrt{n}$, where n is sample
size. There are problems, however, aside from possible
costs. If there are a considerable number of extraneous
environmental factors contributing to variability, then
each must be proportionately distributed across the range
of experimental outcomes else a spurious result could
appear. If randomization leaves only a small chance of not
achieving a proportional allocation, say p = 0.05 for each
extraneous factor, then for k independent factors, the
chance of obtaining a poorly distributed one is kp. At
k = 20, kp = 1.0. Plot experiments appear to be particularly exposed to this problem. It may be difficult to
assess p and costly to reduce it.
    In fact, in the drainage density experiment alluded to
above, no attempt was made to obtain homogeneous topography
or climate. A sample of n = 204 units was analyzed (by
Wood and Snell, 1957) for each areal datum. The successive
areas were defined by circles of increasing radius at each
site, so extraneous factors were distributed identically
across the mean result for each areal datum. The high
degree of confirmation of the predicted result is evidence
for the success of this plan.

Prescreening of experimental units is an attempt to select
ones which are highly homogeneous with respect to extraneous
factors. The factors that guide the pairing of catchments
is a good case of this (an example is given below).

Blocking represents an attempt to group experimental units
on the basis of prior information so that they are substantially more homogenous within groups than between them.
Experimental variation is then sought between groups. The
entire subject of statistical experimental design has

arisen from this approach and includes numerous strategies
for overcoming variability in background factors by
assigning experimental units or treatments proportionally
(cf. Cochran and Cox, 1957). Bovis (1978) has presented
an example of the approach to this Commission. He analyzed
measurements of surficial soil movement in montane, subalpine and alpine zones of the Colorado Rocky Mountains.
A randomized design with replication within strata based on
major vegetation cover type was the basis for the analysis
of variance.

Adjustment of the experimental results using concomitant
information represents an *ex post facto* attempt to separate
the experimental effect from confounding factors. This may
be done in several ways, but the concomitant information
must be independent of the experimental effects. Bovis did
not obtain an unequivocal result in his first analysis, so
he restratified his sample using field observations at plot
scale of factors supposed to act as controls on soil movement. He then obtained a clear demonstration of spatial
variability in soil movement between densely vegetated
sites (tundra meadow and dense forest) and sparsely covered
ones (dry, montane forest and alpine tundra).
    A more formal adjustment is provided by the analysis of
covariance. Significant displacements in the values of
paired data, induced by extraneous effects, can be removed
by adjustment from regression, or experimental effects may
be examined directly in the analysis of variance. Scarf
(1972) illustrated this in a study of the effects of land
use changes on hill catchments in New Zealand.

RECONSIDERATION OF EXPERIMENTAL CRITERIA

According to a strict interpretation of the criteria presented in the introduction to this paper, there is little
doubt that only type I studies constitute true experiments
(cf. Ahnert, 1980). However, the purpose of an experiment
is

> '... to obtain results about a limited set of relation-
> ships... in such a way that they may be used to identify
> or formulate general rules or laws that govern the
> operation of the processes involved.' (ibid. p.6)

All of the foregoing methods contribute to the elimination
of extraneous variability so that a specific putative
relation may be examined.
    In the statistical methods, cases selected for study
may be subjected to some direct treatment (ie., vegetation
manipulation on plots): often, they are not. 'Treatment'
may be imposed by the choice of physical sites, as in the
stratification by vegetation zones in Bovis's study. To
the extent that control is gained over the variability of
the parameters of interest - by purposive selection of
sites or of cases, or by the conduct of focused trials -
the statistical methods discussed here certainly conform
to the spirit of experimentation.

Indeed, proper statistical design must be considered in any experiment (steps (iv) and (v)), so that sound statistical design is synonymous with sound experimental procedure. There is no doubt that Sir Ronald Fisher, in giving the principles of (statistical) experimental design, recognized that he was establishing principles for sound and efficient scientific experimentation. I conclude that successfully controlled statistical experiments, even in the absence of direct, physical 'treatment' of the experimental sites, may be accepted as proper experiments.

Let us now review type I experiments. An experiment in the classical sense involves examining parameters one or two at a time in isolation. So far as analysis of environmental systems goes, that is a prescription for looking at a dysfunctional case: in order to obtain analyzable simplicity, we will have removed most of the essential variation upon which the real system relies for its stability. For the study of long term effects, which tend to be pervasive in geomorphology, it is hard to see how to eliminate variability in the prime driving force - weather - in any case. We must either abandon field experimentation or modify our approach.

A viable approach may be to examine the fully complex operating system and to consider the role of individual features or components by controlling them one at a time. That is, essentially, what is achieved in most full-scale geomorphological 'experiments'. In such an experiment, most extraneous environmental factors remain normally variable, and so some scheme must be developed to measure or otherwise discount their effects. (The time scale of the experiment may, of course, mean that some factors are either relaxed or fixed, hence not of concern). I will describe an example to illustrate this approach.

In 1954 a drainage diversion was completed in the Coast Mountains of British Columbia from the Fraser basin to Kemano River for hydroelectric power generation. The mean flow of Kemano River has been trebled (in several steps, as additional generating capacity has been added) without increasing the maximum flood or the sediment delivery to the river (the flood flow restriction is a requirement of the government's licence for the project). We have here a situation in which one feature of alluvial river regime - duration of nonflood flows - has been controlled and changed. We may attempt to predict the effect on the channel from available regime 'theory', so that the study becomes a test of the regime concept of alluvial river behaviour in a cobble-gravel channel. For example, provided that the regime channel type is not changed, we expect geometrical changes to follow equations 2, so we might expect w_s to increase by $\sqrt{Q_f/Q_i} \sim \sqrt{3} = 1.7$. Since this relation appears to be the most robust feature of alluvial regime theory, this test represents a strong single factor experiment. However, other factors that may affect channel size remain uncontrolled: in particular, prior trends of channel change and the possibility for simultaneous change due to other effects must be considered. We approach these possibilities by making measurements upstream from the powerhouse outfall to investigate extraneous variability,

and by examining the trend of river channel changes prior
to the project. Some results are given in figure 33.1.
The comparison of these data must be done statistically: at
present, we have compared only mean behaviour. The latest
results do not confirm our prediction, but we do not know
whether the adjustment has run its full course (that is, we
have no very good idea of the time scale of the experiment).
It is also possible that some flow substantially greater
than mean flow should be made the basis of prediction.

Our earliest record (from air photography) is for 1938,
so the experiment already spans 43 years. Clearly, I was
not around to plan it, so it is truly an 'inadvertent
experiment'. That does not compromise the results, the key
to which lies in appropriate interpretation of the sequence
of events and their unambiguous analogy with a significant
experimental proposal before analysis begins.

Execution of this experiment requires that the form of
a type I experiment be realized satisfactorily, but that
type III statistical control methods be used to isolate and
identify the relations of interest. Most field experiments
in geomorphology - indeed in all field science - are apt to
be of this form. Note that we have abandoned none of the
criteria for an experimental investigation: criterion (iv)
has, however, been satisfied by a hybrid set of procedures.
It is not clear, at the outset, that analyzable simplicity
(and a successful experiment) necessarily will be achieved.

NON-EXPERIMENTAL STUDIES

Most field studies of geomorphological processes cannot be
considered to be 'experimental' in any of the senses
discussed above. Indeed, the type II 'experiments' of
Slaymaker et al. (1980) - selection of a particular landform and sequential monitoring of changes - do not qualify
because there is no attempt to exert experimental control
over the subject by any means. Do such case studies have
a diminished value? So far as the establishment of
scientific concepts goes, they cannot provide the unequivocal evidence that an experiment does; nonetheless, they
are an essential part of the complete enquiry that develops
conceptual understanding.

A complete geomorphological enquiry begins with the
enunciation of some concept for study, suggested by prior
work, casual observation, or as the result of regularities
observed during descriptive field work. Case studies serve
to enrich the context and permit conceptual models for the
matter of interest to be explored more fully. Most sediment
yield studies, both at plot scale and at small watershed
scale, fall into this category. They serve to increase our
knowledge of sediment yield and its various antecedents,
but they cannot provide unequivocal confirmation of a functional relation between sediment transfer and any particular
antecedent because all the other contributing factors
remain uncontrolled. The group of studies reported in
Gregory and Walling (1974) falls into this frame.

Such work is of value in itself, as well, for the insight that it provides into the stability of environmental

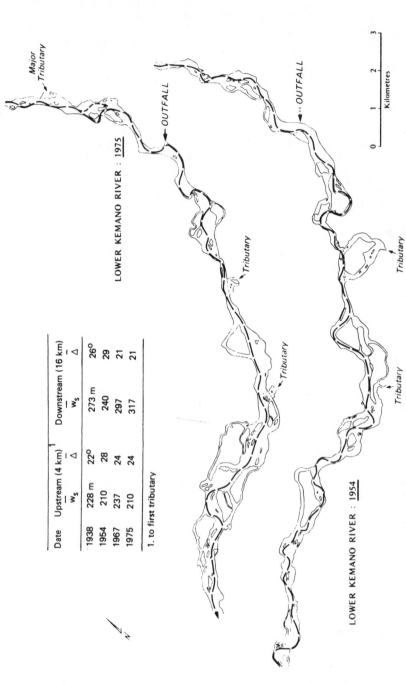

Date	Upstream (4 km)[1]		Downstream (16 km)	
	$\overline{w}_s$	$\overline{\Delta}$	$\overline{w}_s$	$\overline{\Delta}$
1938	228 m	22°	273 m	26°
1954	210	28	240	29
1967	237	24	297	21
1975	210	24	317	21

1. to first tributary

LOWER KEMANO RIVER : 1954

LOWER KEMANO RIVER : 1975

Figure 33.1. Kemano River in 1954, immediately before the inception of the diversion, and in 1975. The dashed line represents the (inferred) talweg position. The river reaches tidewater at the limit of the 1954 map. The data represent reach means derived from measurements at 200 m intervals along the talweg. Δ is the talweg deflection in the 200 m distance (evaluated without regard for sign). For more details of the diversion see Kellerhals, Church and Davies (1979).

systems. Geomorphologists should pursue the establishment
of a class of 'representative studies', modelled after the
hydrological 'representative basins' (Toebes and Ouryvaev,
1970) but not necessarily restricted to catchment units.
Luna Leopold (1962) has, in fact, proposed such a set of
studies at catchment scale in the form of the 'Vigil network'. Criteria for measurements, data management and
analysis must be as strict in representative studies as in
experiments (cf. Leopold and Emmett, 1965), though the
measurements may be simple, since the same public scrutiny
as is applied to experimental results must be possible if
the work is to be of value. There need not, however, be
the antecedent notion of critically testing some concept
to motivate the work. This forces us to discriminate
clearly between measurements made in an observational or
case study context, and those made under experimental
control.

THE USES OF EXPERIMENTS

Eventually, it must be decided whether concepts are capable
of discrimination or confirmation by an experiment. In
many cases they will not be, perhaps for lack of technique
to obtain control or to make critical operational defini-
tions or measurements. In geomorphology, as in all field
sciences, work in which there is an essential historical
element, or in which a specific place on the earth's surface
is an essential concomitant, cannot be reduced to experimental context in any event.

   For concepts amenable to experimental treatment, we
consider two classes of experiments. The initial experiment
is exploratory: it is "an attempt to secure acceptable
scientific evidence of a relation (or relations) between
events" (Flueck, 1978; p 386). A confirmatory experiment
is "an attempt at securing an 'independent' confirmation of
the results that already have been discovered and reported
by an exploratory experiment" (ibid.; 387).

   In an exploratory experiment, the design may change as
the experiment progresses (cf. Bovis, 1978) and the final
form of the underlying conceptual model may even be arrived
at *ex post facto*. However, the evidence submitted in
support of the relation that has been discovered must meet
proper experimental criteria:

 i) data must have been collected and analyzed under
    adequate control;
 (ii) analyses must be accompanied by reasonable statements
    of the probability of observing the actual outcome;
 (iii) hence, the data must be sufficient to discriminate the
    possible outcomes,
 (iv) the experiment must be publicly reported and the
    basic data available for scrutiny.

These requirements serve to establish confidence in the
results.

   Confirmatory experiments begin from well articulated
concepts. Hence, they should be more narrowly focused and
should not be modified during progress. The scheme of
analysis should be fixed beforehand and carried out only

after the data are all in. There are several ways to approach such an experiment:

1) employ a split sample technique in an initially exploratory experiment, and use samples after the first for confirmation;
2) proceed sequentially: after obtaining data in an exploratory experiment, return to the experimental site/arrangement and obtain further data for confirmation (perhaps with some modifications in procedure);
3) perform a new experiment.

Methods (1) and (2) are economical, but have the obvious disadvantage of possibly preserving bias. In earth science, method (3) is always to be preferred, and should be followed at a new site in order to overcome any possibility of geographical bias. Indeed, it is best if the investigators and experimental design are changed as well, so long as the essential experimental focus remains the same.

The context of exploratory and confirmatory experimentation provides the opportunity to employ properly the discretionary power of inferential statistics. In an exploratory experiment, we wish to be reasonably sure that we do not ignore a relation that may prove interesting: accordingly, we set the α-level for statistical detection of a significant result rather low ($\alpha < 0.10$, perhaps). In confirmatory experiments, however, we wish to ensure that weak evidence is not allowed to sustain a questionable assertion: $0.05 > \alpha > 0.01$ is a more appropriate range.

Unfortunately, the sociology of science makes the pursuit of confirmatory experiments rather rare. There is no prize for being second to report a result. Hence, some of our more fundamental conclusions stand on unconfirmed experiments. It is remarkable that our understanding of resistance to flow in alluvial channels rests largely on experiments on sand grain roughness in pipes conducted nearly fifty years ago by Nikuradse (1933) and by Colebrook and White (1937). The data have been much adjusted, but the experiments have not, so far as I know, been repeated with more appropriate arrangements.

At the other extreme, however, the basic water yield experiment in experimental catchments has been repeated many times because of suspicion that results cannot be transferred from one environment to another. Unfortunately, this calls into question the experimental validity of the entire enterprise. This theme is worth pursuing.

EXPERIMENTAL CATCHMENTS

Because we have incomparably more experience with (hydrological) catchment experiments than with any other full scale experimental arrangement within the purview of earth science, and because the focus of interest of many geomorphologists lies in transformations of landscape that are conveniently studied within the same framework, it is useful to examine directly some aspects of catchment experimentation. Rodda (1976) has recently provided a historical perspective and critique.

It is instructive to review the earliest of all watershed experiments to be concluded, that at Wagon Wheel Gap,
Colorado, between 1910 and 1926 (Bates and Henry, 1928).
This was a paired experiment to examine the effect on runoff of forest cutting and burning. The criteria for choice
of the basins were as follows:
- that the selected basins be practically contiguous in
  order that differences in the amount and timing of
  precipitation be minimized;
- that the basins be situated on identical geological
  structure, should have similar ranges of elevation,
  and should be as nearly alike as possible in aspect
  and physiography;
- that the vegetation be representative of the region
  (rather than optimum).
This amounts to a set of requirements to permit the establishment of experimental control. The results remain
amongst the most conclusive ever obtained.

It is possible to consider an experiment in which one
of the criteria listed above is deliberately selected to
be different between the two basins. This contrast re-
places the active treatment of one of the basins. The
Swiss Emmental experimental basins, initiated in the 1890s
and still extant, followed this rubric: surface cover was
deliberately varied. Now, however, the conventional calibration between watersheds before treatment cannot be
carried out. Accordingly, special precautions must be taken
in the design of the measurement programme and during
analysis to ensure that extraneous sources of variability
do not compromise the results. In particular, the study
usually must be prolonged in order that short term inconsistencies in the behaviour of the experimental units, introduced principally by the vagaries of weather but possibly
by other means, may assuredly be smoothed out. Further, it
becomes desirable for the analysis of data to be performed
blind. This variant is interesting, because many recent
geomorphological 'experiments' have been set up in this
way (cf. examples by Arnett, Oxley, and Weyman, for diverse
ends, in Gregory and Walling, 1974). One is usually left
with an interesting case study because of the constraint of
inadequate length of observations.

Two other aspects of the interpretation of catchment
experiments, both inherently difficult, deserve mention.
The application of a treatment to a limited parcel of
terrain surrounded by a large region of contrasting cover
type may certainly provoke a distinctive response from the
treated area, but it may not be equivalent to the response
we would see if it were not isolated, even if scale differences are taken into account. At plot scale, this
problem takes the familiar form of dominance by edge
effects: at larger scale, it represents the 'oasis' phenomenon. The second problem is that treatment of the land
surface nearly always includes vegetation manipulation:
unless the treatment is repeated (a feature of some experiments), the subsequent behaviour of the treated area will
be characterized by transient effects for a more or less
prolonged period as floral succession begins. This makes
experimental interpretation of the observations very

difficult. Even repeated treatment may be accompanied by progressive, irreversible effects that yield trends in data.

The methodological adequacy of catchment experiments has been aired thoroughly in recent years (cf. Hewlett *et al.*, 1969, for a detailed discussion, or Rodda, *op. cit.*). In the litany of criticisms, only two appear to stem from fundamental circumstances (the others remain interesting for what they imply about the requirements for adequate experimental design). These are that results are not transferable, and that results cannot be extrapolated to different scale. The first of these objections seems to derive from a view of catchment research as a means to characterize the behaviour of a complete environmental system. It should be evident that this is not an experimental purpose (though it may still be a very good purpose) and should not be an objection to properly structured experiments on catchments. We have considered the second objection: it is given special point by Sugawara (1972). It appears amenable to study, even though it may be a very difficult problem (cf. Chery, 1967, for a consideration of similitude requirements for hydrological purposes in catchments).

A particular contribution that geomorphologists may be able to make is to further develop hierarchical experimental designs using plots, hillsides and catchments simultaneously (cf. Slaymaker, 1972, for a single basin design). Toebes and Ouryvaev (1970) recommend this as the preferred method of design in order to avoid problems of non-experimental changes in the correlation between paired catchments. Careful hierarchical design might permit assessment of the representativeness of response to treatment. 'Inadvertent experiments' - as in catchments undergoing urbanization - must also be treated in this way, since there will almost never be a pair available. Ultimately, it seems essential to adopt this design in order to understand sediment transfers in the landscape because of large scale-related changes.

Finally, it must be noted that catchment experiments are inevitably expensive. The pattern of recent years for calibration periods to become shorter and shorter is in part a response to this, and in part a response to the growing pressures for immediate results. Two things may be said about this. First, a short calibration period destroys the resolution of the experiment, hence its justification. It is noteworthy that many of the most valuable results continue to come from long-established sites where detailed recent histories of catchment behaviour are available (cf., for example, results from Coweeta Experimental Forest reported by Swank in Chapter 20). Second, since proper conduct of catchment experiments is expensive, they should be designed as multiple purpose experiments as often as possible and be deliberately reviewed before they are initiated to determine that they will contribute strong inferences. We have come full circle to conceptual considerations.

CONCLUSIONS

In this paper, I have described criteria for a scientific experiment and shown that, amongst geomorphological 'experiments', only type I of Slaymaker *et al*. (1980) - intentional, controlled interference in the landscape - might strictly conform to them. As a potentially economical variant upon type I experiments, the use of scaled models has been too little considered heretofore. Consideration of scale models also confronts directly the difficult issue of extrapolating experimental results in the landscape.

It is argued that, with proper statistical design, type III experiments - spatially stratified sampling schemes - may also qualify as true experiments. The parallel that is drawn between statistical design and experimental procedure leaves little doubt that this is the case.

In fact, the traditional notion of a type I experiment - that only one or two parameters are allowed to vary under control, all other factors held constant - is almost never achieved in the field, and would produce a dysfunctional environmental context in any case. Instead, one or two factors are varied under control, whilst many other factors remain normally variable. An experiment mounted on this basis constitutes a sort of hybrid between types I and III. However, the success of such a venture often cannot be guaranteed beforehand.

Type II 'experiments' - purposive measurements of selected landscape factors - do not qualify as scientific experiments since there is no effective control over the subject of interest. They remain valuable, however, as case studies which enrich the context of knowledge out of which concepts for experimental test emerge. In many cases, these studies might usefully be characterized as 'representative studies': the observations and results must then be subject to the same criteria as experiments.

Attention is drawn to the distinctive roles of exploratory and confirming experiments, which have been too little appreciated in earth science. In particular, narrowly focused, well designed confirmatory experiments that permit strong inferences to be made (in the sense of Platt, 1964) are much too rare to permit us the judgement that geomorphology is making satisfactory scientific progress.

Catchment studies, of which there is a long tradition in the cognate field of hydrology, bring most of these issues to focus. These studies are of special interest to geomorphologists, for they suggest an obvious rubric for experiments to study central problems of sediment transfers in the landscape. Such experiments are typically I/III hybrids and, unless adequate attention is given to statistical establishment of control over normally varying factors, including provision for adequate resolution of the intended principal effects, they fail as experiments because unequivocal conclusions cannot be drawn. Often, they also fail as significant case studies for the same reason.

Much of geomorphology falls outside the purview of experimental method. In particular, parts of the subject that deal with physiographic description of landscape or

terrain analysis, and with the history of particular landscapes must use other methods. However, that part which is
concerned with the principles by which landforms develop in
general (so-called 'functional geomorphology' of Ahnert, or
'dynamic geomorphology' of Strahler) should seek to use the
experimental method. Experimentation in the landscape,
with typically long time scales and a wide range of spatial
scales, is expensive and difficult. The conduct of two or
three such experiments, or of a closely related set of
experiments, would in most cases constitute the work of a
career. Neither the difficulties of experimentation, nor
requirements of scientific method, have been widely
appreciated in geomorphology. It should be the main task
of this Commission to rectify the latter problem.

ACKNOWLEDGEMENTS

An article by John A. Flueck, of Temple University, prompted
some of the major themes in this paper. My own attempts at
field experimentation in river channels are supported by
the Natural Science and Engineering Research Council of
Canada. Dr. Olav Slaymaker invited this outburst upon the
Commission.

REFERENCES

Ahnert, F., 1980, A note on measurements and experiments in
    geomorphology, *Zeitschrift für Geomorphologie*, N.F.,
    Supplementband 35, 1-10.

Ashmore, P.E., 1982, Laboratory modelling of gravel braided
    stream morphology, *Earth Surface Processes and
    Landforms*, 7, 201-225.

Bates, C.G. and Henry, A.J., 1928, Forest and stream-flow
    experiment at Wagon Wheel Gap, Colorado: Final
    report on completion of the second phase of the
    experiment, *Monthly Weather Review*, Supplement 30.

Bovis, M.J., 1978, Soil loss in the Colorado Front Range:
    sampling design and areal variation, *Zeitschrift
    für Geomorphologie*, N.F., Supplementband 29, 10-21.

Bryan, R.B., Yair, A. and Hodges, W.K., 1978, Factors
    controlling the initiation of runoff and piping in
    Dinosaur Provincial Park badlands, Alberta, Canada,
    Zeitschrift für Geomorphologie, N.F., Supplementband
    29, 151-168.

Chamberlin, T.C., 1897, The method of multiple working
    hypotheses, *Journal of Geology*, 5, 837-848.

Chery, D.L., jr., 1967, A review of rainfall-runoff,
    physical models as developed by dimensional analysis
    and other methods, *Water Resources Research*, 3,
    881-889.

Church, M. and Mark, D.M., 1980, On size and scale in geo-
    morphology, *Progress in Physical Geography*, 4,
    342-390.

Cochran, W.G. and Cox, G.M., 1957, *Experimental Designs*, (John Wiley, New York).

Colebrook, C.F. and White, C.M., 1937, Experiments with fluid friction in roughened pipes, *Proceedings of the Royal Society, Series A*, 161, 367-381.

Dunne, T. and Dietrich, W.E., 1980, Experimental study of Horton overland flow on tropical hillslopes, *Zeitschrift für Geomorphologie*, N.F., Supplementband 35, 1. Soil conditions, infiltration and frequency of runoff, 40-59; 2. Hydraulic characteristics and hillslope hydrographs, 60-80.

Flueck, J.A., 1978, The role of statistics in weather modification experiments, *Atmosphere-Ocean*, 16, 60-80.

Gregory, K.J. and Walling, D.E., (eds.) 1974, *Fluvial processes in instrumented watersheds*, Institute of British Geographers, Special Publication no.6.

Hewlett, J.D., Lull, M.W. and Reinhart, K.G., 1969, In defence of experimental watersheds, *Water Resources Research*, 5, 306-316.

Kellerhals, R., Church, M. and Davies, L.B., 1979, Morphological effects of interbasin river diversions, *Canadian Journal of Civil Engineering*, 6, 18-31.

Leopold, L.B., 1962, The Vigil Network, *Bulletin of the International Association of Scientific Hydrology*, 7(2), 5-9.

Leopold, L.B. and Emmett, W.W., 1965, Vigil network sites: a sample of data for permanent filing, *Bulletin of the International Association of Scientific Hydrology*, 10(2), 12-21.

Nikuradse, J., 1933, Strömungsgesetze im rauhen Rohren, *Verein Deutsche Ingenieur Forschung*, 361, 43; Also, *United States National Advisory Committee for Aeronautics, Technical Memorandum* 1292, (1950).

Platt, J.R., 1964, Strong inference, *Science*, 146, 347-353.

Rodda, J.C., 1976, Basin studies, in: *Facets of Hydrology*, ed. Rodda, J.C., (Wiley-Interscience, London), 257-297.

Scarf, F., 1972, Hydrological effects of cultural changes at Moutere experimental basin, in: *Symposium on the results of research in representative and experimental basins, Wellington, N.Z., 1-8 December, 1970*, International Association of Scientific Hydrology Publication no.97, 170-186.

Schumm, S.A., 1977, *The Fluvial System*, (John Wiley, New York).

Slaymaker, O., 1972, Patterns of present sub-aerial erosion and landforms in mid-Wales, *Institute of British Geographers Transactions*, 55, 47-68.

Slaymaker, O., Dunne, T. and Rapp, A., 1980, Geomorphic experiments on hillslopes, *Zeitschrift für Geomorphologie*, N.F., Supplementband, 35, v-vii.

Southard, J.B., Boguchwal, L.A. and Romea, R.D., 1980, Test of scale modelling of sediment transport in steady unidirectional flow, *Earth Surface Processes*, 5, 17-23.

Strahler, A.N., 1958, Dimensional analysis applied to fluvially eroded landforms, *Bulletin of the Geological Society of America*, 69, 279-300.

Sugawara, M., 1972, Difficult problems about small experimental basins and necessity of collecting informations on large basins, in: *Symposium on the results of research in representative and experimental basins, Wellington, N.Z., 1-8 December, 1970*, International Association of Scientific Hydrology Publication no.96, 393-397.

Toebes, C. and Ouryvaev, V., 1970, *Representative and experimental basins - an international guide for research and practice*, UNESCO Studies and Reports in Hydrology no.4.

Wood, W.F. and Snell, J.B., 1957, *The dispersion of geomorphic data around measures of central tendency and its applications*, United States Army Quartermaster Research and Development Center, Research Study Report GA-8.

ABSTRACTS OF ARTICLES

CH. 2. CONTROLS ON OVERLAND FLOW GENERATION

M.G. ANDERSON, D. BOSWORTH & P.E. KNEALE

The study attempts to delimit those conditions under which
it may prove possible to make empirical estimations of overland flow responses, and those conditions demanding a much
greater detail of investigation for such predictions to be
successful. It is demonstrated that such a distinction cannot be made according to scale criterion alone. At the
'larger' scale, topography is shown to be an important determinant regarding the feasibility of estimating overland
flow using unit hydrograph techniques, while at the plot
scale the detailed group of topographic variables is shown
to be of overriding importance.

CH. 3. SPATIAL VARIATION OF SOIL HYDRODYNAMIC PROPERTIES IN
        THE PETITE FECHT CATCHMENT, SOULTZEREN, FRANCE -
        PRELIMINARY RESULTS

B. AMBROISE, Y. AMIET & J.-L. MERCIER

The necessity of having a good knowledge of the spatial
heterogeneity of soil hydrodynamic properties is demonstrated. A sampling method is suggested whereby 'genetically
homogenous units' are identified using geomorphologic and
pedologic maps. A mathematical model is presented which
calculates a relative hydraulic conductivity curve based on
the soil water retention curve. Good agreement was obtained
between model results and field determination. The use of
retention curves as a basis for the prediction of hydraulic
conductivity greatly reduces the amount of soil sampling and
analysis required.

CH. 4. PIPEFLOW AND PIPE EROSION IN THE MAESNANT EXPERI-
        MENTAL CATCHMENT

J.A.A. JONES & F.G. CRANE

The research has involved three major areas of enquiry: the
spatial extent, connectivity and general nature of the pipe
networks; the relative importance of pipeflow to stream
discharge; and the factors responsible for generating pipeflow. In addition, the use of bedload traps has shown that
pipeflow is a significant agent of erosion. The research
demonstrates that pipes are a major source of runoff in this
catchment. Great variation between pipes is revealed
suggesting that categories of flow regime need to be expanded
considerably beyond a simple division into ephemeral and
perennial.

CH. 5. FLOODPLAIN RESPONSE OF A SMALL TROPICAL STREAM

S. NORTCLIFF & J.B. THORNES

The runoff dynamics of a small tropical stream are shown to be dependent on activity on the floodplain rather than to hillslope hydrology. Quickflow is almost entirely the result of saturation overland flow from floodplain areas immediately adjacent to the channel. Despite seasonal contrasts in precipitation, the floodplain retains sufficient moisture in the dry season for its runoff response to remain constant throughout the year. The floodplain is fed by two dominant mechanisms, lateral inflow from the channel, and groundwater inflow (as opposed to subsurface stormflow) from beneath the interfluves. Inputs from overbank flow are important in the wet season.

CH. 6. THE PATTERN OF WASH EROSION AROUND AN UPLAND STREAM HEAD

MIKE McCAIG
School of Geography,
University of Leeds, UK.

Results from sediment traps set up around a Pennine stream head have shown that surface wash volumes and sediment transport rates are closely related to the area drained per unit contour length at the sampling point. Mapping area drained (a) and slope gradient (s) together as the variable $\ln(a/s)$ showed a complex pattern in the study catchment due to the occurence of natural pipe systems. Relationships between $\ln(a/s)$, wash frequency, wash volume and sediment transport have been examined and maps produced. The results show that frequently wet drainage line areas (high $\ln(a/s)$) which occupy only 2 - 7% of the catchment area account for around 80% of the erosion accomplished by surface wash beyond the headward extent of the perennial stream channel. The $\ln(a/s)$ variable has also been incorporated into a hydrologically based simulation model of erosion in and around the stream head. The model has been used to investigate the influence of different catchment hydrological characteristics on the pattern of erosion.

CH. 7.   RUNOFF AND SEDIMENT TRANSPORT DYNAMICS IN CANADIAN
         BADLAND MICRO-CATCHMENTS

R.B. BRYAN & W.K. HODGES

The major components of runoff and sediment transport in
badland micro-catchments are described. Rainsplash plays
an important role in sediment entrainment but is not significant in sediment transport. The dominant mode of sediment
transport on all lithologies and surfaces is surface or subsurface runoff. The timing, duration and volume of runoff
discharge are critical, and vary widely on different lithologic surfaces and with changes in antecedent moisture conditions. Despite exhibiting very complex patterns of water
and sediment transport, the results provide a good basis for
understanding the variable response of mesoscale catchments
to typical storm events.

CH. 8.   RUNOFF AND SEDIMENT PRODUCTION IN A SMALL PEAT-
         COVERED CATCHMENT: SOME PRELIMINARY RESULTS

T.P. BURT & A.T. GARDINER

Runoff and sediment production in a small peat-covered
catchment on the blanket peat moorlands of the Southern
Pennines, U.K. is discussed with reference to the contrasts
between the heavily-eroded deep peat of the catchment interfluves and the shallower peat of the sloping Cotton Grass
moorlands. Observations of overland flow, infiltration
capacity, and soil moisture status all suggest that surface
runoff is dominant, this being confirmed by hydrograph
analyses. Clear spatial differences exist in the contributing area of surface runoff in the two types of peat moorland which are reflected in the production of sediment at
the two sites. Suspended sediment transport rates are consistently higher, and better related to stream discharge,
at the heavily-eroded peat subcatchment.

CH. 9.   THE HYDROLOGY AND WATER QUALITY OF A DRAINED CLAY
         CATCHMENT, COCKLE PARK, NORTHUMBERLAND.

ADRIAN C. ARMSTRONG

Close spaced drains and mole drainage treatments are compared to the natural soil water regime on a sloping clay
site. The dominant natural water movement on this site is
down the slope, within the plough layer, and close spaced
drains serve only to intercept this flow. Mole drainage
however lowers the water table in the soil and reduces surface flow. The undrained areas discharge as much runoff by
plough layer and surface layer flow as is removed by artificial drainage. There was no evidence that artificial
drainage increased the nutrient losses from the site. In
an arable situation, artificial drainage has only small

effects on the hydrological output from the site as a whole,
although it has profound effects on the soil water regime
within the site. However, with a damaged grass sward, the
almost complete elimination of surface runoff by artificial
drainage results in a marked reduction in the total runoff
from drained areas.

CH. 10.   RAPID SUBSURFACE FLOW AND STREAMFLOW SOLUTE
          LOSSES IN A MIXED EVERGREEN FOREST, NEW ZEALAND

          M.P. MOSLEY* & L.K. ROWE
          Forest Research Institute,
          Christchurch, New Zealand.

Measurements of subsurface flow during rain storms and under
controlled experimental conditions indicate that there is
rapid movement of water along preferred pathways ('macropores') at rates of up to 2 cm s^{-1}, and that average subsurface flow velocities are about 0.25-0.3 cm s^{-1}. Streamflow in the study area is generated largely by subsurface
flow, both during storm period and baseflow conditions.
Solute concentrations in streamflow (which is considered to
have chemical characteristics similar to those of subsurface
flow) are about 11 mg l^{-1} and in precipitation are about
4.5 mg l^{-1}; there is a net loss of solutes of about
60 kg $ha^{-1}year^{-1}$. In contrast, net losses of suspended and
bed load sediment are in the order of 750-1000 kg $ha^{-1}year^{-1}$.
It is suggested that the hydrological importance of rapid
subsurface flow is responsible for this contrast; in that
the rapidity of slope runoff (1) causes the low concentration
of solutes and (2) by generating large streamflow peaks,
promotes bank erosion and transport of sediment by the
streams.

* Present address: Ministry of Works and Development,
  Christchurch, New Zealand.

CH. 11.   HYDROLOGY AND SOLUTE UPTAKE IN HILLSLOPE SOILS ON
          MAGNESIAN LIMESTONE: THE WHITWELL WOOD PROJECT

          S.T. TRUDGILL, R.W. CRABTREE, A.M. PICKLES &
          K.R.J. SMETTEM
          Department of Geography,
          University of Sheffield, UK.
          T.P. BURT
          Department of Geography,
          Huddersfield Polytechnic, UK.

The hydrological and solutional processes for a wooded hillslope on Magnesian Limestone are described. The routeways
and travel times of soil water flow are described in relatio
to precipitation inputs and the influence of soil structure.
The nature of solute uptake by mobile soil water is related
to the hydrological processes. The implications of soil
water movement and solute uptake are assessed with respect
to temporal variations in stream solute load and to solutional denudation on hillslopes.

CH. 12. SURFACE AND SUBSURFACE SOURCES OF SUSPENDED
SOLIDS IN FORESTED DRAINAGE BASINS IN THE
KEUPER REGION OF LUXEMBOURG

A.C. IMESON, M. VIS & J.J.H.M. DUYSINGS

The forested drainage basins of the Keuper region of Luxembourg are characterised by relatively high rates of erosion
and sediment transport. Excluding the river channels, it
is possible to distinguish between surface and subsurface
sources of suspended solids. Both sources are widely distributed throughout the catchment studied. The subsurface
sources are probably located in the upper part of the B
horizon where clay is dispersed and transported by throughflow to topographic depressions which drain to the catchment rivers. Surface sources of suspended solids are formed
where areas of soil are exposed by zoögenic processes. On
certain occasions sediment from these two sources arrive at
different times in the rivers to produce a double peaked
turbidity trace. The amount of interflow supplying sediment
from the subsurface sources can be calculated. The relationship between sediment supply and turbidity is discussed for
the Schrondweilerbaach for a number of flood events. A comparison is made between the forested Keiwelsbaach and cultivated Mosergriecht drainage basins. For the flood considered more than 36% of the suspended solids are supplied
from subsurface sources to the Keiwelsbaach as opposed to
8% to the Mosergriecht.

CH. 13. SOURCES OF VARIATION OF SOIL ERODIBILITY IN
WOODED DRAINAGE BASINS IN LUXEMBOURG

P.D. JUNGERIUS & H.J.M. van ZON

A large part of the sediment reaching streams from wooded
slopes in Luxembourg is derived from bare parts of the
forest floor which are exposed to splash. Sediment yeild
is therefore partly a function of the size and the erodibility of the exposed areas. Both characteristics are related
to soil fauna activities. The erodibility of these areas
was investigated in 5 drainage basins. Variation in erodibility could to some extent be explained if the drainage
basins are divided into landscape units which represent different ecosystems. Although soil fauna is largely dependent
on the ecological conditions of the landscape units they
form an important second source of variation in soil erodibility, at times overruling the importance of the first
source. This is mainly because soil animals bring material
to the surface from soil horizons with different resistance
to erosion. The effect of the animals is to reduce the
differences between the landscape units. This is partly because moles increase the erodibility of the surface material,
whereas worms decrease it.

CH. 14.   MICROEROSION PROCESSES AND SEDIMENT MOBILIZATION
IN A ROAD-BANK GULLY CATCHMENT IN CENTRAL OKLAHOMA

M.J. HAIGH

6.9 million ha of road-banks are to be found in the USA.
These features are prone to erosion and are a significant
sediment source. A case study of the erosion of a typical,
small, road-bank gully catchment is described. Erosion has
been monitored for 5 years and a soil loss of 520 t ha^{-1}
year^{-1} has been documented. Microrill activity contributes
significantly to the surface roughness of the debris slope.
Wash deposits at the slope foot retain a relatively high
proportion of mobilized sediments in the 0.2 - 0.06 mm size
range. Annual ground loss totals are closely related to
annual precipitation and the number of freeze-thaw cycles.

CH. 15.   WATER AND SEDIMENT DYNAMICS OF THE HOMERKA
CATCHMENT

W. FROEHLICH & J. SLUPIK

Studies of the slope and channel subsystems undertaken within the 18km^2 Homerka catchment in the Flysch Carpathians of
Poland are described. Water and solute discharges and sediment loads demonstrate a positive linear relationship with
catchment area. Specific discharge and denudation rates
exhibit an inverse hyperbolic relationship with basin area.
These relationships must be taken into account when analysing
the relationship between the response of a sub-catchment
and that of the whole drainage basin.

CH. 16.   SOURCES OF SEDIMENT AND CHANNEL CHANGES IN
SMALL CATCHMENTS OF ROMANIA'S HILLY REGIONS

D. BALTEANU *et al*

The hilly regions of Romania cover one third of the country
and supply 41% of the average annual runoff. Erosion rates
are high and values of suspended sediment yield in excess of
25t ha year^{-1} have been recorded. Three small catchments
in the southeast Carpathians have been studied and present-
day processes have been documented using repeated aerial and
terrestrial photographs over a period of 10-17 years. The
photogrammetric evidence has been correlated with topographic
surveys, reservoir accumulation data and periodic measurements of reference markers. Information on rates of gully
growth, volumes of material eroded and rates of reservoir
sedimentation has been assembled.

CH. 17. LANDSLIDING, SLOPE DEVELOPMENT AND SEDIMENT YIELD
IN A TEMPERATE ENVIRONMENT: NORTHEAST ROMANIA

I. ICHIM, V. SURDEANU & N. RADOANE

Sediment yield in Romania is greatest in spring as gullies
and channels are cleared of material during prolonged runoff from frontal rains. The period from May to early July
is more critical for slope wash erosion but this is more
localised. In the mountain forests gullies and river beds
are important sediment sources and slopes develop basal convexities. In the hill regions slope erosion exceeds stream
removal and concave footslopes develop. Many areas are influenced by landslides, especially after deforestation, and
very high sediment yields are produced.

CH. 18. DEVELOPMENT OF FIELD TECHNIQUES FOR ASSESSMENT
OF RIVER EROSION AND DEPOSITION IN MID-WALES, UK

G.J. LEEKS

Techniques employed by the Insitute of Hydrology in a 5 year
programme of research to generate and analyse a data-base
on river erosion and deposition in Mid-Wales are described.
These include topographic surveys using Infra-red Electronic
Distance Measurement (E.D.M.) equipment and theodolites,
sediment sampling on shoals, the measurement of suspended
sediment and bed load transport, and sediment tracing using
artificially enhanced magnetic susceptability.

CH. 19. SUSPENDED SEDIMENT PROPERTIES AND THEIR GEO-
MORPHOLOGICAL SIGNIFICANCE

D.E. WALLING & P. KANE
Department of Geography,
University of Exeter, UK.

Concern for non-point pollution and contaminant transport
within the fluvial system has directed attention to the
significance of suspended sediment properties in influencing
sediment-associated transport. It is suggested that the
geomorphologist can profitably broaden his own perspective
to include the properties of the sediment as well as the
magnitude of the transported loads. The potential significance of studies of sediment properties in catchment
studies are illustrated by examples taken from an ongoing
study of suspended sediment and solute transport within the
$1500km^2$ Exe basin, Devon, UK. Attention is given to the
particle-size characteristics of suspended sediment, the
relationship of the nature of the material leaving the basin
to that of the source material, the 'fingerprinting' of
sediment sources and the elucidation of processes operating
within the basin sediment system.

CH. 20. DYNAMICS OF WATER CHEMISTRY IN HARDWOOD AND PINE
ECOSYSTEMS

WAYNE T. SWANK & W.T. SCOTT SWANK

Solutional weathering is an important form of fluvial denudation on forested landscapes in the eastern United States.
This paper describes the seasonal changes in water chemistry
during the passage of water through one mixed hardwood and
two different pine forests. The greatest change in ion
concentrations occurs during the growing season when plant
uptake and other biological processes are most active.
Patterns of cation changes through compartments are generally
similar for the three forest types, but there are a few exceptions. Patterns are most different for ions such as NO_3
and SO_4 which are mediated by microbial transformations.
The forest canopy and litter-soil interface are major compartments of ion exchange. The presence, absence, and type
of forest vegetation can substantially alter the chemical
composition of water before it reaches the underlying saprolite and bedrock. These alterations may influence weathering
rates and solutional denudation.

CH. 21. VARIABLE SOLUTE SOURCES AND HYDROLOGIC PATHWAYS
IN A COASTAL SUBALPINE ENVIRONMENT

T.M. GALLIE & H.O. SLAYMAKER

Observations of runoff generation within a 2 ha subalpine
catchment indicate that water flowpaths are non-uniform in
time and space. Spatial sampling of solutes confirms that
surface, soil and groundwater movements and mixing are complex phenomena. It is concluded that soil-vegetation complexes are rational sampling units for hydrologic process
studies in small watersheds, that unsaturated, transient
saturated and permanently saturated soil water zones are
useful solute sampling units, and that channel-based sampling programs should supplement, rather than supercede,
slope-based sampling programmes.

CH. 22. SOME IMPLICATIONS OF SMALL CATCHMENT SOLUTE STUDIES
FOR GEOMORPHOLOGICAL RESEARCH

I.D.L. FOSTER & I.C. GRIEVE

The role of small drainage basin studies in geomorphological
research is highlighted with reference to a small catchment
experiment in Midland England. Detailed chemical analysis
relating to precipitation, through-fall, soil water and river
water samples permits evaluation of predominant weathering
processes, and calculation of rates of process operation.
Comparison of such studies with other environments enables
broad regional controls on denudation rates to be identified.

CH. 23. HYDROCHEMICAL CHARACTERISTICS OF A DARTMOOR
HILLSLOPE

A.G. WILLIAMS, J.L. TERNAN & M. KENT

A hydrological model for a Dartmoor hillslope is proposed.
This involves overland flow and four pathways through the
soil. Local hydrological pathways are of great importance
in controlling weathering processes. When the solute composition of the waters was plotted on stability diagrams a
clear contrast between the four interflow pathways was evident. Weathering of silicate minerals on the slope crest
was tending to gibbsite, whilst the other pathways tended
to kaolinite. A deep interflow pathway was very similar in
composition to the spring waters. These results indicate
that weathering is probably rapid and controlled by rain/
soil water interactions.

CH. 24. MAGNITUDE AND FREQUENCY CHARACTERISTICS OF SUS-
PENDED SEDIMENT TRANSPORT IN DEVON RIVERS

B.W. WEBB & D.E. WALLING
Department of Geography,
University of Exeter, UK.

Relatively few catchment experiments have investigated the
magnitude and frequency of sediment transport by rivers.
Continuous monitoring of suspended sediment concentrations
in tributaries of the Exe Basin, Devon, UK, including 9
years for the River Creedy and 5 years for the Rivers Dart
and Barle, has provided an opportunity to establish medium-
term magnitude and frequency properties of suspended solids
removal and to evaluate contrasts in sediment transport
within a river system. Results reveal significant contrasts
between tributaries in specific sediment yields, annual
regime of sediment loads, the timing of suspended solids
removal and the role of extreme events in sediment transport, which reflect variations in catchment scale and sediment availability.

CH. 25.   THE RELATIONSHIP BETWEEN SOIL CREEP RATE AND
CERTAIN CONTROLLING VARIABLES IN A CATCHMENT
IN UPPER WEARDALE, NORTHERN ENGLAND

EWAN W. ANDERSON & NICHOLAS J. COX

Soil creep has been monitored at 20 plots in a small catchment near Rookhope in upper Weardale, northern England, and
measurements made of several associated controlling variables. The observational design is aimed to produce measurements from a variety of different slope, soil, moisture
and vegetation conditions, and to allow comparison of different instruments. Models predicting soil creep rate as a
function of controlling variables have been obtained using
statistical analysis. The strong interdependence of controlling variables implies that models with only one predictor are the most satisfactory. Moisture and related
variables are the most important controls of creep in the
catchment. The relationship between creep and slope gradient is weak and negative.

CH. 26.   PATTERNS OF HILLSLOPE SOLUTIONAL DENUDATION IN
RELATION TO THE SPATIAL DISTRIBUTION OF SOIL
MOISTURE AND SOIL CHEMISTRY OVER A HILLSLOPE
HOLLOW AND SPUR

T.P. BURT, R.W. CRABTREE & N.A. FIELDER

The pattern of soil moisture distribution in a hillslope
hollow and its adjacent spurs is discussed in relation to
observations of the distribution of iron in soil profiles
up the hollow, and to the pattern of micro-weight loss experienced by rock tablets emplaced in the soil at various
locations over the hillslope. The results suggest that the
hollow is the main focus of solutional denudation. Soils
there remain wet for the greatest length of time and acid
water is rapidly transmitted down the hollow due to the
continued presence of the saturated wedge.

CH. 27. SOME RELATIONSHIPS BETWEEN DEBRIS FLOW MOTION AND
         MICRO-TOPOGRAPHY FOR THE KAMIKAMIHORI FAN,
         NORTHERN JAPAN ALPS

S. OKUDA & H. SUWA

The dynamic characteristics of debris flows have been observed at Kamikamihori Fan with an automatic recording system
and micro-topographical changes in the fan have been investigated by repeated surveys. Observations on changes in
front velocity and the longitudinal flow path have been
analysed using a simple one dimensional model of mass point
motion. Travel distance has been correlated with hydrological and physical factors controlling the flow properties.
Topographic changes on the fan resulting from scouring and
deposition have been evaluated and characteristic depositional forms are described.

CH. 28. PRECISE MEASUREMENT OF MICROFORMS AND FABRIC OF
         ALLUVIAL CONES FOR PREDICTION OF LANDFORM
         EVOLUTION

MASASHIGE HIRANO
Department of Geography,
Osaka City University.
TAKAYUKI ISHII
Department of Geography,
Osaka Kyoiku University.

The slope profiler devised by Ishii (1980) was used to measure the detailed form of alluvial cones or fans produced
chiefly by debris flow under a humid climate. The fabric
of surface materials on alluvial cones was also recorded,
partly by conventional photogrammetry using wire nets, and
partly by sketching at the time of profile measurement, in
order to analyse the process of cone formation. In addition,
subsurface profiles along longitudinal sections of the cones
were obtained in an open trench wherever possible. Detailed
maps obtained at cones in the Ashio Mountain, Kanto, Japan
are shown. The formation of these cones is discussed in relation to the meteorological records and to the subsurface
profile. In the case of a fan at the foot of Mt. Yakedake,
an active volcano near Kamikochi, Central Japan, mapping of
microforms focussed on the portion near the outlet of a
gully which supplies debris flows frequently. Results are
discussed in connection with prediction of the orientation
and travel distance of the next debris flow.

CH. 29.   AN ANALYSIS OF SEDIMENT TRANSPORT BY DEBRIS FLOWS
IN THE JIANGJIA GULLY, YUNNAN

KANG ZHICHENG & ZHANG SHUCHENG

The Jiangjia Gully in northeast Yunnan is a large-scale debris flow gully.   It has a drainage area of 47.1 km² and the
trunk gully is 12 km long.   Debris flows occur frequently
and a study of their behaviour has been undertaken.   The
annual supply of solid material by debris flows has been
calculated and a method of relating the total potential
supply of solid material and the active life of a debris
flow to the longitudinal slope of the gully bed has been
developed.   Debris flow control schemes must be based on a
sound knowledge of the factors governing debris flow formation.

CH. 30.   FLOW PROCESSES AND RIVER CHANNEL MORPHOLOGY

R.D. HEY & C.R. THORNE

The use of process-response models in the study of channel
form is discussed.   Study of flow processes in gravel-bed
rivers must include consideration of secondary flows, boundary shear stresses, flow resistance, bed load transport,
bank erosion, bar deposition, and meander mechanisms.   The
application of conceptual, mathematical and empirical models
to the dynamic modelling of channel morphology is reviewed.

CH. 31.   THE INFLUENCE OF VEGETATION ON STREAM CHANNEL
PROCESSES

A.M. GURNELL & K.J. GREGORY
Department of Geography,
University of Southampton, UK.

Vegetation exercises an influence upon stream channel processes in at least three main ways and these are analysed
by reference to empirical data obtained from the Highland
Water Catchment in the New Forest, Hampshire.   The significance of vegetation has been investigated at three scales;
firstly, in relation to drainage network dynamics and discharge within a subcatchment; secondly, in relation to soil
moisture variations and soil permeability on a single hillslope; and, finally, in relation to the components of the
main drainage network, with particular emphasis on channel
form and change within the Highland Water itself.

CH. 32.   STREAM RESPONSE TO FLASH FLOODS IN UPLAND SCOTLAND

A. WERRITTY

Three years' observations on a short reach of Dorback Burn, Cairngorm, Scotland demonstrate great complexity in the pattern of channel adjustments to flow regime in gravel-bed rivers. The paper examines the scale and pattern of channel adjustment, the geomorphic significance of flash floods in such rivers, and the pattern of recovery following a flash flood. The sequence of competant flows and the state of the channel prior to a given flood event are shown to be important controls of channel adjustment.

CH. 33.   ON EXPERIMENTAL METHOD IN GEOMORPHOLOGY

M.J. CHURCH

A properly structured experiment exhibits several distinctive characteristics: i) a conceptual model of the processes or relations of interest which will be supported or refuted by the experiment; ii) specific hypotheses that will be confirmed or controverted by the experimental results; iii) definition of explicit properties of interest and operational statements of the mesurements to be made; iv) a formal schedule of measurements to be made under conditions controlled to ensure that remaining variability be predictable under the research hypotheses; v) a specified scheme for analysis of the measurements that will discriminate the possibilities in (ii), and vi) a data management system designed for this purpose. Statistical methods are used in order to demonstrate the covariability of observations, to account for incidental variability introduced by extraneous conditions, and to judge whether or not the data support the hypotheses. Proper statistical design must be considered for this. In the classification of geomorphological experiments of Slaymaker *et al.*, types I (direct manipulation of landscape) and III (stratified observations) qualify as genuine experiments. Scale change is an important tool to extend the range of type I experiments. Geomorphological enquiry begins with the development of concepts from prior work. Case studies enrich the context and permit conceptual models to be explored more fully. Eventually, it must be decided whether the concepts are susceptible of experimental test: if so, an exploratory experiment is designed and executed. Its successful completion should be followed by confirming experiments, preferably in distinct settings. Most geomorphological 'experiments' can be described at best as 'case studies'. Examples of work on river channel behaviour and on sediment mobilization are used to illustrate the nature of experimental enquiry in geomorphology. A hybrid I/III-type experiment turns out to be most feasible. Special attention is given to catchment experiments, of which there is a long tradition in the cognate science of hydrology, since they appear especially suitable for many geomorphological enquiries.